Aquatic Ecology Series

Volume 8

Editor
Jef Huisman
Institute for Biodiversity and Ecosystem Dynamics, University of Amsterdam,
Amsterdam, The Netherlands

More information about this series at http://www.springer.com/series/5637

Stefan Schmutz • Jan Sendzimir
Editors

Riverine Ecosystem Management

Science for Governing Towards a Sustainable
Future

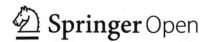 Springer Open

Editors
Stefan Schmutz
BOKU - University of Natural Resources
and Life Sciences
Institute of Hydrobiology and Aquatic
Ecosystem Management (IHG)
Vienna, Austria

Jan Sendzimir
BOKU - University of Natural Resources
and Life Sciences
Institute of Hydrobiology and Aquatic
Ecosystem Management (IHG)
Vienna, Austria

Aquatic Ecology Series
ISBN 978-3-030-10340-8   ISBN 978-3-319-73250-3   (eBook)
https://doi.org/10.1007/978-3-319-73250-3

Printed on acid-free paper

This Springer imprint is published by the registered company Springer International Publishing AG part of
Springer Nature.
The registered company address is: Gewerbestrasse 11, 6330 Cham, Switzerland

Contents

Chapter 1
Challenges in Riverine Ecosystem Management

Jan Sendzimir and Stefan Schmutz

This book is dedicated to those interested in the natural and social sciences and elements of governance that will support the sustainable management of rivers and aquatic ecosystems. Since elements of nature and society interact to determine the integrity and trajectory of these systems, they are referred to hereafter as social-ecological systems (SESs). This introduction opens the door to these topics in four steps. It begins by explaining why a book dedicated to river management and science is needed at this point. In the second part, it outlines the history of some of the major developments that challenge the integrity of SESs worldwide. In the third part, it describes several of the principal tools used to study as well as manage SES. Tools to measure the degree of degradation of an SES include indicators of biological integrity, ecosystem health, and resilience. Tools to assess and manage the trajectory of an SES include the DPSIR and adaptive management. The introduction closes by outlining the structure of the book through the progression of its chapters.

1.1 Justification of Book

Rivers are among the most threatened ecosystems of the world. For more than a century, river science has evolved to define these threatening trends and the mechanisms that cause them. What has emerged, while still incomplete, is a picture of imposing complexity, especially for managers, policy makers, and any concerned citizens interested in addressing these threats. This book surveys the frontier of scientific research and provides examples to guide management toward a sustainable future of riverine ecosystems. Principal structures and functions of the biogeosphere

J. Sendzimir (✉) · S. Schmutz
Institute of Hydrobiology and Aquatic Ecosystem Management, University of Natural Resources and Life Sciences, Vienna, Austria
e-mail: jan.sendzimir@boku.ac.at; stefan.schmutz@boku.ac.at

© The Author(s) 2018   1
S. Schmutz, J. Sendzimir (eds.), *Riverine Ecosystem Management*, Aquatic Ecology
Series 8, https://doi.org/10.1007/978-3-319-73250-3_1

of rivers are explained; key threats are identified, and effective options for restoration and mitigation are provided.

Rivers increasingly suffer from pollution, water abstraction, river channelization, and damming. Fundamental knowledge of ecosystem structure and function is necessary to understand how human activities interfere with natural processes and what interventions are feasible to rectify this. The specifics of such management leverage points become clear through elucidation of cause-effect relationships, especially how socioeconomic drivers create pressures on rivers and how those pressures alter ecosystem functions and impact fauna and flora.

Modern water legislation strives for sustainable water resource management and protection of important habitats and species. However, decision-makers would benefit from more profound understanding of ecosystem degradation processes and of innovative methodologies and tools for efficient mitigation and restoration. This becomes especially important for threats where current policies are ineffective, and both policy and management must support research that identifies solutions. The book provides best-practice examples of sustainable river management from on-site studies, European-wide analyses, and case studies from other parts of the world. It will be of interest to researchers (graduate and post-graduate) in the fields of aquatic ecology, river system functioning, conservation and restoration, to institutions involved in water management, and to water-related industries.

The current wealth of textbooks on river ecology extensively describes structures and functions of riverine ecosystems but gives less attention to river management (Cushing et al. 1995; Giller and Malmqvist 1998; Naiman and Bilby 1998; Allan and Castillo 2007; Dudgeon 2008; Likens 2010). By contrast our book directly targets riverine ecosystem management by examining the formulation and application of policy and providing sufficient depth of river ecology to inform competent decision-making in governance.

1.2 Past and Future Trends

Riverine ecosystems have been systematically modified on increasingly large scales since the invention of irrigation, perhaps as much as 7000 years ago (Mays 2008). However, their historic degradation has been accelerated periodically by surges of economic and/or technological power as empires and technologies erupted and expanded. The most recent surges were powered by coal (late nineteenth century) and oil (post WWII). The harnessing of fossil fuels increased our capacity to mechanically move material by over four orders of magnitude enabling society to engineer and reshape the contours of rivers and the surrounding landscapes on unprecedented scales. Fossil energy drove the massive industrialization and globalization of Western Society that witnessed an unprecedented acceleration of the degradation processes in rivers and lakes worldwide since 1950. Riverscapes were reshaped to accommodate intensive agriculture and industrial uses as well as high-density habitation. However, industrial technologies also amplified access to energy sources other than fossil fuels, especially hydropower. On average, humanity has constructed one 45 m high dam every day for the past 140 years (Bai et al. 2015).

The pace and scale of dam construction and other forms of river modification are reflected in the scale of impacts on aquatic flora and fauna. The greatest acceleration of biodiversity loss due to human activities in human history has occurred since 1970 (Millennium Ecosystem Assessment 2005). The drivers causing loss of biodiversity and, hence, of ecosystem services are either steady, show no evidence of declining over time, or are increasing in intensity. By aggregating the trends of some 3000 wild species, the Living Planet Index has documented a 40% decline in average species abundance between 1970 and 2000. The more rapid decline (50%) of inland water species underscores their greater vulnerability, being closer to the workings and by-products of human enterprise, while both marine and terrestrial species declined by about 30%. The concomitant loss of biodiversity and ecosystem services has been driven by both steady and episodic changes to habitat (land use change and geo-engineering), climate, overexploitation of resources (water, soil, biomass), and pollution. Geo-engineering of rivers has systematically channelized rivers for transport and to increase drainage during high-water events and separated the channel from the floodplain to protect water-sensitive row crops and zones for high-density habitation, commerce, and industry and dammed rivers for hydropower (Zarfl et al. 2014) as well as for water storage as a hedge against drought. Damming rivers currently stores the equivalent of 15% of global annual river runoff (Likens 2010). As a result 48% of rivers (expressed as river volume) globally is moderately-to-severely impacted by either flow regulation, fragmentation, or both. Impacts could double should all planned dams be constructed by 2030 (Grill et al. 2015).

1.2.1 Future Trends in River Engineering

The threat of climate change challenges society to decrease its reliance on carbon as an energy base for the economy (IPCC 2014). Most scenarios of paths to a low-carbon future foresee electricity increasingly replacing fossil fuels in all sectors. Furthermore, renewable power technologies such as hydropower and offshore wind will play an increasing role in electricity generation (Riahi et al. 2012). As the prospect of worldwide carbon pricing becomes realistic, fossil fuels, especially coal, look increasingly suspect as energy sources, and hydropower becomes increasingly attractive. This is especially so in areas with expanding economies and extensive unexploited river reaches, such as China, which currently is building 130 major dams in its southwest (Lewis 2013) and has constructed more than half the new dams built since 1950 worldwide (Wang and Chen 2010). This construction boom has been driven in part by investment policies that have been naively uncritical and optimistic. Authorizing new dam construction has been facilitated by a history of underestimating construction costs by development banks (Ansar et al. 2014). These drivers are projected to increase dam construction globally over the next several decades (Fig. 1.1)

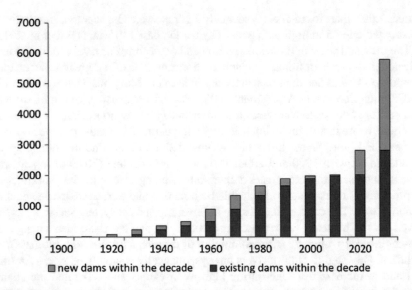

Fig. 1.1 Global pace of hydropower dam construction of existing hydropower dams (Lehner et al. 2011) and outlook for hydropower dams which are under construction or planned (Zarfl et al. 2014) (© Aquatic Sciences—Research Across Boundaries, A global boom in hydropower dam construction, 77/1, 2014, p. 162, Christiane Zarfl. With permission of Springer)

Surges of economic growth made it relatively easy to justify and ignore the impacts of riverine degradation. However, replacing lost riverine ecosystem services with economic and technological services may have seemed feasible when riding the updraft of a growing economy. But it becomes increasingly difficult in a world of increasing economic and ecological turbulence. When even the monumental riverscape engineering of the past century cannot prevent floods and droughts from disrupting communities and economies, the expenses of losing ecosystem services and of repairing and fortifying such an engineering system can no longer be ignored, and the search for alternative management paradigms becomes more attractive (Sendzimir et al. 2007). Indeed, more recent economic assessment that accounts more thoroughly with ecological considerations can be used to justify dam removal (Gowan et al. 2006; Lejon et al. 2009).

The future is never easy to predict, and this challenge is only compounded by the unprecedented levels of change anticipated over the coming century in nature, e.g., climate, and in human society, e.g., economy, demographics, and technology. While previous generations often migrated away from extreme challenges, that luxury no longer exists. There is no "away" to migrate to or to dispose pollution in. Novel levels of uncertainty only raise the challenge of improving the science and technology of managing rivers further. And the first step to make room for innovative ideas is to honestly admit that despite considerable advances, current science is not sufficient to deal with all of the anticipated uncertainty. This book reviews the current science useful to river management and then considers on what basis society can "learn its way into an uncertain future." It begins with assessing the level of riverine degradation and builds on that information to consider ways to mitigate the damage and restore the function of environmental flows and ecosystem services in riverine systems.

1.3 Managing River Systems

1.3.1 Assessing Degradation

"You cannot manage what you cannot measure." (Deming 2000)

For more than half a century, management science has striven to base decisions primarily on experiment-driven data, not opinion, a trend in business management greatly influenced by Deming's philosophy (Hunter 2015). Management based on conventional, tradition-based intuition or opinion has often been the default option when measurement proves difficult. Efforts to measure are often stymied by resource (time, money) limitations and system complexity. However, since 1970 different "metrics" have been developed to measure ecosystem change as input to policy decisions about environmental management and restoration.

Biotic Integrity
In 1972 a national mandate to measure the status of aquatic ecosystems in the United States was provided by the goal of the Clean Water Act: "...restoration and maintenance of the chemical, physical, and biological integrity of the Nation's waters." For these purposes the term *integrity* "implies an unimpaired condition or quality or state of being complete" (Watershed Science Institute 2001). To put this mandate in practice, Karr (1981) developed an Index of Biotic Integrity (IBI) to assess the health condition of an aquatic ecosystem by multiple metrics representing quantifiable attributes of biotic communities. Depending on the types of metrics used, those indices integrate the concepts of biodiversity, functional traits, invasive species, fitness, and population dynamics.

The underlying assumption is that the employed metrics react to human pressures in a predictable way. Individual metrics are compared with reference values that roughly equal pristine or best available conditions and are then integrated into an index. The index represents a numeric estimate of how far the current condition deviates from the expected condition. It is commonly expressed as a verbal scoring system, e.g., high, good, or bad status, that is easy to understand by decision-makers and thus has been frequently introduced in legislative acts related to aquatic ecosystem management, e.g., ecological status assessment of the Water Framework Directive (WFD) in Europe. A number of different IBIs worldwide follow the same principal of a multi-metric index but vary according to the context of targeted biotic communities, the definition of reference conditions, the scoring method, and the used metrics (examples for fish-based IBIs, see Roset et al. 2007).

Ecosystem Health

In assessing the status of ecosystems, ecosystem health (EH) is an index that reflects evidence from more than just the natural sciences. It integrates data and analysis from the natural, social, and health sciences, often as input for collaborative decision-making that incorporates human values and perceptions (Muñoz-Erickson et al. 2007). This expands the scope of assessment from ecosystems out to the wider context of the surrounding society and its culture and economy. Assessing the health of *social-ecological systems* (SESs) demands integrating science inputs and societal values and thereby unpacking some of causes of the pressures behind the drivers that impact ecosystems. When the IBI measures how far a system has moved from "pristine" conditions, the parameters defining those conditions and the change away from them are assessed using natural science. EH might use the same or very similar measurements but adds the perceptions and values of people who live in that social-ecological system and who may be the sources of the drivers of change as well as the recipients of the impacts of those changes.

In general the health of a social-ecological unit is reflected in how its composition, organization, and functions remain relatively stable and sustainable over time (Costanza 1992; Rapport 1998). EH bridges natural, social, and health sciences not so much to provide the definitive scientific basis for policy nor to offer predictive descriptions of causation. Rather it offers a theoretical framework with related monitoring methods (Bertollo 1998) that can be practically applied for case-by-case assessments in real-world settings (Wilcox 2001).

Both measures (IBI and EH) require a reference condition to measure change from, whether it is defined by policy, e.g., for the WFD, or by historical research of pristine conditions, or is complemented by stakeholder opinions (EH). These different applications allow us to distinguish between short-term human impacts and long-term environmental changes. However, if riverine SESs are dynamic, then there may be no fixed and stable condition to refer to, no undisturbed point of origin. For example, rivers are physically dynamic. River channels can move laterally, as much as 750 m per year in the case of the pre-engineered Kosi River, which flows from Nepal into Bihar, India (Smith 1976). In the face of such dynamism, integrity measures based solely on a stable reference condition become suspect. This challenge became apparent as examples of sudden, nonlinear, and sometimes irreversible change in aquatic ecosystems emerged in the last decades of the twentieth century (Jackson 1997; Jackson et al. 2001; Scheffer 2004; Scheffer and Van Nes 2007). After decades of apparently stable, clear water conditions, a single summer storm could cause a shallow lake to "flip" and become turbid, irreversibly, for years afterward (Scheffer 2004). To assess how SES responds dynamically to extreme events, new measures had to be developed to provide a conceptual, and potentially a quantifiable, basis for research and policy for aquatic ecosystems.

Resilience

How can we assess the response of riverine SES to the impacts of slow processes (degradation, accumulation of pollutants) as well as extreme events? One measure developed by engineers to assess the performance of river infrastructure is *engineering resilience*, measured in terms of the time required to return to an optimal state after an extreme event such as a flood. However, if aquatic systems can exhibit very different states, to name but two examples, clear or turbid, and remain in either state for extended periods of time, then perhaps the key question is not "What is the reference (optimal) condition?" but "What is the potential for the SES to move to an undesirable condition?" The fact that movement from one stability domain to another can be surprising (difficult to anticipate), rapid, and very difficult to reverse at best makes this a critical question for managers. *Ecological resilience* has been developed as a concept (Holling 1973) to help explore that potential for SES to remain in a "stability domain" (state) or move to another one. Where riverine restoration is an issue, the question can become: "What is the potential for a riverine SES to move from an undesirable to a desirable stability domain?" The resilience concept relates that potential to a system's capacity to absorb disturbance and recover afterward. That potential to change state rises as those capacities are lost.

Despite several decades of research, it has proven extremely difficult to measure this potential for movement between stability domains, i.e., regime change. One measure, referred to as a critical slowing down (CDS), has shown promise to reflect that an SES is close to a "tipping point," e.g., a point beyond which the SES moves inexorably to a new regime or stability domain. This proximity to a tipping point may be indicated when the system recovers slowly from relatively small perturbations, e.g., when the water column concentrations of nutrients like phosphorus or nitrogen are very slow to recover to average values following sudden spikes (Scheffer 2004; Scheffer and van Nes 2007; Scheffer et al. 2009). Measures like a critical slowing down (CDS) have been found in enough cases to be interesting but not often enough to be general, and there is even more so for a number of other indicators (for an overview, see Dakos et al. 2015). However, even if a more reliable measure could be found, that might not serve science or management very well. Quinlan et al. (2015) warn that:

> Measuring and monitoring a narrow set of indicators or reducing resilience to a single unit of measurement may block the deeper understanding of system dynamics needed to apply resilience thinking and inform management actions.

It is for these reasons that resilience has been applied mostly as a heuristic to help define and explore issues in ecology and natural resource management (Quinlan et al. 2015). However, resilience has also been used as a concept within planning processes and adaptive management exercises (Roux and Foxcroft 2011; Namoi CMA 2013). Resilience can be understood as a system's capacity to "... retain its basic function and structure by absorbing the impact of disturbance and/or recovering and rebuilding post-disturbance" (Namoi CMA 2013). Such a definition is too general to measure precisely (Cabell and Oelofse 2012). A very wide diversity of variables has been used not as direct measurements but as indicators of separate factors that individually and collectively contribute to this capacity in different contexts. Social science applications have assessed various human capacities to cope or adapt in the face of shock or stress as indicators of resilience. These capacities have been variously defined in terms of robustness and vulnerability (Pasteur 2011; Barrett and Constas 2014), response to poverty (Mancini et al. 2012), capacity to learn and innovate (Carpenter et al. 2001), and capacity to organize and develop collaborative networks and adaptive institutions (Atwell et al. 2010; McKey et al. 2010) (for a comprehensive summary, see Quinlan et al. 2015).

1.3.2   Integrating Assessment, Policy, and Action

Development of tools to assess the state and trajectory of an aquatic SES has deepened our appreciation for their complexity and dynamism. This is especially so from the perspective of managers who must contend with a history of changes that have proven difficult or impossible to reverse. The practical potential of such tools is realized when they are applied to develop and guide the implementation of policies to manage such systems. This book considers several frameworks, such as DPSIR and adaptive management, which have been developed to integrate such tools both for research and as part of decision-support processes.

DPSIR

A wealth of cause-effect relations can influence the trajectory of an SES. Clarifying those relations can make management of an SES more flexible and adaptive. To this end several major management agencies (OECD 1993; EEA 1995) developed Driver-Pressure-State-Impact-Response (DPSIR) as a more detailed framework of relationships linking five categories that describe influences and reactions of systems (Fig. 1.2). DPSIR has been used extensively to analyze ecological and social factors influencing the resilience of aquatic SES in the face of anthropogenic pressure. For example, under the aegis of the Water Framework Directive, it has been applied to improve protection of groundwater, inland surface waters, estuaries, and coastal waters (Borja et al. 2006). It has also been used to assess the pressure of alien species (UKTAG 2013) as well as to support the design of an integrated river basin management plan by identifying the structure of environmental problems in a river basin

Fig. 1.2 DPSIR framework (After EEA 2003)

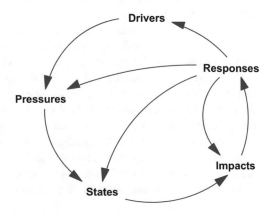

(Kagalou et al. 2012). Gari et al. (2015) conclude that two factors explain the widespread use of DPSIR, especially in the realm of policy-related science: "...it structures the indicators with reference to the political objectives related to the environmental problem addressed; and ... it focuses on supposed causal relationships in a clear way that appeals to policy actors (Smeets and Weterings 1999)." However, the use of DPSIR has been complicated by discrepancies in its application, such as the placement of the same variables in different categories (Gari et al. 2015).

As applied by the EEA (2003), the categories of the DPSIR framework are described as follows. Driving forces are created by the patterns of production and consumption that emerge from the intertwined social, demographic, and economic developments of society. These forces of society's metabolism drive the pressures that impact SES, e.g., emissions of chemical, physical, and biological substances and agents and shifts in land use and land cover. In response to these pressures, the state of an SES can shift physically (temperature), biologically (fish stocks), and/or chemically (atmospheric CO_2, water column nitrogen). Impacts resulting from shifts in ecosystem state are reflected in diminished functioning of the environment, e.g., lower human or ecosystem health, resource availability, and/or biodiversity. Any or all such impacts can precipitate responses to mitigate or adapt, which can emerge at the level of individuals and groups at different levels of organizations (Gari et al. 2015).

Management decisions to hold steady or change course benefit from precise measurements, but such choices grow out of many critical decisions that come beforehand. What should be measured, how, to answer what questions or policy dictates, and whose perspective should be included in the discussion? These are among a plethora of decisions that face river managers. With regard to measurements, who decides how to define the space and time dimensions of the reference condition? What is the baseline in time against which one measures change (degradation or progress) in ecosystem properties? For example, radically different conclusions can be drawn from the number of salmon found in 2002 in the Northwestern US Columbia River basin depending on when one sets the baseline. The baseline's date can inspire optimism (200% increase since 1930) or pessimism (90% decline since 1866) (Olson 2002). To shape sound research as well as policy, management must account for the false optimism inherent in such a *shifting baseline syndrome* (Pauly 1995), which can be reversed if management can integrate ecological restoration within the larger social context, restoring habitat connectivity, local fish populations, as well as local fisheries (McClenachan et al. 2015).

Constructive and effective engagement with these questions can help build a comprehensive overview and a flexible approach that managers need to deal with uncertainty. However, the global decline of river socio-ecosystems reflects a history of management that did not meet these challenges but defaulted to convention and tradition based on previous knowledge and historical relationships. Historically, river management regimes have evolved as complex webs of relationships that reinforce each other and create a momentum carrying them down a development path. In this way, a river system advances along a trajectory determined by complex feedbacks of interacting actors, policies, technologies, and concepts (Sendzimir et al. 2007).

Sometimes such feedbacks reinforce one another in ways difficult to change. When such histories of relations eliminate novelty based on new information or innovation, then management becomes *path dependent* (David 1988; Arthur 1994; Page 2006), i.e., locked into previous decisions, and it loses the initiative to adapt to changes (Barnett et al. 2015). For example, if the history of investment in the science and technology of dam and dike infrastructure makes it unthinkable to open such barriers as part of managing for droughts or floods, this constricts the range of options for research and policy. It is as if the way forward for science or managers can only proceed along a narrow set of rails. These constraints hamper our attempts to experiment by moving laterally. This inertia from path dependence can be especially challenging for managers who seek to experimentally develop policies to address uncertainty arising from the dynamism of nature and/or society. In response to such challenges, decades of experimentation have produced a range of tools to engage these twin challenges and make decision-making and policy formulation more flexible and comprehensive (Gunderson et al. 1995). This book reports on the opportunities afforded by these new approaches under the general rubric of adaptive management and governance.

1.3.3   Adaptive Management and Governance

The challenge of understanding and managing complex systems like aquatic ecosystems is compounded by their dynamism. Initial success at restoring ecosystem integrity often cannot be sustained (Scheffer 2004). So often have initial policy successes collapsed and remained so, despite all efforts at restoration, that the dysfunctional inertia following these surprising reversals has come to be known as *policy resistance* (Sterman 2000, 2002). Attempts to control disturbances (flood, fire, and pests) have often led to larger and more profound disruptions. For example, policies to constrain flood volumes within channels bounded by dikes have not stemmed the trend of increasing flood damages (Sendzimir et al. 2007; Gleick 2002; Pahl-Wostl et al. 2007).

The possibility that path dependence gives rise to policy resistance has provoked a search for ways to improve how we make science-based decisions, a search that has driven experimentation to integrate science and policy in one decision-making process. If ongoing change in ecosystems and society can render any inflexible policy obsolete, then management must dynamically adapt as a counter to perennial uncertainty. Adaptiveness requires the sustained capacity to learn and to flexibly manage. For 40 years a variety of separate experimental lineages [e.g., policy exercises (Toth 1988a, b), adaptive management (Gunderson et al. 1995), group model building (Vennix 1995; Senge 1990), soft systems methodology (Checkland 1989)] have worked in parallel to develop decision-making processes that address the challenge of learning while managing. Within this book we report on one such process, known as adaptive management

Fig. 1.3 Adaptive management: cyclic learning—decision process (After Magnuszewski et al. 2005)

(AM), which offers a framework to integrate the knowledge, methods, and operations of the research, policy, and local practice communities. It has been developed over four decades of experimental applications to understand and manage crises of collapsed fisheries, agriculture, forestry, and rangeland grazing (Holling 1978; Walters 1986; Gunderson et al. 1995). In addition to incorporating multiple perspectives, AM increases adaptive capacity by shifting decision-making processes from linear (crisis–analysis–policy) to a cyclic process (Fig. 1.3). This process structures learning and iteratively integrates how we modify assessment and policy formulation, implementation, and monitoring in order to track and manage change in the world (Magnuszewski et al. 2005).

The search for durable solutions to crises in ecosystems and society has repeatedly expanded the scope of inquiry outward from science to develop policies based on a broader base of experience and practice. Initial experiments (Holling 1978) acknowledged government and local practice but focused mostly on bridging disciplines within science. Subsequent experiments worked to include government (Walters 1986), local practitioners (Light and Blann 2000). However, managing aquatic ecosystems proceeds over time scales (decades) that far exceed those of individual projects or individual management campaigns. To make ecosystems sustainable, the adaptive potential raised by AM must be sustained over periods long enough to institutionalize adaptive and sustainable practices.

This drive to build long-term ecosystem sustainability proposed *adaptive governance* as a framework that would foster AM while addressing social aspects neglected in initial AM experiments (Gunderson et al. 2016). Specifically, it should create a workspace where formal and informal institutions can collaborate to understand and manage complex issues in social-ecological systems (Schultz et al. 2015). Adaptive governance would be distinguished by its capacity to increase the importance of learning and to bridge previously separate levels: formal/informal, scales of administration (polycentricity), in ways that embrace cross-scale interactions in ecosystems and society (Chaffin et al. 2014; Chaffin and Gunderson 2015).

1.4 Structure of the Book

Science can expand knowledge along two fronts defined by depth and breadth of information. This book sacrifices some depth of detail in order to better describe the breadth, e.g., the diversity of knowledge from different disciplines and their interconnections. This may disappoint specialists, but it best serves managers interested in practical insights from a wide spectrum of important aspects of riverine ecosystem management. Overall this book is designed to provide a general understanding of socio-ecological river systems that is grounded by specific examples from problem-oriented research. Given how global change may manifest as increasing variability in natural and/or social systems, governance toward a sustainable future of riverine ecosystems will greatly depend on integrating knowledge across disciplines.

The book is structured to guide the reader from a broad understanding of the structure and function of riverine social-ecological systems to an appreciation of human impacts and, finally, to interventions to manage such evolving systems. This starts with a basic knowledge of ecosystem structure and function. It then expands to include the consequences of human impacts as well as interventions to mitigate and restore these systems and the management tools required to realize them.

The foundations of understanding riverine structure and function are established in Part I. It introduces key system elements and characteristics of riverine ecosystems such as hydrology, morphology, connectivity, sediment, floodplain, riverscape, and water quality. Against this background understanding of riverine ecosystem functioning under natural conditions, the effects of human impacts and biotic responses are described. River management requires assessing these impacts, which begins with careful definition of the baseline or reference conditions against which change is measured. On this basis, one can analyze the dynamics generated by biotic responses as well as the potential effects of human intervention.

Understanding the history of human impacts and identifying tipping points of ecosystem degradation are important for setting up management objectives (Chaps. 15 and 16). The effects of pressures are described in this book in the way they affect key abiotic system elements and associated biota. Hydromorphological processes shape river channels, determine flow patterns, and define available habitat (Chap. 3). Channelization as a result of agriculture, urbanization, and infrastructure development including hydropower and navigation results in habitat degradation, disruption of river continuity, floodplain decoupling, river bed incision, and flow alteration (Chaps. 3–9). River restoration strives for improving in-stream habitat quality, recoupling floodplains, provision of flood retention areas, reestablishment of river continuity, and sustainable sediment management. Dams and water abstraction for irrigation, hydropower production, drinking water, and other purposes reduces discharge, alters flow regime, disrupts river continuity, and results in habitat loss (Chaps. 3–9).

Part II focuses on the management of riverine ecosystems and provides insights into state-of-the-art methodologies of integrated river basin management including international and EU water legislation (Chaps. 15 and 17), the concept of adaptive management (Chap. 16), challenges in managing international rivers (Chap. 18), and supporting methodologies and concepts such as ecosystem services (Chap. 21) and ecological monitoring and assessment (Chap. 19). The last Part III provides more detailed case studies of problem-related research with a focus on large rivers (Danube River, Chaps. 24 and 25), species conservation (sturgeon, Chap. 26), floodplain management (Tisa River, Chap. 28), and bioassessment and fisheries in developing countries (Burkina Faso, Chap. 27).

References

Allan D, Castillo M (2007) Stream ecology: structure and function of running waters. Springer, Cham, 452pp

Ansar A, Flyvbjergb B, Budzierb A, Lunn D (2014) Should we build more large dams? The actual costs of hydropower megaproject development. Energy Policy 69:43–56

Arthur WB (1994) Increasing returns and path dependence in the economy. The University of Michigan Press, Ann Arbor

Atwell RC, Schulte LA, Westphal LM (2010) How to build multifunctional agricultural landscapes in the US corn belt: add perennials and partnerships. Land Use Policy 27(4):1082–1090

Bai X et al (2015) Plausible and desirable futures in the Anthropocene: a new research agenda. Glob Environ Chang 39:351–362. https://doi.org/10.1016/j.gloenvcha.2015.09.017

Barnett J, Evans LS, Gross C, Kiem AS, Kingsford RT, Palutikof JP, Pickering CM, Smithers SG (2015) From barriers to limits to climate change adaptation: path dependency and the speed of change. Ecol Soc 20(3):5. https://doi.org/10.5751/ES-07698-200305

Barrett C, Constas M (2014) Toward a theory of resilience for international development applications. Proc Natl Acad Sci USA 111:14625–14630

Bertollo P (1998) Assessing ecosystem health in governed landscapes: a framework for developing core indicators. Ecosyst Health 4(1):33–51

Borja A, Galparsoro I, Solaun O, Muxika I, Tello EM, Uriarte A, Valencia V (2006) The European water framework directive and the DPSIR, a methodological approach to assess the risk of failing to achieve good ecological status. Estuar Coast Shelf Sci 66:84e96

Cabell JF, Oelofse M (2012) An indicator framework for assessing agroecosystem resilience. Ecol Soc 17(1):18

Carpenter S, Walker B, Anderies JM, Abel N (2001) From metaphor to measurement: resilience of what to what? Ecosystems 4:765–781

Chaffin BC, Gosnell H, Cosens BA (2014) A decade of adaptive governance scholarship: synthesis and future directions. Ecol Soc 19(3):56. https://doi.org/10.5751/ES-06824-190356

Chaffin BC, Gunderson LH (2015) Emergence, institutionalization and renewal: rhythms of adaptive governance in complex social-ecological systems. J Environ Manag 165:81–87

Checkland PB (1989) Soft systems methodology. Hum Syst Manag 8(4):273–289

Costanza R (1992) Toward an operational definition of ecosystem health. In: Costanza R, Norton B, Haskell B (eds) Ecosystem health: new goals for environmental management. Island Press, Washington, DC, pp 239–256

Cushing C, Cummins K, Minshall G (eds) (1995) River and stream ecosystems. Elsevier, Amsterdam, 817p

Dakos V, Carpenter SR, van Nes EH, Scheffer M (2015) Resilience indicators: prospects and limitations for early warnings of regime shifts. Philos Trans R Soc B 370:20130263

David PA (1988) Path-dependence: putting the past into the future of economics. Institute for Mathematical Studies in the Social Sciences, Stanford

Deming W (2000) The new economics: for industry, government, education. MIT Press, Cambridge, 240 pp

Dudgeon D (ed) (2008) Tropical stream ecology. Academic Press, Cambridge, 316p

EEA (1995) Europe's environment: the Dobris assessment. European Environmental Agency, Copenhagen, 8pp

EEA (2003) Environmental indicators: typology and use in reporting. European Environment Agency, Copenhagen, 20pp

Gari SR, Newton A, Icely JD (2015) A review of the application and evolution of the DPSIR framework with an emphasis on coastal social-ecological systems. Ocean Coast Manag 103:63–77

Giller PS, Malmqvist B (1998) The biology of streams and rivers. Oxford University Press, Oxford, 304p

Gleick PH (2002) Water management: soft water paths. Nature 418(6896):373–373

Gowan C, Stevenson K, Shabman L (2006) The role of ecosystem valuation in environmental decision making: hydro-power relicensing and dam removal on the Elwha River. Ecol Econ 56:508–522. https://doi.org/10.1016/j.ecolecon.2005.03.018

Grill G, Lehner B, Lumsdon AE, MacDonald GK, Zarfl C, Reidy Liermann C (2015) An index-based framework for assessing patterns and trends in river fragmentation and flow regulation by global dams at multiple scales. Environ Res Lett 10:015001

Gunderson L, Cosens B, Garmestani AS (2016) Adaptive governance of riverine and wetland ecosystem goods and services. J Envir Mgmt 183:353–360

Gunderson LH, Holling CS, Light SS (eds) (1995) Barriers and bridges to the renewal of ecosystems and institutions. Columbia University Press, New York

Holling CS (1973) Resilience and stability of ecological systems. Annu Rev Ecol Syst 4(1):1–23

Holling CS (ed) (1978) Adaptive environmental assessment and management. Wiley, New York

Hunter J (2015) Myth: if you can't measure it, you can't manage it. The W. Edwards Deming Institute Blog. URL: http://blog.deming.org/2015/08/myth-if-you-cant-measure-it-you-cant-manage-it/. Accessed 20 Oct 2015

IPCC (2014) Climate change 2014: impacts, adaptation, and vulnerability. Part B: regional aspects. Contribution of working group II to the fifth assessment report of the intergovernmental panel on climate change [Barros VR, Field CB, Dokken DJ, Mastrandrea MD, Mach KJ, Bilir TE, Chatterjee M, Ebi KL, Estrada YO, Genova RC, Girma B, Kissel ES, Levy AN, MacCracken S, Mastrandrea PR, White LL (eds.)]. Cambridge University Press, Cambridge, pp 688

Jackson JBC (1997) Reefs since Columbus. Coral Reefs 16(suppl):S23–S32

Jackson JBC, Kirby MX, Berger WH, Bjorndal KA, Botsford LW, Bourque BJ, Bradbury RH, Cooke R, Erlandson J, Estes JA, Hughes TP, Kidwell S, Lange CB, Lenihan HS, Pandolfi JM, Peterson CH, Steneck RS, Tegner MJ, Warner RR (2001) Historical overfishing and the recent collapse of coastal ecosystems. Science 293:629–638

Karr JR (1981) Assessment of biotic integrity using fish communities. Fisheries 6:21–27

Kagalou I, Leonardos I, Anastasiadou C, Neofytou C (2012) The DPSIR approach for an integrated river management framework. A preliminary application on a Mediterranean site (Kalamas River-NW Greece). Water Resour Manag 26(6):16pp

Lehner B, Liermann CR, Revenga C, Vörösmarty C, Fekete B, Crouzet P, Döll P, Endejan M, Frenken K, Magome J, Nilsson C, Robertson JC, Rödel R, Sindorf N, Wisser D (2011) High-resolution mapping of the world's reservoirs and dams for sustainable river-flow management. Front Ecol Environ 9:494–502

Lejon A, Malm Renöfält B, Nilsson C (2009) Conflicts associated with dam removal in Sweden. Ecol Soc 14(2):4

Lewis C (2013) China's great dam boom: a major assault on its rivers. Posted 04 November 2013 on Climate Energy Science & Technology Sustainability Water Asia. URL: http://e360.yale.edu/feature/chinas_great_dam_boom_an_assault_on_its_river_systems/2706/

Light S, Blann K (2000) Adaptive management and the Kissimmee River restoration project. (Unpublished manuscript)

Likens GE (ed) (2010) River ecosystem ecology. A global perspective. Academic Press, London, 424p

Magnuszewski P, Sendzimir J, Kronenburg J (2005) Conceptual modeling for adaptive environmental assessment and management in the Barycz Valley, Lower Silesia, Poland. Int J Environ Res Public Health 2(2):194–203

Mancini A, Salvati L, Sateriano A, Mancino G, Ferrara A (2012) Conceptualizing and measuring the 'economy' dimension in the evaluation of socio-ecological resilience: a brief commentary. Int J Latest Trends Finance Econ Sci 2:190–196

Mays LW (2008) A very brief history of hydraulic technology during antiquity. Environ Fluid Mech 8:471–484

McClenachan L, Lovell S, Keaveney C (2015) Social benefits of restoring historical ecosystems and fisheries: alewives in Maine. Ecol Soc 20(2):31. https://doi.org/10.5751/ES-07585-200231

McKey D, Rostain S, Iriarte J, Glaser B, Birk JJ, Holst I, Renard D (2010) Pre-Columbian agricultural landscapes, ecosystem engineers, and self-organized patchiness in Amazonia. Proc Natl Acad Sci USA 107(17):7823–7828

Millennium Ecosystem Assessment (2005) Ecosystem and human well-being: biodiversity synthesis. World Resources Institute, Washington, DC

Muñoz-Erickson TA, Aguilar-González B, Sisk TD (2007) Linking ecosystem health indicators and collaborative management: a systematic framework to evaluate ecological and social outcomes. Ecol Soc 12(2): 6. [online] URL: http://www.ecologyandsociety.org/vol12/iss2/art6/

Naiman R, Bilby R (eds) (1998) River ecology and management: lessons from the pacific coastal ecoregion. Springer, New York, 732p

Namoi CMA (2013) Namoi catchment action plan 2010–2020. Supplementary document 1, the first step – preliminary resilience assessment of the Namoi Catchment. http://bit.ly/1927lHq. Accessed 10 Nov 2015

OECD (1993) OECD core set of indicators for environmental performance reviews. Organization for Economic Cooperation and Development, Paris, 93pp

Olson R (2002) LA TIMES, sunday opinion section. http://www.shiftingbaselines.org/op_ed/. 17 Nov 2002

Page SE (2006) Path dependence. Q J Polit Sci 1(1):87–115

Pahl-Wostl C, Sendzimir J, Jeffrey P, Aerts J, Berkamp G, Cross K (2007) Managing change toward adaptive water management through social learning. Ecol Soc 12(2):30. [online] URL: http://www.ecologyandsociety.org/vol12/iss2/art30/

Pasteur K (2011) From vulnerability to resilience: a framework for analysis and action to build community resilience. Practical Action Publishing, Warwickshire

Pauly D (1995) Anecdotes and the shifting baseline syndrome of fisheries. Trends Ecol Evol 10(10):430

Quinlan A, Berbes-Blazquez M, Haider LJ, Peterson GD (2015) Measuring and assessing resilience: broadening understanding through multiple disciplinary perspectives. J Appl Ecol 53:677–687. https://doi.org/10.1111/1365-2664.12550

Rapport DJ (1998) Defining ecosystem health. In: Rapport D, Costanza R, Epstein P, Gaudet C, Levins R (eds) Ecosystem health. Blackwell, Malden, pp 18–33

Riahi K, Dentener F, Gielen D, Grubler A, Jewell J, Klimont Z, Krey V, McCollum D, Pachauri S, Rao S, van Ruijven B, van Vuuren DP, Wilson C (2012) Energy pathways for sustainable development: chapter 17. In: Johansson TB, Nakićenović N (eds) Global energy assessment: toward a sustainable future. Cambridge University Press, Cambridge

Roset N, Grenouillet G, Goffaux D, Pont D, Kestemont P (2007) A review of existing fish assemblage indicators and methodologies. Fish Manag Ecol 14:393–405

Roux D, Foxcroft L (2011) The development and application of strategic adaptive management within South African National Parks. Koedoe 53:105

Sendzimir J, Magnuszewski P, Flachner Z, Balogh P, Molnar G, Sarvari A, Nagy Z (2007) Assessing the resilience of a river management regime: informal learning in a shadow network in the Tisza River Basin. Ecol Soc 13(1):11. [online] URL: http://www.ecologyandsociety.org/vol13/iss1/art11/

Senge P (1990) The fifth discipline: the art and practice of the learning organization. Doubleday/ Currency, New York

Scheffer M (2004) Ecology of shallow lakes. Springer Science & Business Media, Berlin

Scheffer M, van Nes EH (2007) Shallow lakes theory revisited: various alternative regimes driven by climate, nutrients, depth and lake size. Hydrobiologia 584(1):455–466

Scheffer M, Bascompte J, Brock WA, Brovkin V, Carpenter SR, Dakos V, Held H, van Nes EH, Rietkerk M, Sugihara G (2009) Early-warning signals for critical transitions. Nature 461:53–59. https://doi.org/10.1038/nature08227

Schultz L, Folke C, Österblom H, Olsson P (2015) Adaptive governance, ecosystem management, and natural capital. Proc Natl Acad Sci USA 112:7369–7374

Smeets E, Weterings R (1999) Environmental indicators: typology and overview. Technical report No 25. EEA, 19pp

Smith DG (1976) Effect of vegetation on lateral migration of anastomosed channels of a glacier meltwater river. Geol Soc Am Bull 87(6):857–860

Sterman J (2000) Policy resistance, its causes, and the role of system dynamics in better avoiding it. In: Business dynamics: systems thinking and modeling for a complex world. McGraw Hill, Boston

Sterman J (2002) All models are wrong: reflections on becoming a systems scientist. Syst Dyn Rev 18(4):501. https://doi.org/10.1002/sdr.261

Toth FL (1988a) Policy exercises objectives and design elements. Simul Games 19(3):235–255

Toth FL (1988b) Policy exercises procedures and implementation. Simul Games 19(3):256–276

UKTAG (2013) Guidance on the assessment of alien species' pressures. U. K. Technical Advisory Group Water Framework Directive, 20pp

Vennix J (1995) Group model building: facilitating team learning using system dynamics. Wiley, Chichester, 316 pp

Walters CJ (1986) Adaptive management of renewable resources. Macmillan, New York

Wang Q, Chen Y (2010) Status and outlook of China's free-carbon electricity. Renew Sust Energ Rev 14(3):1014–1025

Watershed Science Institute (2001) Index of biotic integrity (IBI). Watershed Condition Series: Technical Note 2. ftp://ftp.wcc.nrcs.usda.gov/wntsc/strmRest/wshedCondition/IndexOfBioticIntegrity.pdf. Retrieved 20 Oct 2015

Wilcox BA (2001) Ecosystem health in practice: emerging areas of application in environment and human health. Ecosyst Health 7(4):317–325

Zarfl C, Lumsdon AE, Berlekamp J, Tydecks L, Tockner K (2014) A global boom in hydropower dam construction. Aquat Sci. https://doi.org/10.1007/s00027-014-0377-0

Part I
Human Impacts, Mitigation and Restoration

Chapter 2
Historic Milestones of Human River Uses and Ecological Impacts

Gertrud Haidvogl

2.1 Introduction

History has been acknowledged for 20 years as an important research element for river management that has been applied, for example, to define reference conditions and assess the level of degradation. The evolution of river uses and related ecological conditions, especially in recent decades, has been utilized to show the impact of humans on these ecosystems. Integrating a historical perspective into river management can, however, go beyond these targets (see, e.g., Haidvogl et al. 2014, 2015; Higgs et al. 2014). Just as present river management decisions will influence future conditions, paths trodden by users in the past have a bearing on today's ecology. Sound long-term studies of the natural and societal drivers shaping historical river changes can thus support our understanding of the present situation and identify trajectories of change. In long-term studies taking into account the dynamics of natural forces—in particular climate change and subsequent altered hydrologic and temperature conditions—as well as social dynamics (e.g., decision-making processes, main energy sources and technologies, superordinated practices and values) can reveal distinct overarching patterns of river use and management. This can contribute to developing future strategies and plans with lower ecological impacts.

This chapter describes major milestones of human river uses and ecological impacts. With some brief mention of Asian river case studies, it highlights especially examples, which are representative of industrialized countries of Europe and North America. In Europe, larger environmental changes of aquatic ecosystems occurred already in ancient and medieval times. European colonists spread practices and techniques of river uses to other areas of the industrialized world after they reached

G. Haidvogl (✉)
Institute of Hydrobiology and Aquatic Ecosystem Management, University of Natural Resources and Life Sciences, Vienna, Austria
e-mail: gertrud.haidvogl@boku.ac.at

© The Author(s) 2018
S. Schmutz, J. Sendzimir (eds.), *Riverine Ecosystem Management*, Aquatic Ecology Series 8, https://doi.org/10.1007/978-3-319-73250-3_2

regions, which have previously only been influenced by indigenous people (e.g., Humphries and Winemiller 2009). In the global North, the main milestone of historical river uses and subsequent ecological impacts was certainly the shift from agrarian to industrialized societies in the eighteenth and nineteenth centuries. Accordingly, preindustrial and industrialized rivers exhibit large differences in their ecological functioning as well as in the intensity of human impacts. Shifting from the preindustrial to the industrial mode of living resulted from the change of the prime energy source. While the former depended on wood, the latter requires exploitation of fossil energy, first coal, and, shortly before and particularly after World War II, oil (Sieferle 2006). Fossil fuels offered among others new transport means and possibilities for trading as well as unprecedented options to modify riverine environments. Fossil fuels enabled the systematic channelization of rivers and supported their damming or stocking of nonnative species on a global scale. Industrialization loosened the century-long tight connection of major parts of societies from their local and regional environmental resources and gave way to new practices of exploiting riverine ecosystem services.

2.2 Historical River Uses and Resulting Impacts

2.2.1 General Patterns of River Uses

Rivers provide ecosystem services that have attracted humans for millennia (see Chap. 21). Archaeological and later written evidence provide proof that river uses and necessary technical infrastructures existed already in ancient times, especially in arid zones. The Sadd-el-Kafara Dam on the Nile built some 30 km south of Cairo about 4500 years ago is considered as one of the oldest constructions of its kind (Hassan 2011). Major rivers such as the Nile, the Euphrates, the Indus, and the Jangtsekiang enabled cultures to develop and shaped their economy and culture.

In Europe, the Greek and Roman civilizations started influencing rivers, especially in urban areas to which water was delivered by aqueducts. With the collapse of the Roman Empire, technologically supported water uses diminished quickly in areas colonized by Romans. For several hundred years, they were replaced by rather local and small-scale river uses except for Spain, where the Muslims introduced water wheels and mills after the seventh century (Downs and Gregory 2004; Hassan 2011).

Outside of Europe, continuing technological progress and practices of river use as well as possible ecological effects linked to demographic and economic development can be deduced from the dams built, e.g., in Japan during the European "Dark Ages." The World Commission on Large Dams lists 20 dams higher than 15 m, which were built between 130 and 1492 CE. Most of these (i.e., 14) existed in Japan, and one each in India and Afghanistan. In Europe, by the Early to Late Middle Ages only one dam erected in 130 CE in Spain remained. Larger dam construction started only during the Late Middle Ages: In the present Czech Republic between the thirteenth and fifteenth centuries, three facilities were erected to create fishponds (ICOLD—International Commission on Large Dams 2016).

In most European countries and in North America, rivers served a large variety of human uses up until the beginning of the industrial era in the late eighteenth and nineteenth centuries. In preindustrial times most parts of society depended on local and regional environmental resources, often brought to them by rivers. This constituted their strategic importance. Different societal demands on rivers had to be harmonized to minimize adverse impacts on riverine services. Rivers and brooks were the main source of kinetic energy. They were the main transport routes, either for shipping goods or for transporting wood via rafts, sometimes with goods on it. In the case of very small brooks, wood was driven as loose logs, often during seasonal flooding. Although drinking water came often from groundwater wells, surface waters were sources, too. Surface water was a direct resource for many activities. It was used for cleansing and served many commercial purposes that had an adverse effect on water quality for drinking and cleaning. In urban areas and settlements, any local stream received the waste and wastewater from dwellers. It has to be noted, however, that the latter was rather limited as long as a majority of people depended on wells and their limited water quantities. Wastewater volumes significantly increased starting in the late nineteenth and twentieth centuries, as rapidly increasing urban populations required larger-scale and more sophisticated water management. As a result, central water pipelines supplying individual buildings and their households were built. Aquatic biodiversity is an essential component of ecosystem services, and riverine animals and plants played an important role for local food provision. Fish were central to the diets of many regions, especially for settlers along coastal rivers, but also in Christian countries in continental areas. Frogs, mussels, and even beavers were also used as food and, in the latter case, for fur. Floodplain forests helped to meet the heavy demands for wood as a basic energy source for preindustrial societies.

Growing demand from increasing human populations and the expanding economies of growing settlements and towns intensified all these preindustrial river uses. At the onset of industrialization around the beginning of the nineteenth century, human river uses have been maximized as far as possible in large areas of the Western world. But the exploitation of the various riverine ecosystem services was still limited to the local and regional scales, and finding compromises to mitigate adverse effects of one type of use on the other remained a prerequisite.

"Industrialized rivers" differ fundamentally from preindustrial ones. The shift from wood to fossil fuels enabled river engineers to carry out large-scale systematic regulation projects for navigation or flood protection especially on dynamic large rivers. New technologies produced and conducted electricity from hydropower plants to cities and factories, making electricity production spatially independent from the place of use. Travel times decreased and trade volumes increased with the rise of ships and railways driven by fossil fuels (first coal, then petroleum). Preindustrial patterns of river use and resulting ecological impacts ceased to exist. No longer did local and regional rivers serve all purposes that depended on water. For example, drinking and process water could be brought into cities from distant

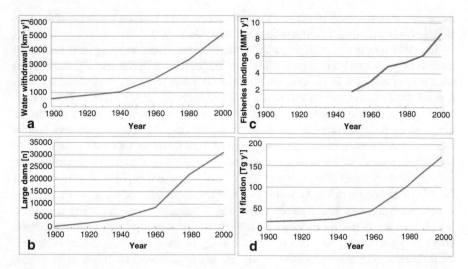

Fig. 2.1 The evolution of human pressure on rivers and freshwater systems in the twentieth century; the examples demonstrate the often exponential increase after World War II. (**a**) Global water withdrawals 1900–2000 (estimation for 2000); (**b**) Number of large dams (higher than 15 m) 1900–2000; (**c**) Fisheries landings from inland waters 1950–2000; (**d**) Global inputs of anthropogenically fixed nitrogen; adapted from Gleick 1993 (**a**), ICOLD 2007 (**b**), Allan et al. 2005 (**c**), Vitousek 1994 (**d**) and Strayer and Dudgeon 2010

rivers and springs, increasing the environmental imprint of urban centers in spatial terms (see, e.g., Billen et al. 2012). Large quantities of fish could be imported from the sea to continental consumers in reasonable times, thus eliminating the need to protect local stocks. Also, food supply based on improving transport started to affect watersheds on a global scale far away from the places of consumption (Vörösmarty et al. 2015).

The industrialization of rivers happened gradually and with increasing pace (see Fig. 2.1). Along with human uses, the resultant ecological impacts increased exponentially, especially after the 1950s. Until the late nineteenth century, often features from the preceding preindustrial period prevailed. For instance, defying elimination by fossil fuels, water mills had grown and become more complex so as to drive sophisticated machinery, to cool water, to improve power generation, to irrigate agricultural land, and/or to secure water supply (Downs and Gregory 2004). But generally, in the nineteenth century and thereafter, shifts in technology, cultural practices, administration, and policy reflected their new roles in river management, especially in European and North American countries. Management of river risks entered a new era. Active flood protection based on dikes became more and more common. It replaced preindustrial strategies of passive flood protection, which aimed at measures to keep damages to goods and lives as low as possible but not at preventing flooding at all (see Chap. 28). Technological and administrative innovations shifted the perspective of the industrial societies toward river ecosystems. The increasing capacity to substitute for river ecosystem services, regardless of distance, eliminated the need to harmonize a large variety of different uses (Jakobsson 2002). This resulted in maximizing one or two river uses, often those that did not adversely affect each other, e.g., power and transport. Other river uses, often fish and fisheries, were given up in favor of the preferred river services. After the 1970s, the negative effects of human impacts on ecological conditions received more and more attention, and river restoration projects have been started. This went hand in hand with thorough scientific observations of the links between human river alterations and biodiversity as well as animal and plant stocks often enabling for the first time to trace ecological changes based on direct field observations.

The following examples of human river uses and ecological impacts can be taken as fairly general, especially for the industrialized world, although with few exceptions (see Zarfl et al. 2015) no global or even continental overviews on the historical development of river uses and ecological impacts exist.

2.2.2 Milestones of Dam Building

Dams are one example of the increasing pressure on river services. Mostly, dams were built to gain hydropower, but they supported also the creation of fishponds or, in dryer areas, irrigation of agricultural land. The number of weirs increased throughout the High and Late Middle Ages and thereafter. For instance, in England, where the oldest comprehensive report exists in the form of the Domesday Book from 1086, 5642 mill weirs were recorded for this time. For France it is assumed that in the beginning of the twelfth century 20,000 dams were operated. Two centuries after, the number had risen to 40,000, and by the end of the fifteenth century (i.e., the end of the Middle Ages), 70,000 dams had been constructed (Braudel 1986). Certainly, the increase in numbers followed the expansion of populations, especially in cities with the increasing wealth of urban dwellers. Bork et al. (1998) added an environmental argument (so-called *Wassermühlenthese*, i.e., "water mill thesis") to the rising number of mills. According to their historical and paleographic study of German landscapes north of the Alps, in the fourteenth century, land-use change, especially forest clearing for the benefit of arable land, meadows, and pastures, reduced transpiration and caused rising groundwater levels. This made springs more abundant and their increasing runoffs were a suitable basis to construct mill weirs. From the turn of the eighteenth to the nineteenth century, it is estimated that in Europe the number of weirs amounted to 500,000–600,000 (Braudel 1986). One can assume, however, that this estimate relates only to larger weirs, while the total number was much higher. For example, a case study of an Austrian alpine river catchment (Möll River in Carinthia) showed that in the 25 communities located along this approx. 80-km-long river and its tributaries, 750 hydropower facilities existed (Haidvogl and Preis 2003, unpublished dataset).

It is evident that already preindustrial weirs—though small compared to modern dams—had modified ecological conditions. They acted as sediment traps and altered channel morphology not least due to their tremendous number. In small, anabranching streams in the mid-Atlantic region of North America, no significant amounts of sediment accumulated before European colonization in the seventeenth century. After European settlers had built thousands of milldams between the seventeenth and nineteenth centuries, 1–5 m of slack water sedimentation had covered the floodplains and the present meandering river channels incised in these sediments (Walter and Merritts 2008).

The impacts of weirs, especially on fish migration, have been known and addressed for centuries (see Chaps. 6 and 9). In preindustrial times, when harmonizing various river uses on local scales was a necessity, finding compromise was key. Although neither historical observations nor fishery records have been kept, this is evident from water legislation. A Scottish statute of 1214 demanded, for instance, openings in dams, and all barrier nets had to be lifted on Saturdays to allow salmon runs (*Salmo salar*, Hoffmann 1996). A fishing decree from 1545 for the Austrian Traisen River, a right-hand tributary of the Danube, provides similar protections for potamodromous fish species (Raab 1978). For tributaries of Alpenrhein (Rhine upstream of Lake Constance), fish passes were planned already in the sixteenth century. Along the Ill River, such a technical facility should have re-enabled migration of lake trout, which was interrupted by a dam to withdraw drinking and process water for the commune Feldkirch. This dam replaced an older and lower construction that was destroyed by a flood in 1566. Some decades later, the manorial lords upstream raised an official complaint because their main fishing target was missed. A fish bypass was suggested as possible solution but never built due to the technical problems of such a construction in the schistose rocks (Zösmair 1886). A fish pass was however realized on the Albula River, a tributary of Hinterrhein in the Swiss canton Graubünden, after millers erected a new dam in the 1680s and interrupted lake trout migration. The passage had a length of 6 m and a width of 1.5 m (Bundi 1988).

In the late nineteenth and especially in the twentieth century, the number of dams rose exponentially around the globe, first in the North and then in the South (see Chaps. 1 and 6). They continued to serve century-long functions especially as hydropower producers and for irrigation. New technologies and machinery built with ever-cheaper steel and powered by fossil energy helped to create concrete edifices of 100 m height and more. Together with the necessary means to transform mechanic energy into electricity and to transmit this electricity over large distances, large manufacturers and railways and urban administration soon started to benefit. After World War II, electricity use rose, not least with domestic demand for household appliances. In arid regions, dams and reservoirs secured irrigation of agricultural land. A summary on dam construction in the twentieth century demonstrates the increasing pace of large dam building after 1950 (Rosenberg et al. 2000). By 1900, several hundreds of large dams (i.e., equal or higher than 15 m; International Commission on Large Dams) existed. Up until 1950, the total global number newly built per decade was less than 1000. During the 1950s, almost 3000 new dam projects were implemented. In the 1970s, the number peaked at more than new 5400 facilities. In the 1990s, still almost 2000 new constructions occurred globally. In the 2000s and 2010s, the number further decreased, but, e.g., Zarfl et al. (2015) assume

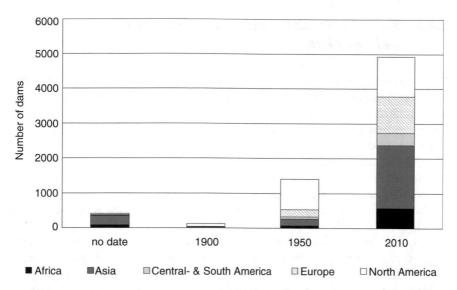

Fig. 2.2 Increase of number of large dams in different geographical areas 1900, 1950, 2009 (Europe includes Russia; based on GRanD-database, Global Water System Project © 2010, Lehner et al. 2011a, b)

that in the 2020s thereafter new dam construction will resume (see also Lehner et al. 2011a) (Fig. 2.2).

The ecological effects of modern dams are manifold (see, e.g., Poff and Hart 2002; see Chap. 6). They reduce velocity and often create almost stagnant waters of varying size; they change water temperature, which influences bioenergetics and vital rates of organisms. Downstream movement of water and sediment is influenced and reduced with adverse effect on river and riparian habitats, and biogeochemical cycles are modified. Dams hamper migration of fish and other aquatic organisms and exchange of nutrients up- and downstream. Almost half of present larger dams are used for irrigation (International Commission on Large Dams; http://www.icold-cigb.org/, Accessed 18 Jul 2016). Water abstraction via dams and reservoirs caused some of the most striking examples of environmental degradation in the last decades. For example, after a severe drought in 1946, the former USSR initiated large-scale dam constructions to redistribute available water resources. The Aral Sea is a prominent case for environmental degradation as it suffers from reduced water inflow due to water abstraction in the main tributaries since the 1960s (Micklin 2007).

2.2.3   River Channelization to Secure Transport and Land Use

There is no direct link between historical river transport and land use in floodplains. However, these two human river system uses have to be perceived as connected, as both required channelization measures. Accordingly, activities evolved centuries ago to prevent erosion of cultivated land and, in particular, to secure navigation. Initial measures focused on stabilizing riverbeds and riverbanks, while flood protection dikes to avoid inundation of settlements became more typical only in and after the nineteenth century with industrialization and subsequent population growth and spread of settlements into floodplains.

Transport has been an essential function of rivers for millennia. It characterizes virtually all rivers that attracted settlement. Just as with hydropower use, its intensity grew with rising population and trading of agrarian, preindustrial societies. Generally, river transport was cheaper and, often, even safer than that on roads, though it was at the same time slower. In addition, navigation, rafting, or log driving was affected by yearly natural cycles, especially low- and high-flow periods or freezing, as typical for alpine and continental regions (e.g., Pounds 1979). Hence, it was the main option for trading bulky goods and, in particular, wood (Pounds 1979; Möser 2008, Sieferle 2008). To support smooth navigation, riverbanks were often fixed and obstacles such as boulders removed manually from rivers or blasted, for instance, on the Austrian Danube in the late eighteenth century (Petts et al. 1989; Hohensinner et al. 2013).

To complement the transport network offered by natural waterways, artificial canals were introduced. In Europe, the first attempts to construct artificial shipping canals date back to Roman Times, e.g., in the Netherlands (Corbulo, Drusus canals) and France (Vella et al. 1999), or to the Early Middle Ages, when Charlemagne projected the Fossa Carolina in 793 (see, e.g., Brolsma 2011; Leitholdt et al. 2012). Charlemagne's plan was far beyond the technologies available at that time, and the canal remained a 3-km-long fragment. In Asia, the approx. 1770-km-long Beijing–Hangzhou Grand Canal was built as strategic waterway before the end of the thirteenth century. It linked five river basins and transferred water from Yangtze to North China Plain (Gregory 2006). By 1411, the Grand Canal was further developed and fed, among others, by water of the Lower Yellow River's main channel, which was stabilized to provide continuous flow (Overeem et al. 2013). To avoid a northward breach of the Yellow River and subsequent damage to the canal, a continuous levee was built on the north bank of the Yellow River and completed in 1494. On the southward banks, breaches diverted water toward the distributaries of the Huai River as flood control measures.

In Europe, in the seventeenth century, first projects in the Netherlands or in France (Canal du Midi) initiated a canal building boom that continued for the next two centuries (Brolsma 2011). Projects became much more ambitious, e.g., as proposals for connections between major European rivers such as Danube, Elbe, or Oder show (see, e.g., Vogemont 1712). Inland canal building continued well into the twentieth century. For example, as the connection between the Black and the North Sea was envisioned by Charlemagne already in the eighth century, the Rhine-Main-Danube Canal became reality only in the 1990s. It can be assumed that, since their completion, artificial canals have supported the dispersal of aquatic animals, in particular fish, to new river systems. The nase (*Chondrostoma nasus*), for instance, entered French rivers via shipping canals at the latest in the second half of the nineteenth century. The expansion of this species is confirmed for the 1860s for the Rhine and a new canal system in north-eastern France. Its occurrence was soon after observed in the Seine, then in the Upper Loire and Rhone basins where it arrived within less than 40 years (Nelva 1997).

In the 1830s, steam-driven railways started to operate, and railway connections intensified quickly in Europe as well as in North America (see, e.g., Pounds 1979 for Europe). Navigation was forced to react to the growing competition, usually by increasing ecological pressures on rivers. Since the first decades of the nineteenth century, the sophistication of steam technology also powered ships, freeing them from the need for tow roads and teams on the riverbanks. Compared to the wooden ships, their requirements for space in the river channel were much stricter, e.g., regarding homogenous and larger river cross sections. Steam ships soon increased in size, boosting the pressure for straightening and channelizing rivers with well-known ecological consequences (see Chap. 3).

While river channelization for navigation dates back centuries, flood protection is more typical for industrialized rivers. In the late nineteenth and twentieth century, previously not intensively used floodplains were newly colonized as urban areas. In the Middle Ages and Early Modern Period, neither settlements nor agricultural lands were protected from floods. It was rather common to adapt land use as much as possible to flooding, e.g., by preferences for elevated terrain and lower water depths during floods. This has been proven for arable land in the Austrian Danube flood-plains in the Machland or for settlements in Vienna (see, e.g., Haidvogl 2008; Haidvogl et al. 2013). Large-scale flood protection measures—often implemented in conjunction with hydropower dams and waterway improvement for shipping—resulted in hydraulic disconnection of areas that previously had been flooded regularly. Cutoff from normal river channel flows as well as, even more importantly, flood pulses, floodplain waters stagnated and filled with sediments and organic matter, raising floodplain elevation and finally drying up (see, e.g., Hohensinner et al. 2004).

2.2.4   Water Supply from Rivers: Increasing Imprint on Urban Hinterland

Rivers were essential water resources in particular for various commercial purposes. In urban areas, they became centers of economy. Washers, tanners, dyers, beer brewers, or slaughterhouses, for example, used them likewise for cleaning and washing. Often this resulted in serious conflicts in water demand between water polluters and other commercial ventures requiring clean water (see, e.g., Billen et al. 1999). Drinking water was often withdrawn from local groundwater sources, although surface waters were used as well, as the example of urban centers such as St. Petersburg shows (Kraikovsky and Lajus 2010). In the nineteenth century, population growth and urbanization increased the pressure on drinking and process water supply. As characteristic for the industrial period, the growing metropoles were driven by the declining quantity and/or quality of water supply to cross the boundaries of their local and regional river catchments. Via water pipelines, they tapped sources far away and transferred also their ecological imprint to other more suitable regions. Prime European examples include Paris (Barles 2012) and Vienna (Gierlinger et al. 2013). Enlarged water supplies often resulted in an enormous growth of water use per capita, sometimes continuing until present times. The Greek capital Athens, for instance, started to search for new water resources outside of the immediate urban surroundings in the 1830s. Since then, water supply infrastructures to tap distant sources have been expanded gradually. At present, Athens controls a significant amount of water reserves of two Greek river basins and no attempts have been made to decrease per capita demand of urban population (Stergiouli and Hadjibiros 2012). A similar historical trend can be observed for Barcelona, with the exception of successful recent efforts to reduce urban water consumption (Tello and Ostos 2012). In mid-nineteenth century, Boston pipelines brought water 20 miles from Lake Cochituate after the local wells became so polluted that they could no longer be used without danger to the lives of urban dwellers. In the 1860s, the city incorporated several communities to extend and secure its water resources. Bostonians used in the 1860s 100 gallons per person per day (approx. 380 L) in contrast to 3–5 gallons (approx. 11–19 L) when water came from wells (Vörösmarty et al. 2015). New York abstracted water from a tributary of the Hudson after erecting the New Croton Dam that was the world's largest masonry dam at its completion in 1906 (Vörösmarty et al. 2015).

As a general historical tendency, more drinking and process water increased the volume of wastewater released into rivers. Newly built centralized sewage systems initiated point-source pollution, built in urban areas since the late nineteenth century, to fight against hygienic nuisance and infectious diseases, such as cholera.

2.2.5 Pollution of Rivers and Its Legacies

Waste—for long historical periods mostly of organic origin—increased the nutrient load in aquatic ecosystems. Centuries ago, smaller and mid-sized rivers suffered certainly more than large ones because of their lower dilution capacity. Medieval castles and monasteries had often a direct connection between their latrines and local rivers (Hoffmann 1996). Already in the beginning of the fourteenth century, Paris effluents had turned the Seine into an infectious and foul canal (Mieck 1981). The quantities of waste were however considerably smaller before the 1900s. For example, human and animal excreta were considered as valuable nutrient resources since agrarian societies depended solely on natural fertilizers for grain production. Only in the late nineteenth century did it become a general habit to flush and dispose, respectively, human and animal excreta. By then, Justus von Liebig's discoveries of the role of nutrients for plants, his invention of a phosphorous fertilizer in 1843, as well as the import of guano and "Chile saltpetre" by steam ship navigation improved the fertilizer sources for European agriculture. Sewage farms collecting in particular organic waste from towns had their heyday in the first decades of the twentieth century. However, the invention of the Haber–Bosch process in 1910 relieved farmers for the first time in history fully from their dependence on natural fertilizers. This had far-reaching consequences for rivers. For example, for the Seine, it was demonstrated that in 1817 when 716,000 dwellers and 16,500 horses lived within the urban borders the amount of nitrogen released into the Seine was negligible. The larger part was returned to the agricultural lands that provided the city in turn with food and feed. A hundred years later, in 1913 when 2,893,000 inhabitants and 55,000 horses lived in Paris, 3100 tons of nitrogen were released annually into the river via central sewers, which were built in the meantime. Still, however, the larger proportion of nitrogen was collected for agriculture, mostly in the large sewage farms along the Seine banks downstream of Paris (i.e., 9100 tons/year; Barles 2007).

On a global scale, Green et al. (2004) compared the change of riverine nitrogen fluxes of the preindustrial era and nowadays. The largest preindustrial flux was found for the Amazon exceeding a load of 3.3 million MT N/year at the river mouth. At present, the largest amounts are closely linked to industrialized areas, e.g., continental Europe, North America, as well as Southern and Southeast Asia. As for nitrogen, eutrophication as a result of excessive phosphorus input became an increasing problem for rivers in the second half of the twentieth century (see, e.g., Liu et al. 2012).

While organic river pollution can produce effects over the short- and midterm, other types of historical pollution will remain for decades and even centuries. The current release of toxic and hazardous substance into rivers and their long-term legacies are widely recognized. For example, chloride pollution in the Rhine is expected to persist for several centuries, forcing France to face a salinity problem on its Alsace aquifer (see Vörösmarty et al. 2015). The long-term legacies of historical events are, however, only slowly getting the scrutiny of river ecologists and managers (but see, e.g., EEA 2001, 2013).

Pollution with heavy metals from mining and ore processing has been relevant throughout history. Several studies exist, for instance, for the mining of mercury in support of large-scale gold and silver exploitation and production since the sixteenth century in Europe and America. Recently, Torkar and Zwitter (2015) investigated the long-term effects of the Slovenian mercury mine in Idrija and the resulting pollution of Idrijca River on fish. Polluted sediments were swept downstream and finally accumulated in the sediments of the northern part of the Gulf of Trieste (Gosar 2008; Foucher et al. 2009). According to Nriagu (1994), the annual loss of mercury in the silver mines of Spanish America averaged 612 tons per year between 1580 and 1900. Total losses of mercury to the environment in the Americas within this period amounted to 257,400 tons. Approximately 60–65% was released to the atmosphere, but large quantities of mercury were deposited on terrestrial and riverine ecosystems from where they may be reemitted. Concerning most of the mercury now sequestered in the sediments of aquatic systems—mainly in marine sediments, Camargo (2002) concluded, however, that the high mercury concentrations currently reported in the global environment are a consequence of the huge pollution caused by human activities during the twentieth century.

The long-term legacies of past sediment pollution have been recognized for the Danube catchment where the risk of accidental release and remobilization of hazardous substances stored in the soils from past industrial activities or waste disposal was identified. An inventory of accident risk spots was elaborated. By 2009, a total of approximately 650 such spots were reported in the flood-prone zones of the entire river basin and 620 were evaluated. Here, a hazardous equivalent of 6.6 million tons has been identified as a potential danger (ICPDR 2009).

2.2.6   Land-Use Change, Hydrology, and Erosion

Land-use change was an indirect but nevertheless severe human impact to preindustrial streams. The large-scale medieval shift from forests to arable land in Europe triggered more rapid surface runoff and erosion, reduced evapotranspiration, and increased the discharges of rivers. Bork et al. (1998) investigated land-use change and its environmental effects for Germany north of the Alps based on palynological and pedological data and demonstrated its strong imprint. Around 650 CE, 93% of the total area was covered by woods (697,500 km^2 out of a total of 750,00 km^2). By 1310, the proportion of woods had diminished to 15% only (i.e., 112,500 km^2) mostly in favor of arable land and grassland. At present, forests cover about one third, arable land 38%, and grassland about 24%. Other land-use types were always of minor importance. Assuming that mean annual precipitation was similar for all periods and amounted to 700 mm per year, total annual surface runoff more or less doubled from 115 mm in 650 to 245 mm in 1310. At present, total annual surface runoff is assumed to be around 220 mm. Although Bork et al. (1998) did not specifically investigate the effects of altered surface runoff on river discharge, they conclude that changed evapotranspiration and interception had an effect. The *Wassermühlenthese* mentioned above clearly points to this link between surface runoff and springs' and rivers' discharges.

Land-use and land-cover change clearly correlated with erosion rates. From the seventh to the end of the tenth century (max. proportion of arable land 20%), for instance, in all of Germany north of the Alps, an annual rate of up to about 9 million tons eroded into river channels. During the first half of the fourteenth century, when forests covered only 15%, the share of arable land had risen to more than 50% (about 55% in 1313–1318), and extreme precipitation events were frequent, annual erosion reached 1900 Mio tons between 1313 and 1318. They peaked at 13,000 Mio tons in 1342, when a 1000-year recurrence flood hit large areas of central Europe. In the second half of the fourteenth century, erosion rates declined together with less wet climate and an increasing proportion of forests recolonizing arable land. The latter was abandoned due to declining human population after the first wave of plague in 1347–1353. Only in the 1780s and in the following decades a new increase to 200 Mio tons per year was noticed—a resurgence due to expansion of arable land and a new period of intensified and more frequent rainstorms (see Bork et al. 1998; but also Lang et al. 2000; Dreibrodt et al. 2010; Dotterweich and Dreibrodt 2011; Brázdil et al. 2005). In total, it is assumed that half of the total erosion that can be observed in Germany between the seventh and the twentieth century took place from 1310 to 1342 (Lang et al. 2000).

In North America and Australia, European settlers introduced new land-use practices that increased erosion. However, changes in sedimentation rates and river morphology date back to native population influences (Overeem et al. 2013). In New Zealand, increase of sediment loads started in the North Island rivers already with the Maoris, and similar trends are associated with cultivation practices of the Native American population. Along the Waipaoa River in New Zealand, sediment yields increased by 140% after Polynesians had arrived between 1250 and 1300 CE. They settled mainly along coastal areas and kept erosion and sediment yield increase comparatively low. This differed from European settlers arriving in the eighteenth century. Their land-use change affected lower and upper catchments and sediment yields increased by 660% (Overeem et al. 2013).

A direct link between land-use change, soil erosion rates, and alluvial sediments is hard to prove. Dating is usually difficult due to the reworking of sediment layers in rivers (Dotterweich 2008; Dreibrodt et al. 2010). Few case studies have investigated, however, the link between increased alluvial sedimentation, land-use change, and extreme precipitation events (Dotterweich 2008; Lang 2003; Lang et al. 2000). Giosan et al. (2012) demonstrated that long-term land-use change in the Danube catchment contributed in the Holocene and, in particular, over the last 1000 years to the evolution of the Danube delta. Human impacts vs. long-term historical climate and subsequent hydrology changes were examined as possible drivers of increased sediment storage rates, and Giosan et al. (2012) found that land-use change was the main factor. Sedimentation rates increased, in particular, after land clearance, affecting also the lower Danube at larger scales during the last two centuries (see also McCarney-Castle et al. 2012). Maselli and Trincardi (2013) found similar trends when comparing the Ebro, Rhona, Po, and Danube. They found two main phases of delta growth. One synchronous increase happened during Roman times under relatively warm climatic conditions, a second during the Little Ice Age. The latter shows, however, slight temporal differences since delta growth coincides temporally mainly for the Ebro, Rhone, and Po (between the sixteenth and twentieth centuries), whereas in the Danube delta growth was found mostly in the nineteenth century and thereafter. Alterations of morphological river types and subsequent habitat change affected riverine fish assemblages as it was shown by Pont et al. (2009) for the Drome River, a tributary of the French Rhone.

2.2.7 Fisheries: Intended and Unintended Dispersal of Nonnative Species

Most human uses and their ecological impacts changed aquatic biota indirectly via habitat modification. Fishing was one exception as it altered stocks directly. Also, until the twentieth century, the appearance of nonnative species was caused mainly by deliberate introductions by fisheries management (however, cf. nonnative fish distribution as a consequence of artificial shipping canals above). Only during the twentieth century, the unintended dispersal of nonnative and invasive fish species and other aquatic animals and plants via transport means increased drastically.

It is evident that fishing put direct stress on the targeted fish populations and changed species assemblages already centuries ago. A remarkable recorded example of medieval overexploitation is the Alpine Zellersee in Austria. After the 1360s, fishermen delivered each year 27,000 whitefish (*Coregonus* sp.) and 18 lake trout (*Salmo trutta*) to the archbishop of Salzburg, taking themselves even more for their own use. Only some decades later the whitefish population collapsed. Pike (*Esox lucius*) was stocked to replace it. When predating pikes had soon diminished trout stocks, only then did the fishing communities decide to reduce fishing pressure (Freudlsperger 1936).

Particularly subjected to overexploitation were diadromous fish because of their predictable spawning runs during which large amounts could be caught. For example, archaeological sturgeon remains from the southern Baltics demonstrate a decrease of average size of specimen and a decline of the percentage in total consumption from 70% in the seventh, eighth, and ninth centuries to only 10% in the twelfth and thirteenth centuries CE. Benecke (1986) clearly attributed this change to overfishing. Weirs built since the High Middle Ages in Europe supported overexploitation (Hoffmann 1996).

Such evidence for declining fish populations are rare for the medieval and even for modern periods. Due to lack of written historical sources that enable tracing depletion of certain fish species and their stocks, it is hardly possible to directly quantify losses before the twentieth century. Nevertheless, some indications help explain the preindustrial decline of fish. As mentioned already above, the latter can be concluded indirectly from fishing laws that were issued in Europe since the thirteenth century (Hoffmann 1996). The laws aimed first at protecting juveniles by regulating minimum lengths or weights of individuals, by forbidding harmful fishing gear, or by defining closed seasons. In contrast, habitat protection is rather a practice of the nineteenth century and afterward.

While overexploitation of fish in the medieval and early modern period took place especially in European countries, North America and Australia followed this pattern after the colonization of European settlers. Travelers' accounts describe the wealth of freshwater fish, e.g., in the Ohio River which was said to have been inhabited by enormous numbers of pike, walleye, catfish, buffalo fish, suckers, drum, and sturgeon as well as small fish such as sand darters, chub, riffle darters, and minnows (Trautman 1981 cited from Humphries and Winemiller 2009). Massive exploitation with a variety of fishing nets, dams to support fishing, as well as milldams hampering fish migration soon raised concern of overexploitation. As in Europe, also in North America fishing regulations followed. The number of fishing days per week was reduced, fishing gear regulated, and closed seasons defined, for example, in Massachusetts in 1710, in Connecticut in 1715, or in Rhode Island in 1735 (Humphries and Winemiller 2009). Sturgeon fishes (*Acipenser oxyrinchus*, *A. brevirostris*), salmon, or shad (*Clupea sapidissima*) were among the fish stocks which have been overfished so heavily that fishing them in the seventeenth and eighteenth centuries was stopped several times after few years of fishing because stocks were too low (Lichter et al. 2006). In North America, the settlers also established a lively beaver trading industry. Hunting beavers began in the early seventeenth century. Between 1630 and 1640, 80,000 individuals were caught annually. By 1900, this species was more or less extinct in North America (Naiman et al. 1988; cited from Humphries and Winemiller 2009). In the late nineteenth and twentieth century, river channelization, flood protection dikes, hydropower dams, and pollution added to the adverse effects of fish overexploitation in most of Western rivers. It is assumed that in Europe 13 fish species have gone extinct since 1700 (Kottelat and Freyhof 2007). A large number of fish species is threatened, especially less tolerant species requiring specific habitats.

Purposeful and unintended species introduction contributed to large-scale changes in fish assemblages. Fish pond networks and fish breeding programs were established to ensure a sufficient and steady supply of a resource that is naturally only seasonally available. Historical records confirm this started in Western Europe in the eleventh century and spread eastward in the twelfth and thirteenth centuries (Hoffmann 1996). Originally, different kinds of cyprinids were raised in the ponds because they could tolerate consistently warm temperatures. Soon, carp (*Cyprinus carpio*), a fish species native to the middle and lower Danube watershed, became the main species as they tolerate longer land transport, have a high fecundity, and grow relatively fast. The earliest traces mark the spread of carps to the upper Danube, the Elbe, or the Rhine in the eleventh and twelfth centuries and to the Maas, Seine, or upper Rhone in the late twelfth and thirteenth centuries. The dispersal into central Bohemia, Southern Poland, the Loire, and southern England happened in the Late Middle Ages. From ponds, carp reached natural waters and had colonized suitable habitats in most of Central, Western, and Northern Europe by 1600 (Hoffmann 1996).

It can be assumed that with the transfer of carp also other species were unintentionally spread and colonized new river systems. Evidence suggests that Bitterling (*Rhodeus amarus*) was introduced to many rivers of Central and Western Europe in a first wave already in the High and Late Middle Ages (1150–1560) together with carp (Damme et al. 2007). It is not possible to trace the origin of tench (*Tinca tinca*) in sixteenth century Spain where it occurred together with carp (Clavero and Villero 2014).

In contrast to many other domesticated animal and plant species, which were transferred purposefully between the continents after the discovery of the Americas, the so-called Columbian Exchange hardly affected riverine environments in the

Early Modern Period, i.e., the sixteenth and seventeenth century (Crosby 1972). A few—though delayed—exceptions are ornamental fish or species that were introduced to help fighting mosquitos. The goldfish (*Carassius auratus*) was brought to Portugal in 1611. In England and France, it was imported in eighteenth century (Copp et al. 2005). The mosquitofish (*Gambusia holbrooki*) was introduced in Europe in the 1920s (Vidal et al. 2010).

Introduction of nonnative fish species and the large-scale spread of invasive fish are clearly attributed to industrialized rivers. With railways, fresh fish could be imported in unprecedented quantities to continental areas. In Vienna, for example, the import of fish from the North Sea started in 1899 when a German steam fish trading company opened its first stand on the Viennese fish market. Only due to these imports the yearly amounts sold on the market could increase from 600 to 2250 tons between 1880 and 1914, securing fish as nutrition for the heavily growing population (Jungwirth et al. 2014). Concurrently, local fish stocks in the Danube exhibited a clear downward trend as they started to be seriously affected by systematic channelization measures for navigation and partly for flood protection.

Although fisheries can be seen as victims of the industrialization of rivers, fishermen eagerly adapted new technologies, thereby contributing seriously on their own to the change of riverine fish assemblages. They began artificial fish breeding and stocking and often the efforts of European fishermen targeted North American fish species since they were considered faster growing and sometimes also better adapted to channelized habitats. Intentional fish translocations happened on a continental as well as an intercontinental scale. In Europe, for instance, catfish (*Silurus glanis*) or pike-perch (*Sander lucioperca*) were introduced in Western Rivers in the nineteenth century (see, e.g., Copp et al. 2005). Modern steam ships enabled relatively easy exchange between the continents, first and foremost between Europe and North America. Rainbow trout (*Oncorhynchus mykiss*)—native to North American and North Asian streams of the Pacific—was one of the main species. In the USA, its artificial breeding for stocking of native and nonnative environments started in the 1870s (Halverson 2010). Import to Europe followed soon after in the 1880s. Brook trout (*Salvelinus fontinalis*), brown bullhead (*Ameiurus nebulosus*), pumpkinseed (*Lepomis gibbosus*), or smallmouth bass (*Micropterus dolomieu*) were other target species. Some of the nonnative species introduced in Europe established self-sustaining populations, e.g., rainbow trout or brook trout (Copp et al. 2005).

2.3   Conclusions

The historical evolution of river uses and resulting ecological impacts exhibit clear temporal patterns. It is evident that human alterations have been numerous for millennia. Preindustrial effects were mostly local and regional, and human practices, such as passive flood protection, were designed to adapt to, not control, the dynamics of rivers. This relates, for instance, to ancient Egypt and likewise to European preindustrial practices of flood protection that depended on measures to mitigate flood damages (see Chap. 28). In intensely populated regions, such adaptive practices at local scales could aggregate up to larger-scale effects. Characteristic for preindustrial rivers is that local aquatic environmental resources were essential for societies. Since substitution by trade was not yet possible, harmonizing a variety of uses was indispensable. This helped keep ecological impacts at low levels. Preindustrial societies, nevertheless, initiated long-term changes of river ecosystems that might influence them even in the present era. Land-use change and erosion as well as weirs as sediment traps are prime cases. Although reliable and detailed records are scarce, it seems that subsequent examples include stocking of nonnative fish species and unintended expansion of fish and other species, for instance, via shipping canals built in and after the seventeenth century contributed to early modifications of aquatic biota and biotic communities. One should note that, in contrast to (well-studied) marine systems (see, e. g., Jackson et al. 2001), in rivers overexploitation, primarily of aquatic animals, was soon followed by effects of other human uses on habitat conditions.

Industrialization had large-scale effects on river uses and their impacts on morphology, hydrology, and aquatic biota. The use of fossil energy enabled intensification of uses with unprecedented ecological consequences. Well into the twentieth century, deteriorating water quality and hydromorphological degradation were perceived as a necessary evil to foster economic development. Riverine impairment peaked in response to a combination of intensifying factors: increasing resource exploitation and use, a rising density of machinery in industry and private households, intensified agriculture driven by an ever-increasing number of machines, as well as fertilizers and pesticides.

As a response in the late 1980s and 1990s, river restoration projects were planned and implemented. Especially in densely populated areas and centers of economic production, rivers and their biotic communities often have been degraded so drastically that restoration toward a natural status appears impossible within any foreseeable political time frame (see, e.g., Hughes et al. 2005; Dufour and Piégay 2009). In addition, some external factors, namely, climate and thus hydrology and temperature, changed naturally as well as due to human impacts for more than a century. This further prevents restoration of presumed pristine conditions. While this might confine the role of history in defining reference conditions, historical investigation of rivers can nevertheless add valuable insights into their trajectories and help explaining the origins of present conditions.

References

Allan JD, Abell R, Hogan Z, Revenga C, Taylor BW, Welcomme RL, Winemiller K (2005) Overfishing of inland waters. Bioscience 55:1041–1051

Barles S (2007) Feeding the city: food consumption and flow of nitrogen, Paris, 1801–1914. Sci Total Environ 375:48–58

Barles S (2012) The seine and Parisian metabolism: growth of capital dependencies in the 19th and 20th centuries. In: Castonguay S, Evenden MD (eds) Urban waters: rivers, cities and the production of space in Europe and North America. Pittsburgh University Press, Pittsburgh, pp 94–112

Benecke N (1986) Some remarks on sturgeon fishing in the southern Baltic region in medieval times. In: Brinkhuizen DC, Clason AT (eds) Fish and archaeology. British Archaeological Reports, Oxford, pp 9–17

Billen G, Garnier J, Deligne C, Billen C (1999) Estimates of early-industrial inputs of nutrients to river systems: implication for coastal eutrophication. Sci Total Environ 243:43–52

Billen G, Garnier J, Barles S (2012) History of the urban environmental imprint: introduction to a multidisciplinary approach to the long-term relationships between western cities and their hinterland. Reg Environ Chang 12:249–253

Bork H-R, Bork H, Dalchow C, Faust B, Piorr H-P, Schatz T (1998) Landschaftsentwicklung in Mitteleuropa (Wirkungen des Menschen auf Landschaften). Klett-Perthes, Gotha

Braudel F (1986) Aufbruch zur Weltwirtschaft. Sozialgeschichte des 15.-18. Jahrhunderts, vol 3. Kindler, München

Brázdil R, Pfister C, Wanner H, Storch HV, Luterbacher J (2005) Historical climatology in Europe – the state of the art. Clim Chang 70:363–430

Brolsma JU (2011) A brief history of inland navigation and waterways - the development of the waterway infrastructure in the Netherlands. RWS Centre for Transport and Navigation, Delft

Bundi M (1988) Von Fischpässen und -treppen einst und heute. Bündner Kalender 1988:70–73

Camargo JA (2002) Contribution of Spanish-American silver mines (1570–1820) to the present high mercury concentrations in the global environment: a review. Chemosphere 48:51–57

Clavero M, Villero D (2014) Historical ecology and invasion biology: long-term distribution changes of introduced freshwater species. Bioscience 64:145–153

Copp GH, Bianco PG, Bogutskaya NG, Erős T, Falka I, Ferreira MT, Fox MG, Freyhof J, Gozlan RE, Grabowska J, Kovac V, Moreno-Amich R, Naseka AM, Penaz M, Povz M, Przybylski M, Robillard M, Russell IC, Stakenas S, Sumer S, Vila-Gispert A, Wiesner C (2005) To be, or not to be, a non-native freshwater fish? J Appl Ichthyol 21:242–262

Crosby AW (1972) The Columbian exchange: biological and cultural consequences of 1492. Greenwood Press, Westport Conn

Damme DV, Bogutskaya N, Hoffmann RC, Smith C (2007) The introduction of the European bitterling (Rhodeus Amarus) to west and Central Europe. Fish Fish 8:79–106

Dotterweich M (2008) The history of soil erosion and fluvial deposits in small catchments of Central Europe: deciphering the long-term interaction between humans and the environment – a review. Geomorphology 101:192–208

Dotterweich M, Dreibrodt S (2011) Past land use and soil erosion processes in Central Europe. PAGES News 19:49–51

Downs P, Gregory K (2004) River Channel management: towards sustainable catchment hydrosystems. Taylor & Francis, New York

Dreibrodt S, Lubos C, Terhorst B, Damm B, Bork HR (2010) Historical soil erosion by water in Germany: scales and archives, chronology, research perspectives. Quat Int 222:80–95

Dufour S, Piégay H (2009) From the myth of a lost paradise to targeted river restoration: forget natural references and focus on human benefits. River Res Appl 25:568–581

EEA – European Environment Agency (2001) Late lessons from early warnings: the precautionary principle 1896–2000. EEA, Luxembourg

EEA – European Environment Agency (2013) Late lessons from early warnings: science, precaution, innovation. EEA, Luxemburg

Foucher D, Ogrinc N, Hintelmann H (2009) Tracing mercury contamination from the Idrija mining region (Slovenia) to the Gulf of Trieste using Hg isotope ratio measurements. Environ Sci Technol 43:33–39

Freudlsperger H (1936) Kurze Fischereigeschichte des Erzstiftes Salzburg. Mitteilungen der Gesellschaft für Salzburger Landeskunde 76:81–128

Gierlinger S, Haidvogl G, Gingrich S, Krausmann F (2013) Feeding and cleaning the city: the role of the urban waterscape in provision and disposal in Vienna during the industrial transformation. Water Hist 5:219–239

Giosan L, Coolen MJL, Kaplan JO, Constantinescu S, Filip F, Filipova-Marinova M, Kettner AJ, Thom N (2012) Early anthropogenic transformation of the danube-black sea system. Sci Rep 2:1–6

Gleick PH (ed) (1993) Water in crisis. A guide to the world's freshwater resources. Oxford University Press, New York

Gosar M (2008) Mercury in river sediments, floodplains and plants growing thereon in drainage area of Idrija mine, Slovenia. Pol J Environ Stud 17(2):227–236

Green PA, Vörösmarty CJ, Meybeck M, Galloway JN, Peterson BJ, Boyer EW (2004) Pre-industrial and contemporary fluxes of nitrogen through rivers: a global assessment based on typology. Biogeochemistry 68:7–105

Gregory KJ (2006) The human role in changing river channels. Geomorphology 79:172–191

Haidvogl G (2008) Von der Flusslandschaft zum Fließgewässer: die Entwicklung ausgewählter österreichischer Flüsse im 19. und 20. Jahrhundert mit besonderer Berücksichtigung der Kolonisierung des Überflutungsraums. Dissertation Universität Wien

Haidvogl G, Preis S (2003) Anthropogene Nutzungen und Eingriffe in und an der Möll um 1830 am Beispiel von zwei ausgewählten Abschnitten. Un-published dataset. Institut für Hydrobiologie und Gewässermanagement, Universität für Bodenkultur, Vienna

Haidvogl G, Guthyne-Horvath M, Gierlinger S, Hohensinner S, Sonnlechner C (2013) Urban land for a growing city at the banks of a moving river: Vienna's spread into the Danube island Unterer Werd from the late 17th to the beginning of the 20th century. Water Hist 5:195–217

Haidvogl G, Lajus D, Pont D, Schmid M, Jungwirth M, Lajus J (2014) Typology of historical sources and the reconstruction of long-term historical changes of riverine fish: a case study of the Austrian Danube and Northern Russian rivers. Ecol Freshw Fish 23:498–515

Haidvogl G, Hoffmann R, Pont D, Jungwirth M, Winiwarter V (2015) Historical ecology of riverine fish in Europe. Aquat Sci 77:315–324

Halverson A (2010) An entirely synthetic fish: how rainbow trout beguiled America and overran the world. Yale University Press, London

Hassan F (2011) Water history for our times. IHP essays on water history 2. Unesco, Paris

Higgs E, Falk DA, Guerrini A, Hall M, Harris J, Hobbs RJ, Jackson ST, Rhemtulla JM, Throop W (2014) The changing role of history in restoration ecology. Front Ecol Environ 12:499–506

Hoffmann RC (1996) Economic development and aquatic ecosystems in medieval Europe. Am Hist Rev 101:631–669

Hohensinner S, Habersack H, Jungwirth M, Zauner G (2004) Reconstruction of the characteristics of a natural alluvial river-floodplain system and hydromorphological changes following human modifications: the Danube river (1812-1991). River Res Appl 20:25–41

Hohensinner S, Lager B, Sonnlechner C, Haidvogl G, Gierlinger S, Schmid M, Krausmann F, Winiwarter V (2013) Changes in water and land: the reconstructed Viennese riverscape from 1500 to the present. Water Hist 5:145–172

Hughes FMR, Colston A, Mountford JO (2005) Restoring riparian ecosystems: the challenge of accommodating variability and designing restoration trajectories. Ecol Soc 10. http://www.ecologyandsociety.org/vol10/iss1/art12/

Humphries P, Winemiller K (2009) Historical impacts on river fauna, shifting baselines, and challenges for restoration. Bioscience 59:673–684

ICOLD – International Commission on Large Dams (2007) Dams and the world's water. An educational book that explains how dams help to manage the world's water. ICOLD, Paris

ICOLD – International Commission on Large Dams (2016) http://www.icold-cigb.net/GB/World_register/general_synthesis.asp?IDA=212. Accessed 18 Jul 2016

ICPDR – International Commission for the Protection of the Danube River (2009) Danube river basin management plan. ICPDR, Vienna

Jackson JBC, Kirby MX, Berger WH, Bjorndal KA, Botsford LW, Bourque BJ, Bradbury RH, Cooke R, Erlandson J, Estes JA, Hughes TP, Kidwell S, Lange CB, Lenihan HS, Pandolfi JM, Peterson CH, Steneck RS, Tegner MJ, Warner RR (2001) Historical overfishing and the recent collapse of coastal ecostystems. Science 293:629–637

Jakobsson E (2002) Industrialization of rivers: a water system approach to hydropower development. Knowl Technol Policy 14:41–56

Jungwirth M, Haidvogl G, Hohensinner S, Waidbacher H, Zauner G (2014) Österreichs Donau. Landschaft – Fisch – Geschichte. Institut für Hydrobiologie und Gewässermanagement. Universität für Bodenkultur, Wien

Kottelat M, Freyhof J (2007) Handbook of European freshwater fishes. Eigenverlag, Cornol

Kraikovsky A, Lajus J (2010) The neva as a metropolitan river of Russia: environment, economy and culture. In: Tvedt T, Coopey R (eds) A history of water, series II, 2, rivers and society: from early civilizations to modern times. I.B. Tauris, London, pp 339–364

Lang A (2003) Phases of soil erosion-derived colluviation in the loess hills of South Germany. Catena 51:209–221

Lang A, Preston N, Dickau R, Bork H-R, Mäckel R (2000) LUCIFS – examples from the Rhine catchment. PAGES News 8:11–13

Lehner B, Reidy Liermann C, Revenga C, Vörösmarty C, Fekete B, Crouzet P, Döll P, Endejan M, Frenken K, Magome J, Nilsson C, Robertson JC, Rödel R, Sindorf N, Wisser D (2011a) Global reservoir and dam database, version 1 (GRanDv1): dams, revision 01, NASA Socioeconomic Data and Applications Center (SEDAC), Palisades, New York. https://doi.org/10.7927/H4N877QK. Accessed 4 Oct 2017

Lehner B, Liermann CR, Revenga C, Vörösmarty C, Fekete B, Crouzet P, Döll P, Endejan M, Frenken K, Magome J, Nilsson C, Robertson JC, Rödel R, Sindorf N, Wisser D (2011b) High-resolution mapping of the world's reservoirs and dams for sustainable river-flow management. Front Ecol Environ 9:494–502

Leitholdt E, Zielhofer C, Berg-Hobohm S, Schnabl K, Kopecky-Hermanns B, Bussmann J, Härtling JW, Reicherter K, Unger K (2012) Fossa Carolina: the first attempt to bridge the Central European watershed – a review, new findings, and geoarchaeological challenges. Geoarchaeology 27:88–104

Lichter J, Caron H, Pasakarnis TS, Rodgers SL, Squiers TS Jr, Todd CS (2006) The ecological collapse and partial recovery of a freshwater tidal eco system. Northeast Nat 13:153–178

Liu C, Kroeze C, Hoekstra AY, Gerbens-Leenes W (2012) Past and future trends in grey water footprints of anthropogenic nitrogen and phosphorus inputs to major world rivers. Ecol Indic 18:42–49

Maselli V, Trincardi F (2013) Man made deltas. Sci Rep 3:1926

McCarney-Castle K, Voulgaris G, Kettner AJ, Giosan L (2012) Simulating fluvial fluxes in the Danube watershed: the 'little ice age' versus modern day. The Holocene 22:91–105

Micklin P (2007) The Aral Sea disaster. Annual review earth. Planet Sci 35:47–72

Mieck I (1981) Die Anfänge der Umweltschutzgesetzgebung in Frankreich. Francia 9:332

Möser K (2008) Prinzipielles zur Transportgeschichte. In: Sieferle RP (ed) Transportgeschichte. Der Europäische Sonderweg, vol 1. LIT, Berlin, pp 39–78

Naiman RJ, Johnston CA, Kelley JC (1988) Alteration of North American streams by beaver. Bioscience 38:753–762

Nelva A (1997) La pénétration du Hotu, Chondrostoma nasus nasus (Poisson Cyprinidé), dans le réseau hydrographique français et ses conséquences. Bull Fr Pêche Piscic 344-345:253–269

Nriagu JO (1994) Mercury pollution from the past mining of gold and silver in the Americas. Sci Total Environ 149:167–181

Overeem I, Kettner AJ, Syvitski JPM (2013) 9.40 impacts of humans on river fluxes and morphology A2. In: Shroder John F (ed) Treatise on geomorphology. Academic Press, San Diego, pp 828–842

Petts GE, Möller H, Roux AL (eds) (1989) Historical change of large alluvial rivers: Western Europe. Wiley, Chichester

Poff NL, Hart DD (2002) How dams vary and why it matters for the emerging science of dam removal. Bioscience 52:659–668

Pont D, Piégay H, Farinetti A, Allain S, Landon N, Liébault F, Dumont B, Richard-Mazet A (2009) Conceptual framework and interdisciplinary approach for the sustainable management of gravel-bed rivers: the case of the Drome River basin (S.E. France). Aquat Sci 71:356–370

Pounds N (1979) An historical geography of Europe. 1500-1840. Cambridge University Press, Cambridge

Raab A (1978) Die traditionelle Fischerei in Niederösterreich, mit besonderer Berücksichtigung der Ybbs, Erlauf, Pielach und Traisen. Dissertation, Universität Wien, Wien

Rosenberg DM, McCully P, Pringle CM (2000) Global-scale environmental effects of hydrological alterations: introduction. Bioscience 50:746–751

Sieferle RP (2006) Das Ende der Fläche – zum gesellschaftlichen Stoffwechsel der Industrialisierung. Böhlau, Köln

Sieferle RP (ed) (2008) Transportgeschichte. Der Europäische Sonderweg, vol 1. LIT, Berlin

Stergiouli ML, Hadjibiros K (2012) The growing water imprint of Athens (Greece) throughout history. Reg Environ Chang 12:337–345

Strayer DL, Dudgeon D (2010) Freshwater biodiversity conservation: recent progress and future challenges. J N Am Benthol Soc 29:344–358

Tello E, Ostos J (2012) Water consumption in Barcelona and its regional environmental imprint: a long-term history (1717–2008). Reg Environ Chang 12:347–361

Torkar G, Zwitter Ž (2015) Historical impacts of mercury mining and stocking of non-native fish on ichthyofauna in the Idrijca River basin, Slovenia. Aquat Sci 77:381–393

Trautman MB (1981) The fishes of Ohio: with illustrated keys. Ohio State University Press, in collaboration with the Ohio Division of Wildlife and the Ohio State University Development Fund

Vella C, Leveau P, Provansal M, Gassend JM, Maillet B, Sciallano M (1999) Le canal de Marius et les dynamiques littorales du golfe de fos. Gallia 56:131–139

Vidal O, García-Berthou E, Tedesco PA, García-Marín J-L (2010) Origin and genetic diversity of mosquitofish (Gambusia holbrooki) introduced to Europe. Biol Invasions 12:841–851

Vitousek PM (1994) Beyond global warming: ecology and global change. Ecology 75:1861–1876

Vogemont L (1712) Teutschlands vermehrter Wohlstand, oder Vorstellung einer grundmässigen Einrichtung der Handlung, wie nemblich solche in Teutschland, durch Schiffreichmachung derer Flüsse, zu wegen gebracht werden könne. Johann Georg Schlegl, Wien

Vörösmarty CJ, Meybeck M, Pastore CL (2015) Impair-then-repair: a brief history & global-scale hypothesis regarding human-water interactions in the anthropocene. Daedalus 144:94–109

Walter RC, Merritts DJ (2008) Natural streams and the legacy of water-powered mills. Science 319:299–304

Zarfl C, Lumsdon A, Berlekamp J, Tydecks L, Tockner K (2015) A global boom in hydropower dam construction. Aquat Sci 77:161–170

Zösmair J (1886) Die Geschichte der Fischerei in der Ill. Feldkircher Zeitung 26(97):1–3

Chapter 3
River Morphology, Channelization, and Habitat Restoration

Severin Hohensinner, Christoph Hauer, and Susanne Muhar

3.1 River Channels as One Piece in the Puzzle

Authorities and planners involved in river restoration projects often tend to focus on the hydromorphological state of a short river reach or certain aquatic habitats where the pending deficits are most evident. Nevertheless, for long-term and sustainable restoration, one should also consider flood dynamics and other interlinked processes at larger spatiotemporal scales, ideally at the catchment scale. Moreover, restoring river morphology also calls for the consideration of the dynamic processes of the whole fluvial system, including the adjacent floodplains, with its diverse interactions between the physical environment (morphology, flow, sediment, etc.) and the riverine coenoses (compare EU Water Framework Directive 2000).

Various concepts in river morphology and ecology address fluvial systems as hierarchical arrangements that integrate typical geomorphic and ecological features over a range of spatial scales. Such well-established schemes are, e.g., the Hierarchical Framework of Stream Habitats (Frissell et al. 1986), the Hydrosystem Approach (Petts and Amoros 1996), the Hierarchical Patch Dynamics Model (Wu and Loucks 1995), the River-Scaling Concept (Habersack 2000), or the Riverine Ecosystem Synthesis (Thorp et al. 2006). They have in common that riverine structures at the local scale are viewed as habitats nested in larger systems at reach scale or catchment scale.

S. Hohensinner (✉) · S. Muhar
Institute of Hydrobiology and Aquatic Ecosystem Management, University of Natural Resources and Life Sciences, Vienna, Austria
e-mail: severin.hohensinner@boku.ac.at; susanne.muhar@boku.ac.at

C. Hauer
Christian Doppler Laboratory for Sediment Research and Management, Institute of Water Management, Hydrology and Hydraulic Engineering, University of Natural Resources and Life Sciences, Vienna, Austria
e-mail: christoph.hauer@boku.ac.at

© The Author(s) 2018
S. Schmutz, J. Sendzimir (eds.), *Riverine Ecosystem Management*, Aquatic Ecology
Series 8, https://doi.org/10.1007/978-3-319-73250-3_3

According to the River Styles Framework, introduced by Brierley et al. (2002), an organism existing in a local habitat is exposed to controls and biophysical fluxes associated with larger spatial entities. These entities exist as a nested hierarchy that builds up from "hydraulic units" as the smallest up to larger "geomorphic units", "river reaches" and "landscape units" and, finally, up to the catchment and ecoregion as the largest spatial scales. These fluvial features can be seen as physical templates that provide the setting in which ecological processes operate and shape riverine coenoses.

Focusing on the ecological functions and the associated biocoenoses of these different spatial entities, aquatic ecologists generally apply the terms micro-, meso-, or macrohabitats. Confusingly, to date, no consistent definition exists that includes both the geomorphological and the ecological perspectives. A microhabitat, roughly corresponding to "hydraulic units," refers to a particular site used by an individual for specific behaviors (e.g., spawning). It can be described by a combination of distinct hydraulic and physical factors such as flow velocity, depth, substrate type, and vegetation cover. Depending on the species (fish, invertebrates, macrophytes, algae, etc.) and the life stage, microhabitats may range from near zero to a few meters. Mesohabitats, typically encountered at the scale of "hydraulic" and "geomorphic units," denote discrete patches of a river channel defined by similar physical characteristics. Such habitats include shallow riffles, deep pools, runs showing high flow velocities, or sediment bars. Depending on the river type, mesohabitats commonly extend over a few square meters but may also cover some hundreds of square meters. While microhabitats refer to sites of individual organisms, mesohabitats can be seen as the area, where aquatic communities and/or specific life stages with similar habitat requirements live (spawning sites, juveniles, adults, etc.). Macrohabitats, spatially best associated with "geomorphic units" or river reaches, typically comprise several mesohabitats shaped by the particular hydromorphological conditions of the respective river reach, branch, or water body (e.g., lotic main channel of an anabranched river, lentic one-side connected backwater, stagnant dead arm). Accordingly, longitudinal continuity and lateral hydrological connectivity and, thus, the distribution and migration possibilities of aquatic organisms are key features for defining macrohabitats.

The different fluvial features—or habitats from the ecological point of view—including those in the adjacent floodplains, undergo permanent hydromorphological and ecological changes owing to influences and fluxes, such as flow and sediments, from the reach or catchment scale. Such adaptive processes of riverine features at a certain spatial scale are also pertinent to specific time scales. The evolution of a new river terrace, for example, usually encompasses longer time spans than the formation of a gravel bar. In many cases, the consequences of physical modifications on the fluvial system are not immediately apparent. Rather, they depend on system-inherent thresholds of response and manifold legacy effects.

Understanding the complex spatiotemporal nature of river landscapes is an essential prerequisite for sustainable and integrative river restoration. However, under daily pressure to balance short-term demands with scarce financial means, the consideration of such complex process-response systems is a challenging task for planners and authorities as well (see Chaps. 15 and 16).

3.2   River Types: Complex Diversity or Confusing Variety?

River systems in the industrialized world today have largely lost their original characteristics. Primarily evident is the disappearance of channel patterns of preindustrial rivers. Such patterns range from deeply incised bedrock channels (gorges) in the headwaters to alluvial anastomosing rivers in the lowlands close to the estuary. Over decades, a confusing number of river classification schemes have been developed to address the various river types from scientific, administrative, or restoration perspectives. In addition, even the terms used to describe specific river types are not applied in a consistent manner in scientific literature. For example, the terms "braiding", "anabranching" or "anastomosing" are sometimes used in a broader sense to describe rivers that show bifurcations in general and in a closer sense in order to explicitly address certain channel styles (Kondolf et al. 2003; Eaton et al. 2010).

Generally, the various classification systems can be distinguished between form-based and process-based schemes. In the first case, rivers are categorized by means of several channel characteristics, such as sinuosity, number of braids, typical forms of cross sections, width-depth ratios, type of substrate, channel slope, etc. (e.g., classification according to Rosgen 1994, 1996). Such descriptive schemes can be used to characterize a channel system in detail; however, it does not provide much information about the underlying fluvial processes, neglects the history of the landscape system, and is of limited value in predicting future channel changes. Accordingly, from the perspective of river management, so-called process-based classification schemes are more useful. They offer a useful framework for assessing potential channel dynamics based on how current forms are shaped by controlling geomorphic processes (e.g., Schumm et al. 1984; Church 1992; Simon et al. 2007). Here, quantitative empirical models provide the best foundation to analyze river forms and to assess the adequacy of management strategies. Based on the early work of Leopold and Wolman (1957), meanwhile, numerous classification systems have emerged that extended our understanding about the relationship between fluvial forms and geomorphic processes. Most schemes are based on critical thresholds with respect to discharge and channel slope (i.e., stream power), sediment volume, and median grain size (see Chap. 8). Other schemes also include bank resistance, the influence of riparian vegetation, and more complex control factors (e.g., Osterkamp 1978; Ferguson 1987; Van den Berg 1995; Yalin and da Silva 2001). The classification of rivers as straight, meandering, and braided originally introduced by Leopold and Wolman (1957) has therefore been substantially expanded.

Today, we understand the complex morphological diversity of rivers as a continuum of fluvial patterns that evolved as a consequence of the given boundary conditions, such as upstream catchment size and its vegetation cover, lateral valley confinement, valley slope, flow regime and sediment type, and transport of material. Channel geometry, patterns, and dimensions reflect the ongoing adjustment to fluctuating flow and sediment yields (bedload/suspended load) and, consequently, the balance of erosional and depositional processes. Here, the concept of stream

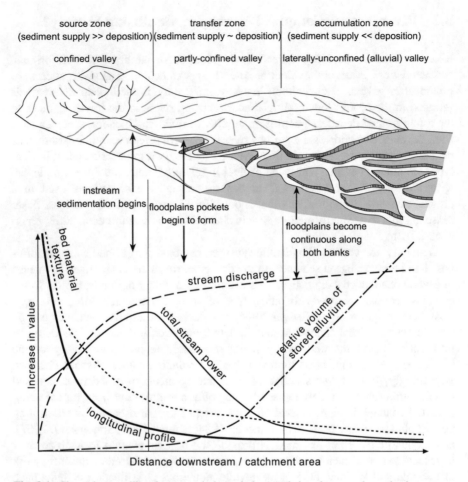

Fig. 3.1 Channel controlling factors and channel characteristics along a schematic river course (source: © 2013 by Kirstie A. Fryirs and Gary J. Brierley; reproduced with the permission of John Wiley & Sons)

power, the product of discharge and channel slope, provides a useful tool to describe the capacity of a river to mobilize and transport material. Comparing stream power and sediment load combined with sediment size helps to identify potential channel adjustments (compare Lane 1954; see Chap. 8).

In an ideal world, the hereinafter described typical sequence of channel patterns (river types) would be identified along a river's course from up- to downstream depending on the abovementioned channel controls (compare Figs. 3.1 and 3.2). In reality, depending on the individual geomorphological setting, rivers may also develop channel forms in mountainous regions that typically would be expected along their lower courses.

In alpine or mountainous headwaters, bedrock-confined rivers that have to follow a narrow and steep valley floor are typical. Stepped-bed profiles with cascades and

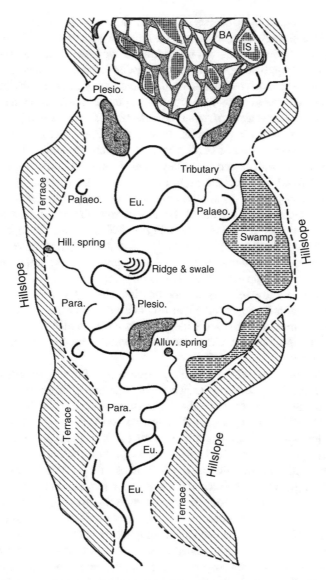

Fig. 3.2 Basic geomorphological features of an idealized river corridor and surface water bodies showing different intensities of hydrological connectivity: *Eu* eupotamal/eurhithral (main channel and lotic side arms), *Para* parapotamal/pararhithral (abandoned braids), *Plesio* plesiopotamal/plesiorhithral (dead arms close to the main channel), *Palaeo* palaeopotamal/palaeorhithral ("oxbow lakes"—abandoned meander bends remote from the main channel), *L* lateral or riparian lake, *BA* bar, *IS* vegetated island (Based on Amoros et al. 1987; modified according to Ward et al. 2000)

pools in combination with coarse sediment load up to the size of boulders are characteristic elements of such rivers. Denudation processes, gully erosion, and channel incisions prevail, and, accordingly, the steep headwaters can be referred to as the sediment supply zones of river systems. In broader valleys, braided rivers carrying coarse gravel may stretch over the whole valley floor. Flashy flow regimes combined with an excess of bedload provide the pulsing power and material to build such river types. Bar-braided rivers almost devoid of vegetated islands indicate a predominance of turnover processes. In island-braided rivers, fluvial dynamics enable at least the evolution of small, vegetated islands on temporally stable gravel bars. As the valley widens and the valley sides do not yet confine the whole river section, small floodplain pockets begin to form. Because discharge increases progressively with catchment area, total stream power typically peaks along a river course in that section downstream of the headwaters where sufficient flow acts on sufficiently steep slopes (Brierley and Fryirs 2005). Here, the upstream zone, characterized by prevailing sediment supply, commonly passes into the sediment transfer zone, where erosional and depositional processes are approximately balanced. If the transport capacity of the river is sufficient or in case of reduced bedload input, e.g., due to a low relief landscape that is tectonically stable, less braided or even sinuous channels may evolve that oscillate between both sides of the valley. Today, such channel patterns are widespread in alpine valleys. However, in most cases, they are products of channelization programs in the nineteenth or early twentieth century.

Further downstream, where the valley bottom significantly widens or the river course enters spacious alluvial plains, we usually find fluvial forms that probably refer to the most common river type worldwide. These show an extraordinary morphological diversity: anabranching rivers (Huang and Nanson 2007). They range from dynamic high- and medium-energy rivers to low-energy systems dominated by accumulation processes. Such river types can be considered as transition forms between braiding and meandering rivers, because they feature characteristics of both. Wandering gravel-bed rivers, as the high-energy variant of anabranching rivers, are mostly located along the sediment transfer zone and may constitute the beginning of the sediment accumulation zone, where the coarse bedload is deposited (Desloges and Church 1987). They usually exhibit a complex channel network with one or two dominant bar-braided or island-braided arms. Individual branches are separated by larger vegetated islands that may show the same terrain elevation as the adjacent floodplain and, thus, divide the flow up to the bankfull stage. Individual channels show independent patterns and may meander, braid, or remain relatively straight (Nanson and Knighton 1996). Wandering gravel-bed rivers are characterized by intensive lateral and vertical turnover processes, driven by a highly variable flow regime and high loads of coarse bed material. Large woody debris or ice jams that block flow and back water up in individual river arms contribute to the fluvial dynamics. Extreme flows can ram accumulations of such materials through river arms and channels, shaping them as they tear off vegetation and substrate. Channel avulsions, the rapid formation of new river arms by incisions in the floodplain terrain, intersecting larger islands, or reclaiming abandoned arms are typical geomorphic processes.

In downstream sections located already in the sediment accumulation zone, anabranching rivers with mixed loads or sand beds emerge. The substrate of the riverbed and, accordingly, channel patterns are closely interlinked with the geological configuration of the respective reach and the sediment load of large tributaries, especially where they meet, e.g., confluences. Deposition of suspended material starts as channel flow slows with decreasing slope and material accumulated in the current hits critical thresholds. This favors the formation of cohesive riverbanks, which facilitates the development of typical meandering rivers. Such systems show a higher bank resistance, and channels primarily migrate laterally or shift downstream. To distinguish between mildly (sinuous) and sharply curving (meandering) rivers, many authors apply a sinuosity index of more than 1.3 or 1.5 (Schumm 1977; Thorne 1997). The sinuosity index indicates the ratio (quotient) between the length of a river course and that of the valley axis or, sometimes instead of the latter, the linear distance between the upper end and lower end. Once the meander bends become too tortuous and shift close to each other, they are cut off, and a new straighter channel emerges, while the former meander loop remains as a an "oxbow lake" (compare "Palaeopotamon" in Fig. 3.2). Meandering rivers still feature flow velocities, i.e., shear stress, that accommodate the formation of distinct river arms and lateral channel adjustments to instream aggradations. The lower the channel slope, the more instream accretion will occur, and the capability of a river to adapt to these deposition processes will be reduced. Under such conditions, a specific low-energy variant of anabranching channel patterns, so-called anastomosing rivers, with very low gradients and stream power associated with stable cohesive banks, will emerge (Knighton and Nanson 1993). Their individual channels are often sinuous and exhibit almost no lateral migration. However, anastomosing channels have insufficient energy relative to bank strength to allow adjustments to instream deposition of mostly suspended material; hence avulsion is more likely to occur. Flooding overtops riverbanks and builds floodplains by vertical accretion of cohesive fine-grained material. The deposits are typically rich in organic material (Nanson and Knighton 1996). Though anastomosing rivers are typical features of the sediment accumulation zone close to the estuary, they can also emerge further upstream in river sections that are wider and unconfined as a consequence of tectonic depressions where the channel gradient and stream power are significantly reduced (compare Fig. 3.2 at the upper margin).

River deltas or estuaries feature environments very different from the rest of the river system. Transport capacity finally is disrupted, and sediment deposition generally constitutes the principal formative process. Delta areas are transition zones between riverine and maritime environments. They reflect structuring influences from both the ocean, such as waves, tides, and saltwater influx, and the river, such as discharge of freshwater and fluvial sediments. Because sea level provides the ultimate base level of the whole fluvial system, the channel gradient of the upstream river section—and over the long term that of the entire channel network—is directly tied to the elevation of the sea.

The described general framework of morphological river types would be only encountered along an ideal longitudinal profile that shows a concave shape with a steep upper section close to the source and a progressively decreasing gradient toward the delta. In reality, however, landscapes are heterogeneous mixes, and the evolution of distinct channel patterns along a river's course depends on the regional and local geological basement, tectonic processes, climate conditions, and vegetation cover. In addition, confluences of large tributaries may alter the flow and sediment regime and, accordingly, channel patterns of the main stem. That's why one can encounter typical meandering river sections or even anastomosing reaches upstream of gorges or braided sections. In order to identify the causes for the confusing variety of river types, principles of hydraulic geometry have been used to derive empirical relationships between channel width, depth, slope, sediment size, flow velocity, and external controls such as catchment size and flow (Leopold and Maddock 1953). Generally, rivers on steeper slopes or systems that transport large volumes of coarse bedload with braided channels tend to develop wider and shallower channels than comparable meandering or straight river reaches (Parker 1979). Similarly, rivers with a flashier discharge regime and relatively high peak flows tend to develop wider channels (Brierley and Fryirs 2005). Recent approaches for river classification strive for a basin-wide analysis that also integrates land cover and human modifications. The usage of a hierarchical framework that nests successive scales of physical and biological conditions allows a more holistic understanding of fluvial processes in the whole basin (Buffington and Montgomery 2013).

3.3   A Shifting Balance of Form and Motion

The biodiversity of riverine ecosystems is closely related to habitat composition and habitat development, which are primarily controlled by natural fluvial disturbances (Ward 1998; Tockner et al. 2006). Along the river continuum, patterns of fluvial processes are closely related to the respective morphological river type and may gradually or abruptly change. Bar-braided or island-braided river reaches are subject to permanent turnover processes driven by their flashy regime and abundant sediment influx. Rapid lateral channel adjustments, a tendency toward vertical aggradation, and noncohesive riverbanks that can be easily reworked facilitate the permanent adaptations of existing channels and formation of new braids. Anabranching rivers, i.e., wandering gravel-bed rivers, are also characterized by intensive lateral and vertical turnover processes that boost the formation of new bars and vegetated islands. In contrast to typical braiding rivers, associated floodplains and larger islands feature significant vertical accretions with coarser material at the base and sand or suspended material in the upper soil layer. In such river sections, channel avulsions are typical phenomena (compare Sect. 3.2). The further downstream a river's course one goes, the more the aggradation processes predominate. Meandering and anastomosing channels in lowlands are subject to instream deposition of sediments that often occurs at point bars and to vertical accretion of suspended load in the floodplain. Both river types have in common a fine-grained, cohesive bank material which limits the potential for the balance of flow/deposition to reshape channel. In contrast, sand channels with insufficient cohesive sediment to form resistant banks are particularly sensitive to flow variability and may easily be reshaped by altered flow conditions or sediment supply (Osterkamp and Hedman 1982).

At the first glance, one may conclude that different forms of channel behavior are bound to certain river types. Instead, morphological river types, i.e., channel patterns, are always products of prevalent fluvial dynamics that also depend on regional differences in climate, lithology, terrain relief, and land cover. In this context, vegetation significantly affects fluvial dynamics and, accordingly, channel patterns in several ways. On the catchment or sectional scale, type and areal extents of the vegetation cover influence the flow regime and local erosion and denudation (areal degradation) processes that, in turn, directly affect sediment availability in the basin (e.g., Allan 2004; Blöschl et al. 2007). On the local scale, riparian vegetation enhances bank resistance and counteracts bank erosion and channel migration but may also boost fluvial dynamics in form of large woody debris (e.g., Gurnell et al. 1995; Corenblit et al. 2007). In the latter case, extreme flows that dislodge vegetation, creates debris masses that can increase the erosive force of a high water event.

Natural river systems never remain in a morphologically static state. Rather they undergo permanent adjustments to internal changes of the system, e.g., when one channel changes in response to alterations in a confluent channel, and to external shifts, such as modified sediment supply or land cover change in the basin. From a temporal perspective, river adjustments reflect cumulative responses to recent events and deferred responses to previous events (Brierley and Fryirs 2005). Thereby, natural channel adjustments are superimposed by human-caused disturbances that additionally boost or curb fluvial dynamics. The geometry of a river channel reflects the balance or unbalance, respectively, of erosional and depositional processes that configure the riverbed and the banks. Generally, rivers seem to "strive" for a state of dynamic equilibrium ("regime status") between the imposed external controls such as valley slope, discharge, and sediment load on the one side and channel responses to those controls, including width, depth, velocity, reach slope, and sediment size, on the other side (Allan and Castillo 2007). While valley slope—from the human perspective—generally remains the same, the flow regime and, in particular, sediment supply are more sensitive and respond to natural or human influences over shorter time frames. This relationship between external controls and channel adjustments is described by "Lane's Law" stating that stream power approximately relates to sediment load (Lane 1954, 1955):

$$Q_S \times D_{50} \sim Q_W \times S$$

Q_S = sediment discharge, D_{50} = median grain size, Q_W = water discharge, S = channel slope; Fig. 3.3.

Stream power, the product of discharge and channel slope, describes the capacity of a river reach to mobilize and transport material. When stream power, i.e., discharge, decreases due to flow regulation or water withdrawal, some of the delivered material can't be transported further downstream, and aggradation processes will transform the channel. The same channel adjustments will occur during unchanged flows, when the sediment supply increases or the material becomes

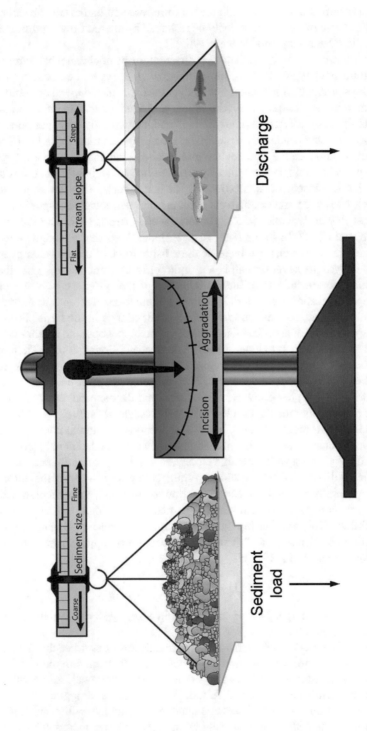

Fig. 3.3 Channel degradation and aggradation as a consequence of the balance between sediment transport (volume and median grain size) and stream power (discharge and channel slope) (Adapted from Pollock et al. 2014; published by Oxford University Press on behalf of the American Institute of Biological Sciences, based on data from Lane 1955)

coarser. On the contrary, dams that retain large shares of bedload generally lead to significantly reduced sediment volumes and smaller sediment sizes in downstream river stretches (see Chap. 6). Lane's Law illustrates that, in this case, stream power is too high for the available sediment load and the river will start to compensate its deficit by eroding the riverbed. Channel degradation downstream of the dam is the consequence. The modification of the channel will last as long as a new balance is not attained and, finally, a new type of channel pattern will emerge. For example, as a consequence of bedload reduction, formerly braiding river reaches may transform to sinuous single-channel rivers (Marti and Bezzola 2004).

Lane's Law and other studies in fluvial morphology assume a kind of equilibrium between external controls and channel geometry or habitat composition (e.g., Mackin 1948; Glova and Duncan 1985; Arscott et al. 2002). Because natural rivers are never totally static, such an equilibrium would be best referred to as a "state of dynamic equilibrium" in which one fluvial process, e.g., erosion, is compensated by a counteracting evolution (in this case aggradation). If fluvial systems did not remain in a kind of equilibrium, they would gradually—or even rapidly if system-inherent thresholds are exceeded—transform to a new morphological state (river type). However, some authors argue that fluvial systems are rarely in dynamic equilibrium, because rivers have to respond to a complex disturbance regime of periodic, episodic, and stochastic events that superimpose themselves on each other. Accordingly, rivers operate in a state of perpetual nonlinear adjustment, rather than oscillating around an equilibrium state (Thorne 1997; Brierley and Fryirs 2005). That way, many rivers show a tendency to develop a recognizable average behavior (Knighton 1998).

Changes in the geomorphological configuration of a river reach can significantly affect its capacity to support the ecological functions and habitat availability of a fluvial system. Likewise, riverine ecosystems, in particular, depend on disturbances that regenerate single parts of the system on a regular basis. Assuming unchanged climate conditions, riverine habitats and their associated biocoenoses undergo ecological successions toward a certain terminal stage that—without further disturbances—would persist (Bravard et al. 1986; Amoros and Roux 1988). Under human undisturbed conditions, periodic and/or stochastic disturbances counteract the general trajectory toward matured terrestrial habitats, rejuvenating the various riverine habitats (Ward 1998; Ward and Tockner 2001). Over the long term, such processes promote morphologically and ecologically differentiated habitat patches, fundamentally determining the competitive interactions at species and community level (Huston 1979, 1994; Hughes 1997). Though an individual habitat may vanish due to disturbances, over lengthier periods and larger areas, in such a "shifting habitat mosaic," the proportions of the differently developed habitat patches are supposed to remain relatively constant as long as the controlling factors do not significantly change (Stanford et al. 2005). Given the hierarchical nature of fluvial systems (compare Sect. 3.1), the "hierarchical patch dynamics" concept emphasizes that higher levels of system organization impose structural and functional constraints on lower levels and its potential ecological processes (Wu and Loucks 1995).

From the landscape perspective of a biocoenosis, e.g., a spatially heterogeneous environment with patches differing in resource quality and quantity, persistence, and connectivity provides the opportunity for a greater biodiversity than under more uniform and stable conditions (Allan and Castillo 2007). Likewise, riverine species have to adapt to the habitats that shift in space and time and, thus, to the underlying disturbance regime. Because individual species show varying habitat preferences and migration capabilities, they respond to landscape heterogeneity and changes in the habitat mosaic in different ways (Wiens 2002). For example, fish diversity generally peaks in intensely connected habitats, while amphibian diversity is higher in habitats with low connectivity (Tockner et al. 1998). This example shows that a high frequency of disturbance does not necessarily result in a higher riverine biodiversity. Once the disturbance regime significantly exceeds the resilience capacity of riverine species, biodiversity will diminish. According to the "intermediate disturbance hypothesis," a moderate level of disturbance potentially may increase diversity enabling the coexistence of species with divergent recruitments (Connell 1978; Ward and Stanford 1983; Fox 2013). In this context, several studies indicate that island-braided and, in particular, anabranched reaches generally show higher diversities than bar-braided, meandering, or anastomosing river sections (e.g., Stanford et al. 1996; Gurnell and Petts 2002).

3.4 Channelized Rivers

One can already say that the mighty . . . stream can never be regulated so as the proud human spirit would like to (Wiletal 1897).

Other than remote human impacts, such as land cover changes or mining in the catchment, river channelization measures comprehensively alter the fluvial morphology of a river reach in the most direct form. Dependent on the objectives of a river training program, various types of hydraulic measures are applied, each associated with specific forms of human interference in the physical configuration and ecological functions of fluvial systems. Construction of dams that present a severe local intervention with remote up- and downstream impacts on fluvial systems is often—but not necessarily—accompanied by channelization measures of longer river reaches (see Chap. 6).

River channelization in general pursues two major aims—the improvement of navigability and flood control. Besides that, river straightening was also seen as a means to increase flow speed and to discharge pollutants. In Europe and North America, owing to the advent of steam navigation in the nineteenth century, several river engineering programs aimed at the improvement of the shipping conditions of medium and large rivers (Sedell and Froggatt 1984; Gore and Petts 1989; Alexander et al. 2012; for human impacts on fluvial systems in earlier periods see Chap. 2). Because load drafts of new steam vessels constantly increased, the water depth along navigable waterways had to be adapted simultaneously. In many rivers, deepening of the channel was achieved by a significant constriction of channel width that in most cases was accompanied by a straightening of the whole river section. This was specifically a major concern in braided or anabranching river sections, where flow was divided into several branches (Wex 1873, 1879). Because they were generally deeper, navigability in sinuous or meandering, single-channel rivers in lowlands was generally easier. However, such systems often had insufficient flood conveyance capacity (N.N. 1853; De Marchis and Napoli 2008). If flood control is the major concern, channelization primarily strives for straightening and/or widening (resectioning) a river reach in order to amplify the conveyance capacity of the channel and to reduce shear stress (Brookes 1988).

Independent from its main purpose, channelization fundamentally modified channel patterns and fluvial dynamics, e.g., when a meandering or braided river section was transformed into a straight, uniform channel. In alluvial reaches, besides the main river arm, the whole riparian ecosystem is affected by channelization's hydraulic measures. Former lotic side arms were cut off and transformed to one side connected backwaters or were totally separated from the main channel. Accordingly, braided and, in particular, anabranching rivers are subject to the most severe impairments with respect to the channel patterns (Gurnell et al. 2009; Tockner et al. 2010). Specifically, in alluvial reaches, river channelization programs were also designed to prevent lateral erosion of floodplain terrain and to gain new arable land. In order to boost terrestrialization processes in cutoff river arms and in low-lying areas of the floodplain, embankments and closure dams were often designed to function as sediment traps and to facilitate deposition of material even during smaller floods. Applying this technique enabled the reclamation of large areas of new land within a few years to decades (Hohensinner et al. 2011). Because navigability was still constricted during periods of reduced discharge, later in many large rivers, additional groynes and training walls for low flow situations were installed. In the twentieth century, channelization measures were often coupled with the construction of reservoirs and hydropower plants, which guaranteed sufficient channel depths for larger vessels. Though flood protection levees are generally not constructed for purposes of river training, they also severely affect fluvial systems in various respects. Levees that are directly located along riverbanks are often accompanied by massive embankments to prevent undercut erosion. In contrast to flood protection levees in the hinterland, such dykes both morphologically and hydrologically constrain river dynamics.

The history of river channelization highlights that most hydraulic measures were designed to fulfill multiple purposes at once in order to facilitate several forms of human uses in fluvial systems (Winiwarter et al. 2012). It also shows that single hydraulic constructions, e.g., a closure dam to cut off a side arm, may impair a fluvial system in multiple ways. Some river engineering measures that are commonly applied—at least at first glance—only affect the channel itself. Transversal protection structures that are installed perpendicular to the water course, such as ground sills on the channel bottom or higher check dams, are generally applied for stabilizing the riverbed and preventing further channel incision. Both types reduce stream power and, consequently, sediment transport capacity in the upstream river reach. Energy dissipation, the conversion of the kinetic energy of flowing waters into other, less hazardous, forms, such as thermal or acoustical energy, is primarily limited to sites just below the transversal hydraulic structures. On the other hand, dredging measures aim for lowering the river bottom and are usually conducted to keep waterways navigable or to increase flood conveyance capacity. Though these measures are performed directly in river channels, they potentially also affect larger parts of riparian systems. Water level changes evoked by the transversal structures may significantly influence the groundwater table or surface water bodies in the adjacent floodplain.

During the past two centuries, river regulation measures caused dramatic "regime shifts" for most European braiding, multi-channel, and transitional rivers (Petts 1989; Tockner et al. 2010). In the Alps, channel patterns commonly shifted from formerly braiding to a single-channel river type. As a consequence, the total length of braiding reaches decreased in France and Austria by 70% and 95%, respectively (Muhar et al. 1998; Habersack and Piégay 2008). River engineering measures not only modify the physical configuration of the channelized river section itself; they indirectly also affect the subsequent up- and downstream reaches. Even if only one of the flow-dependent variables (slope, depth, width, and roughness) is affected by the measures, feedback effects will promote adjustments toward a new morphological state (Brookes 1988). In case of channel narrowing, often applied for the purpose of land reclamation, flow velocity and sediment transport capacity increase, eventually causing bed erosion. Nevertheless, the main cause for amplified bed degradation is channel straightening. Particularly in sinuous or meandering rivers, where the new cutoff is much shorter and steeper, stream power significantly increases, and the riverbed may incise by several meters within a year or several years (Knighton 1998; Kesel 2003). Starting from the upper end of a straightened river section, retrograde erosion that progressively encroaches upstream is a typical response process that may affect large parts of a whole river system (Simon 1989). The mobilized material is transported downstream as far as stream power allows, meaning that large volumes will be deposited just downstream of the straightened section. Here, the opposite adjustment process can be observed: aggradation reduces channel slope, channel width may substantially increase, and new channel patterns may emerge (Brookes 1987; Gregory 2006). Well-documented examples from the Danube River and its tributaries in the nineteenth century show that river straightening programs in alpine tributaries led to marked aggradations and bed modifications in the Danube River, even 150 km downstream of the "improved" section (Schmautz et al. 2000). Once an alluvial Danube section was straightened, downstream aggradation and bed transformation causing severe obstacles for navigation forced the regulation authorities to advance channelization continually downstream until the next gorge section of the Danube was reached (Hohensinner 2008; Hohensinner et al. 2014). However, new problems arose in the alluvial reaches downstream of the gorge, and, finally, they were forced to channelize the whole Austrian Danube section (Schmautz et al. 2002).

Today, distinct channel incisions induced by river "training" (channel engineering) in combination with reduced sediment supply from upstream river sections present a major concern in the industrialized world (Gore and Petts 1989; Stanford et al. 1996). Typical consequences for the biota are the reduction of original instream habitat complexity and habitat availability in increasingly uniform riverbeds (e.g., Toth 1996; Lau et al. 2006). Accordingly, pronounced differences in species composition and abundance can be found compared to more natural sites. Since straightened and constrained river channels generally show higher flow velocities, aquatic communities have to adapt to the altered hydraulic conditions. Fish species and benthic invertebrates preferring moderate or lower flow velocities are largely replaced by rheophilic communities (Jurajda 1995; Jansen et al. 2000). These modifications are referred to as the "rhithralization effect", the shift of a riverine coenoses toward upstream communities (Jungwirth et al. 2000). Higher flow velocities generally result in greater grain sizes of the substrate. Another typical response is riverbed armoring, where the top layer of the bed substrate shows coarser sediment fractions than in the underlying layer. In river sections with negligible bedload transport, such truncated bed dynamics may lead to the clogging of the pore volume of the substrate ("hyporheic interstitial") with silt. Such "colmations" of the riverbed severely impair the exchange processes between the river and the aquifer (Boulton 2007; see Chap. 8).

Apart from the main channel, in alluvial reaches, channelization also affects the hydromorphological configuration and ecological functions of the whole riparian system. Direct forms of impairment include the hydrological separation of the water bodies in the floodplain from the main stem and the promotion of terrestrialization. As already mentioned, the "improvement" of wetlands for better human usage is also an important goal of channelization leading to a drastic reduction of aquatic and semiaquatic habitats. Besides, significantly lowered water levels in the river comprehensively lower downward percolation (infiltration) rates and thereby decrease aquifer recharge in the floodplain. This lowers the resilience of riverine communities to drought. Cutoff side arms and lowered groundwater tables significantly reduce lateral hydrological connectivity, i.e., the various surface and subsurface exchange processes, such as sediments, nutrients, water temperature, or organisms, between the river and the diverse floodplain biotopes (Amoros et al. 1987; Amoros and Bornette 2002).

Accordingly, the stimulating effects of the "flow pulse" at discharges below bankfull and the "flood pulse" at higher stages that in undisturbed condition boost primary production even in remote floodplain areas as a fundamental basis for riverine biodiversity decrease (Junk et al. 1989; Puckridge et al. 1998; Tockner et al. 2000). Moreover, reduced lateral connectivity is reflected by the truncation of the network of potential migration pathways for aquatic organisms. Rheophilic fish species with a preference for lentic conditions in connected backwaters during certain periods in the adult stage, in particular, depend on such lateral migratory pathways between lotic and lentic habitats (e.g., for reproduction, as feeding grounds, or winter refuge; Schiemer and Waidbacher 1992).

Ongoing vertical accretion of sediments during floods further heightens the elevation of the floodplain terrain. As a consequence, besides a lateral decoupling of the floodplain habitats from the river, increasingly a vertical decoupling between the river level (water/groundwater table) and the floodplain terrain is a typical phenomenon (Amoros and Bornette 2002). Historical analyses from Austrian Danube floodplains show that the average depth down to the groundwater table below the terrain surface increased by 63–88% at mean flow situations since the early nineteenth century (Hohensinner et al. 2008). Vertical decoupling of fluvial systems considerably modifies site conditions for riparian vegetation, which is one major cause for the extensive decline of early successional stages and softwood assemblages in the industrialized world (Egger et al. 2007; Mosner 2012; Reif et al. 2013). Today softwood communities are severely endangered and are specifically protected by the EU Flora-Fauna-Habitat Directive.

The brief discussion of potential consequences of channelization shows that channel adjustment to local or sectional hydraulic constructions most likely affects much longer river sections or may even concern the whole river system. Accordingly, in applying such measures, a much larger spatial and temporal scale has to be considered. However, this also applies in the case of ecologically oriented restoration programs.

Given the diverse forms of hydraulic measures and the general lack of basic data, it is difficult to provide scientifically rigorous information about the worldwide or continental impacts on fluvial systems due to channelization. According to a rough estimate, worldwide, approximately 500,000 km of waterways have been altered for navigation (Tockner and Stanford 2002). Even more speculative are estimates about riverine wetlands that are affected by channelization, because the consequences of local channelization measures and those of wetland reclamation or remote impacts, such as altered flow regime and sediment supply due to the construction of dams or land cover changes in the basin, are superimposed upon each other (see Chap. 15).

3.5 Assessing the Hydromorphological State of Rivers

In several European countries, long traditions exist for assessing the morphological conditions of rivers to provide an overall survey of habitat quality. Formerly, such assessments were particularly related to hydraulic engineering activities and river inventories (e.g., Werth 1987; Raven et al. 1997). These studies focused primarily on morphological conditions of rivers and streams, while at the same time, key elements of the physical environment of fluvial systems, like flow and sediment regime, were not or scarcely addressed. In general, hydromorphological assessment is based on the assumption of a strong relationship between the physical environment and aquatic organisms/biocoenoses of riverine ecosystems (Karr 1981; Muhar and Jungwirth 1998). Thus, those hydromorphological attributes are investigated, mapped, and evaluated, which determine the habitat functions of running waters. The methods of such assessments are diverse, depending on the main aims and objectives, ranging from large-scale surveys at the basin scale to local-scale habitat assessment (see Table 3.1).

Since the EU WFD requires the assessment of hydromorphological quality as an essential part in supporting the ecological status of rivers, numerous methods have been revised and further developed (Boon et al. 2010; Belletti et al. 2014; Poppe et al. 2016). Most of them follow the scheme of the WFD by addressing "hydromorphological quality elements" (EU 2000): (1) hydrological regime (e.g., quantity and dynamics of water flow and connection to groundwater bodies), (2) morphological conditions (e.g., river depth and width variation, structure, and substrate of the river and the riparian zone), and (3) river continuity (regarding

Table 3.1   General aims of hydromorphological assessments

Tasks of large-scale surveys at the basin-wide scale
• Overview of the physical status quo of river systems (e.g., according to the EU WFD), overall identification and documentation of habitat improvement/degradation
• Quantification (intensity of impacts, e.g., river engineering measures, artificial barriers, degree of natural highly impacted river stretches, etc.)
• Basis data for supra-regional planning (e.g., establishment of a large-scale river conservation/restorationnetwork)
• Tool for strategic decisions in early stages of project development
Tasks of local-scale habitat assessment
• Detailed habitat investigation in context with biological studies (auto-/synecological studies)
• Identification and assessment of altered habitat conditions due to anthropogenic impacts and the effects on biota
• Monitoring and evaluation of river restoration

migrating species as well as sediment regime). They mainly focus on (field) investigations, frequently supplemented by analyses of remote sensing data (e.g., orthophotos) at reach scale, describing channel characteristics and mesohabitat conditions. Depending on the specific method, respectively, on national guidelines of the EU member states, they follow a predefined scheme to define investigation units; e.g., in Austria, the hydromorphological status assessment is always related to a 500 m river stretch at all rivers with a catchment area of more than 10 km^2 (BMLFUW 2015). The currently applied assessment methods are basically compliant with the EU CEN standards on hydromorphological assessment comprising also a largely comparable set of assessment categories and parameters (see Table 3.2; CEN 2004; Boon et al. 2010).

Such surveys provide a wealth of useful information, but, with some exceptions, they tend to focus on forms rather than processes, typically evaluating hydromorphological degradation on how the characteristics of a river reach differ from "reference" conditions, based on "pristine" sites located elsewhere or how the reach looked at some time during the past. Recently developed studies aimed to go beyond this scheme, to enhance the survey methods to better integrate physical processes as driving forces for the occurrence and reshaping of river channels and instream habitats.

Summarizing, hydromorphological assessment is a key foundation for river basin management and should build on the growing understanding of geomorphological processes (Montgomery and Buffington 1998; Kondolf et al. 2003; Brierley and Fryirs 2005) and integrate biological knowledge with regard to habitat requirements of aquatic species at different spatial scales. In particular, the following issues are crucial:

- To choose methods, harmonized with the specific aims, objectives, and thus spatial scale.
- To identify adequate assessment attributes and evaluation algorithms.

Table 3.2 Assessment categories, features, and attributes comprising a standard hydromorphological assessment according to EN 14614 (From Boon et al. 2010)

Assessment categories	Generic features	Examples of attributes assessed
Channel		
Channel geometry	Planform	Braiding, sinuosity
		Modification to natural planform
	Longitudinal section	Gradient, long-section profiles
	Cross section	Variations in cross section shown by depth, width, bank profiles, etc.
Substrates	Artificial	Concrete, bed-fixing
	Natural substrate types	Embedded (boulders, bedrock, etc.)
		Large (boulders and cobbles)
		Coarse (pebble and gravel)
		Fine (sand)
		Cohesive (silt and clay)
		Organic (peat, etc.)
	Management/catchment impacts	Degree of siltation, compaction
Channel vegetation and organic debris	Structural form of macrophytes	Emergent, free-floating, broad-leaved submerged, bryophytes, macro-algae
	Leafy and woody debris	Type and size of feature/material
		Weed cutting
Erosion/deposition character	Features in channel and at base of bank	Point bars, side bars, mid-channel bars and islands (vegetated or bare)
		Stable or eroding cliffs, slumped or terraced banks
Flow	Flow patterns	Free-flow, rippled, smooth
		Effect of artificial structures (groynes, deflectors)
	Flow features	Pools, riffles, glides, runs
	Discharge regime	Off-takes, augmentation points, water transfers, releases from hydropower dams
Longitudinal continuity as affected by artificial structures	Artificial barriers affecting continuity of flow, sediment transport, and migration for biota	Weirs, dams, sluices across beds, culverts
Riverbanks/riparian zone		
Bank structure and modifications	Bank materials	Gravel, sand, clay, artificial
	Types of revetment/bank protection	Sheet piling, stone walls, gabions, rip-rap
Vegetation type/ structure on banks and adjacent land	Structure of vegetation	Vegetation types, stratification, continuity
	Vegetation management	Bank mowing, tree felling
	Types of land use, extent, and types of development	Agriculture, urban development

(continued)

Table 3.2 (continued)

Assessment categories	Generic features	Examples of attributes assessed
Floodplain		
Adjacent land use and associated features	Types of land use, extent, and types of development	Floodplain forest, agriculture, urban development
	Types of open water/wetland features	Ancient fluvial/floodplain features (cutoff meanders, remnant channels, bog)
		Artificial water features (irrigation channels, fish ponds, gravel pits)
Degree of (a) lateral connectivity of river and floodplain (b) lateral movement of river channel	Degree of constraint to potential mobility of river channel and water flow across floodplain Continuity of floodplain	Embankments and levees (integrated with banks or set back from river), flood walls, and other constraining features Any major artificial structures partitioning the floodplain

- To enhance the methodological approach by comprehensively including the adjacent floodplains/wetlands in assessing the physical environment of river landscapes.
- Far more consideration has to be given to physical processes to better understand the current conditions and the causes of alterations (human uses, restoration measures, etc.) and responding effects (Belletti et al. 2014).

3.6 Conclusion

Addressing the hydromorphological state of riverine ecosystems with profound understanding requires consideration of larger spatial scales. Channel geometry and fluvial dynamics are not solely determined by local geomorphological framework conditions. Rather they are the product of influxes from the upstream catchment. Over the long term, both sediment transport and discharge, on the one side, and the local/sectional setting (e.g., geology, topography), on the other side, lead to the formation of certain channel patterns. However, the typical sequence of morphological river types along a river's course from constrained upstream gorges over braided, anabranched, and meandering rivers to, finally, anastomosing lowland rivers can be rarely found in nature. Tectonic barriers or depressions and large tributaries may interrupt that typical sequence and foster channel patterns that would normally not be expected at a respective site. Changes in upstream sediment delivery and altered discharge regimes trigger local channel adjustments. Even downstream hydromorphological changes may affect channel geometry in upstream sections due to retrograde soil erosion.

Accordingly, channelization measures do not only affect the physical configuration and dynamic fluvial processes at a respective river reach. Rather they influence much longer river sections or even the whole river system, including the tributaries. Human interventions into riverine environments always call for consideration of unintended side effects and potential long-term legacies that may cause new problems at upstream or downstream sections. What seems to be clear for river channelization does also apply to restoration measures. Locally implemented river restoration projects may also influence the up- and downstream fluvial processes and, thus, the habitat availability and the ecological state of longer river sections.

References

Alexander JS, Wilson RC, Green WR (2012) A brief history and summary of the effects of river engineering and dams on the Mississippi river system and delta. U.S. Geological Survey Circular 1375

Allan JD (2004) Landscapes and riverscapes: the influence of land use on stream ecosystems. Annu Rev Ecol Evol Syst 35:257–284

Allan JD, Castillo MM (2007) Stream ecology. Structure and function of running waters, 2nd edn. Springer, Dordrecht

Amoros C, Bornette G (2002) Connectivity and biocomplexity in waterbodies of riverine floodplains. Freshw Biol, Special Issue Riverine Landscapes 47:761–776

Amoros C, Roux AL (1988) Interactions between water bodies within the floodplains of large rivers: function and development of connectivity. In: Schreiber K-F (ed) Connectivity in landscape ecology, vol 29. Münstersche Geographische Arbeiten, Münster, pp 125–130

Amoros C, Roux AL, Reygrobellet JL, Bravard JP, Pautou G (1987) A method for applied ecological studies of fluvial hydrosystems. Regul Rivers Res Manag 1:17–36

Arscott DB, Tockner K, Van der Nat D, Ward JV (2002) Aquatic habitat dynamics along a braided alpine river ecosystem (Tagliamento River, Northeast Italy). Ecosystems 5:802–814

Belletti B, Rinaldi M, Buijse AD, Gurnell AM, Mosselman E (2014) A review of assessment methods for river hydromorphology. Environ Earth Sci 73:2079–2100

Blöschl G, Ardoin-Bardin S, Bonell M, Dorninger M, Goodrich D, Gutknecht D, Matamoros D, Merz B, Shand P, Szolgay J (2007) At what scales do climate variability and land cover change impact on flooding and low flows? Hydrol Process 21:1241–1247

BMLFUW – Bundesministerium für Land- u. Forstwirtschaft, Umwelt und Wasserwirtschaft (2015) Leitfaden zur hydromorphologischen Zustandserhebung von Fließgewässern. Vienna

Boon PJ, Holmes NTH, Raven PJ (2010) Developing standard approaches for recording and assessing river hydromorphology: the role of the European committee for standardization (CEN). Aquat Conserv Mar Freshwat Ecosyst 20:55–61

Boulton AJ (2007) Hyporheic rehabilitation in rivers: restoring vertical connectivity. Freshw Biol 52:632–650

Bravard JP, Amoros C, Pautou G (1986) Impacts of civil engineering works on the succession of communities in a fluvial system: a methodological and predictive approach applied to a section of the Upper-Rhone River. Oíkos 47:92–111

Brierley GJ, Fryirs KA (2005) Geomorphology and river management. Applications of the river styles framework. Wiley, Chichester

Brierley GJ, Fryirs KA, Outhet D, Massey C (2002) Application of the river styles framework as a basis for river management in New South Wales, Australia. Appl Geogr 22:91–122

Brookes A (1987) Recovery and adjustment of aquatic vegetation within channelisation works in England and Wales. J Environ Manag 24:365–382

Brookes A (1988) Channelized rivers. Perspectives for environmental management. Wiley, Chichester

Buffington JM, Montgomery DR (2013) Geomorphic classification of rivers. In: Shroder J, Wohl E (eds) Treatise on geomorphology, Fluvial geomorphology, vol 9. Academic Press, San Diego, pp 730–767

CEN – Comité Européen de Normalisation (2004) Water quality – guidance standard for assessing the hydromorphological features of rivers. BS EN 14614:2004. BS 6068-5.36:2004, Brussels

Church M (1992) Channel morphology and typology. In: Calow P, Petts GE (eds) The rivers handbook, vol 1. Blackwell, Oxford, pp 126–143

Connell JH (1978) Diversity in tropical rain forests and coral reefs. Science, New Series 199 (4335):1302–1310

Corenblit D, Tabacchi E, Steiger J, Gurnell AM (2007) Reciprocal interactions and adjustments between fluvial landforms and vegetation dynamics in river corridors: a review of complementary approaches. Earth Sci Rev 84:56–86

De Marchis M, Napoli E (2008) The effect of geometrical parameters on the discharge capacity of meandering compound channels. Adv Water Resour 31:1662–1673

Desloges JR, Church M (1987) Channel and floodplain facies in a wandering gravel-bed river. In: Ethridge FG, Flores RM, Harvey MD (eds) Recent developments in fluvial sedimentology, Special publication number 39. Society of Economic Paleontologists and Mineralogists, Tulsa, pp 99–109

Eaton BC, Millar RG, Davidson S (2010) Channel patterns: braided, anabranching, and single-thread. Geomorphology 120:353–364

Egger G, Drescher A, Hohensinner S, Jungwirth M (2007) Riparian vegetation model of the Danube River (Machland, Austria): changes of processes and vegetation patterns. In: Proceedings of the 6th Internat. Symposium on Ecohydraulics. CD-Edition, Christchurch, pp 18–23. Feb 2007

EU (2000) Directive 2000/60/EC of the European parliament and of the council of 23 October 2000 establishing a framework for community action in the field of water policy. Brussels

Ferguson RI (1987) Hydraulic and sedimentary controls of channel planform. In: Richards K (ed) River channels: environment and process. Blackwell, Oxford, pp 129–158

Fox JW (2013) The intermediate disturbance hypothesis should be abandoned. Trends Ecol Evol 28:86–92

Frissell CA, Liss WJ, Warren CE, Hurley MD (1986) A hierarchical framework for stream habitat classification: viewing streams in a watershed context. Environ Manag 10:199–214

Fryirs KA, Brierley GJ (2013) Geomorphic analysis of river systems: an approach to reading the landscape. Wiley-Blackwell, Chichester

Glova GJ, Duncan MJ (1985) Potential effects of reduced flows on fish habitats in a large braided river, New Zealand. Trans Am Fish Soc 114:165–181

Gore JA, Petts GE (1989) Alternatives in regulated river management. CRC Press, Boca Raton

Gregory KJ (2006) The human role in changing river channels. Geomorphology 79:172–191

Gurnell AM, Petts GE (2002) Island-dominated landscapes of large floodplain rivers, a European perspective. Freshw Biol 47:581–600

Gurnell AM, Gregory KJ, Petts GE (1995) The role of coarse woody debris in forest aquatic habitats: implications for management. Aquat Conserv Mar Freshwat Ecosyst 5:143–166

Gurnell AM, Surian N, Zanoni L (2009) Multi-thread river channels: a perspective on changing European Alpine river systems. Aquat Sci 71:253–265

Habersack H (2000) The river-scaling concept (RSC): a basis for ecological assessments. In: Jungwirth M, Muhar S, Schmutz S (eds) Assessing the ecological integrity of running waters. Kluwer Academic Publishers, Dordrecht, pp 49–60

Habersack H, Piègay H (2008) River restoration in the Alps and their surroundings: past experience and future challenges. In: Habersack H, Piégay H, Rinaldi M (eds) Gravel-bed rivers VI. Elsevier, Philadelphia, pp 703–735

Hohensinner S (2008) Rekonstruktion ursprünglicher Lebensraumverhältnisse der Fluss-Auen-Biozönose der Donau im Machland auf Basis der morphologischen Entwicklung von 1715–1991 (Reconstruction of historical habitat conditions of the Danube river/floodplain biocoenosis based on the morphological development from 1715–1991 (Machland, Lower/Upper Austria)). PhD thesis at the University of Natural Resources and Life Sciences, Vienna

Hohensinner S, Herrnegger M, Blaschke AP, Habereder C, Haidvogl G, Hein T, Jungwirth M, Weiß M (2008) Type-specific reference conditions of fluvial landscapes: a search in the past by 3D-reconstruction. Catena 75:200–215

Hohensinner S, Jungwirth M, Muhar S, Schmutz S (2011) Spatio-temporal habitat dynamics in a changing Danube River landscape 1812–2006. River Res Appl 27:939–955

Hohensinner S, Jungwirth M, Muhar S, Schmutz S (2014) Importance of multi-dimensional morphodynamics for habitat evolution: Danube River 1715–2006. Geomorphology 215:3–19

Huang HQ, Nanson GC (2007) Why some alluvial rivers develop an anabranching pattern. Water Resour Res 43:W07441

Hughes FMR (1997) Floodplain biogeomorphology. Prog Phys Geogr 21:501–529

Huston MA (1979) A general hypothesis of species diversity. Am Nat 113:81–101

Huston MA (1994) Biological diversity: the coexistence of species on changing landscapes. Cambridge University Press, New York

Jansen W, Böhmer J, Kappus B, Beiter T, Breitinger B, Hock C (2000) Benthic invertebrate and fish communities as indicators of morphological integrity in the Enz River (South-West Germany). In: Jungwirth M, Muhar S, Schmutz S (eds) Assessing the ecological integrity of running waters. Developments in hydrobiology, vol 149. Kluwer Academic Publishers, Dordrecht, pp 331–342

Jungwirth M, Muhar S, Schmutz S (2000) Fundamentals of fish ecological integrity and their relation to the extended serial discontinuity concept. Hydrobiologia 422/423:85–97

Junk WJ, Baylay PB, Sparks RE (1989) The flood pulse concept in the river flood-plain systems. Can Spec Publ Fish Aquat Sci 106:110–127

Jurajda P (1995) Effect of channelization and regulation on fish recruitment in a flood plain river. Regul Rivers Res Manag 10:207–215

Karr JR (1981) Assessment of biotic integrity using fish communities. Fisheries 6:21–27

Kesel RH (2003) Human modifications to the sediment regime of the Lower Mississippi River flood plain. Geomorphology 56:325–334

Knighton AD (1998) Fluvial forms and processes. A new perspective. Arnold, London

Knighton AD, Nanson GC (1993) Anastomosis and the continuum of channel pattern. Earth Surf Process Landf 18:613–625

Kondolf GM, Montgomery DR, Piégay H, Schmitt L (2003) Geomorphic classification of rivers and streams. In: Kondolf GM, Piégay H (eds) Tools in fluvial geomorphology. Wiley, Chichester, pp 171–204

Lane EW (1954) The importance of fluvial morphology in hydraulic engineering. Hydraulic laboratory report, no. 372, United States Department of the Interior, Bureau of Reclamation, Engineering Laboratories, Commissioner's Office. Denver, Colorado

Lane EW (1955) The importance of fluvial morphology in hydraulic engineering. Proceedings of the American Society of Civil Engineers, 81, paper no. 745

Lau JK, Lauer TE, Weinman ML (2006) Impacts of channelization on stream habitats and associated fish assemblages in East Central Indiana. Am Midl Nat 156:319–330

Leopold LB, Maddock T Jr (1953) The hydraulic geometry of stream channels and some physiographic implications. United States Geological Survey Professional Paper, P 0252

Leopold LB, Wolman MG (1957) River channel patterns: braided, meandering and straight. United States Geological Survey Professional Paper, P 0282–B, pp 39–85

Mackin JH (1948) Concept of the graded river. Bull Geol Soc Am 59:463–512

Marti C, Bezzola GR (2004) Sohlenmorphologie in Flussaufweitungen. Mitteilung der Versuchsanstalt für Wasserbau, Hydrologie und Glaziologie der ETH Zürich 184:173–188

Montgomery DR, Buffington JM (1998) Channel processes, classification and response potential. In: Naiman RJ, Bilby RE (eds) River ecology and management. Springer, New York, pp 13–42

Mosner E (2012) Habitat distribution and population genetics of riparian Salix species in space and time – a restoration framework for softwood forests along the Elbe river. PhD thesis at the Philipps-University Marburg, Germany

Muhar S, Jungwirth M (1998) Habitat integrity of running waters – assessment criteria and their biological relevance. Hydrobiologia 386:195–202

Muhar S, Kainz M, Kaufmann M, Schwarz M (1998) Erhebung und Bilanzierung flusstypspezifisch erhaltener Fliessgewässerabschnitte in Österreich (Survey of sections of flowing waters having retained their stream-type specific character). Österreichische Wasser- u Abfallwirtschaft 50:119–127

N.N (1853) Geschichte der Entwicklung der österreichischen Dampfschifffahrt auf der Donau. Deutsche Vierteljahrs Schrift. Jg. 1853(2):163–216

Nanson GC, Knighton AD (1996) Anabranching rivers: their cause, character and classification. Earth Surf Process Landf 21:217–239

Osterkamp WR (1978) Gradient, discharge and particle size relations of alluvial channels in Kansas, with observations on braiding. Am J Sci 278:1253–1268

Osterkamp WR, Hedman ER (1982) Perennial streamflow characteristics related to channel geometry and sediment in Missouri river basin. United States Geological Survey Professional Paper, p 1242

Parker G (1979) Hydraulic geometry of active gravel rivers. J Hydraul Div, Proc Am Soc Civ Eng 105:1185–1201

Petts GE (1989) Historical analysis of fluvial hydrosystems. In: Petts GE, Möller H, Roux AL (eds) Historical change of large alluvial rivers: Western Europe. Wiley, Chichester, pp 1–18

Petts GE, Amoros C (1996) Fluvial hydrosystems. Chapman & Hall, London

Pollock MM, Beechie TJ, Wheaton JM, Jordan CE, Bouwes N, Weber N, Volk C (2014) Using beaver dams to restore incised stream ecosystems. Bioscience 64:270–290

Poppe M, Kail J, Aroviita J, Stelmaszczyk M, Giełczewski M, Muhar S (2016) Assessing restoration effects on hydromorphology in European mid-sized rivers by key hydromorphological parameters. Hydrobiologia 769:21–40

Puckridge JT, Sheldon F, Walker KF, Boulton AJ (1998) Flow variability and the ecology of large rivers. Mar Freshw Res 49:55–72

Raven PJ, Fox P, Everard M, Holmes NTH, Dawson FH (1997) River habitat survey: a new system for classifying rivers according to their habitat quality. In: Boon PJ, Howell DL (eds) Freshwater quality: defining the indefinable? The Stationery Office, Edinburgh, pp 215–234

Reif A, Gärtner S, Zimmermann R, Späth V, Lange J (2013) Developments on alluvial and former alluvial riparian zones along the southern Upper Rhine. Tuexenia Beiheft 6:125–169

Rosgen DL (1994) A classification of natural rivers. Catena 22:169–199

Rosgen DL (1996) Applied river morphology. Wildland Hydrology, Pagosa Springs

Schiemer F, Waidbacher H (1992) Strategies for conservation of a Danubian fish fauna. In: Boon PJ, Calow P, Petts GE (eds) River conservation and management. Wiley, Chichester, pp 363–382

Schmautz M, Aufleger M, Strobl T (2000) Wissenschaftliche Untersuchung der Geschiebe- und Eintiefungsproblematik der österreichischen Donau. Report by order of Verbund – Austrian Hydro Power AG (AHP), Vienna

Schmautz M, Aufleger M, Strobl T (2002) Anthropogene Einflussnahme auf die Flussmorphologie der Donau in Österreich. Österreichische Ingenieur- u Architekten-Zeitschrift 147:171–178

Schumm SA (1977) The fluvial system. Wiley-Interscience, New York

Schumm SA, Harvey MD, Watson CC (1984) Incised channels: morphology, dynamics and control. Water Resources Publications, Littleton

Sedell JR, Froggatt JL (1984) Importance of streamside forests to large rivers: the isolation of the Willamette River, Oregon, U.S.A., from its floodplain by snagging and streamside forest removal. Verhandlungen des Internat Vereins für Limnologie 22:1828–1834

Simon A (1989) A model of channel response in disturbed alluvial channels. Earth Surf Process Landf 14:11–26

Simon A, Doyle M, Kondolf M, Shields FD Jr, Rhoads B, McPhillips M (2007) Critical evaluation of how the Rosgen classification and associated "Natural Channel Design" methods fail to integrate and quantify fluvial processes and channel response. J Am Water Resour Assoc 43:1117–1131

Stanford JA, Ward JV, Liss WJ, Frissell CA, Williams RN, Lichatowich JA, Coutant CC (1996) A general protocol for restoration of regulated rivers. Regul Rivers Res Manag 12:391–413

Stanford JA, Lorang MS, Hauer FR (2005) The shifting habitat mosaic of river ecosystems. Verhandlungen der Internat Vereinigung für Theoretische und Angewandte Limnologie 29:123–126

Thorne CR (1997) Channel types and morphological classification. In: Thorne CR, Hey RD, Newson MD (eds) Applied fluvial geomorphology for river engineering and management. Wiley, Chichester, pp 175–222

Thorp JH, Thoms MC, Delong MC (2006) The riverine ecosystem synthesis: biocomplexity in river networks across space and time. River Res Appl 22:123–147

Tockner K, Stanford JA (2002) Riverine flood plains: present state and future trends. Environ Conserv 29:308–330

Tockner K, Schiemer F, Ward JV (1998) Conservation by restoration: the management concept for river floodplain system on the Danube river in Austria. Aquat Conserv Mar Freshwat Ecosyst 8:71–86

Tockner K, Malard F, Ward JV (2000) An extension of the flood pulse concept. Hydrol Process 14:2861–2883

Tockner K, Paetzold A, Karaus U, Claret C, Zettel J (2006) Ecology of braided rivers. In: Sambrook Smith GH, Best J, Bristow C, Petts GE (eds) Braided rivers. IAS Special Publication/Blackwell Publishers, London

Tockner K, Pusch M, Borchardt D, Lorang MS (2010) Multiple stressors in coupled river-floodplain ecosystems. Freshw Biol 55:135–151

Toth LA (1996) Restoring the hydrogeomorphology of the channelised Kissimmee river. In: Brookes A, Shields FD Jr (eds) River channel restoration: guiding principles for sustainable projects. Wiley, Chichester, pp 369–383

Van den Berg JH (1995) Prediction of alluvial channel pattern of perennial rivers. Geomorphology 12:259–279

Ward JV (1998) Riverine landscapes: biodiversity patterns, disturbance regimes and aquatic conservation. Biol Conserv 83:269–227

Ward JV, Stanford JA (1983) The intermediate-disturbance hypothesis: an explanation for biotic diversity patterns in lotic ecosystems. In: Fontaine TD, Bartell SM (eds) Dynamics of lotic ecosystems. Ann Arbor Science Publishers, Ann Arbor, pp 347–356

Ward JV, Tockner K (2001) Biodiversity: towards a unifying theme for river ecology. Freshw Biol 46:807–819

Ward JV, Tockner K, Arscott DB, Claret C (2000) Riverine landscape diversity. Freshw Biol 47:517–539

Werth W (1987) Ökomorphologische Gewässerbewertung in Oberösterreich (Gewässerzustand kartierungen). Eco-morphological classification of channels in Upper Austria. Österreichische Wasserwirtschaft 39:121–128

Wex G (1873) Über die Wasserabnahme in den Quellen, Flüssen und Strömen, bei gleichzeitiger Steigerung der Hochwässer in den Cultur-Ländern. Zeitschrift des Österreichischen Ingenieur-u. Architekten-Vereines 25:23–30, 63–76, 101–119

Wex G (1879) Zweite Abhandlung über die Wasserabnahme in den Quellen, Flüssen und Strömen, bei gleichzeitiger Steigerung der Hochwässer in den Cultur-Ländern. Zeitschrift des Österreichischen Ingenieur- u. Architekten-Vereines 31:93–99, 125–144

Wiens JA (2002) Riverine landscapes: taking landscape ecology into the water. Freshw Biol 47:501–515

Wiletal I (1897) Die Veränderungen des Donau-Laufes im Wiener Becken. Alt-Wien – Monatsschrift für Wiener Art und Sprache 6:48–51, 65–68

Winiwarter V, Schmid M, Hohensinner S, Haidvogl G (2012) The environmental history of the Danube river basin as an issue of long-term socio-ecological research. In: Singh SJ, Haberl H, Chertow M, Mirtl M, Schmid M (eds) Long term socio-ecological research: studies in society-nature interactions across spatial and temporal scales. Springer, Dordrecht, pp 103–122

Wu J, Loucks OL (1995) From balance of nature to hierarchical patch dynamics: a paradigm shift in ecology. Q Rev Biol 70:439–466

Yalin MS, da Silva AMF (2001) Fluvial processes, IAHR Monograph. IAHR, Delft

Chapter 4
River Hydrology, Flow Alteration, and Environmental Flow

Bernhard Zeiringer, Carina Seliger, Franz Greimel, and Stefan Schmutz

"The water runs the river." This chapter focuses on the river flow as the fundamental process determining the size, shape, structure, and dynamics of riverine ecosystems. We briefly introduce hydrological regimes as key characteristics of river flow. Hydrological regimes are then linked to habitats and biotic communities. The effects of flow regulation as a result of human activities such as water abstraction (irrigation and hydropower), river channelization, land use, and climate change are demonstrated. Finally, methods to assess the environmental flow, the flow that is needed to maintain the ecological integrity, are described, and examples of successful flow restoration presented.

4.1 The Water Cycle and Hydrological Regimes

In temperate zones water received via precipitation is either stored in ice and snow during winter or infiltrates into the groundwater and is released into rivers during summer. Water cycles through stages of evaporation, water storage in the atmosphere, precipitation, (sub)surface runoff, and storage in the ocean. The water cycle and climatic conditions form the boundary conditions for the *hydrological regimes* that define distinct seasonal and daily flow patterns. High altitude rivers receive water mainly from glacial melt during summer with distinct diurnal melting peaks following air temperature warm-up (*glacial regime*) (Fig. 4.1). At lower elevations snow melting in spring causes seasonal peaks (*nival regime*), while periods of high flow and floods due to rainfall can occur at any time of the year (*pluvial regime*).

B. Zeiringer (✉) · C. Seliger · F. Greimel · S. Schmutz
Institute of Hydrobiology and Aquatic Ecosystem Management, University of Natural Resources and Life Sciences, Vienna, Austria
e-mail: bernhard.zeiringer@boku.ac.at; carina.seliger@boku.ac.at; franz.greimel@boku.ac.at; stefan.schmutz@boku.ac.at

© The Author(s) 2018
S. Schmutz, J. Sendzimir (eds.), *Riverine Ecosystem Management*, Aquatic Ecology Series 8, https://doi.org/10.1007/978-3-319-73250-3_4

67

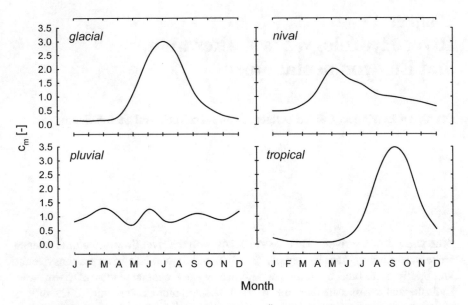

Fig. 4.1 Simple hydrological regimes (glacial, River Ötztaler Ache; nival, River Mur; pluvial, River Stiefing; and tropical, River Niger). The monthly discharge coefficient (c_m) is defined by the ratio of the average monthly discharge and the mean discharge (hydrograph data over several years)

Tropical rivers are characterized by distinct flow cycles related to dry and wet seasons. The *tropical regime* is similar to the pluvial regime, e.g., drought in the dry season and abundant rainfall in the wet season. Depending on the local conditions and position within the catchment, observed flow may represent a mixture of hydrological regimes. Flow regimes are very important to understand the key functions and processes of riverine ecosystems.

Catchments are hydrological units defined as the area collecting the water within a given drainage divide or watershed (a drainage divide is the line that separates neighboring drainage basins). All the catchments for all the tributaries of a river are lumped together to form a *river basin* (e.g., Danube River Basin). The so-called *water balance* of a given catchment or basin is calculated from water gains (precipitation) and losses (evapotranspiration and runoff) including storage phases (soil water, groundwater, ice, snow). The observed discharge (m³/s) at distinct locations within the catchment is determined based on meteorological and biogeophysical factors (see Table 4.1).

The river flow determines the dynamics of the four-dimensional river system (Ward 1989). Sediment and nutrient transport is closely linked to the longitudinal dimension of flow. Floodplain dynamics depend on the lateral hydrological connectivity and flood pulses (Junk et al. 1989). River groundwater interaction represents the vertical dimension of flow dynamics and determines groundwater recharge and groundwater contribution to river flow. The longitudinal, lateral, and vertical flow pattern varies over time representing the fourth dimension of the four-dimensional river system.

Table 4.1 Meteorological and biogeophysical factors determining river flow

Meteorological factors	Biogeophysical factors
– Type of precipitation (rainfall, snow)	– Drainage area
– Rainfall amount, intensity, duration, and distribution over the drainage basin	– Elevation
	– Topography, terrain slope
– Precipitation that occurred earlier and resulting soil moisture	– Basin shape and drainage network patterns
	– Soil type, land use, and vegetation
– Meteorological conditions that affect evapotranspiration and infiltration	– Ponds, lakes, reservoirs, sinks, etc. in the basin, which prevent or delay downstream runoff

4.2 Flow Determines Habitats and Biotic Communities

River flow determines processes that shape and organize the physical habitat and associated biotic communities. Flow variability is a fundamental feature of river systems and their ecological functioning (Poff et al. 1997). The natural flow of a river varies on time scales of hours, days, seasons, years, and longer. Many years of observation from a streamflow gauge are generally needed to describe the characteristic pattern of a river's flow quantity, timing, and variability (Poff et al. 1997). River flow regimes show regional patterns that are determined largely by river size and by geographic variation in climate, geology, topography, and vegetative cover.

The widely accepted natural flow paradigm (*sensu* Poff et al. 1997), where the flow regime of a river, comprising the five key components of variability, i.e., *magnitude, frequency, duration, timing*, and *rate of change*, is recognized as central to sustaining biodiversity and ecosystem integrity (Poff and Ward 1989; Karr 1991; Richter et al. 1997; Rapport et al. 1998; Rosenberg et al. 2000). These components can be used to characterize the entire range of flows and specific hydrologic phenomena, such as floods or low flows, which are critical to the integrity of river ecosystems.

The natural flow regime organizes and defines river ecosystems. In rivers, the physical structure of the environment and, thus, of the habitat is defined largely by physical processes, especially the movement of water and sediment within the channel and between the channel and floodplain. The physical habitat of a river includes sediment size and heterogeneity, channel and floodplain morphology, and other geomorphic features. These features form as the available sediment, woody debris, and other transportable materials are moved and deposited by flow. Thus, habitat conditions associated with channels and floodplains vary among rivers in accordance with both flow characteristics and the type and the availability of transportable materials. Within a river, different habitat features are created and maintained by a wide range of flows (Poff et al. 1997).

Generally, the shaping of hydro-morphological channel and floodplain features (e.g., river bars and riffle-pool sequences) happens continuously. But the dominant, shaping processes occur in episodes of bank-full discharges (see Chap. 3). It is important that these flows are able to move bed or bank sediment and occur frequently enough to continually modify the river channel (Wolman and Miller 1960).

The diversity of instream and floodplain habitat types has stimulated the evolution of species that use the habitat mosaic created by hydrologic variability. For many riverine species, completion of the life cycle requires an array of different habitat types, whose availability over time is regulated by the flow regime (Greenberg et al. 1996).

Aquatic organisms have evolved life history strategies primarily in direct response to natural flow regimes (Bunn and Arthington 2002). The physical, chemical, and biological characteristics of rivers are primarily affected by flow variation as a "master variable." Changes in discharge are a form of disturbance, but a moderate level of hydrological variability enhances biological diversity (*sensu* Connell 1978; Ward and Stanford 1983; Bunn and Arthington 2002). River biota have evolved adaptive mechanisms to cope with habitat changes that result from natural flow variation, and indeed many species rely on regular or seasonal changes in river flows to complete their life cycles (Poff et al. 1997). For detailed discussions of the ecological effects (and knock-on social and economic implications) of hydrological alterations on riverine ecosystems, with impacts ranging from genetic isolation through habitat fragmentation to declines in biodiversity, floodplain fisheries, and ecosystem services, see Ward (1982), Petts (1984), Lillehammer and Saltveit (1984), Armitage (1995), Cushman (1985), Craig and Kemper (1987), Gore and Petts (1989), Calow and Petts (1992), Boon et al. (1992, 2000), Richter et al. (1998), Postel (1998), Snaddon et al. (1999), Pringle (2000), World Commission on Dams (2000), Bergkamp et al. (2000), and Bunn and Arthington (2002).

Bunn and Arthington (2002) propose that the relationship between biodiversity and the physical nature of the aquatic habitat is likely to be driven primarily by large events that influence channel form and shape (principle 1) (Fig. 4.2). However, droughts and low-flow events are also likely to play a role by limiting overall habitat availability. Native biota have evolved in response to the overall flow regime. Many features of the flow regime influence life history patterns, especially the seasonality and predictability of the overall pattern, but also the timing of particular flow events (principle 2). Some flow events trigger longitudinal dispersal of migratory aquatic organisms, and other large events allow access to otherwise disconnected floodplain habitats (principle 3). Catchment land-use change and associated water resource development inevitably lead to changes in one or more aspects of the flow regime resulting in declines in aquatic biodiversity via these mechanisms. Invasions by introduced or exotic species are more likely to succeed at the expense of native biota if the former are adapted to the modified flow regime (principle 4).

4.3 Flow Regulation

The global increase in water demand has resulted in a conflict between using rivers as water and energy sources and the need to conserve rivers as intact ecosystems (Dynesius and Nilsson 1994; Abramovitz 1995; Postel 1995; McCully 1996; World Commission on Dams (2000). This ongoing conflict has stimulated a growing field

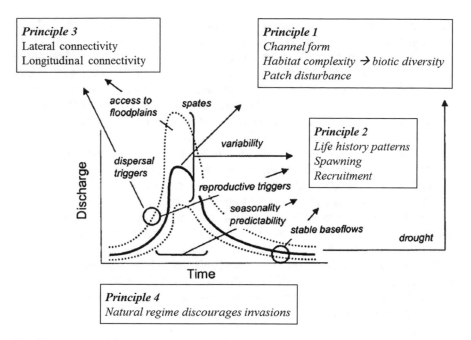

Fig. 4.2 The natural flow regime of a river influences aquatic biodiversity via several interrelated mechanisms that operate over different spatial and temporal scales. The relationship between biodiversity and the physical nature of the aquatic habitat is likely to be driven primarily by large events that influence channel form and shape (principle 1). However, droughts and low-flow events are also likely to play a role by limiting overall habitat availability. Many features of the flow regime influence life history patterns, especially the seasonality and predictability of the overall pattern, but also the timing of particular flow events (principle 2). Some flow events trigger longitudinal dispersal of migratory aquatic organisms, and other large events allow access to otherwise disconnected floodplain habitats (principle 3). The native biota have evolved in response to the overall flow regime. Catchment land-use change and associated water resource development inevitably lead to changes in one or more aspects of the flow regime resulting in declines in aquatic biodiversity via these mechanisms. Invasions by introduced or exotic species are more likely to succeed at the expense of native biota if the former are adapted to the modified flow regime (principle 4) (Bunn and Arthington 2002) (© Environmental management, Basic principles and ecological consequences of altered flow regimes for aquatic biodiversity, 30, 2002, p. 493, Bunn SE, Arthington AH. With permission of Springer.)

of research dedicated to assessing the requirements of rivers for their own water, to enable satisfactory tradeoffs in water allocation among all users of the resource and the resource base itself (the river) (Tharme 2003).

More than half of the world's accessible surface water is already appropriated by humans, and this is projected to increase to 70% by 2025 (Postel 1998). Water resource developments such as impoundments, diversion weirs, interbasin water transfers, run-of-river abstraction, and exploitation of aquifers, for the primary uses of irrigated agriculture, hydropower generation, industry, and domestic supply, are responsible for unprecedented impacts to riverine ecosystems, most of which result from alterations to the natural hydrological regime (Rosenberg et al. 2000). Almost all large river basins are already impacted by large dams (Nilsson et al. 2005).

About 60% of the world's rivers are estimated to be fragmented by hydrologic alteration, with 46% of the 106 primary watersheds modified by the presence of at least one large dam (Revenga et al. 1998, 2000). Dynesius and Nilsson (1994) calculated that 77% of the total discharge of the 139 largest river systems in North America, Europe, and the republics of the former Soviet Union is strongly or moderately affected by flow-related fragmentation of river channels. Moreover, they observed that large areas in this northern third of the world entirely lack unregulated large rivers. EU member countries regulate the flow of around 65% of the rivers in their territories, while in Asia, just under 50% of all rivers that are regulated have more than one dam (World Commission on Dams 2000). Flow regulation through impoundment represents the most prevalent form of hydrological alteration with over 45,000 large dams in over 140 countries (World Commission on Dams 2000); a further 800,000 small dams are estimated to exist worldwide (McCully 1996). The top five dam-building countries (China, United States, India, Japan, Spain) account for close to 80% of all large dams worldwide, with China alone possessing nearly half the world total (World Commission on Dams 2000, cited in Tharme 2003). Dam development is expected to continue, with more than 3700 large hydropower dams alone currently planned or under construction worldwide (Zarfl et al. 2014).

4.4   Human Alteration of Flow Regimes

Human alteration of flow regime changes the established pattern of natural hydrologic variation and habitat dynamics. Modification of natural hydrologic processes disrupts the dynamic balance between the movement of water and the movement of sediment that exists in free-flowing rivers (Dunne and Leopold 1978).

Typical sources of alteration of flow regimes are (after Poff et al. 1997):

- Dam
- Water diversion
- Urbanization, sealing, drainage
- Levees and channelization
- Groundwater pumping

Dams, which are the most obvious direct modifiers of river flow, capture both low and high flows for flood control, electrical power generation (Fig. 4.3), irrigation and municipal water needs, maintenance of recreational reservoir levels, and navigation. Dams capture sediments moving down a river, with many severe downstream consequences (e.g., erosion of fine sediment in the downstream section). The coarsening of the streambed can, in turn, reduce habitat availability for aquatic species living in or using interstitial spaces (Chien 1985). Beside flow regulation as a consequence of dam construction, rivers get fragmented and loose its natural connectivity (see Chap. 6).

Dams also lead to reduction of the magnitude and frequency of high flows, leading to deposition of fines and sealing in gravel and channel stabilization and

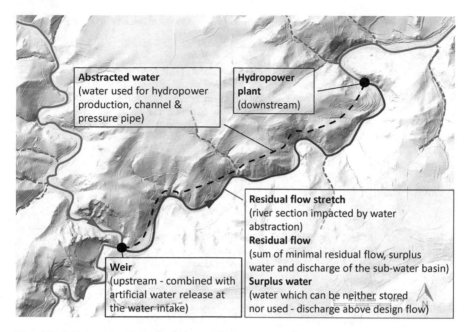

Fig. 4.3 Scheme of a diversion power plant and residual flow stretch (hydropower plant Hohenstein at the River Krems, Austria). Main river (blue solid line), small tributaries (blue dashed line), residual flow stretch (red solid line), and diversion channel (black dashed line)

narrowing. Sealing and land drainage increase the magnitude and frequency of high flows, leading to bank and riverbed erosion and floodplain disconnection. Furthermore, reduced infiltration into soil reduces base flows. Levees and channelization reduce overbank flows, leading to floodplain deposition and channel restriction, causing downcutting and restraining channel migration and formation of secondary channels. Groundwater pumping lowers water table levels and further reduces plant growth. The loss of vegetation leads to streambank stability erosion and channel downcutting.

4.5   Ecological Responses to Altered Flow Regime

In a comprehensive review, Poff and Zimmerman (2010) reported that almost all published research found negative ecological changes in response to a variety of flow alteration (Table 4.2). Only in few instances did values for ecological response metrics increase, indicating shifts in ecological organization, such as increase in non-native species or non-woody plant cover on dewatered floodplains. This also confirms earlier summaries of ecological response to flow regime alterations (Poff et al. 1997; Bunn and Arthington 2002; Lloyd et al. 2003).

Table 4.2 Alterations in flow components and common ecological response (modified after Poff et al. 1997; Poff and Zimmerman 2010)

Flow component	Alteration		Ecological response
Magnitude	Flow stabilization (loss of extreme high and/or low flows)	(a)	Reduced diversity Loss of sensitive species Altered assemblages and dominant taxa Reduced abundance Increase in non-natives
		(r)	Seedling desiccation Ineffective seed dispersal Terrestrialization of flora Lower species richness Encroachment of vegetation into channels Increased riparian cover Altered assemblages
	Greater magnitude of extreme high and/or low flows	(a)	Life cycle disruption Reduced species richness Altered assemblages and relative abundance of taxa Loss of sensitive species
Frequency	Decreased frequency of peak flows	(a)	Aseasonal reproduction Reduced reproduction Decreased abundance or extirpation of native fishes Decreased richness of endemic and sensitive species Reduced habitat for young fishes
		(r)	Shift in community composition Reductions in species richness Increase in wood production
Duration	Decreased duration of floodplain inundation	(a)	Decreased abundance of young fish Change in juvenile fish assemblage Loss of floodplain specialists in mollusk assemblage
		(r)	Reduced growth rate or mortality Altered assemblages Terrestrialization or desertification of species composition Reduced area of riparian plant or forest cover
	Prolonged low flows	(a)	Concentration of organisms Downstream loss of floating eggs
		(r)	Reduction or elimination of plant cover Diminished plant species diversity Desertification of species composition
	Prolonged inundation	(a)	Loss of riffle habitat
		(r)	Change in vegetation functional type Tree mortality

(continued)

Table 4.2 (continued)

Flow component	Alteration		Ecological response
Timing	Shifts in seasonality of peak flows	(a)	Disruption of spawning cues Decreased reproduction and recruitment Change in assemblage structure
	Increased predictability	(a)	Change in diversity and assemblages structure Disruption of spawning cues Decreased reproduction and recruitment
	Loss of seasonal flow peaks	(a)	Disruption of migration cues Loss of accessibility to wetlands and backwaters Modification of food web structure
		(r)	Reduced riparian plant recruitment Invasion of exotic riparian plant species Reduced plant growth and increased mortality Reduction in species richness and plant cover
Rate of change	Rapid changes in river stage	(a)	Drift (washout) and stranding
	Accelerated flood recession	(r)	Failure of seedling establishment

Taxonomic identity of organisms: aquatic (a) and riparian (r)

Taxonomic groups, e.g., fish, macroinvertebrates, and riparian vegetation, show biota-specific responses (abundance, diversity, and demographic parameters) to flow alteration depending on the flow components affected (magnitude, frequency, duration, timing, rate of change). Most of the studies on ecological changes report responses to altered flow magnitude associated with flow stabilization due to water abstraction or water withdrawals for irrigation. For the most part instream taxa react negatively to alteration of flow magnitude. Alterations in flow frequency, referring mainly to decreases in frequency of floods, resulted in negative ecological responses by macroinvertebrates and fish. Riparian communities usually decline in response to flow frequency alteration; but also some increases are indicated (e.g., wood production). Alterations in flow duration, mostly in the form of changes in the duration of floodplain inundation, are primarily associated with decreases in both instream and riparian communities. Similarly, changes in the timing of flows due to loss of seasonal flow peaks reduce both aquatic and riparian communities (Poff et al. 1997; Poff and Zimmerman 2010). The rate of change is an important component of the natural flow regime, commonly altered by hydropeaking, which causes detrimental effects on instream and riparian communities (see Chap. 5).

Fish respond negatively to changes in flow magnitude, whether the flows increase or decrease. Fish metrics decrease sharply in response to reduced flows (see Figs. 4.4, 4.5 and 4.6). Diversity shows a clear decline, especially where changes in flow magnitudes exceed 50%. Therefore, fish are sensitive indicators of flow alteration. Compared to this, macroinvertebrates or riparian species are not such reliable indicators, since they do not consistently respond to changes in flow magnitude. Riparian

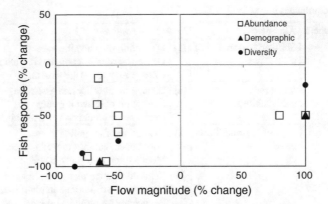

Fig. 4.4 Percent change in fish abundance, demographic parameters, and species diversity (and/or richness) with respect to percent alteration of flow magnitude. Percent change for both fishes and flow magnitude represents alteration relative to a pre-impact or "reference" condition. Alteration in flow magnitude includes changes in peak flow, total or mean discharge, baseflow, or hourly flow (Poff and Zimmerman 2010) (source: Poff and Zimmerman (2010). Ecological responses to altered flow regimes: a literature review to inform the science and management of environmental flows. Freshwater Biology, 55(1), 194–205, reproduced with permission of John Wiley & Sons, Ltd., © 2009 Blackwell Publishing Ltd, Freshwater Biology, 55, 194–205)

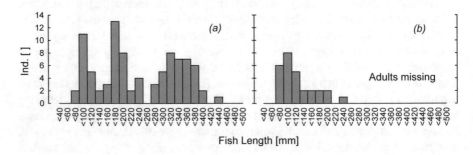

Fig. 4.5 Length distribution of brown trout at River Unrechttraisen (**a**) full water section and (**b**) residual flow section (adapted from Zeiringer 2008b)

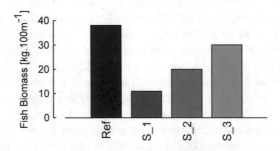

Fig. 4.6 Biomass of brown trout in River Ybbs in full flow section (reference) and residual flow sections, ordered along the river course (adapted from Zeiringer et al. 2010)

Fig. 4.7 Encroachment of vegetation into river channel, example residual flow stretch River Gölsen

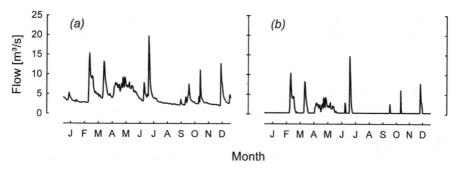

Fig. 4.8 Hydrological effects of water abstraction, (**a**) natural hydrograph, and (**b**) reduced and moderated flow in the residual flow section at the HPP Reichenau/River Schwarza (adapted from Zeiringer 2008a)

responses can be associated with decreases in flood peaks, leading to reduction or elimination of overbank flooding (Poff and Zimmerman 2010) (Fig. 4.7).

Aquatic and riparian species respond to multiple hydrologic drivers, and overlap in their occurrence and impacts often confounds analysis (Poff and Zimmerman 2010). Changes in magnitude of high flows are often accompanied by changes in frequency, and either or both of these may influence biological response (Fig. 4.8). Additionally, other environmental characteristics, like water temperature (Fig. 4.9) or sediment regime (Fig. 4.10), may affect biota independently or in association with flow alteration.

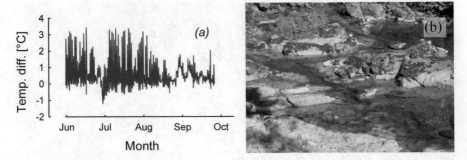

Fig. 4.9 Change of water quality due to water abstraction (**a**) water temperature increase River Mur during summer (adapted from Zeiringer et al. 2008) and (**b**) algae bloom River Lassing

Fig. 4.10 Morphological effects of water abstraction, e.g., reduction of flow velocity and shear stress, change of flow and substrate patterns, silting up of interstitial (clogging), reduced water depth, and reduced wetted width (**a**) River Aschbach and (**b**) and (**c**) River Mur

Poff and Zimmerman (2010) mentioned that there are no studies reported that focus primarily on ecosystem functional responses (e.g., riparian production, nutrient retention), even though many ecological processes are clearly flow dependent (Hart and Finelli 1999; Doyle et al. 2005, cited in Poff and Zimmerman 2010). They emphasized that this absence points to an obvious research gap in the environmental flows research.

4.6 Environmental Flow

Environmental flow (EF) is the quantity or volume of water required over time to maintain river health in a particular state, where the state has to be predetermined or agreed upon based on a trade-off with other considerations (Acreman and Dunbar 2004). Such quanta are captured by a variety of terms, including the environmental flow (regime), instream flow, environmental allocation, or ecological flow requirement, to distinguish these from compensation flows (Gustard et al. 1987, cited in Acreman and Dunbar 2004). The latter have been set for other purposes, such as downstream human uses (e.g., irrigation, hydropower), pollutant dilution, or navigation. The first approaches to quantifying EFs only focused on minimum flow, based on the idea that all river health problems are associated with low flows.

Although there is no generally agreed definition or term (IWMI 2005), it is widely accepted (e.g., Poff et al. 1997; Karr 1991; Bunn and Arthington 2002; Postel and Richter 2003; Annear et al. 2004) that not only the quantity of discharge is decisive but that also the timing and discharge dynamics are key factors for sustaining and conserving native species diversity and ecological integrity of rivers.

4.6.1   The Concept and Definitions of Environmental Flow

The concept of EF historically was developed as a response to the degradation of aquatic ecosystems caused by overuse of water. In this context EF may be defined as the amount of water that is left in an aquatic ecosystem, or released into it, for the specific purpose of managing the condition of that ecosystem (Arthington et al. 2006; Brown and King 2003). Despite the fact that the concept of EF has existed for over 40 years (including other terminology, such as instream flows), there is still no unified definition for it (Moore 2004). This lack of uniform agreement for a definition of EF can be illustrated by looking at a sample of the ways in which it has been defined in the literature by researchers and organizations involved in assessing and implementing the concept all around the world over the last decades. In these definitions of environmental flows, there are always two key aspects of the concept included: the flow regime that should be considered and the level of conservation for the ecosystem that is intended.

Selected definitions of EF:

- Arthington and Pusey (2003) define the objective of environmental flows as maintaining or partially restoring important characteristics of the natural flow regime (i.e., the quantity, frequency, timing, and duration of flow events, rates of change, and predictability/variability) required to maintain or restore the biophysical components and ecological processes of instream and groundwater systems, floodplains, and downstream receiving waters.
- Brown and King (2003) state that environmental flows is a comprehensive term that encompasses all components of the river, is dynamic over time, takes cognizance of the need for natural flow variability, and addresses social and economic issues as well as biophysical ones.
- Dyson et al. (2003) in the IUCN guide on environmental flows define the concept as the water regime provided within a river, wetland, or coastal zone to maintain ecosystems and their benefits where there are competing water uses and where flows are regulated.
- Tharme (2003) defines an environmental flow assessment (EFA) as an assessment of how much of the original flow regime of a river should continue to flow down

it and onto its floodplains in order to maintain specified, valued features of the ecosystem.

- Gupta (2008) defines EFs as discharges of a particular magnitude, frequency, and timing, which are necessary to ensure that a river system remains environmentally, economically, and socially healthy.
- Environmental flows can be described as "the quality, quantity, and timing of water flows required to maintain the components, functions, processes, and resilience of aquatic ecosystems which provide goods and services to people" (Hirji and Davis 2009).

EF is a management concept, and thus it should vary in response to actions or processes that are used and understood by management. Generally, certain human activities create a water demand that requires the development of infrastructure (diversion weirs, dams, etc.). The presence and operation of this infrastructure produces modifications of the natural flow regimes that affects the biophysical conditions of ecosystems. Environmental flows can help to restrict water use, to define the maximum limits of hydrological alteration to maintain a certain biological condition and may appear as a basic tool for the recovery of certain species affected by the modification of aquatic habitats (Navarro and Schmidt 2012). A combination of Arthington and Pusey and Tharme definitions (2003) might consider the most basic and relevant aspects of the concept of environmental flows: environmental flow is the proportion of original flow maintaining or restoring biophysical components, ecological processes, and services of instream and groundwater systems, floodplains, and downstream receiving waters.

4.6.2 Assessing and Implementing Environmental Flows

In many countries a variety of approaches for assessing EF were developed with varying complexity, e.g., look-up tables (preliminary assessment level), desktop analyses and functional analyses (intermediate assessment level), and finally hydraulic habitat modeling (comprehensive assessment level), which we describe in more detail below (see also Table 4.3). Some address just parts or the river system, while others are more holistic (Tharme 2003; Acreman and Dunbar 2004). Currently, there exist at least 200 environmental flow methods classifiable in four major categories according to focus, complexity, and cost and time effectiveness: (1) hydrological methods, (2) hydraulic rating, (3) habitat simulation models, and (4) holistic methodologies (Dyson et al. 2003; Tharme 2003; Arthington et al. 2004; Richter et al. 2006; King et al. 2008).

Hydrological Analyses (also called desktop analyses) are mostly based on simple minimum flow thresholds derived from hydrographs (e.g., mean annual flows, monthly flows, high/low flows, and Q95%) (Barker and Kirmond 1998). For example, the Tennant or Montana method (Tennant 1976) defines EF values as percentage of the average daily discharge or mean annual flow (MQ) with 10% MQ

Table 4.3 Different methods and characteristics of setting environmental flows and choice of method (modified after Acreman and Dunbar 2004; European Commission 2015; Theodoropoulos and Skoulikidis 2014)

Method type	Application range				Pros and cons	Assessment level
	Scoping study, regional planning	Basin-scale planning	Impact assessment (multi-site)	Impact assessment (single-site); River restauration (multi- and single-site)		
Look-up table					Rapid, cheap, not site specific	Preliminary
Desktop					Site specific, limited new data collection, long time series required, use existing ecological data	Intermediate
Functional analysis					Flexible, robust, more focused on whole ecosystem, expensive to collect all relevant data and wide range of experts	
Habitat modeling					Replicable, predictive, expensive to collect hydraulic and ecological data	Comprehensive

considered as minimum flow and 60–00% MQ considered the flow range necessary to provide optimal habitat conditions. More complex hydrological indices are the indicators of hydrologic alteration (IHA) (Richter et al. 1996), the range of variability approach (RVA) (Richter et al. 1997), and the indicators of hydrologic alteration in rivers (IAHRIS) (Martinez and Fernandez 2010). RVA, for example, uses 32 hydrological parameters (their range and variation) as indicators of hydrological alteration (IHA; Richter et al. 1996) to characterize ecologically relevant attributes of the local flow regime and to translate them into defined flow-based management targets. The method suggests a natural flow paradigm including the full range of natural intra- and interannual variation of hydrological regimes and associated characteristics of timing, duration, frequency, and rate of change as critical factors to sustain the integrity of the riverine ecosystem (Richter et al. 1997). Hydrological methods rely primarily on historical hydrological data, requiring flow measurements over long time periods. Although hydrological data collection is resource demanding, the application of such methods itself is time- and cost-effective and simple. Although such methods consider flow dynamics, they only indirectly address requirements of aquatic biota. Therefore, they are not considered appropriate as stand-alone methods, but often are used as initial desktop analyses to assist more complex environmental flow methodologies (Theodoropoulos and Skoulikidis 2014). In fact, these methods lack ecological relevance and sensitivity to individual rivers and are considered as inadequate to provide the data needed to sustain ecological integrity.

Hydraulic Rating methods use simple hydraulic variables and propose EF through the quantifiable relationship between water discharge and instream habitats (Trihey and Stalnaker 1985). Hydraulic rating methods try to incorporate channel-discharge relationships. The generic wetted perimeter method (Reiser et al. 1989, cited in Tharme 2003) is the most applied hydraulic rating approach worldwide. River integrity is directly related to the quantity of wetted perimeter. The modeled relationship between wetted perimeter and discharge is used to determine minimum or preservation flows. The flow events method (FEM; Stewardson and Gippel 2003) evaluates the frequency of hydraulically relevant flow indices (selected by experts) under alternate flow regimes (Acreman and Dunbar 2004). It consists of five steps: After preparing a list of ecological factors affected by flow variation, different flow events and their distribution in time are analyzed. Then hydraulic parameters (e.g., wetted perimeter) at these different flow events are modeled. A comparison and evaluation of different flow management scenarios with regard to ecological consequences leads to the specification of certain flow rules (Stewardson and Gippel 2003). However, these methods have been currently replaced by more sophisticated hydraulic/habitat simulation methods (described below).

Habitat Simulation methods combine flows with habitat availability for selected indicator species and life stages. Waters (1976) invented the concept of weighted usable area (WUA), which was used by the US Fish and Wildlife Service to develop the computer model PHABSIM (Physical Habitat Simulation model, Bovee 1982). Available habitat is weighted by its suitability for certain species under different flow scenarios (Acreman and Dunbar 2004). PHABSIM is embedded into the Instream Flow Incremental Methodology (IFIM; Bovee and Milhous 1978; Reiser et al. 1989) providing a tool for calculating suitable EF. Physical habitat (flow velocity, water depth, substrate) is monitored in the field and/or modeled using mainly 1-D or 2-D hydraulic models or habitat modeling software, such as TELEMAC (Galland 1991), PHABSIM (USGS 2001), CASiMiR (Schneider et al. 2010), and RIVER 2D (Steffler and Blackburn 2002). Habitat preferences for target organisms are retrieved from field observations or literature, and habitat availability is then calculated through the modeling software for different discharges (for more details, see Chap. 7).

Holistic Methodologies require multidisciplinary input and expertise (Tharme 1996, 2000; King et al. 2008; Arthington 1998), address flow requirements of multiple ecosystem components (fish, benthic fauna, macrophytes, riparian vegetation) at various spatial temporal scales, and target a flow regime going beyond simple minimum flow definitions. Examples are the building block methodology (BBM) (Tharme and King 1998; King et al. 2008), the downstream response to imposed flow transformations (DRIFT) (King and Brown 2006), and the ecological limits of hydrologic alteration (ELOHA) (Poff et al. 2010). Field data on a monthly basis are required to construct a flow regime from scratch (bottom-up approaches, BBM, and ELOHA). In contrast, top-down approaches (e.g., DRIFT) are generally scenario based, defining environmental flows as acceptable degrees of divergence from the natural/reference flow regime, being less susceptible to any omission of critical flow characteristics or processes than their bottom-up counterparts (Bunn 1998). More detailed, the building block methodology states that aquatic organisms rely on basic elements (i.e., building blocks) of the flow regime (e.g., low flows, medium flows, and floods). In this method EF is assessed by an expert-based combination of building blocks. The expert panel assessment method (Swales and Harris 1995), the scientific panel approach (Thoms et al. 1996), or the benchmarking methodology (Brizga et al. 2002) tries to evaluate how much a flow regime can be altered before the integrity of the aquatic ecosystem is altered or seriously affected. Also ELOHA is based on the premise that increasing degrees of flow alteration enforce increasing ecological change. The evaluation of the relationship relies on the testing of plausible hypotheses stated by experts. Ecological response variables are most suitable if they react to flow alterations, allow validation using monitoring data, and are esteemed by society (e.g., for fishery) (Poff et al. 2010).

Several modified approaches have also been proposed and implemented, e.g., trying to shift the assessment scale from the micro- to meso-habitat (e.g., Parasiewicz 2007), but their general concept is based on one of the four principles mentioned above. Although progress in environmental flow methodologies is fast and becoming very sophisticated, there still remains a critical need for greater understanding of flow-ecological response relationships and enhanced modeling capacity to support river flow management and ecosystem conservation (Arthington et al. 2010).

While (1) current EF determinations are often prescriptive and not negotiable (i.e., consequences of noncompliance are not discussed) and (2) socioeconomic impacts are not adequately considered (cost-benefit of water resource developments), the DRIFT method (King et al. 2003) tries to incorporate all aspects of the river ecosystem as well as socioeconomic aspects on the basis of scenario assessments. It consists of four modules:

- The biophysical module evaluates changes of the ecosystem (e.g., hydrology, hydraulics, geomorphology, water quality, riparian vegetation, aquatic plants, organisms, etc.) in response to altered flow.
- The socioeconomic module covers all relevant river resources.
- The scenario-building module optimizes flow.
- The economic module considers compensation costs of each scenario.

DRIFT is usually used to build scenarios, but can also be used to set flows for achieving specific objective (e.g., optimizing ecological condition through combinations of dam releases; different timings, magnitudes, and durations; Acreman and Dunbar 2004).

Although many different methodologies exist, it is still a challenge to translate the knowledge of hydrologic-ecological principles into specific management rules (Poff et al. 2003). The selection of the appropriate methodology depends on matching the available resources (e.g., time, money, and data) to the question of concern. Environmental flow assessments should be incorporated into the planning phase of any proposed use of river resources that changes flows, especially hydropower plants. Finally, it has to be kept in mind that each EF assessment, whether calculated by a simple rule of thumb or by a holistic method, has to be evaluated with regard to its biological relevance and effectiveness for the specific river to be assessed. Therefore, the selected EF has to be monitored and, if necessary, adapted accordingly.

Recently, environmental flow assessments have been shifted toward more holistic approaches (Arthington and Pusey 2003; Tharme 2003; King et al. 2008), demanding assessment of the requirements of all ecosystem components through judgment from multidisciplinary teams of scientific experts. Furthermore, at the same time habitat modeling techniques have significantly advanced, offering a greater basis to incorporate data-driven approaches, in the holistic perspective. As a result, habitat modeling applications can now be used to assess the flow requirements of various ecosystem components. This concept is also adopted and incorporated in a three-level (preliminary/intermediate/comprehensive) approach proposed in the EFs Guidance Document of the European Commission (2015), highlighting the need for data-driven holistic environmental flow assessments and using habitat modeling for optimum visualization of the information to stakeholders and water managers (see Table 4.3).

Even though there is no simple choice for which method is the most suitable to assess environmental flow, Acreman and Dunbar (2004) suggest that the main driving force for choice of method is the type of issue to be addressed (i.e., scoping, basin planning, impact assessment, and river restoration). Scoping includes large-scale assessment and national auditing, where the focus encompasses many river basins. Therefore, a rapid method, such as a look-up table, would be most relevant. Basin planning involves the assessment of EFs throughout an entire river basin. Such assessment can be started using look-up tables, but increasing the level of detail assessed requires following up with a desktop approach. Environmental flow assessment often involves impact assessment and mitigation of flow modifications (e.g., dams, abstractions). Where the impact is spread over several sites within a river basin, it may be useful to make initial assessments of the impact around the basin using a desktop method before more specific functional analysis or hydraulic habitat modeling is undertaken as part of a holistic approach (Acreman and Dunbar 2004). The holistic approaches allow assessment of the benefits of any restoration activities (e.g., reduced abstractions, release from reservoirs, structural measures, and morphological river restoration). Some pros and cons useful in selecting different approaches are summarized in Table 4.3.

4.7   Conclusions

Nowadays, hydrological processes forming riverine ecosystems are well understood, and the importance of flow for maintaining the ecological integrity is well perceived. Human uses have altered the hydrological regime of running waters and degraded riverine ecosystems. A number of environmental flow assessment methods have been developed ranging from simple hydrological methods over habitat flow models to more comprehensive methodologies including socioeconomic aspects. While much effort has been dedicated to the development of those methods, the biological effectiveness of environmental flow regulations has been evaluated only in few cases. Further research is necessary to better understand the response of biota and riverine ecosystems to flow restoration by holistic assessments including interactions with river morphology, sediment transport, groundwater, and floodplain dynamics.

References

Abramovitz JN (1995) Freshwater failures: the crises on five continents. World Watch 8:27–35
Acreman MC, Dunbar MJ (2004) Defining environmental river flow requirements? A review. Hydrol Earth Syst Sci Discuss 8:861–876
Annear T, Chisholm I, Beecher H (2004) Instream flows for riverine resource stewardship (revised edn). Instream Flow Council, Cheyenne
Armitage PD (1995) Faunal community change in response to flow manipulation. In: Harper DM, Ferguson AJD (eds) The ecological basis for river management. Wiley, Chichester, pp 59–78
Arthington AH, Pusey BJ (2003) Flow restoration and protection in Australian rivers. River Res Appl 19:377–395
Arthington AH, Brizga SO, Kennard MJ (1998) Comparative evaluation of environmental flow assessment techniques: best practice framework. Occasional paper No. 25/98. Land and Water Resources Research and Development Corporation. Canberra
Arthington AH, Tharme R, Brizga SO, Pusey BJ, Kennard MJ (2004) Environmental flow assessment with emphasis on holistic methodologies. In: Welcomme R, Petr T (eds) Proceedings of the second international symposium on the management of large rivers for fisheries, vol II. RAP Publication 2004/17, FAO Regional Office for Asia and the Pacific, Bangkok, pp 37–65

Arthington AH, Bunn SE, Poff NL, Naiman RJ (2006) The challenge of providing environmental flow rules to sustain river ecosystems. Ecol Appl 16:1311–1318

Arthington AH, Naiman RJ, McClain ME, Nilsson C (2010) Preserving the biodiversity and ecological services of rivers: new challenges and research opportunities. Freshw Biol 55(1):1–16

Barker I, Kirmond A (1998) Managing surface water abstraction. In: Wheater H, Kirby C (eds) Hydrology in a changing environment, vol 1. British Hydrological Society, London, pp 249–258

Bergkamp G, McCartney M, Dugan P, McNeely J, Acreman M (2000) Dams, ecosystem functions and environmental restoration. WCD thematic review—environmental issues II.1. Final report to the world commission on dams. Secretariat of the World Commission on Dams. Cape Town

Boon PJ, Calow P, Petts GE (eds) (1992) River conservation and management. Wiley, Chichester

Boon PJ, Davies BR, Petts GE (eds) (2000) Global perspectives on river conservation: science, policy and practice. Wiley, Chichester

Bovee KD (1982) A guide to stream habitat analysis using the IFIM. US Fish and Wildlife Service Report FWSIOBS-82I 26. Fort Collins

Bovee KD, Milhous R (1978) Hydraulic simulation in instream flow studies: theory and techniques. Instream flow information paper: No. 5. FWS/OBS-78/33. Fish and Wildlife Service, 156p

Brizga SO, Arthington AH, Choy SC, Kennard MJ, Mackay SJ, Pusey BJ, Werren GL (2002) Benchmarking, a 'top-down' methodology for assessing environmental flows in Australian rivers. In: Proceedings of the international conference on environmental flows for river systems, Southern Waters, University of Cape Town, Cape Town

Brown C, King J (2003) Environmental flows: concepts and methods. In: Davis R, Hirji R (eds) Water resources and environment technical note C.1. The World Bank, Washington

Bunn SE (1998) Recent approaches to assessing and providing environmental flows: concluding comments. In: Arthington AH, Zalucki JM (eds) Water for the environment: recent approaches to assessing and providing environmental flows, Proceedings of AWWA forum. Brisbane, Australia, pp 123–129

Bunn SE, Arthington AH (2002) Basic principles and ecological consequences of altered flow regimes for aquatic biodiversity. Environ Manag 30:492–507

Calow P, Petts GE (eds) (1992) The rivers handbook. vol. 1: hydrological and ecological principles. Blackwell Scientific, Oxford

Chien N (1985) Changes in river regime after the construction of upstream reservoirs. Earth Surf Process Landf 10:143–159

Connell JH (1978) Diversity in tropical rain forests and coral reefs. Science 199(4335):1302–1310

Craig JF, Kemper JB (eds) (1987) Regulated streams: advances in ecology. Plenum Press, New York

Cushman RM (1985) Review of ecological effects of rapidly varying flows downstream of hydroelectric facilities. N Am J Fish Manag 5:330–339

Doyle MW, Stanley EH, Strayer DL, Jacobson RB, Schmidt JC (2005) Effective discharge analysis of ecological processes in streams. Water Resour Res 41, W11411, https://doi.org/10.1029/2005WR004222

Dunne T, Leopold LB (1978) Water in environmental planning. W. H. Freeman and Co., San Francisco

Dynesius M, Nilsson C (1994) Fragmentation and flow regulation of river systems in the northern third of the world. Science 266:753–762

Dyson M, Bergkamp M, Scanlon J (2003) Flow: the essentials of environmental flows. IUCN, Switzerland

European Commission (2015) Ecological flows in the implementation of the water framework directive. WFD CIS Guidance Document No. 31

Galland J (1991) TELEMAC: a new numerical model for solving shallow water equations. Adv Water Resour 14:138–148

Gore JA, Petts GE (eds) (1989) Alternatives in regulated river management. CRC Press, Florida

Greenberg L, Svendsen P, Harby A (1996) Availability of microhabitats and their use by brown trout (Salmo trutta) and grayling (Thymallus thymallus) in the river Vojman, Sweden. Regul Rivers Res Manag 12:287–303

Gupta AD (2008) Implication of environmental flows in river basin management. Phys Chem Earth 33:298–303

Gustard A, Cole G, Marshall D, Bayliss A (1987) A study of compensation flows in the UK. Institute of Hydrology report 99, Wallingfort

Hart DD, Finelli CM (1999) Physical-biological coupling in streams: the pervasive effects of flow on benthic organisms. Annu Rev Ecol Syst 30:363–395

Hirji R, Davis R (2009) Environmental flows in water resources policies, plans, and projects: findings and recommendations. The World Bank. Environment and Development series

International Water Management Institute (IWMI) (2005) Environmental flows. Planning for environmental water allocation. Water Policy briefing, 15, pp 1–6

Junk WJ, Bayley PB, Sparks RE (1989) The flood pulse concept in river-floodplain systems. Can Spec Publ Fish Aquat Sci 106:110–127

Karr JR (1991) Biological integrity: a long neglected aspect of water resource management. Ecol Appl 1:66–84

King J, Brown C (2006) Environmental flows: striking the balance between development and resource protection. Ecol Soc 11:26

King J, Brown C, Sabet H (2003) A scenario-based holistic approach to environmental flow assessments for rivers. River Res Appl 19(5–6):619–639

King JM, Tharme RE, De Villiers MS (2008) Environmental flow assessments for rivers: manual for the building block methodology. WRC report no TT 354/08, Cape Town, 364p

Lillehammer A, Saltveit SJ (eds) (1984) Regulated rivers. Universitetsforlaget As, Oslo

Lloyd N, Quinn G, Thoms M, Arthington A, Gawne B, Humphries P, Walker K (2003) Does flow modification cause geomorphological and ecological response in rivers? A literature review from an Australian perspective. Technical report 1 / 2004, CRC for freshwater ecology, ISBN 0-9751642-02

Martinez SMC, Fernández Yuste JA (2010) IAHRIS 2.2. Indicators of hydrologic alteration in rivers. Methodological reference manual & user's manual. http://www.ecogesfor.org/IAHRIS_es.html

McCully P (1996) Silenced rivers: the ecology and politics of large dams. ZED books, London

Moore M (2004) Perceptions and interpretations of environmental flows and implications for future water resource management: a survey study. Masters Thesis, Department of Water and Environmental Studies, Linköping University, Linköping

Navarro SR, Schmidt G (2012) Environmental flows as a tool to achieve the WFD objectives. Discussion paper (in the framework of Service contract for the support to the follow-up of the Communication on Water scarcity and Droughts). Version: Draft 2.0, 11 June 2012

Nilsson C, Reidy CA, Dynesius M, Revenga C (2005) Fragmentation and flow regulation of the world's river systems. Science 308:405–408

Parasiewicz P (2007) The Mesohabsim model revisited. River Res Appl 27:893–903

Petts GE (ed) (1984) Impounded rivers: perspectives for ecological management. Wiley, Chichester

Poff NL, Ward JV (1989) Implications of streamflow variability and predictability for lotic community structure: a regional analysis of streamflow patterns. Can J Fish Aquat Sci 46:1805–1818

Poff NL, Zimmerman JK (2010) Ecological responses to altered flow regimes: a literature review to inform the science and management of environmental flows. Freshw Biol 55(1):194–205

Poff NL, Allan JD, Bain MB, Karr JR, Prestegaard KL, Richter BD, Sparks RE, Stromberg JC (1997) The natural flow regime. Bioscience 47(11):769–784

Poff NL, Allan JD, Palmer MA, Hart DD, Richter BD, Arthington AH, Rogers KH, Meyer JL, Stanford JA (2003) River flows and water wars: emerging science for environmental decision making. Front Ecol Environ 1:298–306

Poff NL, Richter BD, Arthington AH, Bunn SE, Naiman RJ, Kendy E, Acreman M, Apse C, Bledsoe BP, Freeman MC, Henriksen J, Jacobson RB, Kennen JG, Merritt DM, O'Keeffe JH, Olden JD, Rogers K, Tharme RE, Warner A (2010) The ecological limits of hydrologic alteration (ELOHA): a new framework for developing regional environmental flow standards. Freshw Biol 55(1):147–170

Postel SL (1995) Where have all the rivers gone? World Watch 8:9–19

Postel SL (1998) Water for food production: will there be enough in 2025? Bioscience 48:629–637

Postel SL, Richter B (2003) Rivers for life: managing water for people and nature. Island Press, Washington, DC

Pringle CM (2000) River conservation in tropical versus temperate latitudes. In: Boon PJ, Davies BR, Petts GE (eds) Global perspectives on river conservation: science, policy and practice. Wiley, Chichester, pp 371–384

Rapport DJ, Costanza R, McMichael AJ (1998) Assessing ecosystem health. Trends Ecol Evol 13:397–402

Reiser DW, Wesche TA, Estes C (1989) Status of instream flow legislation and practices in North America. Fisheries 14:22–29

Revenga C, Murray S, Abramowitz J, Hammond A (1998) Watersheds of the world: ecological value and vulnerability. World Resources Institute and Worldwatch Institute, Washington, DC

Revenga C, Brunner J, Henninger N, Kassem K, Payne R (2000) Pilot analysis of global ecosystems: freshwater ecosystems. World Resources Institute, Washington, DC

Richter BD, Baumgartner JV, Powell J, Braun DP (1996) A method for assessing hydrological alteration within ecosystems. Conserv Biol 10:1163–1174

Richter BD, Baumgartner JV, Wigington R, Braun DP (1997) How much water does a river need? Freshw Biol 37:231–249

Richter BD, Braun DP, Mendelson MA, Master LL (1998) Threats to imperiled freshwater fauna. Conserv Biol 11:1081–1093

Richter BD, Warner AT, Meyer JL, Lutz K (2006) A collaborative and adaptive process for developing environmental flow recommendations. River Res Appl 22:297–318

Rosenberg DM, McCully P, Pringle CM (2000) Global-scale environmental effects of hydrological alterations: introduction. Bioscience 50(9):746–751

Schneider M, Noack M, Gebler T, Kopecki I (2010) Handbook for the habitat simulation model CASiMiR, Module CASiMiR-Fish, Base Version. 52 pp. Translated by Tuhtan, J. Available from: http://casimir-software.de/ (accessed 21 September 2015)

Snaddon CD, Davies BR, Wishart MJ (1999) A global overview of inter-basin water transfer schemes, with an appraisal of their ecological, socio-economic and socio-political implications, and recommendations for their management. Water Research Commission Technology Transfer report TT 120/00. Water Research Commission. Pretoria

Steffler P, Blackburn J (2002) River 2D: two-dimensional depth averaged model of river hydrodynamics and fish habitat. Introduction to depth averaged modelling and user's manual. University of Alberta, Canada

Stewardson MJ, Gippel CJ (2003) Incorporating flow variability into environmental flow regimes using the flow events method. River Res Appl 19:459–472

Swales S, Harris JH (1995) The expert panel assessment method (EPAM): a new tool for determining environmental flows in regulated rivers. In: Harper DM, Ferguson AJD (eds) The ecological basis for river management. Wiley, Chichester

Tennant DL (1976) Instream flow regimens for fish, wildlife, recreation and related environmental resources. Fisheries 1(4):6–10

Tharme RE (1996) Review of international methodologies for the quantification of the instream flow requirements of rivers. Water law review final report for policy development for the Department of Water Affairs and Forestry, Pretoria. Freshwater Research Unit, University of Cape Town, Cape Town

Tharme RE (2000) An overview of environmental flow methodologies, with particular reference to South Africa. In: King JM, Tharme RE, De Villiers MS (eds) Environmental flow assessments for rivers: manual for the building block methodology. Water Research Commission Technology Transfer report no. TT131/00. Water Research Commission, Pretoria, pp 15–40

Tharme RE (2003) A global perspective on environmental flow assessment: emerging trends in the development and application of environmental flow methodologies for rivers. River Res Appl 19(5–6):397–441

Tharme RE, King JM (1998) Development of the building block methodology for instream flow assessments, and supporting research on the effects of different magnitude flows on riverine ecosystems. Water Research Commission Report No. 576/1/98, 452p

Theodoropoulos C, Skoulikidis N (2014) Environmental flows: the european approach through the water framework directive 2000/60/EC. Proceedings of the 10th International Congress of the Hellenic Geographical Society (in press)

Thoms MC, Sheldon F, Roberts J, Harris J, Hillman T (1996) Scientific panel assessment of environmental flows for the Barwon-Darling river. New South Wales Department of Land and Water Conservation, Sydney

Trihey EW, Stalnaker CB (1985) Evolution and application of instream flow methodologies to small hydropower developments: an overview of the issues. In: Olson FW, White RG, Hamre RH (eds) Proceedings of the symposium on small hydropower and fisheries. Aurora, CO

USGS (2001) PHABSIM for Windows. User's manual and exercises. Open file report 01-340, U.S. Geological Survey.

Ward JA (1982) Ecological aspects of stream regulation: responses in downstream lotic reaches. Water Pollut Manage Rev (New Delhi) 2:1–26

Ward J (1989) The four-dimensional nature of lotic ecosystems. J N Am Benthol Soc 8:2–8

Ward J, Stanford JA (1983) The intermediate-disturbance hypothesis: an explanation for biotic diversity patterns in lotic ecosystems. In: Fontaine F (ed) Dynamics of lotic ecosystems. Ann Arbor Press, Ann Arbor, pp 347–355

Waters BF (1976) A methodology for evaluating the effects of different stream flows on salmonid habitat. In: Orsborn JF, Allman CH (eds) Instream flow needs. American Fisheries Society, Bethseda, pp 254–266

Wolman MG, Miller JP (1960) Magnitude and frequency of forces in geomorphic processes. J Hydrol 69:54–74

World Commission on Dams (WCD) (2000) Dams and development. A new framework for decision-making. The report of the World Commission on dams. Earthscan Publications, London

Zarfl C, Lumsdon AE, Berlekamp J, Tydecks L, Tockner K (2014) A global boom in hydropower dam construction. Aquat Sci 77(1):161–170

Zeiringer B (2008a) Minimum flow study at the river Schwarza/Reichenau. evn naturkraft, p 49

Zeiringer B (2008b) Minimum flow study at the river Unrechttraisen/Mauthof. evn naturkraft, p 54

Zeiringer B, Unfer G, Jungwirth M (2008) Environmental flow study at the river Mur/Pernegg. VERBUND-Austrian Hydro Power AG (AHP), p 146

Zeiringer B, Hinterhofer M, Unfer G (2010) Environmental flow study at the river Ybbs/Opponitz. WIENERGIE, p 162

Chapter 5
Hydropeaking Impacts and Mitigation

Franz Greimel, Lisa Schülting, Wolfram Graf, Elisabeth Bondar-Kunze, Stefan Auer, Bernhard Zeiringer, and Christoph Hauer

5.1 Introduction

Flow is a major driver of processes shaping physical habitat in streams and a major determinant of biotic composition. Flow fluctuations play an important role in the survival and reproductive potential of aquatic organisms as they have evolved life history strategies primarily in direct response to natural flow regimes (Poff et al. 1997; Bunn and Arthington 2002). However, although the organisms are generally adapted to natural dynamics in discharge, naturally caused flow fluctuations may entail negative consequences (e.g., stranding, drift, low productivity), especially if the intensity is exceptionally high or the event timing is unusual (Unfer et al. 2011; Nagrodski et al. 2012). Aside from natural dynamics in discharge, artificial flow fluctuations with harmful impacts on aquatic ecology can be induced by human activities. Hydropeaking—the discontinuous release of turbined water due to peaks of energy demand—causes artificial flow fluctuations downstream of reservoirs.

F. Greimel (✉) · L. Schülting · W. Graf · S. Auer · B. Zeiringer
Institute of Hydrobiology and Aquatic Ecosystem Management, University of Natural Resources and Life Sciences, Vienna, Austria
e-mail: franz.greimel@boku.ac.at; lisa.schuelting@boku.ac.at; wolfram.graf@boku.ac.at; stefan.auer@boku.ac.at; bernhard.zeiringer@boku.ac.at

E. Bondar-Kunze
Institute of Hydrobiology and Aquatic Ecosystem Management, University of Natural Resources and Life Sciences, Vienna, Austria

WasserCluster Lunz Biological Station GmbH, Lunz am See, Austria
e-mail: elisabeth.bondar@boku.ac.at

C. Hauer
Christian Doppler Laboratory for Sediment Research and Management, Institute of Water Management, Hydrology and Hydraulic Engineering, University of Natural Resources and Life Sciences, Vienna, Austria
e-mail: christoph.hauer@boku.ac.at

© The Author(s) 2018
S. Schmutz, J. Sendzimir (eds.), *Riverine Ecosystem Management*, Aquatic Ecology Series 8, https://doi.org/10.1007/978-3-319-73250-3_5

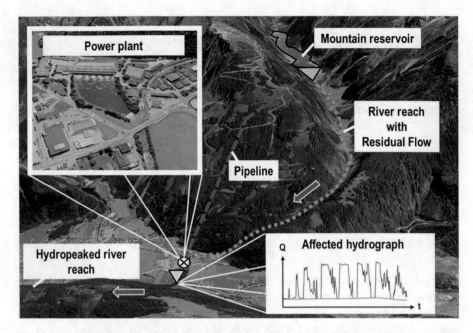

Fig. 5.1 Systematic sketch—high-head storage power plant and discontinuous release of turbined water due to peaks of energy demand (hydropeaking) [background image: Google Inc.—Google Earth 2015 (7.1.5.1557)]

High-head storage power plants usually induce flow fluctuations with very high frequencies and intensities compared to other sources of artificial flow fluctuations (Fig. 5.1). However, run-of-the-river power plants and other human activities may also create artificial hydrographs due to turbine regulation, gate manipulations, and pumping stations.

Hydropeaking frequently occurs in river systems with high river slopes (e.g., alpine regions). Here, storage hydropower plants use the potential energy in water stored at higher elevations for electricity production on demand, which produces significant alterations of the flow regime downstream (e.g., decreased low flow, hydropeaking). As an example, according to the National Water Management Plan for Austria, more than 800 km of river reaches (Fig. 5.2) are likely to be affected by hydropeaking in Austria. Almost all of these reaches are located in the grayling and trout region within the Alpine ecoregion of western Austria (BMLFUW 2010; Illies 1978). Sometimes more than five hydropeaking events (peaks) per day are recorded, but situations in different river systems are highly variable. In addition to hydropeaking, a major part of Austrian hydrographs is affected by so-called hydrofibrillation. The latter show similar frequencies, but much lower intensities than hydropeaking, and are mainly caused by run-off-the-river power plants. Unaffected sub-daily flow regimes can be found primarily on small rivers with a catchment area less than 100 km^2 (Greimel et al. 2015).

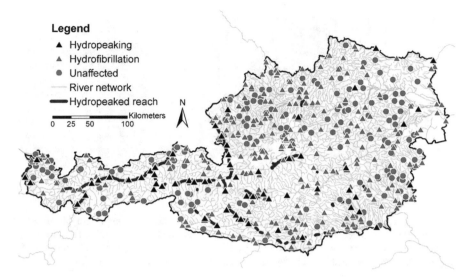

Fig. 5.2 Regulated and unregulated sub-daily flow regimes of Austrian rivers [for method, see Greimel et al. 2015; black triangles, hydropeaking (n = 71); gray triangles, hydrofibrillation (n = 250); circles, unaffected (n = 221); black lines, hydropeaked river reaches according to the National Water Management Plan (data source BMLFUW 2010)]

Sub-daily flow dynamics have to be considered for the integration of scientific knowledge in policy as well as for mitigation measure design to achieve the aims of the European Water Framework Directive. Conceptual models to predict ecological effects of altered sub-daily flow regimes are needed. Detailed ecological knowledge and a quantitative framework incorporating mathematical representations of field and laboratory results on flow, temperature, habitat structure, organism life stages, and population dynamics form the basis to develop these conceptual models (Young et al. 2011).

5.2 Detection and Characterization of Flow Fluctuation Intensity and Frequency

Hydrographs can be used to characterize the hydrological context in rivers. Greimel et al. (2015) developed a method to detect and characterize sub-daily flow fluctuations: flow fluctuations are separated into increase (IC) and decrease (DC) events, which is necessary from an ecological point of view since biota reacts in different ways (e.g., drifting and stranding) to increase and decrease events. To analyze in detail fluctuation conditions for both event types, an event-based algorithm for automated analysis of time series was developed. The algorithm calculates flow (Q) differences of consecutive time steps (ts) of the discrete hydrograph curves (Qts1, Qts2,..., Qtsn) and discriminates between time steps with increasing (IC:

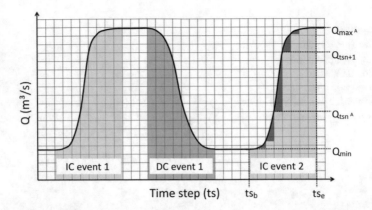

Fig. 5.3 Events definition and relevant values to calculate intensity parameters illustrated at increase event 2 (IC evt. 2): 1 ts $\triangleq$ 900 s or 15 min; time step event beginning (ts_b), time step event ending (ts_e), maximum event flow (Q_{max}), minimum event flow (Q_{min}), flow of a specific time step (Q_{tsn}), flow of subsequent time step (Q_{tsn+1}) (modified from Greimel et al. 2015)

Table 5.1 Event-based intensity parameters: definitions and units (modified from Greimel et al. 2015)

Nr	Parameter	Acronym	Definition	Unit
1	Maximum flow fluctuation rate	MAFR	$Max(abs((Q_{tsn+1}) - (Q_{tsn})))$	m^3/s^2
2	Mean flow fluctuation rate	MEFR	Amplitude/duration	m^3/s^2
3	Amplitude	AMP	$Q_{max} - Q_{min}$	m^3/s
4	Flow ratio	FR	Q_{max}/Q_{min}	
5	Duration	DUR	$ts_e - ts_b$	s

ts_b time step event beginning, ts_e time step event ending, Q_{max} maximum event flow, Q_{min} minimum event flow, Q_{tsn} flow of a specific time step, Q_{tsn+1} flow of subsequent time step, *max* maximum, *abs* absolute, *s* second (1 ts $\triangleq$ 900 s or 15 min)

Qts1 < Qts2) and decreasing flow (DC: Qts1 > Qts2). Continuous time steps with equal trends are defined as a single fluctuation event (Fig. 5.3).

For each event a set of parameters related to fluctuation intensity (Table 5.1) is calculated by the algorithm: the highest flow change within a time step represents parameter (1)—maximum flow fluctuation rate (MAFR). Parameter (2)—mean flow fluctuation rate (MEFR) is calculated by the event amplitude divided by the number of time steps. Parameter (3)—the amplitude (AMP) of an event is defined as the difference between the flow maximum (Q_{max}) and the flow minimum (Q_{min}). Parameter (4)–flow ratio (FR) is defined as (Q_{max})/(Q_{min}). The duration (DUR) of an event (5) is simply the number of continuous time steps with equal flow trend. In addition, timing and daylight condition are determined for every single event.

This method to detect and characterize flow fluctuations using hydrograph curves offers a wide range of applications: intensity, timing, and frequency of flow fluctuations can be detected automatically and in a standardized way. As a consequence the hydrological situation at specific river sections can be compared to each other, and hydrographs can be allocated automatically to different sub-daily flow regimes (see Fig. 5.2). In particular, the contrast between unaffected and artificially affected situations is significant from an ecological point of view. Furthermore, a power plant-specific, longitudinal assessment of hydropeaking intensity and frequency based on multiple hydrograph curves is enabled (see Sect. 5.4).

Summing up, detailed hydrological information forms the basis for scientific analyses since a number of ecologically relevant parameters (see Sect. 5.3) related to unsteady flow hydraulics are determined by flow changes. For example, ramping rates (changes in water surface elevation/discharge per standardized time period, e.g., cm/min) are important for determining the risk of stranding of aquatic organisms in terms of dewatering caused by shutdown of the turbine. Flow velocities for both base and peak flow are important indicators, which determine one of the main physical criteria for habitat suitability of target species at different life stages (e.g., juvenile fish in low velocity habitat along the banks). Similar to flow velocity, the bottom shear stress has to be studied as an indicator for possible sediment dynamics in hydropeaked rivers. In addition to analysis of base and peak flow, bottom shear stress during mean or even extraordinary flooding is a critical determinant of self-forming morphological and sedimentological dynamics. Studies on sediment transport in hydropeaked rivers are required especially for the design of morphological mitigation measures. Here, the sediment regime has not only to be investigated on the reach scale but also at the catchment scale. Furthermore, water temperature fluctuations induced by hydropeaking may be related to cold (summer) or warm (winter) water release from hydropower plants in addition to power plant-related discharge fluctuation. Finally, frequency, periodicity, and timing of hydropeaking constitute essential aspects in the ecological assessment of potential hydropeaking impacts. Ecological effects in reference to several parameters and organisms are discussed in detail below.

5.3 Hydropeaking Impacts on Aquatic Biota

Flow fluctuations induced by hydropeaking operation can have tremendous short- and long-term effects on riverine organisms. Due to increasing hydraulic forces, organisms may get abraded from underlying substrate and drift downstream or must invest significant amounts of energy to avoid downstream displacement during a hydropeaking event. Unintentional drift downstream results in relocation to a possibly less suitable habitat, as well as in physiological, mechanical, or predatory stress. A lateral habitat shift of vagile organisms may help them remain in habitats with suitable hydraulic conditions, but this tactic is linked to a risk of stranding during water level declines. Furthermore, high mechanical stress through increased sediment mobilization and sediment transport can harm organisms or may lead to a decreased primary production (Hall et al. 2015). Besides reducing biomass and abundance, artificial sub-daily flow and water temperature fluctuations may also have negative effects on growth, survival rates, reproduction, and biotic integrity

(Finch et al. 2015; Puffer et al. 2015; Schmutz et al. 2015; Céréghino et al. 2002; Graf et al. 2013; Kennedy et al. 2014; Lauters et al. 1996; Parasiewicz et al. 1998). Furthermore frequent exposure of aerial zones (dewatering) may have negative consequences for the local stream food web (Blinn et al. 1995; Graf et al. 2013; Flodmark et al. 2004). In the following, we discuss the impacts of different hydropeaking-related variables (see Sect. 5.2) on stream biota in detail.

5.3.1 Flow Velocity, Shear Stress, and Sediment Transport

Changes in flow velocity produce higher shear stress, entailing gravel bed movement and, thus, increased fine sediment transport. This may have severe effects on the whole community structure in rivers affected by hydropeaking.

For instance, benthic algae are highly impacted by flow velocities above 10–15 cm/s, because taxonomic composition and nutrient cycling may change (Biggs et al. 1998; Hondzo and Wang 2002). Bondar-Kunze et al. (2016) found in an experimental study that, in an oligotrophic stream ecosystem, daily hydropeaking significantly retarded the development of periphyton biomass with no interference in the relative abundance of the three main algal groups (diatoms, chlorophyta, cyanobacteria) or the photosynthetic activity. The lower biomass could be related to cell abrasion due to a fivefold increase in flow velocity compared to base flow conditions (Biggs and Thomsen 1995). It is also very likely that in the hydropeaking treatment, the colonization with high resistance-to-disturbance taxa such as slow-growing diatoms or low-profile species (short-statured species) took place (Passy and Larson 2011; Smolar-Žvanut and Klemenčič 2013), whereas in the unaffected treatment the typical succession from smaller, low-profile diatoms to larger long-stalked and large-rosette diatoms could occur (Hoagland et al. 1982). But higher trophic levels are also impacted by a pulsed increase of flow velocity due to hydropeaking events.

Hydropeaking-impacted stretches frequently show a reduced macroinvertebrate biomass and a change of community structure and species traits (Céréghino and Lavandier 1998; Graf et al. 2013). Different taxa can withstand different flow velocity thresholds and time spans of being exposed to increased discharge (Oldmeadow et al. 2010; Statzner and Holm 1982; Waringer 1989). Exceeding these taxa-specific thresholds leads to the detachment of the organisms and increased drift. Whether taxa are affected by hydropeaking depends on species traits like morphological and behavioral adaptations (presence of claws/hooks, ability to quickly crawl into the sediments), whereas interstitial taxa are rarely found drifting. Additionally, different life stages show different sensitivities, since juvenile larvae show the strongest tendencies to drift following suddenly increased flow (Fjellheim 1980; Waringer 1989; Limnex 2004).

Similarly, fish larvae and juveniles are particularly affected by hydropeaking due to their preference for shallow habitats with low flow velocities, i.e., habitats that are heavily influenced by hydropeaking. In contrast to adults, the reduced swimming performance of young fish (Heggenes and Traaen 1988) puts them at risk to get drifted downstream. This may entail several consequences. As for other organism groups, the risk to drift following hydropeaking is taxa-specific: postemergence brown trout (*Salmo trutta*) prefer substrate-linked habitats, making them more resistant to drift caused by hydropeaking compared to larval grayling (*Thymallus thymallus*), which start to swim relatively soon within the water column (Auer et al. 2014). Experiments conducted by Schmutz et al. (2013) found a positive relationship between maximum peak flows and drift rates of juvenile graylings. Interestingly, a survey by Thompson et al. (2011) showed that repeated peak events may increase the chances of successful adaptive responses to hydropeaking. Auer et al. (2014) found a decrease of hydropeaking-induced drift during repeated peak events for juvenile graylings.

Reaching certain thresholds of critical flow can additionally induce bed movement and thus suppress periphyton as well as macroinvertebrate biomass through increased drift (Townsend et al. 1997; Biggs and Close 1989; Graf et al. 2013). Further, temporary increases of suspended solid concentration in the water column during peaks followed by fine sediment accumulations between peaks may be another factor depressing periphyton growth. Yamada and Nakamura (2002) observed an inverse correlation between suspended solid concentration and benthic chlorophyll-a concentrations in autumn and winter, which they related to shading effects. However, the amount of the fine sediment load is also an important factor. For example, small deposits of fine sediment on coarse substrata increase habitat heterogeneity, augmenting taxa more tolerant to the movement of fine particles. However, fine particles can crush and bury cells of benthic algae and cyanobacteria (Burkholder 1996) and, hence, potentially also increase taxon richness and evenness via reduced competition with taxa that are strong competitors on a stable substratum (Wagenhoff et al. 2013). During phases of substrate stability (between two hydropeaking events), the importance of invertebrate grazers and, thus, biotic control on periphyton recovery increases (Biggs and Close 1989). Besides periphyton and macroinvertebrates, high shear stress and gravel bed movement can also affect fish communities, e.g., larval brown trout are highly vulnerable due to the preferences for substrate-linked habitats.

5.3.2 Ramping Rate

The ramping rate describes the rapidity of the water level increases or decreases during a peak event, and there is strong evidence that the ramping rate is significantly linked to stream organism responses (Schmutz et al. 2015; Smokorowski 2010).

In contrast to gradual flow increases, fast up-ramping may greatly reduce the time available for seeking shelter, thereby strongly increasing drift rates of aquatic organisms, such as macroinvertebrates (Imbert and Perry 2000). In line with these considerations, further studies (Marty et al. 2009; Smokorowski 2010; Tuor et al. 2014) indicate that unlimited ramping over the long-term can reduce the densities that are sustainable for benthic organisms and therefore affect the food web structure. In these studies, the trophic structure was reduced by one trophic level between macroinvertebrates and fish. Fish had to compensate this lack by increased feeding on baseline taxa. Additionally, there is experimental evidence that drift rates of more vagile organisms, e.g., juvenile graylings, remained unchanged during flow fluctuations with varying up-ramping rates (Schmutz et al. 2013), showing that the effect of reducing up-ramping rates as a mitigating measure most likely is species and life stage specific. At least for juvenile grayling, the risk for drift is higher during nighttime in summer, but could decrease when up-ramping rates were reduced from 3.0 to 0.5 cm/min (Auer et al. 2017).

For fish, the abruptness of the flow decrease (down-ramping rate) seems to be of higher importance than the increase. A fast water level decrease may lead to increased stranding risk for organisms because they may not be able to perform a lateral shift fast enough with a rapidly sinking water level. Several studies observed a positive relationship between stranding and down-ramping rate (Bauersfeld 1978; Hunter 1992; Bradford et al. 1995). For brown trout Halleraker et al. (2003) found a significantly decreased stranding rate, when down-ramping rate was reduced from 60 to 10 cm/h. Recent experiments at the HyTEC facility support a significant relationship between stranding risk and down-ramping rate, depending on species and live stage. For example, stranding of larval graylings during diurnal single-peak experiments vanished at a down-ramping rate of 0.2 cm/min compared to 50% stranding at 2.9 cm/min. A similar relation was identified for larval brown trout, although stranding risk vanished only below 0.1 cm/min. Juvenile grayling could avoid stranding during a down-ramping rate of 3.0 cm/min, and juvenile brown trout, despite their vulnerability as larvae, actually could adapt to a rate of 6.4 cm/min (Auer et al. 2014, 2017). However, there is also evidence that seasonal and daily variation play an important role in terms of stranding risk (see Sect. 5.3.3).

5.3.3 Frequency, Periodicity, and Timing of Hydropeaking

The frequency, periodicity, and timing of hydropeaking events may be crucial parameters for defining mitigation measures for hydropower stations.

Even when single-peak events result in low drift or stranding risk for young fish, cumulative effects due to recurring hydropeaking can have significant impacts on fish populations (Bauersfeld 1978). By contrast, experiments conducted by Friedl and Naesby (2014) showed a kind of temporal adaptation behavior for young graylings. Facing three peak events within 24 h over a period of 21 days, stranding was only detectable during first 9 days. If flow conditions prior to a peak event are stable for more than 24 h, this adaptation seems to vanish. Hunter (1992) reports increased stranding risk of young fish when long stable flow occurred prior to a down-ramping event. However, there is a lack of detailed research and empirical evidence regarding these phenomena.

Additionally, the timing of a peak event is a critical parameter, since the activity of aquatic organisms changes throughout the day. Several studies showed that drift of macroinvertebrates increases during the night when they are more active feeding, i.e., there is a negative correlation between light intensity and the feeding activity of the animals due to predatory pressure during day (Allan 1987; Elliott 1967, 2005; Poff et al. 1991; Schülting et al. 2016). Experiments on larval and juvenile grayling and juvenile brown trout during summer as well as on larval brown trout during winter showed increased stranding during nocturnal experiments (Auer et al. 2014). Other experiments with juvenile graylings showed that three consecutive peak events during daytime could lower stranding rates during subsequent nocturnal peak events. Furthermore, Berland et al. (2004) observed higher stranding of Atlantic salmon parr, and Bradford (1997) found higher side-channel trapping, both during night and summer conditions. During winter conditions other studies showed increased stranding risk for some salmonid species during the daytime (Bradford et al. 1995; Saltveit et al. 2001; Halleraker et al. 2003). Summarizing, behavior seems to be influenced by the photophase as well by seasonally related factors such as water temperature.

5.3.4 Channel Morphology

Hydropeaking effects on aquatic biota also depend on the interaction between hydrology and river morphology. Physical habitat diversity is important to ensure a sufficient availability of different habitats for different life stages of aquatic organisms. Morphological alteration by channelization and bank fixation are common pressures in alpine rivers (Comiti 2012; Muhar et al. 2000) and particularly impact river biota (Arscott et al. 2005; Kennedy and Turner 2011). Hydropeaking reinforces this effect and contributes, e.g., to a selection of specifically rheobiont macroinvertebrate taxa (Bretschko and Moog 1990; Cushman 1985; Graf et al. 2013), while limnophilic taxa tend to decrease.

Besides macroinvertebrates early stages of many rheophilic fish species also prefer lateral habitats with reduced flow velocities (Moore and Gregory 1988) due to lower swimming capacity. If discharge increases during a hydropeaking event, a lateral habitat shift of the organisms is needed, either to avoid higher energy demands for maintaining their position or from getting displaced downstream. On the other hand, temporarily wetted habitats can represent deadly traps as macroinvertebrates, and fishes frequently colonize these refugia during up-ramping phases and subsequently undergo stranding effects during down-ramping periods. Side channels, potholes, or low gradient bars have a greater stranding potential than homogenous channels with steep banks (Hunter 1992). Side channels may trap fish during the down-ramping phase (Bradford 1997), and potholes and low gradient bars also may lead to increased stranding during dewatering (Bauersfeld 1978; Bell et al. 2008; Auer et al. 2017), although they provide better habitats for young fish than channelized rivers (Schmutz et al. 2015). Vanzo et al. (2015) pointed out that heterogeneous river morphology can reduce some negative effects of hydropeaking, but can also cause higher stranding risk due to increased dewatered area following down-ramping. Permanently linked gravel bank structures like bays created by groins can provide temporal habitats that act as refugia during peak phases with lowered flow velocity (Schmutz et al. 2013).

5.3.5 Water Temperature

Surface water temperature in reservoirs is more subject to seasonal variation than the more constant and cold water temperature found in deeper areas. Hypolimnic water release for energy production leads to a decrease in water temperature during peak events in summertime and an increase during wintertime (Ward and Stanford 1979; Maiolini et al. 2007; Zolezzi et al. 2011). Water temperature changes during a hydropeaking event are referred to as thermopeaking. A thermal wave usually occurs shortly after an increase of discharge (Toffolon et al. 2010) and may act as an additional stressor on river biota (Bruno et al. 2013).

This additional thermal stressor can have severe impacts on the periphyton biomass development and community composition. In an experimental study by Kasper (2016), cold thermopeaking led to a decrease in chlorophyll-a (surrogate parameter for biomass) and diatoms remained the dominant species, whereas in the control treatment (no hydro- and thermopeaking), a chlorophyte and diatom community developed. The reason for these patterns can be explained due to higher shear stress, which mitigates the development of high quantities of filamentous green algae, and also to a decrease in temperatures during hydropeaking, which increased the development of diatoms. Therefore thermopeaking affects the quantity and quality of periphyton, which also might affect higher trophic levels (e.g., macroinvertebrates).

Céréghino and Lavandier (1998) found that frequent thermal modifications to stream water can lead to changes in macroinvertebrate growth, flight, and emergence patterns. Following hydropeaking, Carolli et al. (2012) and Bruno et al. (2013) observed in experiments increased macroinvertebrate drift associated with warm and cold thermopeaking. By contrast, results of an experimental study by Schülting et al. (2016) suggest that hydropeaking and cold thermopeaking together have an antagonistic effect on drift for aquatic macroinvertebrates. The findings suggest that macroinvertebrate responses to cold thermopeaking are taxa-specific, but in general lead to reduced drift for most taxa. The underlying mechanisms are still unclear.

Hydropeaking-related effects on fish also depend on water temperature. In general on a seasonal level, lower water temperature during winter lowers activity of Atlantic salmon and brown trout (Saltveit et al. 2001; Halleraker et al. 2003). However, temperatures below 4.5 °C results in a substrate-seeking behavior during daytime, leading to lower stranding during night (Saltveit et al. 2001). On a sub-daily level, thermopeaking as a sudden change in water temperature may also affect fish response. As activity and metabolism are affected by water temperature, fish that face a decrease of water temperature during a flow fluctuation may respond with higher drift and stranding rates. Bradford (1997) could show an increase in stranding for juvenile Chinook salmon (*Oncorhynchus tshawytscha*) when water was 6 °C compared to 12 °C. Preliminary experiments with grayling showed increased drift and stranding during flow fluctuation with decreasing water temperatures (Kaiser 2016).

5.4 Research Application and Hydropeaking Mitigation

5.4.1 Potential Hydropeaking Mitigation Measures

In principle, hydropeaking is a hydrological impact. However, the ecological effects of hydropeaking are linked to the morphological quality of rivers (Hauer et al. 2014; Schmutz et al. 2015), and thus superimposed impacts on the aquatic biota are possible due to river regulation and disturbed sediment regime. Consequently, hydropeaking mitigation measures can be classified into two groups, direct and indirect measures (see Fig. 5.4): direct measures may reduce the hydrological impact from operational measures that modify the power plant operation mode, which produce current costs in terms of an economical loss of profit. The second possibility for a direct reduction of the hydrological impact is to build retention basins that take up the peaks and release the water more smoothly afterward. A further alternative is to divert the water into a side channel or tunnel to be used for a newly built hydropower station downstream where a larger water body (large river, reservoir,

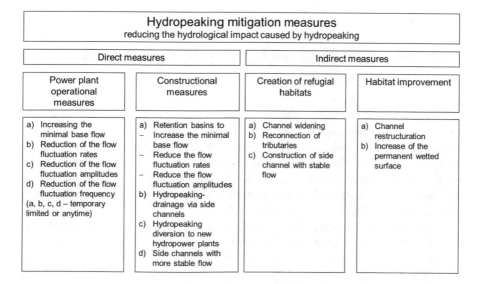

Fig. 5.4 Overview of potential hydropeaking mitigation measures

sea) can better cope with the peaks. These constructional measures primarily entail construction costs and almost no current costs. This also applies to indirect mitigation measures, which reduce the ecological impacts of hydropeaking via adapting the river morphology: the channel width can be enhanced, which leads to decreased water level changes at the widened river section. Tributaries can be reconnected, or side channels with stable flow can be constructed to create refugial habitats. Habitat improvement in general can lead to reduced hydropeaking impacts (Schmutz et al. 2015). Both direct and indirect mitigation measures have the potential to reduce the hydropeaking impact for specific aquatic organism. However, the most substantial improvement can be achieved by taking into account a coordinated river-specific combination of the different mitigation measures (integrative hydropeaking mitigation). Furthermore, the required site-specific design of mitigation has to consider the sediment regime and disturbances of the sediment dynamics in the river stretches impacted due to hydropeaking (Hauer et al. 2014). Identified measures and combinations can be compared with their respective costs to select most effective measures.

The ecological and socioeconomic complexity of hydropeaking mitigation warrants a case-specific quantitative evaluation of measures. A conceptual framework for hydropeaking mitigation is needed that can be transferred to multiple mitigation projects. Bruder et al. (2016) developed such a framework based on current scientific knowledge and on ongoing hydropeaking mitigation projects in Switzerland. The proposed Swiss framework refers to ecological, hydrological, and morphological indicators as well as to aspects of sediment transport. However, detailed knowledge of efficient approaches to mitigate ecological hydropeaking impacts is still rare despite increased interest in research and management in recent decades (Tonolla et al. 2017).

5.4.2   Integrative Hydropeaking Mitigation and Example of Application

In accordance with the abovementioned Swiss framework for hydropeaking mitigation, the concept of integrative hydropeaking management has been developed in Austria. Integrative hydropeaking mitigation requires the consideration of (a) the vulnerability of aquatic organisms; (b) the frequency, intensity, and timing of artificial flow fluctuations and resulting water level changes; (c) the availability and quality of habitats; and (d) the spatial variability of hydrological and morphological impacts (Hauer et al. 2014) (Fig. 5.5).

In general, hydropeaking intensity and ecological impacts diminish downstream. At the scale of a river reach, flow fluctuation rates, in particular, are highly variable due to retention effects and morphological variability (Hauer et al. 2013). A method that allows for the detection of flow fluctuations (Greimel et al. 2015) and the longitudinal development (including retention effects) is described in Greimel

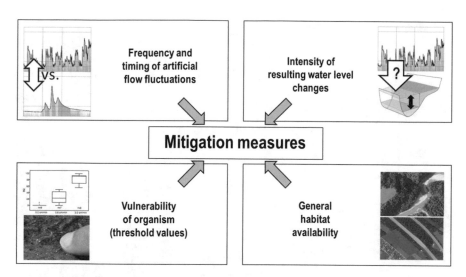

Fig. 5.5 Integrative hydropeaking management: linking abiotic and biotic factors to define and monitor mitigation measures

et al. (2017). Based on data from multiple hydrographs, this method enables the assessment of hydropeaking intensities and frequencies along an affected river reach. This method links flow rate changes to water level changes and, subsequently, to thresholds for harmful impacts (e.g., for ramping rates—see Fig. 5.5) for different species and life stages.

The following example of a hypothetical power plant (Fig. 5.6) should exemplify the application of longitudinal hydropeaking assessment as the basis for integrative hydropeaking mitigation. One way of evaluating different mitigation scenarios is the longitudinal development of maximum flow fluctuation rates of flow decrease events, which are critical for stranding. This approach to assessing the intensity of longitudinal hydropeaking aims to compare the stranding risk for juvenile and larval fish at the actual state ("maximum-intensity scenario") with the risks inherent in mitigation scenarios (reduced scenarios 1 and 2) (Fig. 5.6). Flow fluctuations are tracked downstream of the power plant outlet by analyzing turbine flow data and downstream hydrographs. First, inter-hydrograph models describe the intensity changes between neighboring hydrographs. Then these results are combined in an overall longitudinal assessment schema (Fig. 5.6). The "maximum-intensity scenario" envisions down-ramping the turbine discharge at the rate of 25 m³/s per 15 min (upper dotted line—left axis). This results in water level changes of ca. 2.7 cm/min directly downstream of the turbine (upper continuous line—right axis). During such flow decrease events, retention effects cause a decrease in the event intensity of ca. 10 m³/s per 15 min or 0.9 cm/min at the downstream end of the investigated river reach. Assuming that high stranding risk is designated for flow fluctuation rates over 0.4 cm/min, then under the maximum-intensity scenario, fish stranding appears likely over the entire river reach. The "Reduced scenario 1" limits

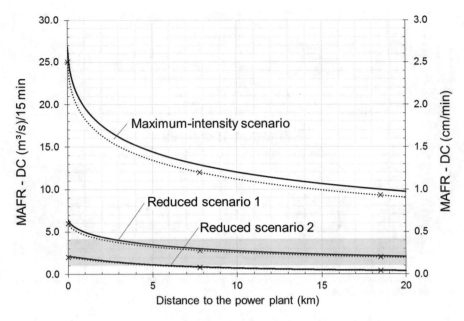

Fig. 5.6 Example of longitudinal assessment of hydropeaking intensity (MAFR-DC: max. Flow fluctuation rate of decrease events; continuous lines refer to the left axis; dotted lines refer to the right axis; the blue box refers to assumed stranding risk of juvenile and larval fish) using multiple hydrographs (marked by crosses) for a hypothetical power plant and evaluation of mitigation scenarios

hydropeaking intensity (e.g., the down-ramping rate) to a maximum turbine flow decrease of 6 m³/s per 15 min or 0.65 cm/min. The maximum turbine flow restriction in "Reduced scenario 2" equals 2 m³/s per 15 min or 0.2 cm/min directly below the power plant. If "Reduced scenario 1" is implemented, it is likely that the stranding risk for juveniles (threshold, 0.4 cm/min) is minimized further than 3 km downstream of the turbines. "Reduced scenario 2" would lead to a minimized stranding risk also for larvae (threshold, 0.1 cm/min) further than 5 km downstream of the turbine.

Besides the maximum flow fluctuation rate of decrease events, the exemplified approach can be applied to flow increase events as well as to other hydrological parameters. That allows a comprehensive description of the hydrological situation downstream of a specific power plant. If the specific timing, intensity, and frequency of artificial flow fluctuations at any point along an affected river reach are known due to these power plant-specific assessments, then potential ecological effects can be evaluated by contrasting the hydrological situation to the vulnerability of aquatic organism and life stages. As a consequence, multiple hydrological mitigation scenarios can be defined by referring to different organism groups and corresponding threshold values. However, it has to be noted that several hydrological mitigation scenarios should be interpreted in the face of the current habitat suitability in the affected river stretch in order to prevent ineffective mitigation scenarios due to low habitat availability. In these cases additional morphological measures may be required. In the last step the hydrological mitigation scenarios could be evaluated, both ecologically and economically, if specific costs are linked to the hydrological mitigation scenarios and/or types of mitigation measures, e.g., retention basins and changing of the power plant operation mode.

5.4.3 Summary and Outlook

Current scientific knowledge allows to develop fundamental conceptual models in order to describe ecological effects of hydropeaking and to predict potential effects of mitigation scenarios. For this purpose, hydrological, morphological, sedimentological, hydraulic, and ecological aspects have to be linked.

As hydropeaking intensity, frequency, and timing are attenuated or changed along a river course downstream of power stations, it is important to develop case-specific assessment schemas (longitudinal hydropeaking assessment). This approach enables (a) to monitor both hydropeaked and unaffected sub-daily flow regimes, (b) to transfer laboratory results (e.g., from stranding experiments), and (c) to model mitigation scenarios. However, in addition to hydrological aspects, sedimentological (sediment transport) and morphological (habitat suitability) issues also have to be considered to describe potential hydropeaking impacts on aquatic organism.

Ecological knowledge has been established to a varying extent for fish, macroinvertebrates, and periphyton. In general, hydropeaking reduces the quality and availability of suitable habitats, which leads to reduced reproduction, survival, and biodiversity. The repeated artificial flow fluctuations and the corresponding variation of related parameters (e.g., flow velocity, water depths, shear stress, water temperature) require a lateral habitat shift of vagile aquatic organism that prompts increased rates of drifting (up-ramping) or stranding (down-ramping). Ecological responses to hydropeaking are species and life stage specific and may affect the entire food web. Additionally, daylight conditions, water temperature, habitat quality, and other seasonal aspects may interact with hydropeaking effects. Some threshold values to draw the line between harmful and harmless peaking have been established, which is an important step to predicting ecological effects and thereby defining mitigation measures. Furthermore, it is evident that more "natural" river morphology can decrease hydropeaking impacts.

In the absence of implemented and validated mitigation measures, current conceptual models should be considered as relatively rudimentary. Existing mitigation concepts should be enhanced and broadly implemented to support the collection of detailed observations and monitoring data. Additionally, further research is necessary to fill knowledge gaps concerning poorly understood hydropeaking effects, such as stranding of invertebrates, substrate clogging, cyprinid species, benthic algae, microorganisms, and riparian vegetation.

References

Allan J (1987) Macroinvertebrate drift in a Rocky Mountain stream. Hydrobiologia 144:261–268

Arscott DB, Tockner K, Ward JV (2005) Lateral organization of aquatic invertebrates along the corridor of a braided floodplain river. J North Am Benthol Soc 24:934–954. https://doi.org/10. 1899/05-037.1

Auer S, Fohler N, Zeiringer B, Führer S, Schmutz S (2014) Experimentelle Untersuchungen zur Schwallproblematik – Drift und Stranden von Äschen und Bachforellen während der ersten Lebensstadien. Forschungsbericht

Auer S, Zeiringer B, Führer S, Tonolla D, Schmutz S (2017) Effects of river bank heterogeneity and time of day on drift and stranding of juvenile European grayling (*Thymallus thymallus* L.) caused by hydropeaking. Sci Total Environ 575:1515–1521. https://doi.org/10.1016/j.scitotenv. 2016.10.029

Bauersfeld K (1978) Stranding of juvenile salmon by flow reductions at Mayfield Dam on the Cowlitz River, 1976 Technical Report 36. Department of Fisheries, Washington

Bell E, Kramer S, Zajanc D, Aspittle J (2008) Salmonid fry stranding mortality associated with daily water level fluctuations in Trail Bridge Reservoir. Oregon North Am J Fish Manage 28 (5):1515–1528. https://doi.org/10.1577/m07-026.1

Berland G, Nickelsen T, Heggenes J, Økland F, Thorstad EB, Halleraker J (2004) Movements of wild Atlantic salmon parr in relation to peaking flows below a hydropower station. River Res Appl 20(8):957–966. https://doi.org/10.1002/rra.80

Biggs BJ, Close ME (1989) Periphyton biomass dynamics in gravel bed rivers: the relative effects of flows and nutrients. Freshw Biol 22(2):209–231

Biggs BJ, Thomsen HA (1995) Disturbance of stream periphyton by shear stress: time to structural failure and differences in community resistance. J Phycol 31(2):233–241

Biggs BJ, Goring DG, Nikora VI (1998) Subsidy and stress responses of stream periphyton to gradients in water velocity as a function of community growth form. J Phycol 34(4):598–607

Blinn W, Shannon JP, Stevens LE, Carder JP (1995) Consequences of fluctuating discharge for lotic communities. J North Am Benthol Soc 14:233–248. https://doi.org/10.2307/1467776

BMLFUW – Nationaler Gewässerbewirtschaftungsplan (2009/2010) http://www. lebensministerium.at/wasser/wasser-oesterreich/plan_gewaesser_ngp/nationaler_ gewaesserbewirtschaftungsplan-nlp/ngp.html. 13 February 2014

Bondar-Kunze E, Maier S, Schönauer D, Bahl N, Hein T (2016) Antagonistic and synergistic effects on a stream periphyton community under the influence of pulsed flow velocity increase and nutrient enrichment. Sci Total Environ 573:594–602

Bradford MJ (1997) An experimental study of stranding of juvenile salmonids on gravel bars and in sidechannels during rapid flow decreases. River Res Appl 13:395–401. https://doi.org/10.1002/ (sici)1099-1646(199709/10)13:5<395::aid-rrr464>3.0.co;2-l

Bradford MJ, Taylor GC, Allan JA, Higgins PS (1995) An experimental study of the stranding of juvenile Coho Salmon and Rainbow Trout during rapid flow decreases under winter conditions. North Am J Fish Manage 15(2):473–479

Bretschko G, Moog O (1990) Downstream effects of intermittent power generation. Water Sci Technol 22:127–135

Bruder A, Tonolla D, Schweizer S, Vollenweider S, Langhans SD, Wüest A (2016) A conceptual framework for hydropeaking mitigation. Sci Total Environ 568:1204–1212. https://doi.org/10. 1016/j.scitotenv.2016.05.032

Bruno MC, Siviglia A, Carolli M, Maiolini B (2013) Multiple drift responses of benthic invertebrates to interacting hydropeaking and thermopeaking waves. Ecohydrology 6:511–522. https:// doi.org/10.1002/eco.1275

Bunn SE, Arthington AH (2002) Basic principles and ecological consequences of altered flow regimes for aquatic biodiversity. Environ Manag 30(4):492–507. https://doi.org/10.1007/ s00267-002-2737-0

Burkholder JM (1996) Interactions of Benthic Algae with their Substrata-9. In: Stevenson RJ, Bothwell ML, Lowe RL (eds) Algal ecology – freshwater benthic ecosystems. Academic, San Diego, pp 253–297

Carolli M, Bruno MC, Siviglia A, Maiolini B (2012) Responses of benthic invertebrates to abrupt changes of temperature in flume simulations. River Res Appl 28:678–691. https://doi.org/10.1002/rra

Céréghino R, Lavandier P (1998) Influence of hypolimnetic hydropeaking on the distribution and population dynamics of Ephemeroptera in a mountain stream. Freshw Biol 40:385–399

Céréghino R, Cugny P, Lavandier P (2002) Influence of intermittent hydropeaking on the longitudinal zonation patterns of benthic invertebrates in a mountain stream. Int Rev Hydrobiol 87:47–60

Comiti F (2012) How natural are Alpine mountain rivers? Evidence from the Italian Alps. Earth Surf Process Landf 37:693–707. https://doi.org/10.1002/esp.2267

Cushman RM (1985) Review of ecological effects of rapidly varying flows downstream from hydroelectric facilities. North Am J Fish Manag 5:330–339. https://doi.org/10.1577/1548-8659 (1985)5<330:ROEEOR>2.0.CO;2

Elliott JM (1967) Invertebrate drift in a Dartmoor stream. Arch Hydrobiol 63:202–237

Elliott JM (2005) Contrasting diel activity and feeding patterns of four instars of Rhyacophila dorsalis (Trichoptera). Freshw Biol 50:1022–1033. https://doi.org/10.1111/j.1365-2427.2005. 01388.x

Finch C, Pine WE, Limburg KE (2015) Do hydropeaking flows alter juvenile fish growth rates? A test with juvenile humpback chub in the colorado river. River Res Appl 31:156–164. https://doi. org/10.1002/rra.2725

Fjellheim A (1980) Differences in drifting of larval stages of Rhyacophila nubila (Trichoptera). Holarct Ecol 3:99–103. https://doi.org/10.1111/j.1600-0587.1980.tb00714.x

Flodmark LEW, Vøllestad LA, Forseth T (2004) Performance of juvenile brown trout exposed to fluctuating water level and temperature. J Fish Biol 65:460–470. https://doi.org/10.1111/j.1095-8649.2004.00463.x

Friedl J, Naesby K (2014) Habitateignung der HyTEC-Versuchsanlage sowie Einfluss von Schwall auf Wachstums- und Konditionsentwicklung von Jungäschen (Thymallus thymallus). Master Thesis. Institut für Hydrobiologie, Gewässermanagement (IHG), BOKU-Universität für Bodenkultur, p 142

Graf W, Leitner P, Moog O, Steidl C, Salcher G, Ochsenhofer G, Müllner K (2013) Schwallproblematik an Österreichs Fließgewässern – Ökologische Folgen und Sanierungsmöglichkeiten. Dateneerhebung und Analyse Benthische Invertebrtaten [WWW Document]. http://hydropeaking.boku.ac.at/

Greimel F, Zeiringer B, Höller N, Grün B, Godina R, Schmutz S (2015) A method to detect and characterize sub-daily flow fluctuations. Hydrol Process 30:2063–2078. https://doi.org/10.1002/ hyp.10773

Greimel F, Zeiringer B, Führer S, Holzapfel P, Fuhrmann M, Höller N, Hauer C, Schmutz S (2017) Longitudinal assessment of hydropeaking intensity and frequency based on multiple hydrograph curves – a method proposal (in prep)

Hall RO, Yackulic CB, Kennedy TA, Yard MD, Rosi-Marshall EJ, Voichick N, Behn KE (2015) Turbidity, light, temperature, and hydropeaking control primary productivity in the Colorado River, Grand Canyon. Limnol Oceanogr 60:512–526. https://doi.org/10.1002/lno.10031

Halleraker H, Saltveit SJ, Harby A, Arnekleiv JV, Fjeldstad H-P, Kohler B (2003) Factors influencing stranding of wild juvenile brown trout (Salmo trutta) during rapid and frequent flow decreases in an artificial stream. River Res Appl 19(5–6):589–603. https://doi.org/10.1002/ rra.752

Hauer C, Schober B, Habersack H (2013) Impact analysis of river morphology and roughness variability on hydropeaking based on numerical modelling. Hydrol Process 27(15):2209–2224

Hauer C, Unfer G, Holzapfel P, Haimann M, Habersack H (2014) Impact of channel bar form and grain size variability on estimated stranding risk of juvenile brown trout during hydropeaking. Earth Surf Process Landf 39(12):1622–1641

Heggenes J, Traaen T (1988) Downstream migration and critical water velocities in stream channels for fry of four salmonid species. J Fish Biol 32(5):717–727. https://doi.org/10.1111/j.1095-8649.1988.tb05412.x

Hoagland KD, Roemer SC, Rosowski JR (1982) Colonization and community structure of two periphyton assemblages, with emphasis on the diatoms (Bacillariophyceae). Am J Bot 69:188–213

Hondzo M, Wang H (2002) Effects of turbulence on growth and metabolism of periphyton in a laboratory flume. Water Resour Res 38(12):1277

Hunter MA (1992) Hydropower flow fluctuations and salmonids: a review of the biological effects, mechanical causes, and options for mitigation. Technical Report No. 119. Department of Fisheries, State of Washington

Illies J (1978) Limnofauna Europeae.- 2. überarbeitete und ergänzte Auflage. G. Fischer Verlag, Stuttgart, New York; Swets & Zeitlinger B.V., Amsterdam

Imbert J, Perry J (2000) Drift and benthic invertebrate responses to stepwise and abrupt increases in non-scouring flow. Hydrobiologia 436:191–208

Kaiser L (2016) Zum Einfluss unterschiedlich temperierter Schwallereignisse auf die Drift und Strandung von Jungäschen – demonstriert an der Versuchsanlage HyTEC Lunz am See. Masterarbeit – Institut für Hydrobiologie, Gewässermanagement (IHG), BOKU-Universität für Bodenkultur, p 92

Kasper V (2016) The effect of thermopeaking on structural and functional characteristics of a microphytobenthos community. Masterarbeit – Institut für Hydrobiologie, Gewässermanagement (IHG), BOKU-Universität für Bodenkultur

Kennedy TL, Turner TF (2011) River channelization reduces nutrient flow and macroinvertebrate diversity at the aquatic terrestrial transition zone. Ecosphere 2:art35. https://doi.org/10.1890/ES11-00047.1

Kennedy TA, Yackulic CB, Cross WF, Grams PE, Yard MD, Copp AJ (2014) The relation between invertebrate drift and two primary controls, discharge and benthic densities, in a large regulated river. Freshw Biol 59:557–572. https://doi.org/10.1111/fwb.12285

Lauters F, Lavandier P, Lim P, Sabaton C, Belaud A (1996) Influence of hydropeaking on invertebrates and their relationship with fish feeding habits in a Pyrenean river. Regul Rivers Res Manag 12:563–573. https://doi.org/10.1002/(SICI)1099-1646(199611)12:6<563::AID-RRR380>3.0.CO;2-M

Limnex (2004) Auswirkungen des Schwallbetriebes auf das Ökosystem der Fliessgewässer: Grundlagen zur Beurteilung. Bericht im Auftrag des WWF, Zürich

Maiolini B, Silveri L, Lencioni V (2007) Hydroelectric power generation and disruption of the natural stream flow: effects on the zoobenthic community. Stud Trent Sci Nat, Acta Biol 83:21–26

Marty J, Smokorowski K, Power M (2009) The influence of fluctuating ramping rates on the food web of boreal rivers. River Res Appl 25:962–974. https://doi.org/10.1002/rra.1194

Moore KMS, Gregory SV (1988) Summer habitat utilization and ecology of cutthroat trout fry (*Salmo clarki*) in Cascade Mountain Streams. Can J Fish Aquat Sci 45:1921–1930

Muhar S, Schwarz M, Schmutz S, Jungwirth M (2000) Identification of rivers with high and good habitat quality: methodological approach and applications in Austria. Hydrobiologia 422:343–358. https://doi.org/10.1023/A:1017005914029

Nagrodski A, Raby GD, Hasler CT, Taylor MK, Cooke SJ (2012) Fish stranding in freshwater systems: sources, consequences, and mitigation. J Environ Manag 103:133–141. https://doi.org/10.1016/j.jenvman.2012.03.007

Oldmeadow DF, Lancaster J, Rice SP (2010) Drift and settlement of stream insects in a complex hydraulic environment. Freshw Biol 55:1020–1035. https://doi.org/10.1111/j.1365-2427.2009.02338.x

Parasiewicz P, Schmutz S, Moog O (1998) The effect of managed hydropower peaking on the physical habitat, benthos and fish fauna in the River Bregenzerach in Austria. Fish Manag Ecol 5:403–417. https://doi.org/10.1046/j.1365-2400.1998.550403.x

Passy SI, Larson CA (2011) Succession in stream biofilms is an environmentally driven gradient of stress tolerance. Microb Ecol 62(2):414–424

Poff NL, DeCino RD, Ward JV (1991) Size-dependent drift responses of mayflies to experimental hydrologic variation: active predator avoidance or passive hydrodynamic displacement? Oecologia 88:577–586. https://doi.org/10.1007/BF00317723

Poff NL, Allan JD, Bain MB, Karr JR, Prestegaard KL, Richter BD, Sparks RE, Stromberg JC (1997) The natural flow regime. Bioscience 47:769–784

Puffer M, Berg OK, Huusko A, Vehanen T, Forseth T, Einum S (2015) Seasonal effects of Hydropeaking on growth, energetics and movement of juvenile Atlantic Salmon (Salmo Salar). River Res Appl 31:1101–1108. https://doi.org/10.1002/rra.2801

Saltveit SJ, Halleraker JH, Arnekleiv JV, Harby A (2001) Field experiments on stranding in juvenile Atlantic Salmon (Salmo salar) and Brown trout (Salmo trutta) during rapid flow decreases caused by hydropeaking. River Res Appl 17(4–5):609–622. https://doi.org/10.1002/rrr.652.abs

Schmutz S, Fohler N, Friedrich T, Fuhrmann M, Graf W, Greimel F, Höller N, Jungwirth M, Leitner P, Moog O, Melcher A, Müllner K, Ochsenhofer G, Salcher G, Steidl C, Unfer G, Zeiringer B (2013) Schwallproblematik an Österreichs Fließgewässern – Ökologische Folgen und Sanierungsmöglichkeiten. BMFLUW, Wien

Schmutz S, Bakken TH, Friedrich T, Greimel F, Harby A, Jungwirth M, Melcher A, Unfer G, Zeiringer B (2015) Response of fish communities to hydrological and morphological alterations in hydropeaking rivers of Austria. River Res Appl 31:919–930. https://doi.org/10.1002/rra.2795

Schülting L, Feld CK, Graf W (2016) Effects of hydro- and thermopeaking on benthic macroinvertebrate drift. Sci Total Environ 573:1472–1480. https://doi.org/10.1016/j.scitotenv.2016.08.022

Smokorowski BKE (2010) Effects of experimental ramping rate on the invertebrate Community of a Regulated River Benthic Invertebrates as test organisms. Proceedings of the Colorado River Basin Science and Resource Management Symposium: 149–156

Smolar-Žvanut N, Klemenčič AK (2013) The impact of altered flow regime on periphyton. In: Maddock I, Harby A, Kemp P, Wood P (eds) Ecohydraulics: an integrated approach. Wiley, Chichester, pp 229–243

Statzner B, Holm TF (1982) Morphological adaptations of benthic invertebrates to stream flow – an old question studied by means of a new technique (Laser Doppler Anemometry). Oecologia 53:290–292. https://doi.org/10.1007/BF00389001

Thompson LC, Cocherell SA, Chun SN, Cech JJJ, Klimley AP (2011) Longitudinal movement of fish in response to a single-day flow pulse. Environ Biol Fish 90:253–261. https://doi.org/10.1007/s10641-010-9738-2

Toffolon M, Siviglia A, Zolezzi G (2010) Thermal wave dynamics in rivers affected by hydropeaking. Water Resour Res 46:W08536. https://doi.org/10.1029/2009WR008234

Tonolla D, Bruder A, Schweizer S (2017) Evaluation of mitigation measures to reduce hydropeaking impacts on river ecosystems – a case study from the Swiss alps. Sci Total Environ 574:594–604. https://doi.org/10.1016/j.scitotenv.2016.09.101

Townsend CR, Scarsbrook MR, Dolédec S (1997) The intermediate disturbance hypothesis, refugia, and biodiversity in streams. Limnol Oceanogr 42(5):938–949

Tuor KMF, Smokorowski KE, Cooke SJ (2014) The influence of fluctuating ramping rates on the diets of small-bodied fish species of boreal rivers. Environ Biol Fish 98:345–355. https://doi.org/10.1007/s10641-014-0264-5

Unfer G, Hauer C, Lautsch E (2011) The influence of hydrology on the recruitment of brown trout in an Alpine river, the Ybbs River, Austria. Ecol Freshw Fish 20:438–448. https://doi.org/10.1111/j.1600-0633.2010.00456.x

Vanzo D, Zolezzi G, Siviglia A (2015) Eco-hydraulic modelling of the interactions between hydropeaking and river morphology. Ecohydrology 9(3):421–437. https://doi.org/10.1002/eco.1647

Wagenhoff A, Lange K, Townsend CR, Matthaei CD (2013) Patterns of benthic algae and cyanobacteria along twin-stressor gradients of nutrients and fine sediment: a stream mesocosm experiment. Freshw Biol 58(9):1849–1863

Ward JV, Stanford JA (1979) Ecological factors controlling stream zoobenthos with emphasis on thermal modification of regulated streams. In: Ward JV, Stanford JA (eds) The ecology of regulated streams. Plenum Press, New York, pp 35–55

Waringer JA (1989) Resistance of cased caddis larva to accidental entry into the drift: the contribution of active and passive elements. Freshw Biol 21:411–420

Yamada H, Nakamura F (2002) Effect of fine sediment deposition and channel works on periphyton biomass in the Makomanai River, Northern Japan. River Res Appl 18:481–493

Young PS, Cech JJ, Thompson LC (2011) Hydropower-related pulsed flow impacts on stream fishes: a brief review, conceptual model, knowledge gaps, and research needs. Rev Fish Biol Fish 21:713–731. https://doi.org/10.1007/s11160-011-9211-0

Zolezzi G, Siviglia A, Toffolon M, Maiolini B (2011) Thermopeaking in Alpine streams: event characterization and time scales. Ecohydrology 4:564–576

Chapter 6
Dams: Ecological Impacts and Management

Stefan Schmutz and Otto Moog

6.1 Introduction

Dam construction goes back in human history for more than 5000 years (e.g., Sadd el-Kafara dam in Egypt for flood protection), but most of the world's existing dams have been built after the Second World War as consequence or basis of economic development. Today, there are about 6000 existing or planned large hydropower dams (>15 m height) worldwide (Zarfl et al. 2014) and an uncountable number of small dams. For example, with more than 5000 mostly small hydropower plants, Austria is one of the countries with the highest density of hydropower dams (about 6 dams per 100 km^2, Wagner et al. 2015). Downstream flows are mainly altered by large dams, e.g., there are 654 reservoirs with storage capacities $\geq$0.5 km^3 (Lehner and Döll 2004). Damming rivers currently stores the equivalent of 15% of global annual river runoff (Likens 2010). As a result, 48% of rivers (expressed as river volume) globally are moderately to severely impacted by either flow regulation, fragmentation, or both.

Besides flow, sediment transport is severely altered by dams. A total of approximately 25–30% of pre-disturbance sediment flux is sequestered by modern impoundments (Fig. 6.1).

Impacts might double should all planned dams be constructed by 2030 (Grill et al. 2015). This is especially so in areas with expanding economies and extensive unexploited river reaches, such as China, which currently is building 130 major dams in its Southwest (Lewis 2013) and has constructed more than half the new dams built since 1950 worldwide (Wang and Chen 2010).

Unfortunately, there is no generally accepted descriptive nomenclature of dams. The term "dam" is often applied to both the physical structure retaining the water and

S. Schmutz (✉) · O. Moog
Institute of Hydrobiology and Aquatic Ecosystem Management, University of Natural Resources and Life Sciences, Vienna, Austria
e-mail: stefan.schmutz@boku.ac.at; otto.moog@boku.ac.at

© The Author(s) 2018
S. Schmutz, J. Sendzimir (eds.), *Riverine Ecosystem Management*, Aquatic Ecology Series 8, https://doi.org/10.1007/978-3-319-73250-3_6

Fig. 6.1 Sediment trapping by large dams. GWSP Digital Water Atlas (Available online at http://atlas.gwsp.org)

the water so retained. For the purposes of this chapter, dam will be used solely to describe the physical structure (e.g., weir), and the term "reservoir" will be used to denote the artificially created water body. This leads to the following definition: "A dam is a barrier to obstruct the flow of water and to create a reservoir." Reservoirs are also called "impoundments." Reservoirs are built for specific community needs:

- Drinking, industrial, and cooling water supply
- Hydropower generation
- Agricultural irrigation
- River regulation and flood control
- Navigation
- Recreation and fisheries

Dams are among the most damaging human activities in river basins, deeply modifying the physiography of watersheds. Reservoirs may look very much like natural lakes; however, the operating regime determined by the purpose for which the reservoirs were created may significantly alter their physicochemical character and biological responses. The peculiar form of a reservoir, its location, and mode of operation may cause considerable, actual variation of the basic limnological behavior. Reservoirs undergo great changes in water quality during the early stages of their formation until a new ecological balance becomes established (Straskraba et al. 1993). Reservoirs follow a succession of (1) physicochemical alteration, (2) modification in the structure and dynamics of primary producers, and (3) changes in the community of consumers, especially invertebrates and fish (Petts 1985). After that, reservoirs may pass over into a kind of stability, but occurrence of floods, dam operation, or other impacts may create new disturbances to the system.

Reservoirs not only affect the inundated river sections but also block upstream fish migration (see Chap. 9) and downstream flow and sediment transport. The magnitude of impact is strongly correlated with the location of the dam, size of reservoir (height of dam, volume of reservoir), and *water residence time*. The average length of time water remains within the boundaries of an aquatic system is one of the key parameters controlling the system's biogeochemical behavior. This time scale, which is generally referred to as the water residence time, is fundamental for multiple and complex processes in reservoirs (Rueda et al. 2006). Furthermore, the dam operation mode determines the seasonal variation of stored water, water level fluctuations, sediment capture and release, as well as daily and seasonal downstream flow patterns.

The main impacts associated with reservoirs are as follows:

- Interruption of river continuity (longitudinal and lateral, fish migration, sediment and nutrient transport)
- Siltation of river bed and clogging of interstitial
- Homogenization of habitats
- Downstream river bed incision
- Alteration of river/groundwater exchange
- Downstream flow and water quality alteration

6.2   Transforming Rivers to Reservoirs

River damming is a process so drastic that it results in the creation of a completely new ecosystem (Baxter 1977). Therefore, the occurrence of environmental impacts is inherent with any impoundment due to fundamental change of the hydrology and morphology of the river. Flow represents the main force behind freshwater ecosystems, and it is responsible for geohydrological structure, matter and energy fluxes, system productivity, and distribution and function of biota (Poff et al. 1997). As a consequence, the alteration of natural flow regimes and morphodynamic patterns has far-reaching impacts (see Chaps. 3 and 4), including production, biodiversity, and changes in functions and services provided by aquatic ecosystems (Nilsson et al. 2005).

Reservoirs differ from natural lakes with respect to hydrological, limnological, and ecological dynamics. Depending on the size and shape of the reservoir, a longitudinal hydrological gradient may develop from the dam (lentic or lacustrine zone) to upstream reaches (riverine zone), showing intermediate characteristics in middle stretches (transition zone, with lentic and lotic features) (Kimmel and Groeger 1984). While fluvial characteristics are maintained to some extent in small reservoirs, e.g., run-of-the-river hydropower plants, lentic conditions prevail in large storage reservoirs (Fig. 6.2).

Coarse sediments settle within the riverine section of the reservoir, while fine sediments (sand, silt) and particulate organic matter (POM) are deposited in the lacustrine zone (Fig. 6.2, Table 6.1). Soon after the reservoir is filled, patterns of thermal/chemical stratification intensify progressively in the water column, and eutrophication may occur due to upstream matter input, decay, and nutrient release from the flooded organic matter (vegetation, litter, and soil) or from pollution. Reservoirs are much more susceptible to eutrophication that rivers due to the higher self-purification capacity of running waters. Consequently, water quality may deteriorate in reservoirs (e.g., thermal stress, low dissolved oxygen, acidification), especially close to the bottom (Agostinho et al. 2008). Depending on a reservoir's characteristics, anoxic water or sediment layers may evolve as a consequence of stratification, deposition, and decomposition of organic material. Stratification (thermocline, light conditions, etc.) changes production (autotrophic, heterotrophic) and the entire food chain (Fig. 6.2).

Dams are often associated with lateral dams or levees, disconnecting the reservoir hydrologically from the floodplains. This results in limited or abandoned inundation and reduced interchange with the groundwater and lowers chances of recolonization, both lateral and longitudinal. As a result, the new ecosystem is colonized by those species that inhabited the original river and are able to adapt to the new conditions (Agostinho et al. 2008). Nonmigratory, eurytopic species dominate the lacustrine zone of reservoirs because they usually have less complex requirements with regard to life-cycle dynamics. Migratory and rheophilic species experience declines in the reservoir due to the lentic environment, spatial fragmentation imposed by the dam, and the loss of critical habitats (e.g., spawning habitats in free-flowing river stretches). Consequently, lentic fish replace lotic species and dominate the reservoir fish communities (Zhong and Power 1996). In terms of richness, fish species tend to remain in environments that preserve the original fluvial characteristics or in those

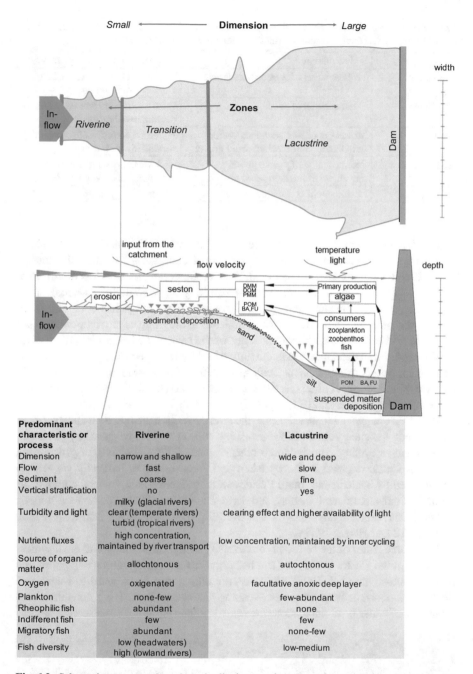

Fig. 6.2 Schematic representation of longitudinal reservoir zones and associated key processes, (**a**) plan view and (**b**) longitudinal profile; *DMM* dissolved mineral matter, *DOM* dissolved organic matter, *PMM* particulate mineral matter, *POM* particulate organic matter, *BA* bacteria, *FU* fungi (adapted after Herzig 1984; Kimmel and Groeger 1984)

Table 6.1 Sedimentation of river sediments as a function of flow velocity and associated dominating benthic invertebrate community (modified after Jungwirth et al. 2006)

Flow velocity	Sedimentation	Benthic invertebrate community
<20 cm s^{-1}	Sedimentation of organic matter and silt	Oligochaeta (Tubifex), Pisidia, Diptera (Chironomidae)
25–50 cm s^{-1}	Sedimentation of sand	Due to instability less favorable for benthic invertebrates
60–90 cm s^{-1}	Movement of fine gravel and sand, sedimentation of medium-sized gravel	Increased diversity of (rheophilic) benthic invertebrates
120–170 cm s^{-1}	Sedimentation of coarse gravel movement of medium-sized gravel	High diversity of (rheophilic and rheobiont) benthic invertebrates
>170 cm s^{-1}	Movement of coarse gravel	Rheophilic organisms

with relatively high habitat heterogeneity, i.e., tributaries and lotic stretches upstream, and it is common to find a gradient of decreasing fish diversity toward dams. If coldwater streams are dammed, the warmer water in reservoirs might favor species adapter to warmer temperatures, resulting in a so-called potamalization effect, i.e., a shift from rhithral to potamal communities (Jungwirth et al. 2003).

Fine sediment deposition in the reservoir leads to clogging of the river bottom (see Chap. 8). This affects aquatic communities of the hyporheic interstitial (Ward et al. 1998), resulting in depauperated fauna dominated by few species (e.g., chironomids, Table 6.1). As a consequence, the ecological status of reservoirs, in particular within the lacustrine section, is often classified as poor or bad (sensu EU Water Framework Directive, Ofenböck et al. 2011, Fig. 6.3).

While the energy and matter fluxes in the riverine section are based on allochthonous matter input, the energy fluxes of the lacustrine section are also triggered by photosynthesis and inner cycling. Allochthonous organic matter (DOM, dissolved organic matter; POM, particulate organic) is directly taken up by consumers or indirectly via detritus decomposed by bacteria and fungi. Besides water residence time, mineral components (DMM, dissolved mineral matter) and light conditions (PMM, particulate mineral matter; POM) regulate algal growth and overall productivity in the lacustrine section (see Fig. 6.2 and Herzig 1984).

In all three zones of the reservoir, biodiversity is highest in the littoral environment as a result of the greater availability and heterogeneity of feeding resources, shelter, and habitats (Agostinho et al. 2008). However, the littoral may be exposed to water level fluctuations, causing frequent stress events to fauna and flora. High magnitude and frequency of water level fluctuations creates "dead zones" along the reservoir shores.

Whenever rivers are turned into reservoirs, the former fluvial habitat is widely lost. The new lentic ecosystem resembles lake-type systems, but, depending on type and dam operation, reservoirs are disturbed by artificial water level fluctuations, drawdowns, and floods. Consequently, dammed rivers are hybrid systems that lose their lotic but gain only partly lacustrine functions.

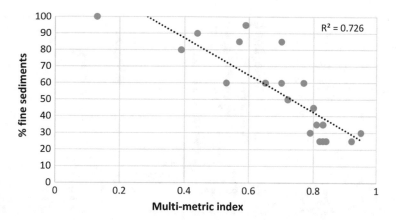

Fig. 6.3 Response of macroinvertebrate communities to increased proportion of fine sediments (akal/psammal/pelal) in three impounded streams (Austria) expressed as multimetric index (1 = high ecological status, 0 = bad ecological status; adapted from Ofenböck et al. 2011)

Sediment trapping by reservoirs is well recognized, with the extent to which sediments are trapped dependent on the morphology of the impoundment, the characteristics (grain size) of the inflowing sediments, the hydrodynamics within the impoundment, and the operating regime of the dam. Smaller impoundments have lower rates as compared to the larger reservoirs. Sediment trapping is cumulative, but as coarser material is trapped in upstream impoundments, the actual trapping rates in downstream impoundments may decline, due to the finer nature of the influent sediment load. This preferential trapping can also lead to nutrients being trapped in different proportions as compared to sediments, due to nutrients' affinity for the finer-grained sediments (Koehnken 2014).

As an example, the dam cascade at the Lancang River (upper Mekong) has the potential to trap most of the sediments (Kummu and Varis 2007, Fig. 6.4). Recent monitoring results suggest that suspended sediment loads in the lower Mekong basin (downstream of China) are now in the range of 44% compared with the historic values (~70 Mt/year compared with ~160 Mt/year; Koehnken 2014).

6.3   Downstream Effects

Although investigated less, downstream impacts are equally or even more damaging to aquatic fauna, given that impoundments affect primarily water flow dynamics, i.e., the main force working in fluvial ecosystems. Impoundments redistribute river discharge in space and time, affecting several hydrological attributes, e.g., flood period, intensity, amplitude, duration, frequency, and, consequently, the structure, dynamics, and functioning of ecosystems located downstream (see Chap. 4). In

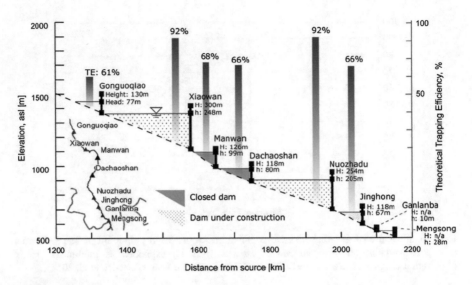

Fig. 6.4 Average theoretical sediment trapping efficiency of reservoirs of the Lancang (Upper Mekong) Cascade in China (Kummu and Varis 2007) (reproduced from Geomorphology, 85/3–4, 2007, Matti Kummu, Olli Varis, Sediment-related impacts due to upstream reservoir trapping, the lower Mekong river, pp. 275–293, with permission from Elsevier. © 2007 Elsevier Inc. All rights reserved)

addition to flow regulation, other important alterations inevitably follow dam construction, such as blockage of migration routes for some fish species and the retention of sediments and nutrients upstream—a process that decreases turbidity as well as nutrient load and suspended material. This last phenomenon imposes limitations on biological productivity in areas downstream, reducing the fertility of wetlands, and affecting their carrying capacity. The loss of fertilizing services of the flood pulse has been documented in several systems worldwide (WCD 2000).

In case of downstream floodplains, negative effects of flow regulation are still more pronounced. The structure and functioning of such ecosystems rely on the alternation of extreme events, e.g., flood and drought (Junk et al. 1989), a dynamic pulsing that disappears with impoundment, because dams usually decrease maximum discharges (absence of seasonal flood pulses) and stabilize or increase minimum discharges. Consequently, hydrological connectivity among environments is considerably modified in space and time. The redistribution of the flooding regime has several direct and indirect effects on fish populations. The decrease in connectivity between the river and lateral floodplain affects riparian communities and reshapes other environments and interface zones that provide important habitats for fish, especially nurseries. In the absence of floods, even if adults successfully reproduce in tributaries, eggs, larvae, and young fish drifting downstream have limited access to lateral habitats, and population recruitment is negatively affected (Agostinho et al. 2008).

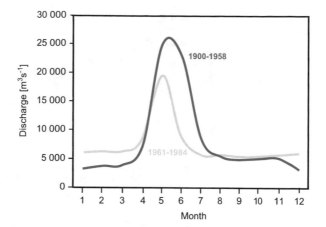

Fig. 6.5 Alteration of discharge as a result of upstream damming in the Volga River (adapted from Górski et al. 2012)

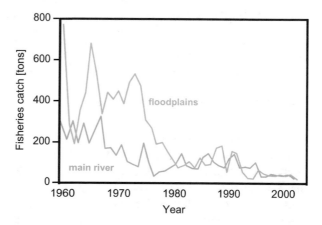

Fig. 6.6 Post-damming commercial fisheries catch decline in the Volga River (adapted from Górski et al. 2012)

In 1960, the Volgograd hydropower dam, the last dam of the Volga–Kama cascade, was completed. After damming, annual maximum peak discharges have decreased, minimum discharges increased, but average discharges remained similar to pre-damming conditions (Fig. 6.5). Moreover, because of riverbed incision of over 1.5 m, a higher discharge is needed to reach bank-full level and to inundate floodplains, which is the largest floodplain in Europe (length 300 km, width 20 km). Commercial fish catches severely decreased after damming, both in the main channel and in the floodplain lakes (Fig. 6.6).

If the continuity of sediment transport is interrupted by dams or removal of sediment from the channel by gravel mining, the flow may become sediment starved (hungry water) and prone to erode the channel bed and banks, producing channel incision (downcutting), coarsening of bed material, and loss of habitat for litophilic species (Kondolf 1997; Fig. 6.7). Riverbed incision reduces the connectivity to floodplain habitats. Together with reduced flood flows (e.g., due to storage) the dimension and quality of floodplain habitats is reduced, affecting the productivity of the entire river-floodplain system. Also, further downstream, lack of sediment may cause habitat degradation due to erosion of river deltas or coastal shores.

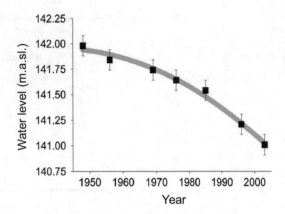

Fig. 6.7 Riverbed incision of the Danube River downstream of the Austrian hydropower cascade in the years 1950–2003 expressed as low water level (*m.a.sl.* meter above sea level) at river 1894.7 km (gauging station Wildungsmauer, Reckendorfer et al. 2005) (© Reckendorfer, W. et al. 2005. The Integrated River Engineering Project for the free-flowing Danube in the Austrian Alluvial Zone National Park: contradictory goals and mutual solutions Archiv für Hydrobiologie, Supplementband "Large Rivers", 155: 613–630, www.schweizerbart.de/series/archiv_Suppl, reproduced with permission from Schweizerbart'sche Verlagsbuchhandlung)

6.4 Other Downstream Impacts

Downstream segments are also subjected to other impacts related to dam operation and water quality of released water. The operation of hydroelectric impoundments tends to follow demands for electricity, creating variable flow regimes. Such irregular discharges called hydropeaking intensify erosive processes downstream and can caused drift and stranding of fish and macroinvertebrates (see Chap. 5).

Downstream release of poor-quality water by turbines and spillways also creates unfavorable conditions, e.g., anoxic hypolimnic water and altered water temperatures. For example, below the Xinanjiang and Danjiangkou dams, spawning of fish was delayed 20–60 days by lower water temperatures (Zhong and Power 1996). Spillflow at high dams may cause oversaturation of oxygen creating the so-called gas bubble disease in fish.

6.5 Mitigation Measures

Reservoirs impose system shifts on running waters, making restoration in the sense of reestablishing pre-damming conditions impossible. Hence, any attempt to improve the ecological condition can be regarded as a mitigation effort to reduce but not to remove the impacts. This is definitely true for the reservoir itself where mitigation measures have to take into account the new boundary conditions of the lacustrine environment. Therefore, mitigation measures mainly focus on the habitat

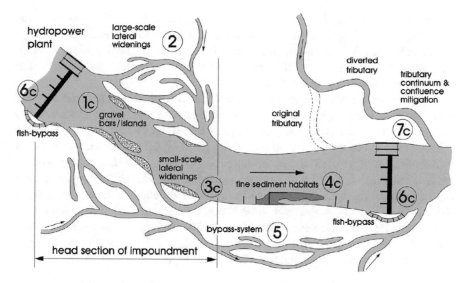

Fig. 6.8 Mitigation measures proposed for Danube reservoirs for improving connectivity and habitat (Jungwirth et al. 2005) (© Jungwirth, M et al. (2005). Leitbild-specific measures for the rehabilitation of the heavily modified Austrian Danube River. Archiv für Hydrobiologie, Supplementband "Large Rivers", 155: 17–36, www.schweizerbart.de/series/archiv_Suppl, reproduced with permission from Schweizerbart'sche Verlagsbuchhandlung)

improvements of the riverine section of reservoirs (head section) and on river sections located up- and downstream of the reservoir (Fig. 6.8). Focus is given to the reestablishment of longitudinal continuity by enabling fish migration through fish passes (see Chap. 9). Downstream mitigation measures involve environmental flow regulations (see Chaps. 4 and 5) and sediment transport by targeted sediment management (see Chap. 8).

6.5.1 Reestablishing Longitudinal Continuity

A common restoration measure for dams is the implementation of fish passes to enable upstream fish passage (for more details on fish passage, see Chap. 9). While fish passes have proven to be effective to pass fish across dams when constructed according to the requirements of migrating species, their role in effectively maintaining populations in dam cascades is still unclear. Even highly efficient fish passes may not be able to pass enough fish upstream when fish have to negotiate multiple dams.

Some studies suggest that ladders are problematic in fish conservation as they lead fish into ecological traps (Pelicice and Agostinho 2008; Pelicice et al. 2015). Migratory fishes travel long distances during the reproductive season in search of habitats suitable for spawning and the development of young. The movement is mostly upstream and, in case of dams, often supported by fish passes. After spawning adults migrate back to their downstream habitats in main rivers. Eggs and/or larvae are then carried downstream by currents but are not further propagated downstream of reservoirs. This leads to lack of recruitment for riverine fish populations downstream of the dams.

Reservoirs themselves often provide unsuitable habitats for juvenile fish and finally represent ecological traps. Four conditions are required to characterize a fish passage/reservoir system as an ecological trap (Pelicice and Agostinho 2008): (1) attractive forces leading fish to ascend the passage; (2) unidirectional migratory movements (upstream); (3) the environment above the passage has poor conditions for fish recruitment, e.g., the absence of spawning grounds and nursery areas; and (4) the environment below the passage has a proper structure for recruitment. When these conditions exist, individuals move to poor-quality habitats, fitness is reduced, and populations are threatened. Based on current and proposed river regulation scenarios, it is concluded that conservation of migratory fish will be much more complicated than previously believed (Pelicice et al. 2015).

6.5.2 Sediment Management

The overarching goal of sediment management should be to make dams transparent to sediment transport as much as possible. Management options include sediment flushing, sediment bypass, and sediment augmentation downstream of reservoirs (Kondolf et al. 2014). Some general guidance relating the size, water inflows and sediment inflows, and applicable mitigation measures was developed by Basson and Rooseboom (1997), who identified a relationship between the capacity of reservoirs and the mean annual water and sediment inflows and appropriate mitigation measures (Fig. 6.9):

Sediment sluicing: The aim of sediment sluicing is to maintain sediment in suspension and move it through the impoundment prior to deposition. Sediment sluicing typically involves a reduction in the water level in the impoundment by opening gates when sediment concentrations are elevated. *Turbidity venting* is similar to sediment sluicing but uses low level gates or deep sluices to enable sediment laden water to "flow" along the bottom of the reservoir to the toe of the dam.

For *sediment flushing*, reservoir levels are reduced to pre-impoundment levels, enabling the "river" to erode deposited sediments. At least twice the mean annual flow is required.

Bypass structures, whether they are tunnels, constructed canals, or existing river channels, can be used to pass high sediment-bearing water and bedload around an impoundment, thus decreasing the trapping of sediment. An advantage is that the seasonality of sediment delivery to the downstream river is maintained (e.g., Lake Miwa, Japan).

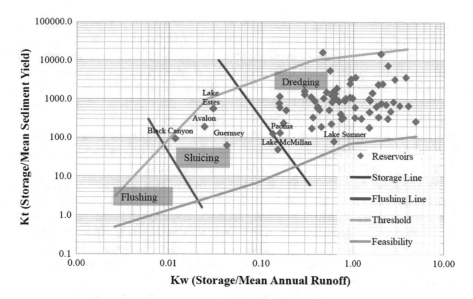

Fig. 6.9 Reservoir management options depending on storage capacity, mean annual runoff, and mean annual sediment load (based on Basson and Rooseboom 1997 and adapted from Kimbrel et al. 2014). Kw is the ratio between reservoir storage capacity and mean annual runoff, and a ratio of >1 means that the reservoir can store more than the mean annual runoff. Kt is the ratio between reservoir storage and mean annual sediment yield. A ratio of 100 means that the mean annual sediment yield can be stored over 100 years until the reservoir is filled (without flushing)

Reintroduction of dredged material: Approaches for reintroducing excavated or recovered material back into the downstream channel are implemented at Danube River downstream of Vienna, Austria, and Rhine River at Iffezheim, Germany.

Most reservoirs have a capacity mean annual flow ratio of between 0.2 and 3 and a life span of 50–2000 years when considering reservoir sedimentation. When the capacity mean annual flow ratio is less than 0.03, sediment sluicing or flushing should be carried out during floods and through large bottom outlets, preferably with free outflow conditions. Flushing is a sustainable operation and a long-term equilibrium storage capacity can be reached. Seasonal flushing for ca. 2 months per year could be used in regions where the hydrology is less variable with capacity mean annual flow ratio up to 0.2. When capacity mean annual flow ratios are, however, larger than 0.2, not enough excess water is available for flushing, and the typical operational model is storage operation. Density current venting can be practiced at these reservoirs as well as dredging to recover lost storage capacity (Basson 2004).

Reservoir flushing is an important mitigation measure for sediment remobilization and, thus, the restoration of natural sediment dynamics, including the formation of type-specific habitats. Nevertheless, reservoir flushing is also associated with immediate negative effects on physicochemical conditions, e.g., turbidity, oxygen deficiency, and hydropeaking, impacting fish directly, e.g., increased drift, gill, and skin injuries, stress, and fish kills, and indirectly, e.g., reduced food supply caused by increased drift and loss of benthic invertebrates, reduced growth, and lost habitats due to sedimentation (Henley et al. 2000; Crosa et al. 2009; Kemp et al. 2011; Jones et al. 2012).

Aquatic organisms have evolved with the dynamics of natural levels of total suspended solids (TSS). The natural level of TSS highly depends on the geo-hydromorphological conditions in the catchment and can be highly variable. The highest natural TSS peaks occur during floods and may affect fish. However, healthy ecosystems are usually able to compensate these effects in the long run. Reservoir flushing can cause TSS concentrations much higher than the natural background concentration and can result, depending on concentration level and duration, in stress or complete elimination of the fish stock.

Newcombe and Jensen (1996) classified the effects of reservoir flushing as (1) lethal effect (high-to-low mortality, high-to-medium habitat degradation), (2) lethal and para-lethal effects (high predatory pressure, prolonged hatching of larvae), (3) sublethal effects (reduction of growth, fitness and feeding, disturbed homing effect, physiologic stress, elevated breath frequency), and (4) behavioral effects (emigration, active/passive drift). While behavioral effects are mainly reversible and limited to the duration of exposure, physiologic changes have a more chronic character.

The intensity of impacts depends mainly on the concentration and duration of exposure, but also the size and texture of particles, water temperature, and chemical and physical conditions. Furthermore, toxic substances, acclimatization, and other stressors and their interaction are considered as relevant. For example, the release and decay of organic matter and resulting oxygen depletion may lead to suffocation of fish and benthic invertebrates. The "ranked effects model" is a tool for quantifying negative effects of suspended solids on fish (Newcombe and Jensen 1996). On the basis of duration (h) and concentration (mg/l) of exposure, a so-called severity of ill index (SEV) is calculated whereby several models are used depending on the species and age class. The resulting index ranges from 0 (no changes in behavior) to 14 (80–100% mortality).

6.5.3   Habitat Improvements in Reservoirs

Mitigation measures generally comprise "instream structures" such as gravel bars, islands, etc., "lateral widenings" of the cross profiles in riverine sections of impoundments, creating artificial habitats in lacustrine section, and "bypass systems" within the alluvial floodplains (Fig. 6.8, Jungwirth et al. 2005).

Due to raised water tables and sedimentation in the impoundments, the drastic loss of originally typical instream structures (gravel bars and islands) strongly affects reproduction and young-of-the-year habitats of the rheophilic fish community. Since the head sections of reservoirs still offer relatively high hydromorphological dynamics, river-type-specific rehabilitation can most likely be achieved here. The best way to recreate near-natural gravel bars and islands is to induce natural formation by lateral widenings of the cross profiles. Further rehabilitation measures comprise removing the embankments, reconnecting the former floodplains including abandoned arms, or forming new side arms (Fig. 6.8).

Fig. 6.10 Example of a bypass system, a near-natural channel to circumvent the impoundment, implemented at hydropower station Ottensheim-Wilhering, Danube, Austria (15 km long, discharge 2.5–20 m³/s, adapted after http://www.life-netzwerk-donau.at, accessed 1.10.2016)

The lacustrine sections of reservoir sections are ecologically heavily degraded due to the loss of fluvial dynamics and intensive sedimentation of fine substrate. Establishing new gravel habitats would soon fail because of strong aggradation with fine sediments during high flow periods. Nevertheless, the construction of artificial, stabilized silt or sand islands at hydraulically appropriate zones along the embankments can provide valuable habitats for various species (Fig. 6.8, for more details for habitat improvements in reservoirs, see Chap. 24).

Bypass systems are designed to connect adjoining impoundments within alluvial floodplains and represent innovative solutions for enhancing longitudinal and lateral connectivity of the fragmented environment as well as for substituting lost fluvial habitat. Depending on the local situation and the ecological objectives, the bypass systems can be established by connecting existing floodplain water bodies or by constructing new artificial channels. The flow of the bypass system should mimic pre-damming flow hydrographs to enable dynamic hydromorphological processes (Fig. 6.10).

References

Agostinho AA, Pelicice FM, Gomes LC (2008) Dams and the fish fauna of the Neotropical region: impacts and management related to diversity and fisheries. Braz J Biol 68:1119–1132

Basson G (2004) Hydropower dams and fluvial morphological impacts – an African perspective. The 10th United Nations Symposium on Hydropower and Sustainable Development. Beijing, pp 1–16

Basson G, Rooseboom A (1997) Dealing with reservoir sedimentation. Water Research Commission, Pretoria

Baxter RM (1977) Environmental effects of dams and impoundments. Annu Rev Ecol Syst 8:255–283

Crosa G, Castelli E, Gentili G, Espa P (2009) Effects of suspended sediments from reservoir flushing on fish and macroinvertebrates in an alpine stream. Aquat Sci 72:85–95

Górski K, van den Bosch LV, van de Wolfshaar KE, Middelkoop H, Nagelkerke LAJ, Filippov OV, Zolotarev DV, Yakovlev SV, Minin AE, Winter HV, De Leeuw JJ, Buijse AD, Verreth JAJ (2012) Post-damming flow regime development in a large lowland river (Volga, Russian Federation): implications for floodplain inundation and fisheries. River Res Appl 28:1121–1134

Grill G, Lehner B, Lumsdon AE, MacDonald GK, Zarfl C, Reidy Liermann C (2015) An index-based framework for assessing patterns and trends in river fragmentation and flow regulation by global dams at multiple scales. Environ Res Lett 10:15001

Henley WF, Patterson MA, Neves RJ, Lemly AD (2000) Effects of sedimentation and turbidity on lotic food webs: a concise review for natural resource managers. Rev Fish Sci 8:125–139

Herzig A (1984) Zur Limnologie von Laufstauen alpiner Flüsse – Die Donau in Österreich. Osterreichische. Wasserwirtschaft 36:95–103

Jones JI, Murphy JFJ, Collins AL, Sear DA, Naden PS, Armitage PD (2012) The impact of fine sediment on Macro-Invertebrates. River Res Appl 28:1055–1071

Jungwirth M, Haidvogl G, Moog O, Muhar S, Schmutz S (2003) Angewandte Fischökologie an Fließgewässern. Facultas Universitätsverlag, Wien, p 547

Jungwirth M, Haidvogl G, Hohensinner S, Muahr S, Schmutz S, Waidbacher H (2005) Leitbild-specific measures for the rehabilitation of the heavily modified Austrian Danube River. Arch Hydrobiol 155:17–36

Jungwirth M, Moog O, Schmutz S (2006) Auswirkung der Stauregelung großer Flüsse auf die aquatische Tierwelt (Fische und Makro-zoobenthos). Limnologie Aktuell 12:79–98

Junk WJ, Bayley PB, Sparks RE (1989) The flood pulse concept in river-floodplain systems. Can Spec Publ Fish Aquat Sci 106:110–127

Kemp P, Sear D, Collins A, Naden P, Jones I (2011) The impacts of fine sediment on riverine fish. Hydrol Process 25:1800–1821

Kimbrel S, Collins K, Randle T (2014) Formulating guidelines for reservoir sustainability plans. Denver

Kimmel BL, Groeger AW (1984) Factors controlling primary production in lakes and reservoirs: a perspective. Lake Reservoir Manage 1:277–281

Koehnken L (2014) Discharge Sediment Monitoring Project (DSMP) 2009–2013 summary & analysis of results. Vientiane

Kondolf GM (1997) Hungry water: effects of dams and gravel mining on river channels. Environ Manag 21:533–551

Kondolf GM, Gao Y, Annandale GW, Morris GL, Jiang E, Zhang J, Cao Y, Carling P, Fu K, Guo Q, Hotchkiss R, Peteuil C, Sumi T, Wang H-W, Wang Z, Wei Z, Wu B, Wu C, Yang CT (2014) Sustainable sediment management in reservoirs and regulated rivers: experiences from five continents. Earths Future 2:256–280

Kummu M, Varis O (2007) Sediment-related impacts due to upstream reservoir trapping, the Lower Mekong River. Geomorphology 85:275–293

Lehner B, Döll P (2004) Development and validation of a global database of lakes, reservoirs and wetlands. J Hydrol 296:1–22

Lewis C (2013) China's great dam boom: a major assault on its rivers. http://e360.yale.edu/feature/chinas_great_dam_boom_an_assault_on_its_river_systems/2706/

Likens GE (2010) River ecosystem ecology: a global perspective. Academic Press, Amsterdam

Newcombe CC, Jensen JOTJ (1996) Channel suspended sediment and fisheries: a synthesis for quantitative assessment of risk and impact. N Am J Fish Manag 16:693–727

Nilsson C, Reidy CA, Dynesius M, Revenga C (2005) Fragmentation and flow regulation of the world's large river systems. Science 308:405–408

Ofenböck T, Graf W, Hartmann A, Huber T, Leitner P, Stubauer I, Moog O (2011) Abschätzung des ökologischen Zustandes von Stauen auf Basis von Milieufaktoren. Vienna

Pelicice FM, Agostinho AA (2008) Fish-passage facilities as ecological traps in large neotropical rivers. Conserv Biol 22:180–188

Pelicice FM, Pompeu PS, Agostinho AA (2015) Large reservoirs as ecological barriers to down-stream movements of Neotropical migratory fish. Fish Fish 16:697–715

Petts GE (1985) Impounded rivers – perspectives for ecological management. Wiley, Chichester

Poff NLR, Allan JD, Bain MB, Karr JR, Prestegaard KL, Richter BD, Sparks RE, Stromberg JC (1997) The natural flow regime; a paradigm for river conservation and restoration. Bioscience 47(11):769–784

Reckendorfer W, Schmalfuss R, Baumgartner C, Habersack H, Hohensinner S, Jungwirth M, Schiemer F (2005) The integrated river engineering project for the free-flowing Danube in the Austrian Alluvial Zone National Park: contradictory goals and mutual solutions. Arch Hydrobiol 155:613–630

Rueda F, Moreno-Ostos E, Armengol J (2006) The residence time of river water in reservoirs. Ecol Model 191:260–274

Straskraba M, Tundisi JG, Duncan A (1993) State-of-the-art of reservoir limnology and water quality management. In: Comparative reservoir limnology and water quality management. Kluwer Academic Publisher, Dordrecht, pp 213–288

Wagner B, Hauer C, Schoder A, Habersack H (2015) A review of hydropower in Austria: past, present and future development. Renew Sust Energ Rev Elsevier 50:304–314

Wang Q, Chen Y (2010) Status and outlook of China's free-carbon electricity. Renew Sust Energ Rev 14:1014–1025

Ward JV, Bretschko G, Brunke M, Danielopol D, Gibert J, Gonser T, Hildrew AG (1998) The boundaries of river systems: the metazoan perspective. Freshw Biol 40:531–569

WCD (2000) Dams and development. A new framework for decision making. Report. World Commission on Dams, London

Zarfl C, Lumsdon AE, Berlekamp J, Tydecks L, Tockner K (2014) A global boom in hydropower dam construction. Aquat Sci 77:161–170

Zhong Y, Power G (1996) Environmental impacts of hydroelectric projects on fish resources in China. Regul Rivers Res Manag 12:81–98

Chapter 7
Aquatic Habitat Modeling in Running Waters

Andreas Melcher, Christoph Hauer, and Bernhard Zeiringer

7.1 Introduction

The understanding behind managing and conserving the environment, including water resources, has an important role in worldwide development strategy. The high priority given to reestablishing and maintaining good ecological status is reflected in multiple national legislations in Europe as well as in the EU Water Framework Directive (WFD). However, despite these emerging institutional protections, water withdrawal and, among other economic uses, continue to claim large fractions of the goods and services provided by aquatic ecosystems in the world's river basins. Consequently, much research and experimentation is needed to reestablish the ecological integrity of aquatic ecosystems, their habitats, and flow conditions.

A. Melcher (✉)
Institute of Hydrobiology and Aquatic Ecosystem Management, University of Natural Resources and Life Sciences, Vienna, Austria

Centre for Development Research, University of Natural Resources and Life Sciences, Vienna, Austria
e-mail: andreas.melcher@boku.ac.at

C. Hauer
Christian Doppler Laboratory for Sediment Research and Management, Institute of Water Management, Hydrology and Hydraulic Engineering, University of Natural Resources and Life Sciences, Vienna, Austria
e-mail: christoph.hauer@boku.ac.at

B. Zeiringer
Institute of Hydrobiology and Aquatic Ecosystem Management, University of Natural Resources and Life Sciences, Vienna, Austria
e-mail: bernhard.zeiringer@boku.ac.at

© The Author(s) 2018
S. Schmutz, J. Sendzimir (eds.), *Riverine Ecosystem Management*, Aquatic Ecology Series 8, https://doi.org/10.1007/978-3-319-73250-3_7

What Is Habitat?

Habitat is where aquatic organisms prefer to live or the living characteristics of a river that aquatic organisms are using. Although habitat is fundamentally a description of what animals use and where animals are found, most ecologists assume that habitat also is what animals need to survive and reproduce. Field experiments sensu "habitat modeling" give the most reliable data what animals need, and ecologists regularly engage in discussions about best available concepts, scales, and whether our habitat studies are properly designed and interpreted. However, in this chapter we assume that habitat is the part of a river that fish or benthic invertebrates and their life stages prefer for a successful survival and reproduction.

Habitat modeling can contribute to meeting the ongoing challenge of wisely balancing demands for the environmental services between society and nature (Bain 1995). This is especially so for those environmental services that sustain the integrity of ecosystems, e.g., *environmental flows* (e-flow). Habitat modeling offers a tool to apply e-flow concepts for science research and management policy. The concept of e-flows is used to mitigate the impacts of altered flow regime, often by assigning compensation flow releases to maintain ecological integrity and a good ecological status (see Chap. 4).

Ideally, attempts to establish or maintain environmental flow regimes will take into consideration the quantity, timing, duration, frequency, and quality of water flows needed to maintain ecosystems and the services they sustain. Prescriptions to reestablish ecologically suitable compensation flows can be based on hydrological metrics, e.g., percentage of average flow and/or hydraulic habitat algorithms. The latter link hydraulic descriptions of rivers with "preference" models of fish life stage responses to microhabitat hydraulics (Linnansaari et al. 2012). There is a growing consensus to combine these approaches, because hydrological metrics characterize temporal variations in the aquatic environment but are poorly suited to analyze spatial variations, whereas the opposite is true for hydraulic habitat models (e.g., Poff 2009; Poff and Zimmerman 2010). Furthermore, approaches have been proposed to model the ecological effects of flow regime on population processes (e.g., growth and survival; Armstrong and Nislow 2012) and dynamics rather than time-averaged population abundance (Shenton et al. 2012).

Hydrologically based methods are still the most widely used approaches internationally (Tharme 2003). This is probably due to their ease of use and low cost, since such methods use only "stream real" or simulated flow data series. A naturally variable regime of flow, rather than just a minimum low flow, is required to sustain freshwater ecosystems (Poff et al. 1997; Bunn and Arthington 2002; Poff 2009), and this understanding has contributed to the implementation of environmental flow management on thousands of river kilometers worldwide (Lobb and Orth 1991; Linnansaari et al. 2012).

The flow regime is regarded by many aquatic ecologists to be a key driver of ecological processes that sustain the integrity of river ecosystems. Flow is a major determinant of the parameters that constitute physical habitat in streams, which in turn, is a major determinant of biotic composition. Consequently, flow dynamics play an important role for aquatic organisms (see Chap. 4). Aquatic organisms have evolved life history strategies primarily in direct response to natural flow regimes (Schmutz et al. 2000; Bunn and Arthington 2002). River discharge typically varies significantly during the annual cycle, depending on climate and catchment conditions. Aside from natural phenomena, long- and short-term flow fluctuations can be altered by human activities. Therefore, flow/habitat relationships have to be established for all relevant species and life stages in order to cover the entire variability of responses to the natural flow.

7.2 Principles of Habitat Modeling

Habitat models allow one to assess the quality and quantity of habitat for a species within the study area or a river reach and provide the basic information required for environmental (flow) assessment. Aquatic habitat suitability models relate suitability to individual maps that are divided into uniform, spatially discrete units, e.g., rasters. These maps are digitally stored as raster-based layers, wherein each raster contains data, such as abiotic topographic descriptors. Current methods assume that the hydraulic measure is directly or indirectly related to habitat quantity for a target species, almost exclusively fish (e.g., Bovee 1982; Reiser et al. 1989), or in some instances the ecological function of the river (e.g., Gippel and Stewardson 1998).

All habitat modeling approaches depend on spatial scales and incorporate biological data based on standardized sampling methods for ecological assessment. These approaches use hydromorphological indicators for habitat assessment, which relies on correlative relations between habitat suitability for biota and hydrological features of river stretches on different scales (e.g., micro- or mesohabitat; Parasiewicz and Walker 2007).

The primary components of the physical habitat in running waters are water depth, velocity, substrate size, and cover, and most habitat models for aquatic organisms are based on these parameters (e.g., IFIM; Bovee 1982). After Jowett (2003) habitat modeling can be generally subdivided into two main categories:

1. *Empirically based habitat suitability models* are based on a description of the abiotic environment that is subsequently linked to the biotic system of flora and fauna that are described based on the concept of their available habitat. Univariate or multivariate functions link abiotic characteristics to habitat suitability. Univariate functions consider individual parameters, while multivariate analysis takes into account the interaction of physical variables and determines species response to cumulative effects of a number of environmental characteristics.
2. *Process-based population or bioenergetic models* describe biological processes based on knowledge of species population dynamics and/or energy budgets for feeding, growth, or other functions. These models can either be linked to the results of a physical habitat model or be directly linked with data describing the physiographic environment. Bioenergetic models are a special type of biological process model where optimal fish or benthic invertebrate location is based on

energy budgets. These models compute how much energy a fish uses as a function of water velocity or turbulence and of food intake. Optimal locations for indicator species are denoted by budget excesses of energy intake over energy loss due to the current.

While a suite of different types in habitat simulation methods can be identified, the general approach to evaluate effects of flow on habitat quantity is the same across different habitat modeling methods. The objective is to establish a relationship between river discharge and, typically, the amount of wetted perimeter and/or the wetted usable area (WUA), and then use this relationship to identify a "critical threshold." Briefly, this means finding a discharge level below which a drastically increasing amount of river bed becomes unsuitable for biota or even dry. A typical application measures the response in hydraulic variables across a number of "representative" cross sections of the river channel over a range of different discharges (measured or simulated using a 1D hydrodynamic model).

In general, a number of state-of-the-art habitat models focusing at different scales (micro, meso, and macro) with various implementing statistics and modeling techniques are available for e-flow assessment. Mostly these are based on the principles of PHABSIM (physical habitat simulation) technique which is used currently all over the world (e.g., Fausch et al. 1988; Harby et al. 2004a, b).

The PHABSIM technique enables the quantitative prediction of suitable physical microhabitat in a river reach for chosen species and life stages under different river flow scenarios, similar to that are mesohabitat models (e.g., MesoHABSIM or MEM (Mesohabitat Evaluation Model); Parasiewicz 2001 and Hauer et al. 2008). Other alternative methods are available, but their predominant emphasis on hydrology does not support a comprehensive assessment of both the hydrological and morphological conditions (e.g., Hauer et al. 2011). A short overview of the implementation on micro- and mesohabitat scale is given in Sect. 3.

Consequently, such indicator-based habitat model consists of several integrative parts (Fig. 7.1) that are linked together:

1. *Biotic habitat modeling*: The aim is to model and assess the biological species occurrence with their physical environment. This includes sampling and analyses of habitat and the morphological characteristics: fish or benthic invertebrate ecological assessment, determination of standardized habitat use, and habitat preference curves for key indicator species and their live stages.

   (a) Standardized biotic sampling of species abundance:

      (i) Microscale: point abundance sampling (e.g., electrofishing, snorkeling)
      (ii) Mesoscale: mesohabitat sampling (e.g., electrofishing)

   (b) Hydromorphological parameter sampling across a range of discharges: water depth and flow velocity, substrate size, embeddedness and stability, cross-sectional geometry, slope, river type, topology, channel or bank stability, sinuosity, width-depth ratio, presence of barriers, land-use activity, geology-lithology, geomorphology, altitude, Froude number, etc.

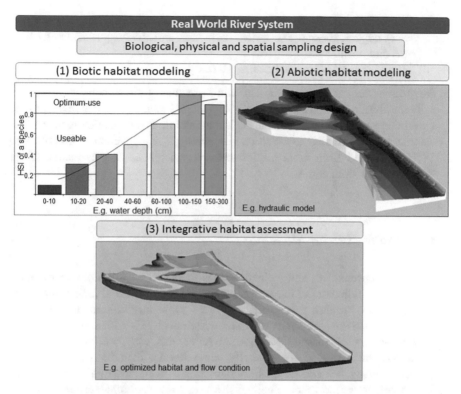

Fig. 7.1 A conceptual example for aquatic habitat modeling in rivers illustrating the process and main steps: (1) biotic and (2) abiotic habitat modeling which lead to (3) integrative habitat assessment. The colors in (1) and (3) show the optimum (green), the useable (brown and yellow), and not useable (red) water depth-based HSI (habitat suitability index) for a certain indicator species (e.g., adult European grayling)

(c) The most important factors are flow (discharge), flow velocity, water depth, substrate, and cover. Within each of these categorical parameters, several to many classes (factors) are included.

2. *Abiotic habitat modeling*: Physical factors, or hydraulic modeling, provide information of changes in the physical habitat as a function of discharge (hydraulic model). The objective is to quantify changes of the physical environment in relation to changes in flow or even morphological adjustments (natural or man-made). This includes physical and spatial measurement (sampling) and analyses of:

(a) *Hydrologic characteristics*: base flow, peak flow and duration, drought events, inter-annual variation of flow, flood and drought regime, spatial variation of discharge, longitudinal variation of cumulative water yield, seasonal variability in runoff, mean and maximum monthly water temperature, drainage area, stream order, branching degree, and distribution.

(b) *Hydraulic characteristics*: local flow velocity, mean cross-sectional velocity, water depths, shear stress, wetted perimeter, surface-subsurface lateral linkage, and turbulence.
(c) *Morphological characteristics*: see biotic part.

3. *Integrative habitat assessment*: The aim is to merge biotic habitat assessment with the abiotic flow and hydromorphological part and, as a management tool, to determine an adapted and suitable environmental flow for aquatic organism. The metrics are determined as reach-related averages, e.g., weighted useable area (WUA) derived from biota-specific habitat preferences and hydraulic 1D or 2D/3D model simulations.

7.2.1   Biotic Habitat Modeling

For multiple reasons, fish and their life stages have proven to be one of the most suitable impact indicators of human activities related to flow and habitat modifications. This is so, because fish populations are significantly affected by all human impact types on rivers, especially by water withdrawals. Fish identification is relatively easy and their taxonomy, ecological requirements, esp. for complex migration patterns and life histories, are generally better known than for other taxa. The longevity of many fish species enables assessments to be sensitive to disturbance over relatively longtime scales. Finally, fish are valuable economic resources and are of public concern. Using fish as indicators confers an easy and intuitive understanding of cause-effect relationships to stakeholders beyond the scientific community. However, also other indicator groups such as macroinvertebrates or macrophytes might be appropriate depending on the questions to be answered.

There are two common technical ways to build these models:

1. Literature review and expert opinion-based habitat suitability models.
2. Empirical and statistical techniques for estimating habitat suitability.

Literature Review and Expert Opinion-Based Habitat Suitability Models
A common habitat suitability modeling technique is based on literature review and expert opinion and generally follows the ideas established in 1980 by the US Fish and Wildlife Service publication "Habitat as a basis for Environmental Assessment." While literature-based models are subject to uncertainty and errors when transforming literature-based habitat studies to a specific river, they are relatively easy to create, because they do not require new collection of detailed field data and can be applied to multiple study areas and allow rapid analyses and modeling designs.

The procedure assigns a weight to each factor (parameter) and a habitat suitability score to each class within this factor. Suitability scores for all habitat factors are then combined to form a single habitat suitability map with a suitability score for each point on the sampling grid.

Combining habitat factors, the two most common methods of combining factors are arithmetic mean and geometric mean models. Further details on these models can be found in the Standards for Development of HSI Models section of the Habitat Evaluation Procedures Handbook.

Empirical and Statistical Techniques for Estimating Habitat Suitability
Sampling design

If presence-absence or species occurrence data are available in a study area, then empirical statistical models can be created by relating the species occurrence data to habitat factors. Sampling is the prerequisite for any related impact assessment and therefore takes a crucially important role for any modeling considerations. Generally, a distinction between (i) qualitative and (ii) quantitative sampling methods can be made. Depending on the scope or aim of a specific project and/or research hypothesis, it can be desirable to quantify the exact number of individuals, e.g., population number or density, in a certain area, or just to gain knowledge of the occurring species and their relative abundances (Bozeck and Rahel 1992). The main advantage of qualitative fishing is the reduced effort compared to that required for quantitative population estimates. In order to achieve the best possible results, several fish sampling methods can be applied and combined such that they are aligned to the methods selected for habitat modeling and e-flow assessments:

Fish data, obtained by *electrofishing,* can be used to assess ecological impacts and the sufficiency of e-flows. Standardized electric fishing procedures are described in detail in the European CEN Directive on Water Analysis—Fishing with Electricity (EN 14011; CEN 2003) for rivers. Fishing procedures and equipment differ depending upon the water depth and wetted width of the sampling site. Point abundance sampling by electrofishing (PASE) is a frequently used sampling method to define fish habitat; however, size selectivity and fish escapement patterns might be of concern (e.g., Persat and Copp 1988; Brosse et al. 1999).

Snorkeling is a prime method for underwater observation and study of fish in flowing waters. Snorkel surveys are widely used to monitor fish populations in streams and to estimate both relative and total abundance (Slaney and Martin 1987). Snorkeling can also be used to assess fish distribution, presence/absence surveys, species assemblages (i.e., diversity), some stock characteristics (e.g., fish length estimation), and habitat use. Snorkeling gear is worn by biologists who, individually or in small teams, survey fish abundance, distribution, size, and habitat use while slowly working in (generally) an upstream direction. This technique is most commonly used to survey juvenile salmonid populations but can also be used to assess other species groups. Snorkel survey programs have been designed and implemented so as to standardize procedures for underwater techniques to survey fish species in streams (Thurow 1994; Greenberg et al. 1996; O'Neal 2007).

Visual observation is an appropriate method to conduct daily surveys of fish species' presence, number of individuals, and habitat size, e.g., in their spawning habitat. Very clear water and shallow habitats are required to count spawning individuals by visual observation. Habitat features, i.e., flow velocity, water depth,

Fig. 7.2 Description of a sampling design on microhabitat scale, using transects to measure hydromorphological parameters at each habitat where fish have been observed (grey dots) and remaining habitat with no fish habitat observed (white dots) (after Schneider et al. 2008 and Melcher et al. 2012)

shading, cover, flow protection, type of structure, substrate, and embeddedness, are recorded at spawning grounds.

As an example, Melcher and Schmutz (2010) monitored spawning habitats of 1250 nase (*Chondrostoma nasus*) in the river Pielach, Lower Austria. Spawning took place in April and fish spawned in shoals on shallow gravel bars that are easy to identify from the river bank. A grid of equally spaced points was laid over the spawning area (grid size 1 m^2; see Fig. 7.2). Additionally, representative sites were sampled with different morphological characteristics within the study area to describe the entire available habitat. Furthermore, point measurements were taken interspersed at 2 m intervals along transects, resulting in hundreds of microhabitat measurements as graphically explained in Fig. 7.2.

Statistical techniques

In general statistical techniques such as generalized linear or generalized additive models (e.g., logistic or Poisson regression), artificial neural networks, classification and regression trees (CARTs), and genetic algorithms can all be used to create a map of a species probability of occurrence at any point of interest (e.g., standardized sampling grid) in a river.

With these models, data observed at each site is typically extracted from a "habitat database" and assembled by occurrence hierarchy; analyzed with statistics packages such as R, SPSS, or SAS; and then fed back into the database to create a table and map storing each probability of occurrence of a certain species.

While empirical models are probably more accurate than rule-based or literature-review based models, they require gathering a good set of field observations for every species and life stage in the linkage area, which can require a considerable amount of resources.

All habitat approaches have a fundamental, sometimes untested, assumption that, e.g., fish species make decisions about how to move along a river using the same rules they use to select habitat. It is reasonable to assume that a species prefers to move through areas that provide food, sufficient water, cover, and reproductive opportunities. But it is important to admit that we never know for sure, e.g., if reproductive individuals were trapped in a river reach by dams and their presence implies that they breed there. Only a small fraction of papers on movement describe the type of movement we are most interested in, namely, why, how, and when animals move between patches of suitable and unsuitable habitat.

Univariate suitability functions

Biological habitat models describe deterministic relations that link biological responses to physical habitat. The models interpret the species presence or abundance in areas with particular characteristics (e.g., depth, velocity) as the measure of the habitat's suitability for any given species. Originally each characteristic was analyzed individually, and algorithms selected *a priori* were used to account for this information. In these univariate suitability functions, the suitability of a habitat is a function of one variable characterizing one physical characteristic of the habitat. Usually, the function gets values between 0 and 1, so that for the least suitable conditions the function has the value 0, and at the most suitable conditions the value 1 (Fig. 7.1).

Three different types of habitat suitability indices are distinguished after Bovee (1986):

Category I indices are based on information other than field observations made specifically for the purpose of suitability index development. They can be derived from life history studies in the literature or from professional judgment. This latter case may involve round table discussions, the Delphi technique (which overcomes some disadvantages of traditional committee meetings), or hybrid techniques such as "habitat recognition," where the experts are taken to a stream and asked to assess the suitability of various habitats.

Category II indices use data collected specifically for habitat studies, based on frequency analysis of the actual habitat conditions used by different species and life stages in a stream. Location of target species may be by one of a number of methods—direct observation (from the bank, snorkeling, or scuba) video, telemetry, trapping/physical capture, or electric fishing. Location of target species is accompanied by measurement of the relevant physical habitat variables at the point of observation.

Category III data combine a category II frequency analysis with additional information on the availability of habitat combinations in the sampling reaches. It has been suggested that this methodology can correct for bias caused by habitat availability in the source stream(s) and thus make indices more generic. It is clear that calculation of preference cannot take into account habitats that are not present in a stream and that occupancy/non-occupancy of low-availability habitat may significantly alter calculated preference. However, so long as care is taken not to undertake surveys on rivers with low physical diversity, calculation of preference provides the best way of removing the complicating problem of differing habitat availability between sites.

Univariate preference curves can be derived using a number of ratio formula or preference indices. The simplest is a ratio where preference (P) = use (U)/availability (A). The preference functions can be delimited to take account of time of day, seasonal, life history, and activity factors. When a physical habitat is described by more than one parameter, a combination of several preference curves has to be made. Several combinatory techniques can be chosen here, for example, to use the minimum value of each preference outcome or to use a mean or sum of all parameters or more elaborated statistical methods (see below).

Multidimensional statistical analyses for biological modeling

Numerous habitat modeling studies have been undertaken over time in North America and Europe, first mainly for salmonids (e.g., Northcote 1984; Shirvell 1989; Wollebaek et al. 2008; Moir and Pasternack 2010) but later for non-salmonids also (e.g., Melcher and Schmutz 2010). Predominantly these studies applied *univariate habitat use* and preference curves. In order to assess anthropogenic alterations on riverine systems, most attention was focused on morphological habitat attributes.

It was recognized that functional processes in riverine environments depend on the interactions of many factors, such as flow velocity and/or riparian vegetation (Melcher and Schmutz 2010). As a result, a more sophisticated analytical toolset is required to quantify the biological consequences of impacted multi-metric environments and to assess fish habitat improvements in river activities. Multidimensional analyses are needed to better identify and understand habitat requirements (Melcher et al. 2012). Until now parametric methods such as classical variance, regression, or discriminant function analyses (Ahmadi-Nedushan et al. 2006) have been the main statistical methods used for habitat modeling. Due to their specific statistical presumptions and requirements, their use is frequently limited in comparison to nonparametric methods (e.g., CHAID tree).

Logistic regression is a multiple regression model used in habitat modeling in the way that the probability of occurrence is regressed against a number of potential habitat characteristics. It requires field observations of habitat characteristics available and utilized by indicator species. Habitat choice is described by the probability of a specific choice occurring along a habitat gradient. Using the stepwise procedure, all significant parameters to describe the habitat are listed (Melcher et al. 2012). The result can be a map that shows the probability of fish occurrence for each location, each area, or each computational grid cell. The probability of occurrence can be converted to an HSI (habitat suitability index score).

Classification trees, often referred to as decision trees, predict the value of a discrete dependent variable with a finite set of values (called classes) from the values of a set of independent variables (called attributes), which may be either continuous or discrete (Breiman et al. 1984; Quinlan 1986). Data describing a real system, represented in the form of a table, can be used to "learn" or automatically construct a decision tree. Decision trees thus constitute a multivariate statistical method of exploration and data analysis by classification. This approach can be applied to predict the presence/absence of fish species from habitat characteristics described by a set of independent variables. The habitat modeling CHAID tree method describes specific fish habitat on microhabitat to reach scales. The method allows one to highlight the importance of and interactions among hydromorphological parameters, e.g., flow velocity or substrate for typical fish habitats (Melcher et al. 2012).

Fuzzy rule-based preference functions

Another approach to evaluate habitat quality is fuzzy rule-based modeling (Jorde et al. 2001; Schneider et al. 2001). Fuzzy modeling allows working with imprecise or "fuzzy" information. This comes with the significant advantage that expert knowledge readily available from experienced fish biologists and supported by local investigations (electrofishing, observation) can easily be transferred into preference data sets by setting up checklists. These lists or so-called fuzzy rule systems (e.g., CASiMiR (computer- aided simulation model for instream flow and riparian), http://www.casimir-software.de) offer a range of possible combinations of relevant physical criteria and let experts define if habitat quality is good or low.

7.2.2   Abiotic Habitat Modeling

All habitat modeling techniques require some information about hydraulic characteristics. The most commonly used method is the "wetted perimeter method" that predicts wetted area of a cross section as a function of discharge at a location (one point) in the river (Tharme 2003). Hydraulic factors may come directly from measurements or from hydraulic models and hydraulic assessment methods (e.g., Harby et al. 2004a, b).

Direct measurements of hydraulic factors: By sampling several times over a range of flows, it is possible to construct an empirical relationship between physical conditions and discharge. Habitat suitability's are calculated for measured flow rates, and habitat suitability's for different discharges are derived by interpolating from the measured range of flow data. Such a method does not require the investigator to accept any underlying requirements or assumptions of a particular hydraulic modeling technique. The predictability of this model is limited to the range of measured discharges.

Hydraulic rating methods, which are also known as habitat retention or hydraulic geometry methods (Tharme 2003), are based on a relationship between some hydraulic parameters of a river (usually wetted perimeter or depth) and discharge (e.g., Jowett 1993, 1997). Leopold and Maddock (1953) described simple power functions that can be used in describing changes in hydraulic variables as a function of discharge. The constants and the exponents in these equations should be empirically developed for each river or region, as the general form of river channels is variable (Linnansaari et al. 2012).

An alternative approach is to use statistically based models of river hydraulics. While empirical models of variation in broad morphometric channel variables have been in existence for many years, it is only recently that these statistically based techniques have been applied to model habitat hydraulics at the reach or "mesohabitat" scale. It has been suggested that at the reach scale, statistical hydraulic models can provide estimates of the frequency distribution of hydraulic variables, when given simple inputs such as mean river velocity, depth, and width (Lamouroux et al. 1995, 1999). Statistical techniques have shown that consistent patterns of such distributions appear among different streams. Based on power laws or multiple measurements, both depth-discharge and width-discharge relationships can be obtained, linking discharge to existing hydraulic distribution patterns. This method requires a wide range of input data from different streams in various catchment areas. Once a "library" of occurring patterns is established, the effort necessary for obtaining depth/width-discharge relations is relatively low. It should be also noted that current models are most suited to rivers with relative natural morphology. Their value lies in their ability to analyze broad trends in habitat hydraulics, rather than the specific description of a particular reach (Linnansaari et al. 2012).

Additionally, *hydrodynamic-numerical models* have been used in habitat modeling for many years. Hydrodynamic-numerical modeling strongly relies on using a range of river stretches with catchment typical hydromorphological characteristics (hydrology, bed substratum, bed structures, degree of braiding, sinuosity of the river course, mean bed width, and bed slope). As a result, a set of model equations enables the simulation of fish habitat conditions in river stretches as a function of flow and morphology. The habitat suitability of selected river sections is assessed mainly in terms of the needs of the life stages of important indicator fish species.

Flow calculations based on the conservation of mass, momentum, and energy provide the foundations of hydrodynamic-numerical modeling. These calculations can be generalized as nonlinear, partial differential equations, e.g., Navier–Stokes, which can be solved in the general case only by approximation with numerical methods. Navier–Stokes equations include no simplifications. That means, if no errors are introduced by the numerical solution, complex flow phenomena can even be correctly calculated to the last detail.

The Navier–Stokes and Reynolds equations can be simplified in various ways, namely, by reducing the dimensionality or through neglect or simplification of terms of output equations. Especially for habitat modeling (e-flow studies), most applications allow a dimension to be neglected only if the flow components in the corresponding direction are negligible (e.g., cross distribution of flow velocity).

In computational fluid dynamics, high-resolution techniques are applicable, e.g., direct numerical solutions (DNS) (Lin and Liu 1998), Reynolds-averaged Navier–Stokes (RANS) (Sinha et al. 1998), large eddy simulation (LES) models (Wu 2004), and smoothed particle hydrodynamics (Monaghan and Kos 1999). Those numerical codes can be used for environmental flow assessments, such as detailed studies on the impacts of turbulence phenomena on macroinvertebrates. For reach-scale environmental flow assessments, however, computational time and the required modeling boundaries make this kind of analysis mostly impracticable.

According to the computed resolution of flow velocity, three-dimensional, two-dimensional, and one-dimensional hydrodynamic-numerical models can be applied for hydraulic and environmental flow studies. The simplest form of numerical modeling is the one-dimensional modelling approach with the assumption that there is no variability of the flow velocity in the vertical and lateral direction in the cross section.

These simplifying assumptions of a one-dimensional model represent often a limit to the applicability of such ecohydraulic studies. Local differences in flow velocity (especially in cross-sectional distribution) cannot be determined with the assumptions of one-dimensional, shallow water equations. To apply simplifications to the 2D-shallow water equations requires two assumptions: (i) the velocities in the x and y directions are taken into account and (ii) the speeds are averaged over the water depth. Two- and three-dimensional models are predictive models in a sense that they also require calibration and validation. An important new development in two- and three-dimensional modeling is the capacity of using a nested grid, i.e., different spatial resolutions in the same model application. This allows for using fine grids in ecologically important areas, while coarser grids can be used in areas that do not require such resolution (Olsen 2000).

Two-dimensional and three-dimensional hydraulic models apply the principles of conservation of mass and momentum on a spatial, computational grid. Two-dimensional horizontal models use depth-averaged velocity, where three-dimensional models have computational layers in the vertical dimension. These models also include empirical or stochastic representation of water turbulence. The model schematization is based on detailed topographic input from the study area in combination with data on bed roughness and boundary conditions for water level and discharge. Velocity and water level measurements are usually used for validation.

A key aspect in numerical (habitat) modeling is the modeler's own considerations of simplifications and assumptions concerning the physics involved. The (habitat) modeler has to decide which numerical approach fits best according to the requirements of the project to describe the abiotic environment. If the near-bottom velocity needs to be addressed (e.g., benthic habitats), a three-dimensional code would be required. If the cross section's variability is important to characterize river morphological characteristics on the meso-unit (mesohabitat scale), then depth-averaged two-dimensional modeling should be selected. For aspects of minimum flow depths in a longitudinal view of the river, the one-dimensional model could deliver the required information for various discharges (e.g., residual flow studies), just to name some aspects of needs for a decision. Here, expert knowledge of this decision-making process is a mandatory in pre-project steps.

7.2.3 Integrative Habitat Assessment

Once the (hydraulic) model has been calibrated and the species-habitat relationships have been established, the two separate components need to be combined into a composite flow-habitat relationship (sometimes referred to as habitat-discharge rating curves). Usually this is done by applying the weighted usable area (WUA) (Bovee 1982) concept used in the PHABSIM family of fish habitat models. The WUA is calculated as an aggregate of the product of a composite suitability index (CSI, range 0.0–1.0). The CSI is calculated as a combination of the separate suitability indices for every single physical parameter. The suitability index for each parameter is evaluated by linear interpolation from an appropriate preference curve to be supplied separately. Velocities and depths are usually taken directly from the hydrodynamic model, while substrate and cover derives from additional mapped data. For quantitative assessment of habitat suitability as a function of flow rate, hydraulic rating curves (flow vs. habitat relationship) for the different key species are generated.

A specific WUA can only be seen as an index because perceived physical area is multiplied by unit-less habitat suitability attributes (Payne 2003). These attributes were originally termed "elective criteria" (Bovee and Cochnauer 1977) under the assumption that species will elect to leave an area when conditions become unfavorable. Electivity is variously expressed as probability of use (or nonuse), preference, suitability, or utilization over the possible range of conditions. Electivity indices range between 0 and 1, have no units, and are most commonly derived from frequency analysis of field observations.

To combine multiple habitat factors into one aggregate habitat suitability model assessment, first it is useful to assign weights to each factor that reflect their relative importance. If a habitat factor is not important for a species, it is assigned a weight of 0%.

Weighting is one of the weakest parts of the models if lacking any underlying theory or hard data. One theoretical issue, for example, is this: When the scores are combined across factors, does the overall score still have the same biological interpretation we established when scoring suitability for each factor? Therefore, habitat assessment should be built on a model that uses weights based on empirical data.

Further criteria to consider are anthropogenic reductions of the mean flow velocity in the cross section along the river stretch; anthropogenic migration obstacles occurring in the natural fish habitat must be passable by fish all year long and stream bed stabilization (river-bottom sills, bank dynamics, and local protections) in context to open substrate und dynamics.

Physical habitat units on the mesoscale (hydromorphological units) are addressed at an intermediate level between microhabitats and reach-scale habitat characteristics and hence are most commonly termed as mesohabitats (Maddock 1999; see also case studies River Ybbs and MEM below). Although, various hydromorphological studies have yet to find consistent numbers of distinctly mesohabitat types for the aquatic environment (e.g., Bain and Knight 1996), variable descriptions of abiotic parameters exist to determine the different habitats on the mesoscale and have served mainly to distinguish between pool, riffle, and run habitats. For characterizing mesohabitats, the Froude number, the water surface slope, the range of water depth and velocities, and the bed material size have been used. Based on the importance of mesohabitats for instream studies and river restoration, various parameters have been developed and implemented as different modeling approaches for mesohabitat description and quantification (e.g., Parasiewicz 2001; Le Coarer 2005; Hauer et al. 2009).

7.3   Managing River Systems Through Habitat Assessment

The following sections describe examples of implementation of habitat modeling at micro- and mesohabitat spatial scales.

7.3.1   Case Study on Microhabitat Scale: E-Flow Study at River Ybbs, Austria

As part of an environmental flow study on the River Ybbs (Zeiringer et al. 2010), a microhabitat modeling approach was carried out as an integrative assessment method to identify ecologically reasonable minimum flows. For the appropriate ecological evaluation of the current situation, quantitative fish surveys (after Haunschmid et al. 2010) and hydromorphological measurements were carried out along the residual flow stretch as well as in unaffected stretches further upstream of the water abstraction inlet for the power plant (see Fig. 7.3). The fishing results formed the basis for deficit analysis and habitat modeling. Within the residual flow stretch, two sections were surveyed and mapped in order to enable numerical modeling of flow. These sections constitute a basic requirement for evaluating the habitat suitability depending on different flow rates. Further, the hydrological conditions along the 30 km long river section were quantified using several water level logger and stage-discharge relationships. Thus, precise residual flow along the river section could be derived over long periods.

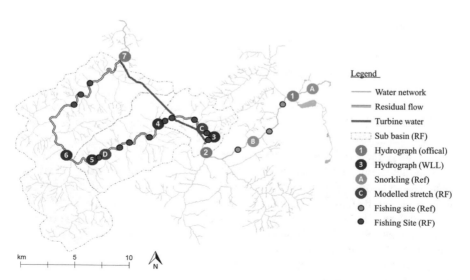

Fig. 7.3 Case study area and study sites are located at River Ybbs in Lower Austria. (WLL = water level logger, Ref = reference/fully discharged river stretch, RF = residual flow)

For the habitat modeling approach, the habitat requirements of the key species in this river section, e.g., brown trout (*Salmo trutta*) and European grayling (*Thymallus thymallus*), for different age stages (0+, 1+, 2+>) were defined. This was done using preference functions (univariate and multivariate, see Fig. 7.4, 2a, b, c) of the indicator species for different seasons (summer and fall) derived from unaffected river stretches further upstream (Fig. 7.3). Fish were observed via snorkeling in the fully discharged sections (Thurow 1994; Greenberg et al. 1996), and the abiotic characteristic (water depth, flow velocity, substrate, and cover) of used and available habitat depended on species and life stage were measured.

The hydraulics were modeled using the software River2D, which is a two-dimensional depth-averaged model of river hydrodynamics and can also be used for fish habitat modeling (Steffler and Blackburn 2002). The habitat suitability of the modeled river stretches was calculated by linking the physical (habitat descriptive) parameters with the habitat requirements of selected indicator species. The fish habitat component of River2D is based on the weighted usable area (WUA). For the quantitative assessment of habitat suitability as a function of flow rate, hydraulic rating curves (flow vs. habitat relationship) for the different key species were generated (Fig. 7.4). This was then combined with a historical flow time series to produce a physical habitat time series and hence a physical habitat duration curve (Maddock 1999).

7.3.2   Example at Mesohabitat Scale: Mesohabitat Evaluation Model (MEM)

The conceptual MEM model was developed and validated by Hauer et al. (2008, 2011) and allows evaluation of six different hydromorphological units (mesohabitats) according to their abiotic characteristics. Three abiotic parameters (flow velocity, water depth, and bottom shear stress) were incorporated into the MEM analysis. For practical purposes, the MEM concept was implemented using a Java software application, which enables MEM evaluation based on one of three different two-dimensional (CCHE2D, River2D, Hydro_AS-2D) and two different three-dimensional models (RSim-3D, SSIIM) (Tritthart et al. 2008).

Recently, the MEM approach was successfully applied for the evaluation of various anthropogenic pressures and the habitat quality assessment of restored river sites in Austria. For example, hydropeaking impact studies (unsteady dynamics in mesohabitat patterns) are significant recent applications. Here, the MEM results were used as a fundamental basis for the discussion of future mitigation measure design (Hauer et al. 2013). Moreover, the MEM concept was used to evaluate the habitat quality of larger river systems like the Drava (Drau) river or the Danube (Fig. 7.5). Here, the distribution of mesohabitats could be linked to the presence of fish guilds, which enables habitat evaluation not only for single fish species, but for groups with similar preferences for the aquatic environment. Using the fish guild

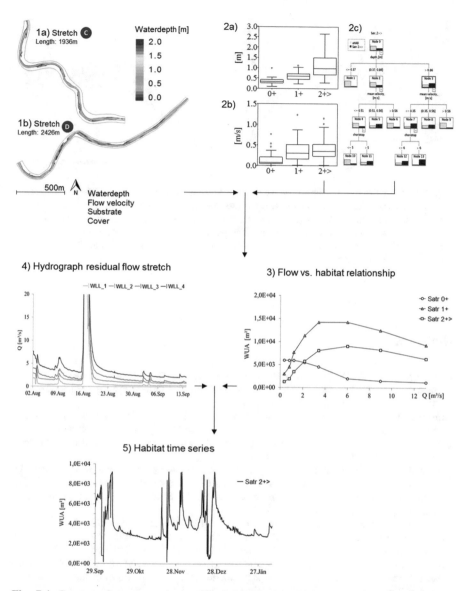

Fig. 7.4 Concept of a case study on microhabitat scale, which combines abiotic and biotic information (after Maddock 1999). The overall physical habitat simulation shows the integration of (1) hydraulic measurement and modeling and (2) biotic habitat suitability criteria to define the (3) flow versus habitat relationship, which is combined with (4) flow time series to produce (5) habitat time series (example of WUA for brown trout older than 2 years), whereas examples for brown trout life stages habitat use are given in 2a for water depth, in 2b for flow velocity, and in 2c as a CHAID tree method selecting specific fish habitat preferences

concept, the model evaluates habitat suitability for spawning, juveniles, subadult, and adult life stages of rheophilic, indifferent, and stagnophilic fish species (see for details: Hauer et al. 2011, 2014).

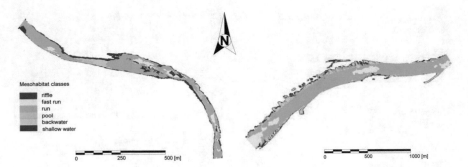

Fig. 7.5 Visualization of the mesohabitat evaluation model (MEM) featuring calibrated mesohabitats under low flow conditions for a restored section of the Drava River (left) and the regulated Danube east of Vienna (right) (Hauer et al. 2011) (reproduced from: Hauer et al. 2011. Variability of mesohabitat characteristics in riffle-pool reaches: Testing an integrative evaluation concept (FGC) for MEM-application. River Research and Applications 27(4):403–430, © 2010 John Wiley & Sons, Ltd.)

References

Ahmadi-Nedushan B, St-Hilaire A, Bérubé M, Robichaud É, Thiémonge N, Bobée B (2006) A review of statistical methods for the evaluation of aquatic habitat suitability for instream flow assessment. River Res Appl 22(5):503–523

Armstrong JD, Nislow KH (2012) Modelling approaches for relating effects of change in river flow to populations of Atlantic salmon and brown trout. Fish Manag Ecol 19(6):527–536

Bain MB (1995) Habitat at the local scale: multivariate patterns for stream fishes. Bulletin Francais de la Peche et la Pisciculture 337–339:165–177

Bain MB, Knight JG (1996) Classifying stream habitat using fish community analysis. In: Leclerc M, Capra H, Valentin S, Boudreault A, Cote Y (eds) Proceedings of second IAHR symposium on habitat hydraulics, Ecohydraulics 2000. Institute National de la Recherche Scientifique – Eau, Ste-Foy, pp 107–117

Bovee KD (1982) Instream flow methodology. US Fish Wildlife Serv. FWS/OBS 82:26

Bovee KD (1986) Development and evaluation of habitat suitability criteria for use in the instream flow incremental methodology. US Fish Wildlife Serv Biol Rep 86(7):1–235

Bovee KD, Cochnauer T (1977) Development and evaluation of weighted criteria, probability-of-use curves for instream flow assessments: fisheries (No. 3). Department of the Interior, Fish and Wildlife Service, Office of Biological Services, Western Energy and Land Use Team, Cooperative Instream Flow Service Group

Bozeck MA, Rahel FJ (1992) Generality of microhabitat suitability models for young Colorado cutthroat trout (Oncorhynchus clarkii pleuriticus) across sites and among years in Wyoming streams. Can J Fish Aquat Sci 49:552–564

Breiman L, Friedman J, Stone CJ, Olshen RA (1984) Classification and regression trees. CRC press, Florida

Brosse S, Guegan JF, Tourenq JN, Lek S (1999) The use of artificial neural networks to assess fish abundance and spatial occupancy in the littoral zone of a mesotrophic lake. Ecol Model 120(2):299–311

Bunn SE, Arthington AH (2002) Basic principles and ecological consequences of altered flow regimes for aquatic biodiversity. Environ Manag 30(4):492–507

CEN (2003) Water quality – sampling of fish with electricity. Document CEN/TC 230, Ref. No. EN 14011:2003 E, 16p

Fausch KD, Hawkes CL, Parsons MG (1988) Models that predict standing crop of stream fish from habitat variables: 1950–1985

Gippel CJ, Stewardson MJ (1998) Use of wetted perimeter in defining minimum environmental flows. Regul Rivers Res Manag 14(1):53–67

Greenberg L, Svendsen P, Harby A (1996) Availability of microhabitats and their use by Brown Trout (Salmon trutta) and grayling (Thymallus thymallus) in the river Vojman, Sweden. Regul Rivers Res Manag 12(2–3):287–303

Harby A, Halleraker JH, Sundt H, Alfredsen KT, Borsanyi P, Johnsen BO, Forseth T, Lund R, Ugedal O (2004a) Assessing habitat in rivers on a large scale by linking microhabitat data with mesohabitat mapping. Development and test in five Norwegian river. In: Proceedings of the fifth international symposium on ecohydraulics, pp 829–833

Harby A, Baptist M, Dunbar MJ, Schmutz S (eds) (2004b) State-of-the-art in data sampling, modelling analysis and applications of river habitat modelling: COST action 626 report. European Aquatic Modelling Network

Hauer C, Unfer G, Schmutz S, Habersack H (2008) Morphodynamic effects on the habitat of juvenile cyprinids (Chondrostoma nasus) in a restored Austrian lowland river. Environ Manag 42(2):279–296

Hauer C, Mandlburger G, Habersack H (2009) Hydraulically related hydro-morphological units: descriptions based on a new conceptual mesohabitat evaluation model (MEM) using LiDAR data as geometric input. River Res Appl 25:29–47

Hauer C, Unfer G, Tritthart M, Formann E, Habersack H (2011) Variability of mesohabitat characteristics in riffle-pool reaches: testing an integrative evaluation concept (FGC) for MEM-application. River Res Appl 27(4):403–430

Hauer C, Schober B, Habersack H (2013) Impact analysis of river morphology and roughness variability on hydropeaking based on numerical modelling. Hydrol Process 27(15):2209–2224

Hauer C, Mandlburger G, Schober B, Habersack H (2014) Morphologically related integrative management concept for reconnection abandoned channels based on airborne LiDAR data and habitat modelling. River Res Appl 30(5):537–556

Haunschmid R, Schotzko N, Petz-Glechner R, Honsig-Erlenburg W, Schmutz S, Unfer G, Bammer V, Hundritsch L, Sasano B, Prinz H (2010) Leitfaden zur Erhebung der biologischen Qualitätselemente, Teil A1–Fische, BMLFUW ISBN: 978-3-85174-059-2 Version Nr. - A1-01j_FIS. Wien

Jorde K, Schneider M, Peter A, Zoellner F (2001) Fuzzy based models for the evaluation of fish habitat quality and instream flow assessment. In: Proceedings of the 2001 international symposium on environmental hydraulics, vol 3, pp 27–28

Jowett IG (1993) A method for objectively identifying pool, run, and riffle habitats from physical measurement. N Z J Mar Freshw Res 27:241–248

Jowett IG (1997) Instream flow methods: a comparison of approaches. Regul Rivers Res Manag 13(2):115–127

Jowett IG (2003) Hydraulic constraints on habitat suitability for benthic invertebrates in gravel-bed rivers. River Res Appl 19(5–6):495–507

Lamouroux N, Souchon Y, Herouin E (1995) Predicting velocity frequency distributions in stream reaches. Water Resour Res 31(9):2367–2375

Lamouroux N, Capra H, Pouilly M, Souchon Y (1999) Fish habitat preferences in large streams of southern France. Freshw Biol 42(4):673–687

Le Coarer Y (2005) "Hydrosignature" software for hydraulic quantification. In: Harby A et al (eds) COST 626 –European aquatic modelling network. Proceedings from the final meeting in Silkeborg, Denmark, 19–20 May 2005. National Environment Research Institute, Silkeborg, pp 199–203

Leopold LB, Maddock T Jr (1953) The hydraulic geometry of stream channels and some physiographic implications. USGS Professional Paper No. 252, pp 1–57

Lin P, Liu PL-F (1998) A numerical study of breaking waves in the surf zone. J Fluid Mech 359:239–264

Linnansaari T, Monk WA, Baird DJ, Curry RA (2012) Review of approaches and methods to assess environmental flows across Canada and internationally. DFO Can Sci Advis Secr Res Doc 39:1–74

Lobb MD, Orth DJ (1991) Habitat use by an assemblage of fish in large warm water stream. Trans Am Fish Soc 120:65–78

Maddock I (1999) The importance of physical habitat assessment for evaluating river health. Freshw Biol 41:373–391

Melcher AH, Schmutz S (2010) The importance of structural features for spawning habitat of nase Chondrostoma nasus (L.) and barbel Barbus barbus (L.) in a pre-Alpine river. River Systems 19(1):33–42

Melcher AH, Lautsch E, Schmutz S (2012) Non-parametric methods–tree and P-CFA–for the ecological evaluation and assessment of suitable aquatic habitats: a contribution to fish psychology. Psychol Test Assess Model 54(3):293–306

Moir HJ, Pasternack GB (2010) Substrate requirements of spawning Chinook salmon (Oncorhynchus tshawytscha) are dependent on local channel hydraulics. River Res Appl 26(4):456–468

Monaghan JJ, Kos A (1999) Solitary waves on a Cretan beach. J Waterw Port Coast Ocean Eng 125(3):145–154

Northcote TG (1984) Mechanisms of fish migration in rivers. In: Mechanisms of migration in fishes. Springer, Boston, pp 317–355

O'Neal JS (2007) Snorkel surveys. In: Salmonid field protocols handbook: techniques for assessing status and trends in Salmon and Trout populations. The American Fisheries Society in Association with State of the Salmon, Bethesda, pp 325–340

Olsen NRB (2000) A three-dimensional numerical model of sediment movements in water intakes with multiblock option. Version 1.1 and 2.0 for OS/2 and Windows. User's manual. Trondheim, Norway

Parasiewicz P (2001) MesoHABSIM: a concept for application of instream flow models in river restoration planning. Fisheries 26:6–13

Parasiewicz P, Walker JD (2007) Comparison of MesoHABSIM with two microhabitat models (PHABSIM and HARPHA). River Res Appl 23(8):904–923

Payne TR (2003) The concept of weighted usable area as relative suitability index. IFIM users workshop 1–5 June 2003 Fort Collins, CO, 14p

Persat H, Copp GH (1988) Electrofishing and point abundance sampling for the ichthyology of large rivers. In: Cowx I (ed) Developments in electrofishing. Fishing New Books, Hull, pp 197–209

Poff NLR (2009) Managing for variability to sustain freshwater ecosystems. J Water Resour Plan Manag 135(1):1–4

Poff NLR, Zimmerman JK (2010) Ecological responses to altered flow regimes: a literature review to inform the science and management of environmental flows. Freshw Biol 55(1):194–205

Poff NLR, Allan JD, Bain MB, Karr JR, Prestegaard KL, Richter BD, Sparks RE, Stromberg JC (1997) The natural flow regime. Bioscience 47(11):769–784

Quinlan JR (1986) Induction of decision trees. Mach Learn 1(1):81–106

Reiser DW, Ramey MP, Wesche TA (1989) Flushing flows. In: Gore JA, Petts GE (eds) Alternatives in regulated river management, pp 91–135

Schmutz S, Kaufmann M, Vogel B, Jungwirth M, Muhar S (2000) A multi-level concept for fish based, river-type-specific assessment of ecological integrity. In: Jungwirth M, Muhar S, Schmutz S (eds) Assessing the ecological integrity of running waters. Kluwer Academic Publishers, Dordrecht, pp 279–289

Schneider M, Jorde K, Zöllner F, Kerle F (2001) Development of a user-friendly software for ecological investigations on river systems, integration of a fuzzy rule-based approach. In: Proceedings environmental informatics 2001, 15th international symposium, informatics for environmental protection

Schneider M, Noack M, Gebler T (2008) Handbuch für das Habitatsimulationsmodell CASiMiR. Schneider & Jorde Ecological Engineering GmbH, Universität Stuttgart Institut Wasserbau, Stuttgart

Shenton W, Bond NR, Yen JD, Mac Nally R (2012) Putting the "ecology" into environmental flows: ecological dynamics and demographic modelling. Environ Manag 50(1):1–10

Shirvell CS (1989) Ability of PHABSIM to predict chinook salmon spawning habitat. Regul Rivers Res Manag 3:277–289

Sinha S, Sotiropoulos F, Odgaard A (1998) Three-dimensional numerical model for flow through natural rivers. J Hydraul Eng 124(1):13–24

Slaney PA, Martin AD (1987) Accuracy of underwater census of trout populations in a large stream in British Columbia. N Am J Fish Manag 7(1):117–122

Steffler P, Blackburn J (2002) River 2D-two-dimensional depth averaged model of river hydrodynamics and fish habitat introduction to depth averaged modeling and user's. Introduction to depth averaged modeling and user's manual. University of Alberta. 120 pp

Tharme RE (2003) A global perspective on environmental flow assessment: emerging trends in the development and application of environmental flow methodologies for rivers. River Res Appl 19(5–6):397–441

Thurow RF (1994) Underwater methods for study of salmonids in the Intermountain West. Gen. Tech. Rep. INT-GTR-307. Odgen, UT: U.S. Department of Agriculture, Forest Service, Intermountain Research Station. 28p

Tritthart M, Hauer C, Liedermann M, Habersack H (2008) Computer-aided mesohabitat evaluation, part II – model development and application in the restoration of a large river. In: Altinakar MS, Kokpinar MA, Darama Y, Yegen EB, Harmancioglu N (eds) International Conference on Fluvial Hydraulics, River Flow 2008, 3.-5.9.2008, Cesme-Izmir

U.S. Fish and Wildlife Service (1980) Habitat as a basis for environmental assessment. In: Habitat evaluation procedures handbook, 101 ESM. USDI Fish and Wildlife Service, Division of Ecological Services, Washington, DC. [online available from https://www.fws.gov/policy/esmindex.html]

Wollebaek J, Thue R, Heggenes J (2008) Redd site microhabitat utilization and quantitative models for wild large brown trout in three contrasting boreal rivers. N Am J Fish Manag 28:1249–1258

Wu TR (2004) A numerical study of braking waves and turbulence effects. PhD thesis, Cornell University

Zeiringer B, Unfer G, Hinterhofer M (2010) Gewässerökologische Restwasserstudie am Kraftwerk Opponitz – Studie zur Restwasserdotation Modul I – fischökologische Untersuchungen, Hydromorphometrie und Restwassermodellierung. Wienstrom GmbH, 162

Chapter 8
The Role of Sediment and Sediment Dynamics in the Aquatic Environment

Christoph Hauer, Patrick Leitner, Günther Unfer, Ulrich Pulg, Helmut Habersack, and Wolfram Graf

8.1 Introduction

The dynamic component in hydrology, sedimentology, and, consequently, river morphology serves as a backbone for the entire river environment (Maddock 1999). In addition to water pollution, the hydro-morphological/sedimentological degradation is one of the main pressures on river systems (Ward and Stanford 1995; Dudgeon et al. 2006). The EU Water Framework Directive (WFD, Directive 2000/60/EC) mentions various aspects of hydro-morphological disturbances that must be addressed by management plans to achieve the aims of a good ecological status or a good ecological potential (Article 3/Article 4). However, to reach these goals, the sediment conditions of a river (e.g., sediment continuum) are not part of the evaluation needs. Here, to achieve "good ecological status," it is assumed that the biotic criteria reflect the hydro-morphological status, while direct assessments of dynamic sedimentological processes are not taken into account (Hauer 2015).

In general, sediments play a decisive role for diversification and composition and, hence, the quality of habitats, especially for the mid- to long-term development of

C. Hauer (✉) · H. Habersack
Christian Doppler Laboratory for Sediment Research and Management, Institute of Water Management, Hydrology and Hydraulic Engineering, University of Natural Resources and Life Sciences, Vienna, Austria
e-mail: christoph.hauer@boku.ac.at; helmut.habersack@boku.ac.at

P. Leitner · G. Unfer · W. Graf
Institute of Hydrobiology and Aquatic Ecosystem Management, University of Natural Resources and Life Sciences, Vienna, Austria
e-mail: patrick.leitner@boku.ac.at; gunther.unfer@boku.ac.at; wolfram.graf@boku.ac.at

U. Pulg
Uni Research Environment, Laboratory for Fresh Water Ecology and Inland Fisheries (LFI), Bergen, Norway
e-mail: ulrich.pulg@uni.no

© The Author(s) 2018
S. Schmutz, J. Sendzimir (eds.), *Riverine Ecosystem Management*, Aquatic Ecology Series 8, https://doi.org/10.1007/978-3-319-73250-3_8

habitat features. According to Leopold et al. (1964), there are eight factors forming the morphological traits of a river: channel width, depth, flow velocity, discharge, channel slope, roughness of channel material, sediment load, and sediment size. Disturbances in any of those factors can alter the general habitat composition of the river and consequently the morphological type of a river. Sediments are both habitat forming (e.g., boulders) and part of morphological structures (e.g., gavel at gravel bars) (Hauer et al. 2014).

Concerning possible impacts of sediment disturbances on the aquatic biota, both the time scale and the form of impact (direct or indirect) are decisive. On the one hand, mid- to long-term indirect impacts are evident due to changes of the physical environment (e.g., changes in sedimentology, loss of spawning sites) as well as short-term, direct (highly dynamic) impacts due to physiological stress (e.g., high turbidity for fish) or risk of abrasion (e.g., for macroinvertebrates). Especially, catchment or reach-scale sedimentological and hydro-morphological disturbances may change the channel shape and/or the habitat composition in the mid- to long-term. Disturbances of the sediment regime are always related to deficits or surpluses in sediment supply and sediment transport (e.g., Brooks and Brierley 1997; Sutherland et al. 2002) which are presented in this book chapter.

Specifically, in alpine regions, the impact of sediment deficits is responsible for riverbed incision and related habitat degradation (Habersack and Piégay 2008). At the same time, increase in sediment load and transport is hardly found in alpine regions but is a major problem in regions with soil erosion due to intensive agriculture or forestry (Leitner et al. 2015; Höfler et al. 2016). Man-made reductions in the sediment load due to torrent controls or retention by hydropower use may have two different consequences, sometimes occurring simultaneously in one and the same river. On one hand, depending on the frequency of floods, the coarsening of substrate due to selective transport leads to fluvial armor or pavement layers (Sutherland 1987). On the other hand, in alpine basins with fine material deposits from the tertiary (marine sediments) below the quaternary gravel layer of the riverbed, the risk of a so-called riverbed breakthrough (Habersack and Klösch 2012) may be realized due to a single flood (e.g., the Salzach River in 2002; Hopf 2006). Another increasingly frequent problem connected to sediment retention is the flushing of reservoirs (see also Chap. 6). During flushing, large amounts of retained suspended load are released in a short period of time, mostly in conjunction with flood events resulting in a surplus of sediments in downstream river sections. Consequently, high loads of mostly fine sediments cause high concentrations of turbidity and can be responsible for losses and mortality of aquatic organisms (e.g., Espa et al. 2015).

Consequences of sediment deficits and impacts on the river are (1) decrease in habitat heterogeneity (Kondolf 1997); (2) risk of river bank erosion (Rinaldi and Casagli 1999); (3) risk of damage to infrastructure, e.g., scouring bridge piers (Jäger et al. 2018); (4) lack of spawning habitats for salmonid fish species (Hauer et al. 2013) and depauperate macroinvertebrate fauna (Graf et al. 2016); (5) decrease in sediment turnover rates and river type-specific sediment quality (Kondolf 1997); and (6) risk of channel avulsion during extreme events (Brizga and Finlayson 1990).

Channel avulsion refers to abrupt changes of the river course leading to a new active channel in the former floodplain.

The aim of this book chapter is to give an overview of the role of sediment and sediment dynamics for the aquatic environment with a special focus on alpine rivers and their fish fauna. We describe how sediment dynamics determine river morphology and habitat-forming processes. Moreover, problems of human-induced sediment increase (e.g., reservoir flushings, intensive agricultural and forestry land use leading to intrusion of fine sediments) and deficits (e.g., deposition by torrent controls and hydropower plants) are targeted with respect to the biotic requirements of macroinvertebrates and fish.

8.2 Sediments and River Morphology

Depending on the morphological river type (Montgomery and Buffington 1997), single grain sizes can be hydraulically habitat forming (e.g., cascade or step-pool type) or just components of a morphological feature (e.g., a gravel bar) that determine the hydraulic patterns of a river (e.g., riffle—pool type) (Hauer et al. 2014). As a decisive variable for channel- and habitat-forming processes, the role of sediments is described in the following subchapters according to their importance in morphological classification, sources in and along river corridors considering river scaling aspects.

8.2.1 River Morphology and Substrate Size

The substrate size and variability in substrate resistance according to the stream power are important agents controlling river morphology (according to Leopold et al. 1964). In this chapter, in contrast to the description of the morphological classification presented in Chap. 3, we use the more sediment size-based classification of Montgomery and Buffington (1997). Here, five different river types for alpine rivers can be distinguished with differences in sediment composition, sediment dynamics, and habitat features:

1. The cascade type is characterized by irregular boulders, local pools, and a large range of particle sizes. Energy dissipation is dominated by continuous tumbling and jet-and-wake flow around and over individual, large clasts (Peterson and Mohanty 1960). The large bedforming material of cascade reaches is immobile during typical flows. Large amounts of bedforming material are mobilized in cascade reaches only during infrequent, hydrologically extreme events with recurrence intervals of 50 up to >200 years (Grant et al. 1990; Phillips 2002). Locally stored gravel and finer grains on the lee sides of flow obstructions (e.g., boulders) are typical sedimentological characteristics of cascade reaches (Montgomery and Buffington 1997). Gravel bed spawning grounds are often small and patchy.

2. Step-pool morphology is characterized by downstream alterations of steps (clasts, wood, and/or bedrock) and plunge pools that develop downstream of each step (Chin 1999; Wohl 2013). Step-pool reaches are most commonly situated along river sections where relatively immobile clasts of coarse sediment can additionally trap wood (Wohl 2013). Energy dissipation is distributed stepwise with high levels at the steps and low dissipation at the outlet of the plunge pools. It is often these outlets which offer good spawning hydraulics and sediment conditions for salmonids. According to Whittaker (1987), step-pool channels reflect a sediment supply-limited system. Potential control variables (reach-scale gradient, discharge, sediment supply and size) for step-pool morphologies have been frequently investigated (e.g., Maxwell and Papanicolaou 2001). Here, in alluvial step-pool systems, particle size was found to determine the step height and discharge as the dominant factor determining the step wavelength (Chartrand and Whiting 2000).

3. Plane-bed reaches are characterized by a lack of gravel bars (e.g., point bars or mid-channel bars), which occur due to a low width-to-depth ratio and a large value of relative roughness (i.e., the ratio of the d_{90} percentile to bank-full depth) (Montgomery and Buffington 1997). Plane-bed channels tend to be intermediate between step-pool and pool-riffle channels regarding gradient slope and grain size (e.g., d_{90}) (Wohl and Merrit 2008). Moreover, the characteristics of plane-bed channels typically in combination with an armored bed surface indicate a transport capacity larger than the sediment supply (Montgomery and Buffington 1997). Hence, supply-limited conditions are found for most discharges (Wohl and Merrit 2008) with some exceptions for high flows (e.g., Sidle 1988). Therefore, a lack of upstream bed-load supply (gravel-to-cobble sized sediments) may be responsible for the development of this specific morphological type. Larger gravel bed spawning grounds are rare and patchily distributed.

4. Riffle-pool channels occur at moderate-to-low gradients and are generally unconfined by valley margins or lateral obstructions (Montgomery and Buffington 1997), with a pool spacing of five to seven times the channel width (Keller and Melhorn 1978). In near-natural river systems, riffle-pool channels contain woody debris leading to forced pool formation with irregular distributions of these local depressions (Lisle 1986). Upstream sediment supply and transport rates cause variable changes in the storage capacity and changes in the channel configuration in low gradient riffle-pool channels (Schumm 1977). High-quality spawning sites for salmonid fish (e.g., brown trout) are usually not limited, especially in the transition zone downstream of the pool and upstream of the riffle crest (Hauer et al. 2013).

5. The low gradient dune-ripple type is associated with sand-bed channels (Montgomery and Buffington 1997). One of the main differences from the plane-bed, riffle-pool, step-pool, and cascade morphological types is that dune-ripple channels exhibit wandering bedforms (Henderson 1963) which are mobile during most water stages. For dune-ripple reaches, bed-load transport occurs even under low flow conditions, caused by the low critical mean flow velocity for the initiation of motion of the fine material predominately consisting of

weathered granite and gneiss [according to Hjülström (1935)]. The occurrence of the dune-ripple type, which is classified as transport limited, is shaped by a high intake of fine sediments from tributaries. Such rivers usually provide poor spawning conditions for gravel-spawning fish species.

8.2.2 Sediment Sources

The sources of sediment are not addressed in the classification of river types and whether these sources are self-formed or relict. Self-formed and relict-non-fluvial streams can be difficult to distinguish in the field. For relict-non-fluvial stream, the off-river sediment supply is low or sediment input only occurs sporadically (Bunte and Abt 2001). In self-formed rivers, however, sediment sources are related entirely to on-site bed material, bank erosion, and upstream fluvial sediments (Andrews 1984). If the sediment sources are not coupled to hillslopes or other partially non-fluvial sources, streams are classified as uncoupled streams (e.g., Trainor and Church 2003). In contrast, coupled streams are determined by sediment supply from relict-fluvial and non-fluvial sources (e.g., Harvey 2001).

8.2.3 Scaling of Sediment Dynamics in the River Environment

Various concepts for scaling river morphology and instream habitats have been developed (e.g., Frisell et al. 1996; Habersack 2000; Maritan et al. 1996; Newson and Newson 2000). From an ecological point of view, the strong dependence of aquatic organisms on abiotic changes in the environment (e.g., sediment turnover, flow fluctuations) has to be emphasized (Hauer 2015). Changes in sediment composition and quantity directly impact aquatic life on various scales. For example, excessive sediment transport rates may change the morphological river type on the reach scale. Consequently, a switch from a riffle-pool morphology to a dune-ripple type can appear due to excessive supply of coarse sand based on impacts of climate change and intensified land use (Hauer 2015). Moreover, the morphological features on the meso-unit scale (decrease in depth variance) as well as the habitat quality at the on-site micro-unit scale can alter. Such local-scale phenomena as, e.g., the loss of interstitial volume and morphological heterogeneity impact macroinvertebrates (Crosa et al. 2010), fish (Pulg et al. 2013; Hauer et al. 2013; Sutherland et al. 2002), and, especially, mussel habitats (Geist and Auerswald 2007). All taxa are strongly influenced by sediment supply at both reach and catchment scales. Therefore, local-scale investigations and research might neglect important aspects of habitat degradation or fail to consider the mid- to long-term evolution and dynamics when mitigation measures are elaborated without considering the driving sedimentological processes at the reach and catchment scales (e.g., reduced sediment supply due to hydropower) (Hauer 2015).

Changes (natural or anthropogenic) of the sediment dynamics on the catchment scale may lead to large-scale disturbances as, e.g., changes in the "sedimentary-link" concept with far-reaching consequences on the instream sediment quality. The sedimentary link concept describes the form of lateral sediment supply from tributaries and its impact on the longitudinal distribution of grain size (Rice and Church 1998). In alpine landscapes, the concept describes the increase in the amount of bed load combined with an increase in the grain size diameter at tributaries followed by a regular downstream fining (Rice 1998; Rice and Church 1998). Unlike alpine river catchments where sediment input from tributaries leads to an increase in the sediment caliber, the "revised" sedimentary link concept for rivers with high sediment input posits a partial decrease in the sediment caliber at tributaries due to the increased deposition of fines (Hauer 2015).

8.3 Sediment Dynamics and Anthropogenic Alterations of the Sediment Flux: What Aquatic Biota Need and How They React to Alterations

Too Little: The Consequences of Sediment Deficits

Rivers exhibiting naturally (downstream of lakes) or anthropogenically reduced sediment supply are "supply-limited" rivers (Montgomery and Buffington 1997). Limited supply leads to continuous armoring of bed surface sediments, a process occurring during ordinary flood events and without extraordinary floods (Fig. 8.1a). In addition to natural bed armoring, human activities can reduce gravel supply and therefore lead to armors. For instance, dams and weirs are responsible for interruptions of the sediment continuum. Further bank stabilization measures reduce lateral sediment supply. In combination, these man-made structures are likely to reduce

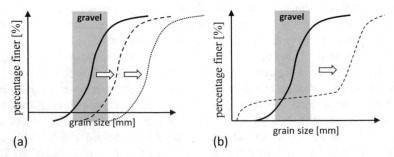

Fig. 8.1 Conceptual schema of mid- and long-term development of spawning gravel in terms of significant (solid line to dashed line) (**a**) lack of sediment supply from upstream reaches in rivers with low concentration of fines and (**b**) lack of sediment supply from upstream with high accumulation of fine sediments in the immobile coarse bed surface (clogging)

gravel supply significantly and can thus increase armoring and intensify flushing out heterogeneous sorted sediments. As a consequence of artificially determined, supply-limited conditions, the resultant deficits in bed-load transport may lead to continuous riverbed incision with the risk of channel avulsion and riverbed break-through during single flood events (Habersack and Klösch 2012). Continuous riverbed incision is the main driver of decoupling floodplains from the required water stage-dependent dynamics of the main river (see Chap. 3).

Beside problems related to riverbed incision and the coarsening of bed surface, increases in fine sediments are known to change grain size distribution and consequently cause degradation of spawning grounds (Sear and DeVries 2008; Pulg et al. 2013), especially in "supply-limited" rivers (Fig. 8.1b). On the one hand, the armoring of the bed surface reduces or prevents cleaning effects of sediment relocations, which naturally generate suitable spawning habitats in the riverbed.

On the other hand, the increase of fines clogs the pore space and can lead to "sustained clogging" (Fig. 8.1b), since the turnover rate is markedly reduced or prevented even in the case of exceptional high flows. In such situations, washed out soil (e.g., from agricultural land use) or fines (e.g., of a glacier environment) may lead to sedimentation of fines on coarse bed material and/or artificially placed gravel with consequent, negative impacts on embryo survival of gravel-spawning fish through suffocation (Reiser 1998; Greig et al. 2005; Pulg et al. 2013).

Too Much: Consequence of an Increased Fine Sediment Yield

Under natural situations, only extraordinary events (e.g., flooding, torrents) produce "too much" sediment. The "excess" sediments generated in extreme events often raise the issue of fine sediments for analysis and/or management of river ecology. In general, in river morphology (Evans and Wilcox 2014) and fish habitat studies (e.g., Pulg et al. 2013), fine sediments are classified as particles <1 mm. Clogging of interstitial space due to clay intrusion called *siltation* degrades macroinvertebrate habitats (e.g., Buddensiek 1995). However, also coarse sand (>1 mm) may impact habitats of macroinvertebrates (Leitner et al. 2015).

Fine sediment intrusion (FSI) is part of the natural sediment and morphological dynamics in most river systems (Smith and Smith 1980). Land-use properties (e.g., Allan 2004) and geological (e.g., Walling 2005) and hydrological catchment-scale characteristics (flood disturbances, frequency, and magnitude of daily glacier melt-off) (e.g., Smith and Smith 1980; Milner and Petts 1994) have often been identified as drivers for natural FSI or clogging of surface and subsurface layers. Aside from glacial rivers, human (anthropogenic) disturbances have greater impacts on the fine sediment dynamics than natural processes. Man-made changes, however, might increase as well as decrease the amount of (fine) sediment load with mostly negative impacts on aquatic ecology in case of increases. For example, hydropower may cause significant alterations of the (fine) sediment regime based on the storage of water and the capture of sediment by dams which cause profound downstream changes in the natural patterns of the hydrologic variation and sediment transport (Poff and Hart 2002). In particular, fine sediment may be trapped in reservoirs and artificially released during controlled events, which may lead to variable meso-unit scale deposition patterns and significant alterations of bed-load transport rates downstream (Wohl et al. 2010). Ecological consequences of reservoir flushing are long-term depletions downstream fish stocks (Espa et al. 2015; Buermann et al. 1995) and short-term impacts on macroinvertebrate communities (Rabení et al. 2005; Crosa et al. 2010).

8.3.1 Ecological Adaptations of Macroinvertebrates to Sediment Dynamics

The faunal structure of benthic macroinvertebrates depends on substrate type, diversity, and spatial patch configuration (Beisel et al. 2000). Habitat conditions of macroinvertebrates are to a large extent determined by flow parameters affecting the macroinvertebrates through hydraulic stress near the bottom (Statzner 1981) which is linked to substrate composition (Percival and Whitehead 1929; Beisel et al. 2000). Accordingly, some species prefer the surface of larger substrates where they feed on biofilms in high current, resulting in a flattened body form (Minshall 1967); others that hide in sand and mud are adapted to temporarily low-oxygen concentrations; those who feed on leaves or wood are restricted to organic matter (Schröder et al. 2013). As a consequence, many species are associated to a certain extent to specific habitats, which are composed of either mineral substrate (e.g., sand, gravel stones) or organic matter (e.g., living plants, dead leaves, deadwood) (see examples in Fig. 8.2a). However, habitat preferences frequently change within the life cycle of invertebrate taxa, indicating the importance of mosaic habitat patterns on a micro-scale (Fig. 8.2b).

In general, benthic invertebrates are adapted to sediment dynamics and natural disturbances (erosion). Animals can usually compensate for infrequent extreme events as floods or ice jams that result in destructive sediment transport. Depending

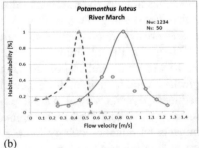

(a)   (b)

Fig. 8.2 (a) Examples of habitat-specific benthic organisms: Perla sp. (macrolithal), Ametropus fragilis (psammal), Nemurella pictetii (fallen leaves), and Lepidostoma basale (deadwood); clockwise from top left; (b) habitat suitability regarding flow velocity of the mayfly Potamanthus luteus in summer (red line, nymphs) and winter (blue line, early instars) at the March River (adapted from Büsch 2014)

on their autecological adaptions (anatomy, strategy) and stage of development (egg, different larval stages, and pupal stage), animals hide in the interstice or go into drift in case of disturbances. Drift is a means of recolonizing denuded downstream habitats and structuring benthic invertebrate communities (Tonkin and Death 2013). However, to preserve stable self-sustaining populations in cases of extreme events, successive downstream drifting has to be compensated by upstream migration by larval stages or by compensation flights by adult insects (Williams and Hynes 1977).

However, anthropogenically induced, long-term alteration of the streambed can result in dramatic shifts of the benthic faunal composition. A coarsening of the bed surface in "supply-limited" rivers can lead to a decrease of macroinvertebrate diversity and/or density for those taxa with habitat preferences for fine sediments comprising certain Oligochaeta, Bivalvia, Diptera, or burrowing Ephemeroptera species. Nevertheless, as many studies show that only a low number of taxa indicate a clear preference for fine substrates (e.g., Minshall 1984; Jowett et al. 1991; Leitner et al. 2015; Graf et al. 2016), the more serious effect in supply-limited river stretches is the clogging of the interstices and embedding of coarse substrate by fines. This phenomenon results in a decline in diversity and abundance of interstices inhabiting sprawlers, such as many Plecoptera and Ephemeroptera species (e.g., Weigelhofer and Waringer 2003).

In particular, anthropogenically induced, fine sediment deposition and siltation in streambeds seriously alters benthic fauna composition and, thus, is becoming a considerable stress for rivers throughout the world. Following Wood and Armitage (1997, Fig. 8.3), increased fine sediment yield affects macroinvertebrates (1) in changing substrate suitability for some taxa (Erman and Ligon 1988; Richards and Bacon 1994), (2) in increasing drift due to sedimentation or substrate instability (Culp and Davies 1985; Rosenberg and Wiens 1978), (3) in limiting respiration by deposition of fine sediments on respiration organs (Lemly 1982) or low-oxygen concentrations in the interstices (Eriksen 1966), and (4) in deteriorating feeding conditions due to effects of increased suspended solids on filter feeders (Aldridge et al. 1987) and in the reduction of the food value of the periphyton (Cline et al. 1982; Graham 1990) as well as prey organisms (Broekhuizen et al. 2001; Yamada and Nakamura 2002; Jones et al. 2012).

Consequently, increased input of fine sediments leads to a decrease in diversity, abundance, and biomass of macroinvertebrates as well as to a shift in community structure (Berkman and Rabeni 1987; Wood and Armitage 1997; Angradi 1999; Leitner et al. 2015). For example, Graf et al. (2016) demonstrated that only Chironomidae and Oligochaeta show a habitat preference for sand or are at least more tolerant to this type of substrate, while other taxa belonging to the orders Ephemeroptera, Plecoptera, and Trichoptera (EPT) show preferences for coarser substrate types and are highly sensitive to siltation.

Briefly, increased fine sediment yield has serious effects on benthic macroinvertebrates in lotic systems, emerging as a steady, often unnoticed, process with a high-risk potential for affecting biodiversity leading to critical ecological degradation.

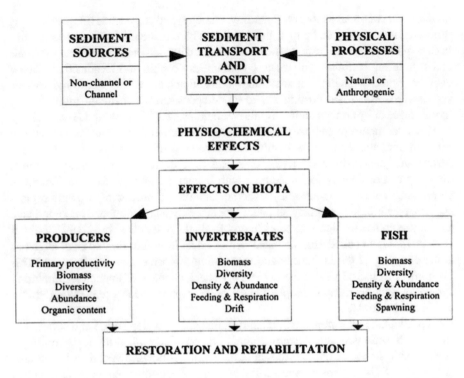

Fig. 8.3 A holistic overview of fine sediment in the lotic ecosystem, after Wood and Armitage (1997) (© Environmental management, Biological effects of fine sediment in the lotic environment, 21(2), 1997, 203–217, Wood, P. J., Armitage, P. D. With permission of Springer)

8.3.2 Ecological Adaptations of Lithophilic Fishes

Sediments play a crucial role in the life cycles of many riverine fish species. This is not surprising since fish fauna had to evolve within the frame of habitat conditions governed by sediment dynamics. Fishes developed strategies or adaptions to cope with dynamic and often stochastically changing sediment conditions. Extensive sediment transport and related relocation are generally destructive events decreasing the survival of incubated egg and juvenile stages of salmonids (e.g., Cattanéo et al. 2002; Lobón-Ceriá and Rincón 2004; Unfer et al. 2011). While the older life stages can actively search for cover, early juvenile stages are exposed to erosive forces that result in high mortality rates. On the other hand, flood events and related sediment relocation reshape the riverbed and refine potential spawning ground for the upcoming spawning period (Poff et al. 1997; Unfer et al. 2011).

For gravel-spawning fish species (lithophilic, most rheophilic species in Europe, such as salmonids and many cyprinids, Fig. 8.4), suitable spawning sediment (bed material composition) and further abiotic components such as water temperature,

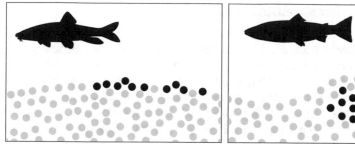

Fig. 8.4 Egg deposition of on-substrate spawners (left, e.g., many cyprinids) and interstitial spawners (e.g., many salmonids)

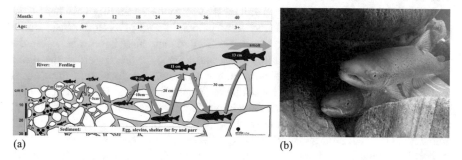

(a)   (b)

Fig. 8.5 (**a**) Habitat use of Atlantic Salmon and brown trout juveniles in relation to grain size distribution in Norwegian salmonid rivers (figure adapted from Pulg et al. 2017). (**b**) Adult Atlantic salmon of approx. 100 cm in length seeking shelter in the river bottom of the boulder-dominated cascade river Nordøla in Western Norway (Photo: Ulrich Pulg).

oxygen concentration, or flow velocity are essential for successful recruitment. Lack of suitable spawning substrate can create bottlenecks for population size and production rates (Pulg et al. 2013). Excessively large grains (large cobble) or armor layers prevent salmonids from redd building (Kondolf 2000), while, on the other hand, high percentages of small grains (fine gravel, sand, silt, clay) do not allow successful reproduction due to reduced permeability and, consequently, insufficient supply of water and oxygen (Sear and DeVries 2008). Besides fines, washout of spawning gravel as well as reduced gravel supply from upstream sources can limit spawning habitats (Barlaup et al. 2008).

Riverine fish depend on substrate also at older life stages (Fig. 8.5). Juveniles of many salmonids spend long periods of their life in the shelter of the sediment, and adults seek shelter on the porous river bottom or behind boulders (Jonsson and Jonsson 2011). Other species (predominately cyprinids) are drifting downstream as larvae and depend on a variable river morphology providing coves, side channels, and oxbows, which are likewise structured by riverbed sediments (Jungwirth et al. 2003).

8.4 Sediment Management Options

Options for sediment management in river catchments are manifold. Basically, they can be divided into (1) structural and (2) nonstructural measures, which can be established on various river scales, including potential consequences (improvements) for downstream river reaches. As an important nonstructural measure, land-use change has to be mentioned. Due to the fact that increased erosion of fines is frequently associated with agricultural land use and intensive forestry (Walling 1990), a reduction of input of erodible soil surfaces provides a management option, especially to prevent clogging of bed sediments (Bakker et al. 2008).

Structural measures on a patch scale (e.g., installation of boulders or deadwood) are useful to create patches of habitats providing the required substratum quality (Hauer 2015). Structural features, such as boulders, have the advantage in that specifically during high (scouring) flows, they provide sheltering habitats in the wake zone accompanied by reduced flow velocities and/or bottom shear stress. Boulder placement or instream use of deadwood can also have effects on the hydraulic conditions and river morphology and, hence, indirectly affect the biota. For example, lateral scour pools with coarse substrate are formed if the flow is vertically or laterally constricted by boulders (Wood-Smith and Buffington 1996).

Examples of structural measures on a larger/local scale are the implementation of river widenings or changes in energy slope (e.g., ramps). Both exhibit local-scale impacts on the sediment transport capacity of rivers. River widenings, in particular, resemble an opportunity to stop riverbed incision, which is often the consequence of a disturbed sediment continuum and channel rectification, specifically in alpine environments. Compared to regulated river sections, local channel widenings increase the hydraulic radius, leading to a decrease in velocity and bottom shear stress (Hauer et al. 2015). In widened river sections, the sediment transport capacity is reduced, which can stop riverbed incision by increasing the aggradation of transported sediments.

Changing the bed (energy) slope is a hydraulic engineering opportunity to influence sediment transport and sediment dynamics when sediment management is required. A large number of artificial transversal obstructions (mainly ramps) are installed to stop ongoing riverbed incision in rectified stretches of alpine gravel bed rivers (DeBene et al. 2016). For this purpose, the bed gradient is reduced between the ramps and the differences in height, and consequently high erosional potential of the flow is controlled by the ramp and the downstream scouring pool (Pagliara 2007). In addition to these technical concepts, by reducing energy slope for channel stability, changes in the bed slope can be explicitly targeted in river restoration (e.g., Habersack et al. 2010) as well as spawning habitat restoration projects (Pulg et al. 2013; Hauer et al. 2015).

Artificial gravel dumping, as an example of structural improvements, is a restoration measure frequently applied below dams (Brown and Pasternack 2008). It affects geomorphic units at meso-scales and thus hydraulic patterns on the microscale (Pasternack 2008). Wheaton et al. (2004) highlight the use of artificial gravel placement as one possible measure to restore or enhance hydro-morphologically suitable spawning habitat conditions for salmonids. For example, in Western Norway, the restoration of anthropogenically impacted (partially destroyed) spawning habitats of Atlantic salmon (*Salmo salar*) was mainly achieved by artificial gravel dumping (e.g., Barlaup et al. 2008) and the restoration of fluvial processes (Fjeldstad et al. 2012). Other restoration techniques include hydraulic structure placement (e.g., single boulders or groins), mainly to create suitable water depths and flow velocities combined with specific sediment sorting, or an "artificial enhancement" of existing spawning gravels by periodic turnovers of spawning substrate to reduce the amounts of aggregated fine sediments at spawning grounds. The problem inherent with all the above mentioned spawning habitat improvement methods (gravel cleaning, gravel dumping, hydraulic adjustments) is that they were designed to increase the habitat suitability for target species during median or low flow conditions (spawning/incubation period, Wheaton et al. 2004) or to reduce the deposition of fine sediments (Pulg et al. 2013). However, the stability and/or scouring depth of spawning substrate during high flow conditions is typically not assessed in spawning habitat restoration design (short- to mid-term time scale) (e.g., DeVries 2008; Lisle 1989; Buffington et al. 2004).

Concerning sediment management actions in relation to hydropower production, many recent studies focus on sediment management techniques in the reservoir (Schleiss et al. 2010). In this context, very often measures removing sediments from the reservoir, such as mechanical and hydraulic dredging (reservoir flushing), are used (Gaisbauer and Knoblauch 2001). Moreover, sediment bypass systems are frequently investigated and described mitigation measures for sediment management in reservoirs. The diversion of sediments through a tunnel (bypassing) can be seen as a preventive and catchment-scale measure against reservoir sedimentation (Boillat and Pougatsch 2000), as it inhibits the input of bed load and part of the suspended load into the reservoir, ensures sediment continuity during floods (Vischer and Chervet 1996), and thus can improve river ecology and sustainability by preventing riverbed erosion downstream the dam (Schleiss and Boes 2011). Turbidity currents are gravity currents driven by the density contrast between sediment-laden fluid and ambient fluid and are an additional sediment management option (Baas et al. 2005). Moreover, dredging of (fine) sediment material is not only important in alpine hydropower reservoirs but also in run-of-river plants in particular. The dredged material needs to be considered in the morpho-dynamic evolution and sediment balance of the reservoir, while the material dumped downstream of the dam yields an important sediment input on the downstream river reach.

8.5   Conclusions and Outlook

Depending on the morphological river type, sediments can be hydraulically habitat forming or just components of a morphological feature that determines the hydraulic patterns of a river. Aquatic biota (e.g., macroinvertebrates, fish) contain different sediment requirements (e.g., morphological adaption) concerning the sediment quantity and distribution in relation to different life stages. Moreover, different reactions in terms of an increased sediment surplus or sediment deficits by a disturbed sediment regime are given. Thus, among the most important issues for sustainable river management in the future are studies on processes and consequently an improved process understanding of sediment dynamics on all river scales. Based on this improved process, understanding restoration measures has to be adjusted to cope with, e.g., increased fine sediments, which are now often trapped in reservoirs. Hence, a holistic view of the river systems and driving abiotic processes has to be targeted for future management—including responsible actors in the present sediment management like water management authorities as well as hydropower companies.

Acknowledgement The financial support by the Austrian Federal Ministry of Science, Research and Economy and the National Foundation of Research, Technology and Development is gratefully acknowledged.

References

Aldridge DW, Payne BS, Miller AC (1987) The effects of intermittent exposure to suspended solids and turbulence on three species of freshwater mussels. Environ Pollut 45:17–28

Allan JD (2004) Landscapes and riverscapes: the influence of land use on stream ecosystems. Annu Rev Ecol Evol Syst 35:257–284

Andrews ED (1984) Bed-material entrainment and hydraulic geometry of gravel-bed rivers in Colorado. Bull Geol Sot Am 95:371–378

Angradi TR (1999) Fine sediment and macroinvertebrate assemblages in Appalachian streams: a field experiment with biomonitoring applications. J N Am Benthol Soc 18:49–66

Baas JH, Mccaffrey WD, Haughton PD, Choux C (2005) Coupling between suspended sediment distribution and turbulence structure in a laboratory turbidity current. J Geophys Res Oceans 110(C11):1978–2012

Bakker MM, Govers G, van Doorn A, Quetier F, Chouvardas D, Rounsevell M (2008) The response of soil erosion and sediment export to land-use change in four areas of Europe: the importance of landscape pattern. Geomorphology 98:213–226

Barlaup BT, Gabrielsen SE, Skoglund H, Wiers T (2008) Addition of spawning gravel – a means to restore spawning habitat of Atlantic Salmon (Salmo salar L.), and anadromous and resident brown trout (Salmo trutta L.) in regulated rivers. River Res Appl 24:543–550

Beisel JN, Usseglio-Polatera P, Moreteau JC (2000) The spatial heterogeneity of a river bottom: a key factor determining macroinvertebrate communities. In: Assessing the ecological integrity of running waters. Springer Netherlands, Cham, pp 163–171

Berkman HE, Rabeni CF (1987) Effect of siltation on stream fish communities. Environ Biol Fish 18:285–294

Boillat JL, Pougatsch H (2000) State of the art of sediment management in Switzerland. Proceedings of the international workshop and symposium on reservoir sedimentation management, pp 35–45

Brizga SO, Finlayson BL (1990) Channel avulsion and river metamorphosis: the case of the Thomson River, Victoria, Australia. Earth Surf Process Landf 15(5):391–404

Broekhuizen N, Parkyn S, Miller D (2001) Fine sediment effects on feeding and growth in the invertebrate grazers Potamopyrgus antipodarum (Gastropoda, Hydrobiidae) and Deleatidium sp (Ephemeroptera, Leptophlebiidae). Hydrobiologia 457:125–132

Brooks AP, Brierley GJ (1997) Geomorphic responses of lower Bega River to catchment disturbance, 1851–1926. Geomorphology 18:291–304

Brown RA, Pasternack GB (2008) Engineering channel controls limiting spawning habitat reha-
  bilitation success on regulated gravel bed rivers. Geomorphology 97:631–654
Buddensiek V (1995) The culture of juvenile freshwater pearl mussels Margaritifera margaritifera L
  in cages: a contribution to conservation programmes and the knowledge of habitat requirements.
  Biol Conserv 74:33–40
Buermann Y, Du Preez HH, Steyn GJ, Harmse JT, Deacon A (1995) Suspended silt concentrations
  in the lower Olifants River (Mpumalanga) and the impact of silt releases from the Phalaborwa
  Barrage on water quality and fish survival. Koedoe 38(2):11–34
Buffington JM, Montgomery DR, Greenberg HM (2004) Basin-scale availability of salmonid
  spawning grave as influenced by channel type and hydraulic roughness in mountain catchments.
  Can J Fish Aquat Sci 61:2085–2096
Bunte K, Abt SR (2001) Sampling surface and subsurface particle-size distributions in wadable
  gravel- and cobble-bed streams for analysis of sediment transport, hydraulics, and stream bed
  monitoring. Gen. Techn. Rep. RMRS-GTR-74. Fort Collins, CO: U.S. Department of Agriculture,
  Forest Service, Rocky Mountains Research Station. 428 p
Büsch M (2014) Hydraulische Habitatpräferenzen ausgewählter Makrozoobenthos-Taxa an der
  March und der Thaya in Niederösterreich und ihre Bedeutung für Restaurierungsmaßnahmen.
  In: Masterarbeit. Universität für Bodenkultur, Wien
Cattanéo F, Lamouroux N, Breil P, Capra H (2002) The influence of hydrological and biotic
  processes on brown trout (Salmo trutta) population dynamics. Can J Fish Aquat Sci 59
  (1):12–22
Chartrand SM, Whiting PJ (2000) Alluvial architecture in headwater streams with special emphasis
  on step–pool topography. Earth Surf Process Landf 25(6):583–600
Chin A (1999) The morphologic structure of step–pools in mountain streams. Geomorphology 27
  (3–4):191–204
Cline LD, Short RA, Ward JV (1982) The influence of highway construction on the
  macroinvertebrates and epilithic algae of a high mountain stream. Hydrobiologia 96(2):149–159
Crosa G, Castelli E, Gentili G, Espa P (2010) Effects of suspended sediments from reservoir
  flushing on fish and macroinvertebrates in an alpine stream. Aquat Sci 72:85–95
Culp JM, Davies RW (1985) Responses of benthic macroinvertebrate species to manipulation of
  interstitial detritus in Carnation Creek, British Columbia. Can J Fish Aquat Sci 42(1):139–146
DeBene A, Diermayr M, Hauer C (2016) Erfahrungen aus dem naturnahen Rampenbau mit
  Berücksichtigung der WRRL am Beispiel des Ischlflusses. Österreichische Wasser- und
  Abfallwirtschaft 11:534–544
DeVries P (2008) Bed disturbance processes and the physical mechanisms of scour in salmonid
  spawning habitat. In: American Fisheries Society Symposium, vol 65. American Fisheries
  Society, Bethesda, pp 121–147
Directive 2000/60/EC (2000) The EU water framework directive – integrated river basin manage-
  ment for Europe. European Commission
Dudgeon D, Arthington AH, Gessner MO et al (2006) Freshwater biodiversity: importance, threats,
  status and conservation challenges. Biol Rev 81:163–182
Eriksen CH (1966) Diurnal limnology of two highly turbid puddles. Verhandlungen des Vereins für
  Limnologie 16:507–514
Erman DC, Ligon FK (1988) Effects of discharge fluctuation and the addition of fine sediment on
  stream fish and macroinvertebrates below a water-filtration facility. Environ Manag 12(1):85–97
Espa P, Crosa G, Gentili G, Quadroni S, Petts G (2015) Downstream ecological impacts of
  controlled sediment flushing in an Alpine valley river: a case study. River Res Appl 31
  (8):931–942
Evans E, Wilcox AC (2014) Fine sediment infiltration dynamics in a gravel-bed river following a
  sediment pulse. River Res Appl 30(3):372–384
Fjeldstad HP, Barlaup BT, Stickler M, Gabrielsen SE, Alfredsen K (2012) Removal of weirs and the
  influence on physical habitat for salmonids in a Norwegian river. River Res Appl 28(6):753–763

Frisell CA, Liss WJ, Warren CE, Hurley MD (1996) A hierachical framework for stream habitat classification: viewing streams in a watershed context. Environ Manag 10:199–214

Gaisbauer H, Knoblauch H (2001) Feststoffmanagement bei Stauanlagen. Österreichische Wasser- und Abfallwirtschaft 53(11–12):265–268

Geist J, Auerswald K (2007) Physicochemical stream bed characteristics and recruitment of the freshwater pearl mussel (Margaritifera margaritifera). Freshw Biol 52:2299–2316

Graf W, Leitner P, Hanetseder I, Ittner LD, Dossi F, Hauer C (2016) Ecological degradation of a meandering river by local channelization effects: a case study in an Austrian lowland river. Hydrobiologia 772(1):145–160

Graham AA (1990) Siltation of stone-surface periphyton in rivers by clay-sized particles from low concentrations in suspention. Hydrobiologia 199:107–115

Grant GE, Swanson FJ, Wolman MG (1990) Pattern and origin of stepped-bed morphology in high-gradient streams, Western Cascades, Oregon. Geol Soc Am Bull 102:340–352

Greig SM, Sear DA, Carling PA (2005) The impact of fine sediment accumulation on the survival of incubating salmon progeny: implications for sediment management. Sci Total Environ 344 (1):241–258

Habersack H (2000) The river scaling concept (RSC): a basis for ecological assessments. Hydrobiologia 422/423:49–60

Habersack H, Klösch M (2012) Monitoring und Modellierung von eigendynamischen Aufweitungen an Drau, Mur und Donau. Österreichische Wasser- und Abfallwirtschaft 64:411–422

Habersack H, Piégay H (2008) River restoration in the Alps and their surroundings: past experience and future challenges. In: Habersack H, Piegay H, Rinaldi M (eds) Gravel-bed rivers VI: from process understanding to river restoration. Elsevier, Amestrdam

Habersack H, Liedermann M, Tritthart M, Hauer C, Klösch M, Klasz G, Hengl M (2010) Maßnahmen für einen modernen Flussbau betreffend Sohlstabilisierung und Flussrückbau–Granulometrische Sohlverbesserung, Buhnenoptimierung, Uferrückbau und Gewässervernetzung. Österreichische Wasser und Abfallwirtschaft 11–12:571–581

Harvey AM (2001) Coupling between hillslopes and channels in upland fluvial systems: implications for landscape sensitivity, illustrated from the Howgill Fells, northwest England. Catena 42:225–250

Hauer C (2015) Review of hydro-morphological management criteria on a river basin scale for preservation and restoration of freshwater pearl mussel habitats. Limnologica 50:40–53

Hauer C, Unfer G, Habersack H, Pulg U, Schnell J (2013) Bedeutung von Flussmorphologie und Sedimenttransport in Bezug auf die Qualität und Nachhaltigkeit von Kieslaichplätzen. KW–Korrespondenz Wasserwirtschaft 4/13:189–197

Hauer C, Blamauer B, Mühlmann H, Habersack H (2014) Morphodynamische Aspekte der Ökohydraulik und Habitatmodellierung im Kontext der rechtlichen Rahmenbedingungen. Österr. Wasser- und Abfallwirtschaft, 56.JG, 66:169–178

Hauer C, Pulg U, Gabrielsen SE, Barlaup BT (2015) Application of step-backwater modelling for salmonid spawning habitat restoration in Western Norway. Ecohydrology 8(7):1239–1261

Henderson FM (1963) Stability of alluvial channels. Trans Am Soc Civ Eng 128:657–686

Hjülström F (1935) The morphological activity of rivers as illustrated by river Fyris. Bulletin of Geological Institute, 25

Höfler S, Hauer C, Gumpinger C (2016) Ökologische Maßnahmen an kleinen und mittelgroßen Fließgewässern -Auswirkungen auf die Qualitätselemente der Europäischen Wasserrahmenrichtlinie und Grenzen der Wirksamkeit – unter besonderer Berücksichtigung der Feinsedimentproblematik. Österreichische Wasser- und Abfallwirtschaft 11:519–533

Hopf G (2006) Die Sanierung der unteren Salzach. – LWF Wissen, 55, 62–66, 8 Abb., Freising (Bayerische Landesanstalt für Wald und Forstwirtschaft)

Jäger E, Hauer C, Habersack H (2018) A novel tool in integrated flood risk management: river-adapted vegetation management VEMAFLOOD

Jones JI, Murphy JF, Collins AL, Sear DA, Naden PS, Armitage PD (2012) The impact of fine sediment on macro-invertebrates. River Res Appl 28(8):1055–1071

Jungwirth M, Haidvogel G, Muhar S, Schmutz S (2003) Angewandte Fischökologie an Fließgewässern. UTB Facultas. 547 pp

Jonsson B, Jonsson N (2011) Migrations. In: Ecology of Atlantic Salmon and Brown Trout. Springer, Dordrecht, pp 247–325

Jowett IG, Richardson J, Biggs BJ, Hickey CW, Quinn JM (1991) Microhabitat preferences of benthic invertebrates and the development of generalised Deleatidium spp. habitat suitability curves, applied to four New Zealand rivers. N Z J Mar Freshw Res 25(2):187–199

Keller EA, Melhorn WN (1978) Rhythmic spacing and origin of pools and riffles. Geol Soc Am Bull 89(5):723–730

Kondolf GM (1997) PROFILE: hungry water: effects of dams and gravel mining on river channels. Environ Manag 21(4):533–551

Kondolf GM (2000) Assessing salmonid spawning gravel quality. Trans Am Fish Soc 129 (1):262–281

Leitner P, Hauer C, Ofenbock T, Pletterbauer F, Schmidt-Kloiber A, Graf W (2015) Fine sediment deposition affects biodiversity and density of benthic macroinvertebrates: a case study in the freshwater pearl mussel river Waldaist (Upper Austria). Limnologica 50:54–57

Lemly AD (1982) Modification of benthic insect communities in polluted streams: combined effects of sedimentation and nutrient enrichment. Hydrobiotogia 87:222–245

Leopold LB, Wolman MG, Miller JP (1964) Fluvial processes in geomorphology. Freeman, San Francisco, CA, 522 pp

Lisle TE (1986) Stabilization of gravel channel by a large streamside obstruction and bedrock bends, Jacoby Creek, northwestern California. Geol Soc Am Bull 97:999–1011

Lisle TE (1989) Sediment transport and resulting deposition in spawning gravels, north coastal California. Water Resour Res 25:1303–1319

Lobón-Cerviá J, Rincón PA (2004) Environmental determinants of recruitment and their influence on the population dynamics of stream-living brown trout Salmo trutta. Oikos 105:641–646

Maddock I (1999) The importance of physical habitat assessment for evaluating river health. Freshw Biol 41(2):373–391

Maritan A, Rinaldo A, Rigon R, Giacometti A, Rodríguez-Iturbe I (1996) Scaling laws for river networks. Phys Rev E53:1510

Maxwell AR, Papanicolaou AN (2001) Step-pool morphology in high-gradient streams. Int J Sediment Res 16(3):380–390

Milner AM, Petts GE (1994) Glacial rivers: physical habitat and ecology. Freshw Biol 32(2):295–307

Minshall GW (1967) Role of allochthonous detritus in the trophic structure of a woodland springbrook community. Ecology 48:139–149

Minshall GW (1984) Aquatic insect-substratum relationships. In: Resh VH, Rosenberg DM (eds) The ecology of aquatic insects. Praeger Publishers, New York, pp 358–400

Montgomery DR, Buffington JM (1997) Channel reach morphology in mountain drainage basins. Geol Soc Am Bull 109:596–611

Newson MD, Newson CL (2000) Geomorphology, ecology and river channel habitat: mesoscale approaches to basin-scale challenges. Prog Phys Geogr 24:195–217

Pagliara S (2007) Influence of sediment gradation on scour downstream of block ramps. J Hydraul Eng 133(11):1241–1248

Pasternack GB (2008) Spawning habitat rehabilitation: advances in analysis tools. In: Sear DA, DeVries P (eds) Salmon spawning habitat in rivers, vol 65. American Fisheries Symposium, Bethesda, pp 321–349

Percival E, Whitehead H (1929) A quantitative study of some types of stream bed. J Ecol 17:282–314

Peterson DF, Mohanty PK (1960) Flume studies of flow in steep, rough channels. J Hydraul Div/Am Soc Civil Eng 86:55–76

Phillips JD (2002) Geomorphic impacts of flash flooding in a forested headwater basin. J Hydrol 269:236–250

Poff NL, Hart DD (2002) How dams vary and why it matters for the emerging science of dam removal an ecological classification of dams is needed to characterize how the tremendous variation in the size, operational mode, age, and number of dams in a river basin influences the potential for restoring regulated rivers via dam removal. Bioscience 52(8):659–668

Poff NL, Allan JD, Bain MB, Karr JR, Prestergaard KL, Richter BD, Sparks RE, Stromberg JC (1997) The natural flow regime. Bioscience 47:769–784

Pulg U, Barlaup BT, Sternecker K, Trepl L, Unfer G (2013) Restoration of spawning habitats of brown trout (Salmo trutta) in a regulated chalk stream. River Res Appl 29:172–182

Pulg U, Barlaup BT, Skoglund H, Velle G, Gabrielsen SE, Stranzl SF, Espedal EO, Lehmann GB, Wiers T, Skår B, Normann E, Fjeldstad HP (2017) Tiltakshåndbok for bedre fysisk vannmiljø: God praksis ved miljøforbedrende tiltak i elver og bekker. Uni Research AS. 180 pages LFI Uni Miljø (296)

Rabení CF, Doisy KE, Zweig LD (2005) Stream invertebrate community functional responses to deposited sediment. Aquat Sci 67(4):395–402

Reiser DW (1998) Sediment gravel bed rivers: ecological and biological considerations. In: Gravel-bed rivers in the environment. Water Research Centre, Colorado, pp 199–228

Rice S (1998) Which tributary disrupt downstream fining along gravel-bed rivers? Geomorphology 22:39–56

Rice S, Church M (1998) Grain size along two gravel-bed rivers: statistical variation, spatial patterns and sedimentary links. Earth Surf Process Landf 23:345–363

Richards C, Bacon KL (1994) Influence of fine sediment on macroinvertebrate colonisation of surface and hyporheic stream sediments. Great Basin Naturalist 54:106–113

Rinaldi M, Casagli N (1999) Stability of streambanks formed in partially saturated soils and effects of negative pore water pressures: the Sieve River (Italy). Geomorphology 26(4):253–277

Rosenberg DM, Wiens AP (1978) Effects of sediment addition on macrobenthic invertebrates in a northern Canadian river. Water Res 12(10):753–763

Schleiss A, Boes R (2011) Dams and reservoirs under changing challenges. Proceedings of the international symposium on dams and reservoirs under changing challenges, 79 Annual Meeting of ICOLD, June 2011, Swiss Committee on Dams, Lucerne

Schleiss A, De Cesare G, Jenzer Althaus J (2010) Verlandung der Stauseen gefährdet die nachhaltige Nutzung der Wasserkraft. Wasser Energie Luft 102(1):31–40

Schröder M, Kiesel J, Schattmann A, Jähnig SC, Lorenz AW, Kramm S, Hering D (2013) Substratum associations of benthic invertebrates in lowland and mountain streams. Ecol Indic 30:178–189

Schumm SA (1977) The fluvial system. Wiley, New York. 338 pp

Sear DA, DeVries P (2008) Salmonid spawning habitat in rivers: physical controls, biological responses, and approaches to remediation, vol 65. American Fisheries Society, Bethesda

Sidle RC (1988) Bed load transport regime of a small forest stream. Water Resour Res 24:201–218

Smith DG, Smith ND (1980) Sedimentation in anastomosed river systems: examples from alluvial valley near Bannf, Alberta. J Sediment Res 50(1):157–164

Statzner B (1981) A method to estimate the population size of benthic macroinvertebrates in streams. Oecologia 51(2):157–161

Sutherland AJ (1987) Static armour layers by selective erosion. In: Thorne CR, Bathurst JC, Hey RD (eds) Sediment transport in gravel-bed rivers. Wiley, Chichester, pp 243–260

Sutherland AB, Meyer JI, Gardiner EP (2002) Effects of land cover on sediment regime and fish assemblage structure in four southern Appalachian streams. Freshw Biol 47:1791–1805

Tonkin JD, Death RG (2013) Macroinvertebrate drift-benthos trends in a regulated river. Fundam Appl Limnol/Archiv für Hydrobiologie 182(3):231–245

Trainor K, Church M (2003) Quantifying variability in stream channel morphology. Water Resour Res 39. https://doi.org/10.1029/2003WR001971

Unfer G, Hauer C, Lautsch E (2011) The influence of hydrology on the recruitment of brown trout in an Alpine river, the Ybbs River. Austria Ecol Freshw Fish 20:438–448

Vischer D, Chervet A (1996) Geschiebe-Umleitstollen bei Stauseen; Möglichkeiten und Grenzen. Mitteilungen der Versuchsanstalt fur Wasserbau, Hydrologie und Glaziologie an der Eidgenossischen Technischen Hochschule Zurich 143:26–43

Walling DE (1990) Linking the field to the river: Sediment delivery from agricultural land. In: Boardman J, Foster IDL, Dearing JA (eds) Soil erosion on agricultural land. Wiley, Chichester, pp 129–152

Walling DE (2005) Tracing suspended sediment sources in catchments and river systems. Sci Total Environ 344(1):159–184

Ward JV, Stanford JA (1995) Ecological connectivity in alluvial river ecosystems and its disruption by flow regulation. Regul Rivers Res Manag 11:105–119

Weigelhofer G, Waringer J (2003) Vertical distribution of benthic macroinvertebrates in riffles versus deep runs with differing contents of fine sediments (Weidlingbach, Austria). Int Rev Hydrobiol 88(3–4):304–313

Wheaton JM, Pasternack GB, Merz JE (2004) Spawning habitat rehabilitation-II. Using hypothesis development and testing in design, Mokelumne river, California, USA. Int J River Basin Manag 2(1):21–37

Whittaker JG (1987) Sediment transport in step-pool streams. In: Sediment transport in gravel-bed rivers. Wiley, New York, pp 545–579

Williams DD, Hynes HBN (1977) Benthic community development in a new stream. Can J Zool 55 (7):1071–1076

Wohl EE (2013) Mountain rivers revisited, American Geophysical Union. Print, 574 pp. ISBN: 9780875903231, Online ISBN: 9781118665572, doi: https://doi.org/10.1029/WM019

Wohl EE, Merritt DM (2008) Reach-scale channel geometry of mountain streams. Geomorphology 93:168–185

Wohl EE, Cenderelli DA, Dwire KA, Ryan-Burkett SE, Young MK, Fausch KD (2010) Large in-stream wood studies: a call for common metrics. Earth Surf Process Landf 35(5):618–625

Wood PJ, Armitage PD (1997) Biological effects of fine sediment in the lotic environment. Environ Manag 21(2):203–217

Wood-Smith RD, Buffington JM (1996) Multivariate geomorphic analysis of forest streams: implications for assessment in landuse impacts on channel conditions. Earth Surf Process Landf 21:377–393

Yamada H, Nakamura F (2002) Effect of fine sediment deposition and channel works on periphyton biomass in the Makomanai River, northern Japan. River Res Appl 18(5):481–493

Chapter 9
River Connectivity, Habitat Fragmentation and Related Restoration Measures

Carina Seliger and Bernhard Zeiringer

9.1 The Importance of Connectivity in Riverine Ecology

For a long time, connectivity conservation focused on interactions and exchanges between terrestrial and, in most cases, homogenous habitat patches. Thereby, rivers have all too often been considered as two-dimensional elements of terrestrial landscapes neglecting their own internal structure and heterogeneity (Wiens 2002). It was therefore not until the early 1980s that the term "river corridors" started to appear in scientific literature only to be then gradually replaced by the term "connectivity" (Amoros and Roux 1988; Pringle 2006) for describing the spatial connections within river systems (Ward 1997; Wiens 2002).

Although knowledge and approaches from terrestrial assessments can also be transferred to aquatic ecosystems, rivers exhibit certain characteristics, which should grant them a special position in connectivity conservation:

1. Riverine systems are characterized by their inherent water-mediated connectivity wherein the river itself represents both habitat and migration corridor (Ward 1989; Wiens 2002). As a consequence, two sites with a low Euclidean distance may indeed show a stream distance of several hundreds of kilometres (Labonne et al. 2008).
2. Connectivity acts on one temporal and three spatial dimensions: longitudinally from headwaters to confluences and the sea, laterally from the main channel to floodplains and vertically from the river towards the hyporheic interstitial and the groundwater (Ward 1989; Jungwirth et al. 2003). The importance of each dimension changes along the river course (Vannote et al. 1980; Ward and Stanford 1995b) and leads to the development of different river concepts (see below).

C. Seliger (✉) · B. Zeiringer
Institute of Hydrobiology and Aquatic Ecosystem Management, University of Natural Resources and Life Sciences, Vienna, Austria
e-mail: carina.seliger@boku.ac.at; bernhard.zeiringer@boku.ac.at

© The Author(s) 2018
S. Schmutz, J. Sendzimir (eds.), *Riverine Ecosystem Management*, Aquatic Ecology
Series 8, https://doi.org/10.1007/978-3-319-73250-3_9

3. Hydrologic connectivity supports the passive downstream transport of matter and energy (Ward and Stanford 1995a; Pringle 2006) but enables a multidimensional dispersal of organisms (Ward and Stanford 1995a; Branco et al. 2014).
4. While terrestrial connectivity often focuses on interactions of homogenous patches, the connection of different habitats is equally or, in aquatic ecology, maybe even more important, since certain species and life stages require diverse habitat patches to complete their life cycle (Jungwirth et al. 2003).

The longitudinal alteration of physical parameters in the downstream direction does not only affect the four dimensionality of rivers but also induces the development of distinct life strategies of organisms living in the river (see Fig. 9.1). Vannote et al. (1980) developed the "River Continuum Concept" (RCC) to highlight the longitudinal, biocoenotic change related to hydro-morphological conditions on a functional basis (i.e. expressed as production/respiration ratio). While the RCC focused on the downstream succession of feeding types, it was criticized for insufficiently considering the lateral and vertical dimensions. Furthermore, due to its limited applicability to anthropogenically disturbed systems, it was followed by the "extended serial discontinuity concept" (ESDC). Developed by Ward and Stanford (1983, 1995b) to describe the longitudinal variation of the four dimensions, the ESDC also allowed the incorporation of anthropogenic alterations. In particular, it considers barriers as well as thermal and flow alterations (e.g. induced by impoundments or water abstractions), which also interrupt the river continuum (Ward and Stanford 1983; Branco et al. 2014) (see Sect. 9.2).

Since both habitats and populations are potentially connected by said four dimensionality (Ward 1989, 1997), ecologists usually differentiate between ecological and landscape/riverscape connectivity. The former deals with the fundamental concept of metapopulation ecology and discusses the impacts of limited genetic exchange between populations (Moilanen and Hanski 2006). Riverscape connectivity, on the other hand, can be further divided into two kinds of connectivity: structural-to characterize relationships between habitat patches (i.e. quantity, location and potential corridors connecting them) (Keitt et al. 1997; Tischendorf and Fahrig 2000; Antongiovanni and Metzger 2005; Segurado et al. 2013)—and functional connectivity, to describe the complex relationships and biological response of individuals or populations to the landscape structure (Tischendorf and Fahrig 2000), which depends on the ecology of the species of concern (e.g. preferences, swimming abilities, requirements) (Bowne et al. 2006).

Aquatic organisms, and especially fish as vagile organisms, evolved in relation to habitat distribution and adapted their life history patterns in response to their connectivity over space and time (Ward 1989; Jungwirth et al. 2000; Schmutz and Mielach 2013). Consequently, all fish species perform targeted "habitat shifts" to exploit a diverse array of habitats (Schmutz et al. 1997; Jungwirth 1998; Mader et al. 1998; Northcote 1998) and to optimize their production and use of resources in response to changing requirements (e.g. for distribution, growth, reproduction, shelter and protection from predators) or changing habitat patches (e.g. due to floods, climate change, etc.) (Northcote 1978, 1998; Lancaster 2000; Wiens 2002). As the

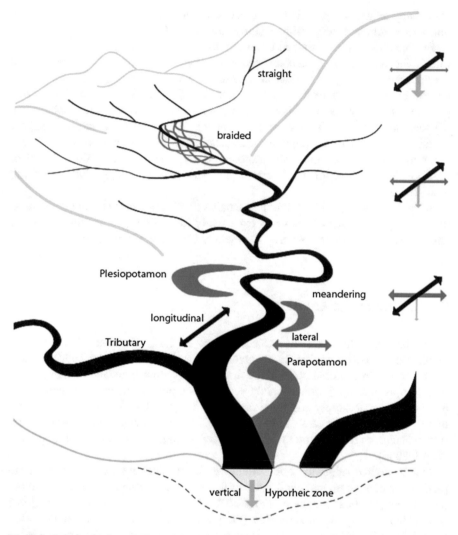

Fig. 9.1 Longitudinal succession of the three spatial dimensions and their importance for aquatic organisms and fish in particular (based on Jungwirth et al. 2000; Amoros et al. 1987)

integrity of fish populations relies to a high degree on the availability and accessibility of spatially separated habitat patches within the river network, fish are good indicators for continuity and connectivity conditions in riverine ecosystems (Jungwirth et al. 2003).

Fish migrations are usually induced by several complex and often interacting factors (Pavlov 1989; Colgan 1993; Lucas and Baras 2001). Examples are water temperature, season, light, discharge, water quality but also internal factors as imprinting and homing effect (Lucas and Baras 2001). While spawning migrations mostly occur towards headwaters, tributaries or floodplains, downstream migrations are usually related to spreading, drift, accessing autumn/winter habitats and return migrations from spawning habitats (Jungwirth et al. 2003).

Fish species can be classified according to their migratory guild as diadromous (inhabiting both seawater and freshwater habitats during certain life stages) and potamodromous (only in freshwater systems). Potamodromous species are characterized by migrations related to (1) spawning, (2) passive drift of larvae and juveniles, (3) age-related habitat changes, (4) flood migrations/catastrophic drift, (5) seasonal habitat shifts (e.g. winter habitats), (6) migrations regarding feeding/nutrition and (7) dispersal migrations. Depending on the migratory distance, long-, medium- and short-distance migrations are distinguished (i.e. >300, 30–300 or <30 km in one direction per year) (Waidbacher and Haidvogl 1998; Jungwirth et al. 2003).

Although many species spawn in inundated floodplains and rely on intact lateral connectivity, this chapter addresses exclusively longitudinal connectivity, leaving lateral connectivity to be discussed in other chapters (e.g. Chaps. 3 and 6).

9.2   River Fragmentation

Rivers belong to the most diverse ecosystems on earth (Bosshard 2015) but are highly threatened by habitat fragmentation (Dynesius and Nilsson 1994; Nilsson et al. 2005). On a global scale, there are currently more than 58,400 large dams (i.e. >15 m in height) which mainly serve the purpose of irrigation, hydropower production, water supply and flood control (ICOLD 2016). These dams fragment more than 60% of all large rivers (i.e. >1000 km in length) with even higher fragmentation rates in Europe where only 28% of large rivers remain free flowing (WWF 2006).

Due to limited migration opportunities in stream networks, disconnections are particularly damaging, making it more difficult or even impossible for fish to avoid barriers (Fagan 2002; Fullerton et al. 2010). Therefore, the dramatic loss in global aquatic biodiversity is not surprising (Pringle et al. 2000; Rosenberg et al. 2000). With 37% of Europe's freshwater fishes threatened and another 4% near-threatened with extinction, they show one of the highest threat levels of any major taxonomic group (Freyhof and Brooks 2011). As it is assumed that 10,000–20,000 freshwater species are already extinct or at risk of extinction (Vörösmarty et al. 2010), the current rates are more than 1000 times the normal background rate (Master 1990). This may explain the steep decline in abundance since the mid-1980s (Latham et al. 2008) for migratory fish species, which are particularly susceptible to fragmentation (Lucas and Baras 2001; Pringle 2006; Ovidio and Philippart 2008). Examples include endangered medium-distance migrants (e.g. *Acipenser ruthenus, Hucho hucho*) and large-distance migratory species (e.g. *Acipenser stellatus, Huso huso*) which became extinct in the upper Danube catchment as a consequence of the closure of the Iron Gate dams (Spindler et al. 1997; Jungwirth et al. 2003).

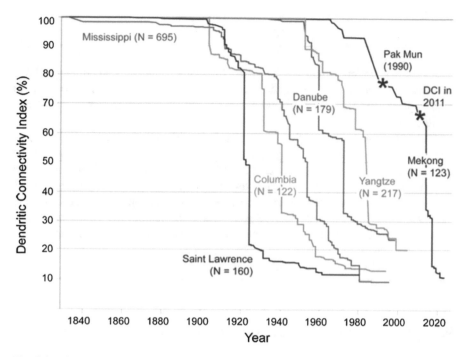

Fig. 9.2 Historic development of the dendritic connectivity index (DCI; Cote et al. 2009) for selected large river basins (Grill et al. 2014) (reprinted from Ecological Indicators, 45, Grill G., Dallaire C.O., Chouinard E.F., Sindorf N., Lehner B., Development of new indicators to evaluate river fragmentation and flow regulation at large scales: A case study for the Mekong River Basin, 148–159, © 2014, with permission from Elsevier Inc. All rights reserved)

The construction of weirs represents the most obvious way to fragment riverine habitats. However, dams may also be associated with other hydro-morphological changes, which then alter the spatial and temporal patch composition and, consequently, connectivity patterns (Wiens 2002). In this context, residual flow sections and impoundments may not only alter habitats but can also contribute to habitat fragmentations by preventing fish migrations through sections with limited water depths and flow velocities (see Chaps. 4 and 6).

The intensity of fragmentation is often expressed by the number of dams within a catchment or per river-km and the mean or maximum length between two barriers (Nilsson et al. 2005; Grill et al. 2014). However, these statements do not allow causal conclusions on the distribution and, consequently, the ecological impact of barriers.

The dendritic connectivity index (DCI; Cote et al. 2009) represents an easy and elegant way to describe catchment fragmentation (see Fig. 9.2). For binary passability ratings, the DCI is calculated as shown below, where n is the number of fragments, l_i is the river length of the fragment i and L is the total length of the entire river network.

$$DCI = \sum_{i=1}^{n} \frac{l_i^2}{L^2} * 100$$

As fragmentation began in the early twentieth century, several large rivers experienced rapid reductions in their connectivity, and the same is expected for many other rivers (e.g. the Mekong) if current hydropower plans are implemented (Grill et al. 2014).

One key disadvantage of the DCI is that it considers the entire river network as equally important and thereby neglects the fact that some sections are more vulnerable to fragmentation than others (see also Sect. 9.3.1). While barriers close to the river mouth disconnect the entire catchment upstream, dams in headwaters make only small proportions inaccessible. To take this into account, Grill et al. (2014) extended the DCI by ecologically relevant weighting factors (e.g. river volume or river classes). Furthermore, to reflect the importance of a segment for migratory fish, the "River Migration Connectivity Index" (RMCI) incorporates the proportion of migratory species potentially visiting a river fragment (Grill et al. 2014).

9.3 Restoration of Longitudinal Continuity

With habitat fragmentation progressing worldwide, ecological research put much effort into conservation measures for maintaining and restoring connectivity of riverine habitats (e.g. Mesa and Magie 2009; Kemp and O'Hanley 2010; O'Hanley 2011; Segurado et al. 2013; Branco et al. 2014). The importance of continuity restoration is reflected in several international and national directives and frameworks (European Commission 1992, 2000; Schmutz and Mielach 2013; ICPDR 2015) and consequently is of high importance in aquatic habitat restoration (Branco et al. 2014).

As it is the case for all environmental impacts, it is recommended to make use of the full management action hierarchy, starting with avoidance, minimization, mitigation and, only if unavoidable, compensation. Ideally, these steps should be considered on a large scale (e.g. catchments or sub-catchments) and support the local decision process. While most new dams already incorporate solutions for sustaining fish migrations, fish pass facilities of existing dams, if present at all, are all too often inoperable, since subsequent retrofitting is both expensive and, due to the large number of impassable dams, time consuming. Therefore, large-scale assessments can also help to identify existing barriers with high priority for continuity restoration. We first examine such large-scale assessments in detail (Sect. 9.3.1) and then discuss state-of-the art solutions for impact avoidance and restoration of individual barriers (Sect. 9.3.2).

9.3.1 Large-Scale Concepts

As proposed by Hartmann et al. (2013), hydropower planning should take place on the system scale and consider multiple parameters (Seliger et al. 2016). The selection of criteria depends on the respective conservation goal (e.g. conservation of overall connectivity or of species-specific migration routes) and can include both qualitative (e.g. habitat quality) or quantitative data (e.g. distance to the next barrier).

Opperman et al. (2015) performed multiple dam building scenarios for three case studies and compared them on the basis of hydropower capacity and impacts on connectivity (defined as the longest connected network in the catchment). They showed that the impacts on connectivity varied considerably between scenarios with the same energy output and that a certain share of the hydropower capacity (or a certain number of barriers) can usually be realized with insignificant connectivity declines. Consequently, large-scale assessments represent one possible approach to significantly reduce the overall impact (Opperman et al. 2015; Seliger et al. 2016; see also Chap. 23).

Also the International Commission for the Protection of the Danube River (ICPDR) developed a two-level approach for guiding sustainable hydropower development in the Danube catchment (ICPDR 2013). Site-specific mitigation measures can be planned after identifying locations with favourable hydropower potential and ecological criteria (ICPDR 2013).

As efficient as large-scale concepts can be for decision-making on new barriers, they can just as well support the planning process for restoring existing dams. Continuity restoration for all existing obstacles in a river might not be feasible or expedient. Due to limited resources (e.g. time, money), it might be worthwhile to identify those barriers where continuity restoration yields the best ecological benefit. This is, of course, only the case if suitable habitats are made accessible. Consequently, the inclusion of habitat quality parameters is highly recommended.

Several examples for the prioritization of barriers for continuity restoration already exist (e.g. O'Hanley and Tomberlin 2005; Mesa and Magie 2009; O'Hanley 2011; Segurado et al. 2013). Also transnational implementations are applicable, as shown by the prioritization index applied for the Danube catchment (ICPDR 2015) assigning decreasing importance from the Danube to the tributaries and including protected areas, length of the reconnected habitat and presence of other hydromorphological pressures. The prioritization index of the ICPDR and many others are based on simple cost-benefit analysis, including a set of assessment criteria to identify barriers of importance. While such scoring-and-ranking systems (e.g. Karle 2005) are easy to apply and comprehend, they assess each barrier independently. On the other hand, detailed GIS analyses (performing "what if"-type assessments) (Dumont et al. 2005; Gough et al. 2012) and optimization models (Kuby et al. 2005; O'Hanley and Tomberlin 2005) which can incorporate cumulative effects remain reserved for specialists (Kemp and O'Hanley 2010).

Although large-scale concepts represent suitable tools for both protecting and restoring aquatic ecosystems, they are rarely applied, and decisions are all too often made on a case-by-case basis. Furthermore, while prioritization concepts provide guidance for efficient continuity restoration, on the long run, all barriers in natural fish habitats should be made passable in both ways.

9.3.2 Fish Migration Aids

Once the barriers for continuity restoration are known, suitable mitigation measures have to be investigated. Since up- and downstream migrations require different settings, they usually cannot be restored by a single facility but rather require independent solutions. Exceptions may apply to certain types, e.g. fish lifts.

While significant knowledge and state-of-the-art measures are already available for restoring upstream migrations (Adam et al. 2005; BMLFUW 2012; Seifert 2012; DWA 2014), efficient solutions for downstream migrations are much less advanced and require further research and practical experience, especially in rivers with diverse fish assemblages (Böttcher et al. 2015).

Overall, while fish passes and bypass systems can reduce the impact of a barrier, they mostly cannot restore connectivity to pristine conditions as limitations might remain for selected species or life stages. Furthermore, barriers are usually related to other pressures, e.g. sedimentation processes in the impoundment with subsequent sediment deficit downstream (see Chap. 8), limited flow velocity in impoundments (see Chap. 6) or insufficient residual flow (see Chap. 4). These pressures, along with poorly executed fish passes, may contribute to migration delays, especially if spawning grounds are separated by several consecutive barriers.

Facilities for up- and downstream migration have to function as an alternative migration corridor. To be accepted as such, their design has to meet the requirements of migratory species (e.g. swimming capabilities, orientation, migration corridors). Therefore, knowledge of the following parameters is essential for the implementation of functional facilities:

1. An important factor for describing migratory capabilities of fish is the swimming speed, which is directly related to body length (i.e. expressed in body lengths per second; DVWK 1996) and depends on species- and age-specific characteristics (e.g. body shape, muscular system) as well as external factors (e.g. water temperature; DWA 2014). The respective pace also depends on the duration it can be sustained. However, in general, the "critical burst swimming speed" (i.e. speed at which drift occurs after 20 s; Clough and Turnpenny 2001) of the weakest swimmer should serve as a benchmark for ecohydraulic planning of fish migration aids (Clough et al. 2004).
2. Although fish use all their senses for orientation, one main parameter is flow (Lucas and Baras 2001). As long as the flow velocity in the fish pass exceeds a species- and age-specific threshold (i.e. from 0.15 to >0.30 m/s), fish show a positive rheoactive orientation (DWA 2014).

3. Fish usually migrate within the main current or, in the case of too high flow velocities, parallel to it. Furthermore, the migration corridor (i.e. surface vs. - bottom-oriented and shoreline vs. open water) depends on species-specific preferences (Seifert 2012). Bypasses have to be directly connected to migration corridors of all relevant species, and attraction flows should enhance their traceability.

Facilities for Upstream Migration

Several guidance documents on planning, construction and operation of fish passes were already developed or are currently under development (Dumont et al. 2005; BMLFUW 2012; Seifert 2012; Schmutz and Mielach 2013, 2015; DWA 2014).

As upstream migrations mostly serve reproduction, facilities have to support at least sexually mature age classes. Three main aspects have to be considered: (1) the perceptibility of the entry, (2) the passability of the fish pass and (3) post-passage effects.

Perceptibility depends to a high degree on the position and attraction flow of the fish pass entry. In general, it should directly link the fish pass to the natural migration corridor of fish and therefore be located close to the barrier, the main current (for hydropower plants, this means close to the turbines) and the shoreline. For oblique weirs, the pointed angle of the weir proved to be advantageous. For bottom-dwelling fish, a continuous connection to the river bottom is required (e.g. by a ramp with rough substrate and a slope <1:2). Success may depend on a combination of several factors: multiple entries or collection galleries to cover wide barriers (>100 m), varying water levels and several species with different migration corridors and/or swimming capabilities.

The attraction flow has to provide a continuous connection between the migration corridors up- and downstream of the barrier. It should be as parallel as possible to the main current (e.g. <30°), cause no turbulences and provide a high impulse of flow (defined as the product of volume and flow velocity; Larinier 2002; Seifert 2012). While the flow velocity is limited by the species' swimming capabilities, the volume can be further increased. At least 1–5% of the turbined flow are required as attraction flow (Larinier 2002; Dumont et al. 2005). In many cases, the operational discharge, which only serves the passability of the fish pass, is too low and has to be enhanced by additional flow introduced into the lowest part of the fish pass. In this case, the installation of attraction flow turbines can reduce energetic losses (Hassinger 2009a; Seifert 2012).

Passability of a fish pass is ensured, if it provides a suitable migration corridor for all relevant species. This is the case when (a) hydraulic conditions do not exceed swimming capabilities, (b) the minimum rheoactive flow velocity is provided, (c) the spatial dimensions and geometry (depth, width and length) allow adult fish of the size-decisive species (i.e. species with highest spatial demands) to pass the entire fish pass and (d) continuous rough substrate supports bottom-dwelling and weaker species by ensuring moderate flow velocities towards the bottom.

With regard to *post-passage effects*, fish should be able to continue their migration (without the risk of downstream drift) and find suitable habitats. As unidirectional connectivity restoration can transform reservoirs into ecological traps (Pelicice and Agostinho 2008), upstream migration facilities have to be combined with downstream solutions (see below).

The selection of measures for upstream continuity restoration depends on the type of the barrier (e.g. function and use), local conditions (e.g. topology, space availability, fish assemblage) and financial resources and includes the following options:

1. The removal of barriers that no longer fulfil their purpose or have lost their functionality should be considered as a sustainable solution that also restores downstream connectivity. However, it requires prior assessment of related consequences (e.g. possible adverse effects on other facilities).
2. Rock ramps and river bottom sills may cover the entire riverbed or only parts of it (e.g. partial ramps). They are usually not hydropower-related but rather used for restoring barriers serving the purpose of flood control. They have the advantage of good perceptibility, provision of several migration corridors (also downstream), low sensitivity to debris (i.e. low maintenance costs) and habitat enrichment for rheophilic species (Gebler 2007). However, the disadvantages of very high construction costs and potentially reduced passability during low flows have to be considered (BMLFUW 2012).
3. Nature-like fish passes became popular in the 1980s in Central Europe and are now successfully built worldwide (Gough et al. 2012). Since nature-like bypass channels or pool-type fish passes mimic a small natural river, they do not only restore connectivity but also provide suitable habitats for reproduction and juvenile age classes. Thus, they can partially substitute the loss of fluvial habitats and can contribute to large-scale restoration if, e.g. installed as bypass system for chains of impoundments (see Chap. 6). One main disadvantage is, however, the high spatial demand and related high costs, especially if land acquisition is necessary.
4. Technical fish passes are usually built in a way that the slope is reduced over defined, constant height differences between pools, which are connected by slots or sluices. A multispecies-efficient representative of this type is the vertical-slot fish pass, but also other types (e.g. technical pool and weir fish pass, Denil fish pass or bristles pass) exist. Although this type does not provide suitable habitat for fish, is often more expensive in construction and requires increased maintenance, its low spatial demands and wide area of application represent major advantages (BMLFUW 2012).
5. The last group includes special constructions, which might only be used under certain conditions or in combination with other measures. As shipping locks show characteristics (i.e. low flow velocity, outside of migration corridor, no continuous attraction flow and functionality) that causes more random than targeted passage of fish, their application is not recommended as alternative passage for particular species (Travade and Larinier 2002; DWA 2014). Fish lifts guide fish into a chamber that is then moved upstream. Under certain circumstances, also trap-and-truck solutions might be feasible. However, all the above-described

solutions have the disadvantage of discontinuous functionality, which is why their application has to be tested on a case-by-case basis.

In general, it can be concluded that upstream fish pass solutions are well developed and, in many cases, proven to successfully restore connectivity-at least for barriers of moderate height in small- to medium-sized rivers of temperate zones. However, there are currently no functional examples for large dams in tropical rivers where vast and diverse species assemblages and seasonal biomass peaks require special solutions (Schmutz and Mielach 2012).

Facilities for Downstream Migration

While measures for continuity restoration started with the construction of upstream fish passes, downstream migration problems were only recognized and addressed more recently (Larinier and Travade 2002). Therefore, solutions are less advanced and require further research and practical experience before they can be considered as state of the art (Böttcher et al. 2015). In any case, facilities supporting both up- and downstream migrations are required for restoring and maintaining healthy fish populations.

In contrast to upstream migrations, hydropower plants usually do not totally block downstream migrations, as fish still may be able to pass through turbines or opened spillways. However, depending on the local characteristics, fish entering these paths might get injured or even killed. Therefore, measures preventing fish from entering harmful plant components and providing alternative migration corridor are required. At the same time, progress continues in developing less harmful turbines that, however, still might not deserve to be called "fish-friendly".

Current measures for restoring connectivity in downstream direction include facilities for (1) improving safe passage, (2) prohibiting transit through harmful hydropower plant components and (3) providing alternative migration routes.

Based on thorough research of the parameters related to turbine injuries (Cada 2001; Larinier and Travade 2002), recommendations for mitigation include: reduce blade numbers, decrease the gap between the blade and its coating and lower the rotation velocity and pressure differences. Furthermore, new turbine concepts were developed. The VLH (very low head; www.vlh-turbine.com) turbine is applicable for heads of 1.4–3.2 m and flows of 10–26 m^3/s. While eels and salmon smolts showed promising survival rates (92.3%), results on other species are still missing. Furthermore, both the Archimedean screw (Schmalz 2010) and the Alden turbine (i.e. applicable for heads from 20 to 30 m and flows >30 m^3/s; Cook et al. 2003) promise high survival rates, whereby the latter has yet to be validated in the field. Another innovative example is the double rotating hydroconnect turbine, which is a gap-free screw with integrated fish lift. It was successfully tested in the rivers Jeßnitz (drop height, 3.3 m; flow, 1 m^3/s) and Sulm (drop height, 5.5 m; flow, 0.4 m^3/s) and allowed injury-free passage of many species and size classes (www.hydroconnect.at). Although its applicability for higher heads and flows still has to be assessed, it has the major advantage of being passable both ways.

If turbines cannot provide a safe passage, turbine entrainment has to be prohibited by physical barriers (e.g. rakes) sufficiently tight to provide effective protection also for small individuals. However, smaller clearances are usually connected with higher energetic losses (DWA 2014). Therefore, also in this field, efforts are made to develop physical barriers with low injury and passage rates as well as low hydraulic losses. Examples are the wedge wire screen and the Opperman screen (Hassinger 2009b). In general, the velocity in front of the screen should not exceed the critical swimming speed (0.25–0.5 m/s). Although many studies investigate the functionality of behavioural barriers, which produce a repulsive stimulus (e.g. with electricity, air bubbles, light, sound), pilot experiments are not yet convincing especially for diverse species assemblages (Gosset and Travade 1999). While louvres and bar racks, which induce a certain flow pattern to guide fish towards a bypass, might represent suitable solutions for small hydropower plants, additional tests are required to prove their efficiency.

When shielding fish from harmful passage routes, alternative migration corridors have to be offered, and fish have to be attracted to enter them. Fish passes for upstream migration usually do not work for downstream migration since fish use other corridors for up- and downstream migration. However, fish can be guided into an existing fish pass via a bypass, which has to fulfil certain criteria to allow a safe transfer. Also spillway passage can be targeted under certain conditions. Finally, also trap and truck is possible, if no other solutions are suitable.

Fish pass facilities should function the whole year, and the assessment of their functionality should include indirect (i.e. measurement of abiotic parameters, e.g. flow velocity) and direct assessments (i.e. monitoring of successful passage, mortality, injuries of fish) (Woschitz et al. 2003).

References

Adam B, Bosse R, Dumont U, Hadderingh R, Jörgensen L, Kalusa B, Lehmann G, Pischel R, Schwevers U (2005) Fischschutz- und Fischabstiegsanlagen – Bemessung, Gestaltung, Funktionskontrolle. Deutsche Vereinigung für Wasserwirtschaft, Abwasser und Abfall e.V, Hennef

Amoros C, Roux AL (1988) Interaction between water bodies within the floodplain of large rivers: function and development of connectivity. Münstersche Geographische Arbeiten 29:125–130

Amoros C, Roux AL, Reygrobellet JL, Bravard JP, Pautou G (1987) A method for applied ecological studies of fluvial hydrosystems. Regul Rivers: Res Manage 1:17–36

Antongiovanni M, Metzger JP (2005) Influence of matrix habitats on the occurrence of insectivorous bird species in Amazonian forest fragments. Biol Conserv 122:441–451

BMLFUW (2012) Leitfaden zum Bau von Fischaufstiegshilfen. Bundesministerium für Land- und Forstwirtschaft, Umwelt und Wasserwirtschaft, Wien

Bosshard P (2015) 30 things you didn't know about rivers. Available http://www.huffingtonpost.com/peter-bosshard/30-things-you-didntknow_b_7812408.html

Böttcher H, Unfer G, Zeiringer B, Schmutz S, Aufleger M (2015) Fischschutz und Fischabstieg – Kenntnisstand und aktuelle Forschungsprojekte in Österreich. Osterreichische Wasser- und Abfallwirtschaft 67:299–306

Bowne DR, Bowers MA, Hines JE (2006) Connectivity in an agricultural landscape as reflected by interpond movements of a freshwater turtle. Conserv Biol 20:780–791

Branco P, Segurado P, Santos JM, Ferreira MT (2014) Prioritizing barrier removal to improve functional connectivity of rivers. J Appl Ecol 51:1197–1206

Cada GF (2001) The development of advanced hydroelectric turbines to improve fish passage survival. Fisheries 26:14–23

Clough SC, Turnpenny WH (2001) Swimming speeds in fish: phase 1. Environment Agency, Bristol

Clough SC, Lee-Elliott IE, Turnpenny AWH, Holden SDJ, Hinks C (2004) Swimming speeds in fish: phase 2 literature review. Environment Agency, Bristol

Colgan P (1993) The motivational basis of fish behavior. In: Pitcher TJ (ed) The behaviour of teleost fishes. Chapman and Hall, London, pp 31–55

Cook TC, Hecker GE, Amaral SV, Stacy PS, Lin F, Taft EP (2003) Final report - pilot scale tests Alden/concepts NREC turbine

Cote D, Kehler DG, Bourne C, Wiersma YF (2009) A new measure of longitudinal connectivity for stream networks. Landsc Ecol 24:101–113

Dumont U, Anderer P, Schwevers U (2005) Handbuch Querbauwerke. Ministerium für Umwelt und Naturschutz. Landwirtschaft und Verbraucherschutz des Landes Nordrhein-Westfalen, Düsseldorf

DVWK (Deutscher Verband für Wasserwirtschaft und Kulturbau E.V.) (1996) Merkblatt 232/1996 - Fischaufstiegsanlagen - Bemessung, Gestaltung, Funktionskontrolle. Deutscher Verband für Wasserwirtschaft und Kulturbau e.V. (DVWK), Bonn

DWA (Deutsche Vereinigung für Wasserwirtschaft, Abwasser und Abfall e. V.) (2014) Merkblatt DWA-M 509 – Fischaufstiegsanlagen und fischpassierbare Bauwerke – Gestaltung, Bemessung, Qualitätssicherung. Deutsche Vereinigung für Wasserwirtschaft, Abwasser und Abfall e.V, Hennef

Dynesius M, Nilsson C (1994) Fragmentation and flow regulation of river systems in the nothern third of the world. Science 266:753–762

European Commission (1992) Council Directive 92/43/EEC of 21 May 1992 on the conservaiton of natural habitats and of wild fauna and flora

European Commission (2000) Directive 2000/60/EC of the European Parliament and the Council of 23 October 2000. Establishing a framework for community action in the field of water policy

Fagan WF (2002) Connectivity, fragmentation, and extinction risk in dendritic metapopulations. Ecology 83:3243–3249

Freyhof J, Brooks E (2011) European red list of freshwater fishes. Publication Office of the European Commission, Luxembourg

Fullerton AH, Burnett KM, Steel EA, Flitcroft RL, Pess GR, Feist BE, Torgersen CE, Miller DJ, Sanderson BL (2010) Hydrological connectivity for riverine fish: measurement challenges and research opportunities. Freshw Biol 55:2215–2237

Gebler J (2007) Rock ramps and nature-like-bypass-channels – design-criteria and experiences. IFAC Working Party - Fish passage best practices, Salzburg. Oct 8–10

Gosset C, Travade F (1999) Etudes de dispositifs d'aide à la migration de dévalaison des salmonidés: barrières comportementales? Cybium 23:45–66

Gough P, Philipsen P, Schollema PP, Wanningen H (2012) From sea to source - international guidance for the restoration of fish migration highways. Regional Water Authority Hunze en Aa's, Veendam

Grill G, Dallaire CO, Chouinard EF, Sindorf N, Lehner B (2014) Development of new indicators to evaluate river fragmentation and flow regulation at large scales: a case study for the Mekong river basin. Ecol Indic 45:148–159

Hartmann J, Harrison D, Opperman J, Gill R (2013) Planning at the system scale – the next frontier of hydropower sustainability. The Nature Conservancy, s.l

Hassinger R (2009a) Energieeffiziente künstliche Erzeugung von Leitströmungen bei Fischwanderhilfen. Österreichische Wasser und Abfallwirtschaft 3–4:32–34

Hassinger R (2009b) Neuartiger Fisch schonender Rechen für Wasserkraftanlagen. Desdener Wasserbauliche Mitteilungen Heft 39:251–258

ICOLD (International Commission on Large Dams) (2016) Available http://www.icold-cigb.org/GB/World_register/general_synthesis.asp

ICPDR (2013) Sustainable hydropower development in the Danube basin - guiding principles. ICPDR - International Commission for the Protection of the Danube River, Vienna

ICPDR (2015) Danube river basin district management plan - update 2015. ICPDR - International Commission for the Protection of the Danube River, Vienna

Jungwirth M (1998) River continuum and fish migration - going beyond the longitudinal river corridor in understanding ecological integrity. In: Jungwirth M, Schmutz S, Weiss S (eds) Fish migration and fish bypasses. Fishing News Books, Oxford, pp 19–32

Jungwirth M, Muhar S, Schmutz S (2000) Fundamentals of fish ecological integrity and their relation to the extended serial discontinuity concept. Hydrobiologia 422(423):85–97

Jungwirth M, Haidvogl G, Moog O, Muhar S, Schmutz S (2003) Angewandte Fischökologie and Fließgewässern. Facultas Universitätsverlag, Wien

Karle KF (2005) Analysis of an efficient fish barrier assessment protocol for highway culverts. Juneau. Final report to Alaska Department of Transportation. FHWA-AK-RD-05-02

Keitt TH, Urban DL, Milne BT (1997) Detecting critical scales in fragmented landscapes. Ecol Soc 1:4

Kemp PS, O'Hanley JR (2010) Procedures for evaluating and prioritising the removal of fish passage barriers: a synthesis. Fish Manag Ecol 17:297–322

Kuby MJ, Fagan WF, ReVelle CS, Graf WL (2005) A multiobjective optimization model for dam removal: an example trading off salmon passage with hydropower and water storage in the Willamette basin. Adv Water Resour 28:845–855

Labonne J, Ravigne V, Parisi B, Gaucherel C (2008) Linking dendritic network structures to population demogenetics: the downside of connectivity. Oikos 117:1479–1490

Lancaster J (2000) Geometric scaling of microhabitat patches and their efficacy as refugia during disturbance. J Anim Ecol 69:442–457

Larinier M (2002) Fishways - general considerations. Bulletin Français de la Pêche et de la Pisciculture 364(Suppl):23–38

Larinier M, Travade F (2002) Downstream migration: problems and facilities. Bulletin Français de la Pêche et de la Pisciculture 364(Suppl):181–207

Latham J, Collen B, McRae L, Loh J (2008) The living planet index for migratory species: an index of change in population abundance. Final Report for the Convention on the Conservation of Migratory Species, London

Lucas MC, Baras E (2001) Migration of freshwater fishes. Blackwell Science, Oxford

Mader H, Unfer G, Schmutz S (1998) The effectiveness of nature-like bypass channels in a Lowland river, the Marchfeldkanal. In: Jungwirth M, Schmutz S, Weiss S (eds) Fish migration and fish bypasses. Fishing News Books, Oxford, pp 384–402

Master L (1990) The imperiled status of North American aquatic animals. Biodivers Netw News 3:1–8

Mesa MG, Magie CD (2009) Prioritizing removal of dams for passage of diadromous fishes on a major river system. River Res Appl 25:107–117

Moilanen A, Hanski I (2006) Connectivity and metapopulation dynamics in highly fragmented landscapes. In: Crooks KR (ed) Connectivity conservation. Cambridge University Press, Cambridge, pp 44–71

Nilsson C, Reidy CA, Dynesius M, Revenga C (2005) Fragmentation and flow regulation of the world's large river systems. Science 308:405–408

Northcote TG (1978) Migratory strategies and produciton in freshwater fishes. In: Gerking SD (ed) Ecology of freshwater fish production. Blackwell Publishing Ltd, Oxford, pp 326–359

Northcote TG (1998) Migratory behavior of fish and its significance to movement through riverine fish passagee facilities. In: Jungwirth M, Schmutz S, Weiss S (eds) Fish migration and fish bypasses. Fishing News Books, Oxford, pp 3–18

O'Hanley JR (2011) Open rivers: barrier removal planning and the restoration of free-flowing rivers. J Environ Manag 92:3112–3120

O'Hanley JR, Tomberlin D (2005) Optimizing the removal of small fish passage barriers. Environ Model Assess 10:85–98

Opperman J, Grill G, Hartmann J (2015) The power of rivers: finding balance between energy and conservation in hydropower development. The Nature Conservancy, Washington, DC

Ovidio BM, Philippart JC (2008) Movement patterns and spawning activity of individual nase Chondrostoma nasus (L.) in flow-regulated and weir-fragmented rivers. J Appl Ichthyol 24:256–262

Pavlov DS (1989) Structures assisting the migrations of non-salmonid fish: USSR. Food and Agriculture Organization of the United Nations, Rome

Pelicice FM, Agostinho AA (2008) Fish-passage facilities as ecological traps in large neotropical rivers. Conserv Biol 22:180–188

Pringle C (2006) Hydrologic connectivity: a neglected dimension of conservation biology. In: Crooks KR (ed) Connectivity conservation. Cambridge University Press, Cambridge, pp 233–254

Pringle CM, Freeman MC, Freeman BJ (2000) Regional effects of hydrologic alterations on riverine macrobiota in the new world: tropical–temperate comparisons. Bioscience 50:807–823

Rosenberg DM, Mccully P, Pringle CM (2000) Global-scale environmental effects of hydrological alterations: introduction. Bioscience 50:746–752

Schmalz W (2010) Untersuchungen zum Fischabstieg und Kontrolle möglicher Fischschäden durch die Wasserkraftschnecke an der Wasserkraftanlage Walkmühle an der Werra in Meiningen - Abschlussbericht. Im Auftrag der Thüringer Landesanstalt für Umwelt und Geologie, Breitenbach

Schmutz S, Mielach C (2012) Scoping of existing research on fish passage through large dams and its applicability to Mekong mainstream dams. Inception paper. 1–25

Schmutz S, Mielach C (2013) Measures for ensuring fish migration at transversal structures - technical paper. International Commission for the Protection of the Danube River, Vienna

Schmutz S, Mielach C (2015) Review of existing research on fish passage through large dams and its applicability to Mekong mainstream dams. Mekong River Commission, Phnom Penh

Schmutz S, Zitek A, Dorninger C (1997) A new automatic drift sampler for riverine fish. Hydrobiol 139:449–460

Segurado P, Branco P, Ferreira MT (2013) Prioritizing restoration of structural connectivity in rivers: a graph based approach. Landsc Ecol 28:1231–1238

Seifert K (2012) Praxishandbuch: Fischaufstiegsanlagen in Bayern. Hinweise und Empfehlungen zu Planung, Bau und Betrieb. Bayerisches Landesamt für Umwelt, München

Seliger C, Scheikl S, Schmutz S, Schinegger R, Fleck S, Neubarth J, Walder C, Muhar S (2016) HY:CON: a strategic tool for balancing hydropower development and conservation needs. River Res Appl 32:1438–1449

Spindler T, Chovanec A, Zauner G, Mikschi E, Kummer H, Wais A, Spolwind R (1997) Fischfauna in Österreich: Ökologie-Gefährdung-Bioindikation-Fischerei-Gesetzgebung. Umweltbundesamt, Wien

Tischendorf L, Fahrig L (2000) How should we measure landscape connectivity? Landsc Ecol 15:633–641

Travade F, Larinier M (2002) Fish locks and fish lifts. Bulletin Français de la Pêche et de la Pisciculture 364:102–118

Vannote RL, Minshall GW, Cummins K, Sedell JR, Cushing CE (1980) The river continuum concept. Can J Fish Aquat Sci 37:130–137

Vörösmarty CJ, McIntyre PB, Gessner MO, Dudgeon D, Prusevich A, Green P, Glidden S, Bunn SE, Sullivan CA, Liermann CR, Davies PM (2010) Global threats to human water security and river biodiversity. Nature 467:555–561

Waidbacher H, Haidvogl G (1998) Fish migration and fish passage facilities in the Danube: past and present. In: Jungwirth M, Schmutz S, Weiss S (eds) Fish migration and fish bypasses. Fishing News Books, Oxford, pp 85–98

Ward JV (1989) The four-dimensional nature of lotic ecosystems. J N Am Benthol Soc 8:2–8

Ward JV (1997) An expansive perspective of riverine landscapes: pattern and process across scales. River Ecosyst 6:52–60

Ward JV, Stanford JA (1983) The serial discontinuity concept of lotic ecosystems. In: Fontaine TD, Bartell SM (eds) Dynamics of lotic ecosystems. Ann Arbor Science, Michigan, pp 29–42

Ward JV, Stanford JA (1995a) Ecological connectivity in alluvial river ecosystems and its disruption by flow regulation. Regul Rivers Res Manag 11:105–119

Ward JV, Stanford JA (1995b) The serial discontinuity concept: Extending the model to floodplain rivers. Regul Rivers Res Manag 10:159–168

Wiens JA (2002) Riverine landscapes: taking landscape ecology into the water. Freshw Biol 47:501–515

Woschitz G, Eberstaller J, Schmutz S (2003) Mindestanforderungen bei der Überprüfung von Fischmigrationshilfen (FMH) und Bewertung der Funktionsfähigkeit. Österreichischer Fischereiverband, Wien

WWF (2006) Free-flowing rivers: economic luxury or ecological necessity? World Wildlife Fund, Zeist

Chapter 10
Phosphorus and Nitrogen Dynamics in Riverine Systems: Human Impacts and Management Options

Gabriele Weigelhofer, Thomas Hein, and Elisabeth Bondar-Kunze

10.1 Introduction

Water chemistry constitutes one key factor for the ecological state of streams and rivers as it determines the composition of the media in which the aquatic organisms live. Among the various chemical substances dissolved in water, phosphorus (P) and nitrogen (N) are particularly important for the management of riverine systems. These two macronutrients are essential components of all organisms and are closely linked to the aquatic carbon cycle, determining both the primary production and the microbial mineralization of organic matter in aquatic systems. The industrialization and intensification of agricultural production during the twentieth century has resulted in the nutrient enrichment and eutrophication of many freshwaters in Europe and the USA, impairing the water quality of rivers, lakes, and aquifers (Grizetti et al. 2011). Among others, eutrophication is responsible for toxic algal blooms, water anoxia, and habitat and biodiversity loss in freshwater ecosystems and poses direct threats to humans by impairing drinking water quality (Smith and Schindler 2009). Nutrient enrichment causes severe problems in coastal zones and can even affect the climate through increased greenhouse gas emissions. Despite current improvements in wastewater treatment from industrial and municipal sources in Europe (Kroiss et al. 2005), phosphorus and nitrogen remain of concern for river managers especially in regions where intensive urban or agricultural land use results in pollution of aquatic systems through diffuse nutrient inputs. Diffuse sources challenge the management of nutrients in riverine systems by requiring a combination of mitigation measures on both the catchment and the reach scale (Mainstone and Parr 2002).

G. Weigelhofer (✉) · T. Hein · E. Bondar-Kunze
Institute of Hydrobiology and Aquatic Ecosystem Management, University of Natural Resources and Life Sciences, Vienna, Austria

WasserCluster Lunz Biological Station GmbH, Lunz am See, Austria
e-mail: gabriele.weigelhofer@boku.ac.at; thomas.hein@boku.ac.at; elisabeth.bondar@boku.ac.at

© The Author(s) 2018
S. Schmutz, J. Sendzimir (eds.), *Riverine Ecosystem Management*, Aquatic Ecology
Series 8, https://doi.org/10.1007/978-3-319-73250-3_10

The following chapter describes dominant input pathways and transformation processes for these two nutrients in streams and rivers and deals with the various human impacts on these processes which impair the nutrient retention function of riverine systems, with specific reference to the Danube River basin. The chapter provides an overview of measures for the mitigation and management of diffuse nutrient inputs in both the river and the riparian zone. Technical treatment of wastewater and point sources is not further addressed in this book chapter and can be found in, e.g., Tchobanoglous et al. (2003). In addition, we address consequences of river restoration measures on the nutrient uptake and release in running waters.

10.2   Historic and Current Emission Situation in the Danube River Basin

Between the 1950s and the 1980s, the emissions of nitrogen and phosphorus into the rivers of the Danube basin increased by more than the twofold as a result of industrialization, urbanization, and intensification of agriculture (Kroiss et al. 2005; Grizetti et al. 2011). Since the 1990s, slight-to-moderate mitigations of nutrient inputs have been achieved by improving wastewater treatment via the implementation of collection systems and new technologies mainly in Germany and Austria and by reducing industrial discharges in the lower Danube countries. However, trends are not consistent throughout European water bodies. In particular, emissions from diffuse sources originating from agricultural areas remain elevated (Kroiss et al. 2005). The significance of diffuse sources for phosphorus and nitrogen inputs to riverine systems in Austria and the Danube basin is shown in Fig. 10.1. While wastewater treatment plants account for about 20–26% of nutrient emissions to Austrian rivers, diffuse inputs via groundwater, soil erosion, and surface runoff (including urban areas) play a key role in delivering nitrogen and phosphorus to riverine systems (BMLFUW 2017). Currently, only 6% of Austrian streams and rivers have a moderate to high risk for failure in water quality due to point sources, but about 25% are threatened in their water quality by diffuse sources.

While point sources are relatively easy to control, the management of diffuse sources requires an integrative approach on multiple levels, comprising (1) minimization of emissions in the catchment (see Chap. 13), (2) nutrient retention in riparian buffer zones and floodplains, and (3) the control of the nutrient cycling within the river channel (Mainstone and Parr 2002). The increasing application of good management practices in agricultural catchments during the last decade has resulted in the mitigation of nutrient loads in rivers, especially as regards nitrogen (Kronvang et al. 2005; Oenema et al. 2005). However, the effects of catchment measures on river water quality are often less effective than expected due to nitrogen and phosphorus legacies in soils and groundwater from past land use activities (Sharpley et al. 2014). Therefore, catchment management needs to be supported by on-site measures in both the riparian zone and the channel. Effective nutrient management within the riverine system, in turn, requires a profound understanding of the various biogeochemical processes phosphorus and nitrogen undergo in running waters.

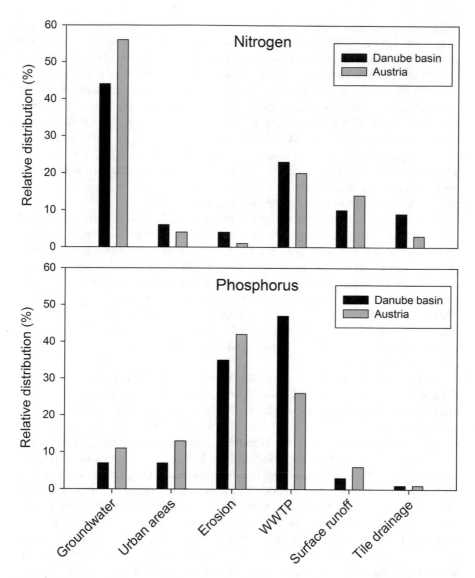

Fig. 10.1   Sources of nitrogen and phosphorus to riverine systems in Austria and the Danube basin (Data from Kroiss et al. 2005; BMLFUW 2017)

10.3   Forms and Sources of Phosphorus and Nitrogen

Phosphorus occurs in aquatic systems in four basic forms: dissolved inorganic phosphorus (usually referred to as soluble reactive phosphorus (SRP) which is immediately bioavailable), dissolved organic phosphorus (DOP; e.g., P-esters), particulate organic phosphorus (POP; in detritus and living biomass), and particulate inorganic phosphorus (PIP; e.g., iron- or aluminum-bound phosphorus on particles) (Allan and Castillo 2007). SRP is biologically available for both aquatic primary producers and microorganisms, while the particulate and soluble organic fractions must undergo chemical or biological transformations to SRP before being bioavailable.

Nitrogen occurs in freshwater ecosystems in three forms: dissolved inorganic nitrogen (DIN), including ammonium (NH_4-N), nitrate (NO_3-N), and nitrite (NO_2-N), dissolved organic nitrogen (DON; e.g., amino acids, polypeptides), and particulate organic nitrogen (PON) (Allan and Castillo 2007). In addition, nitrogen occurs in gaseous forms as dinitrogen gas N_2 and nitrous oxide (N_2O), a potent greenhouse gas. DIN is biologically available for both aquatic primary producers and microorganisms, whereby NH_4-N is preferentially taken up by the aquatic community due to lower physiological costs (Birgand et al. 2007).

Natural sources for phosphorus and nitrogen comprise leaching from terrestrial soils and plant material during decomposition, release of P from weathering rocks, atmospheric deposition of N_2 (precipitation and dry fallout), and biological N fixation through cyanobacteria (Mainstone and Parr 2002; Bernot and Dodds 2005; Birgand et al. 2007; Withers and Jarvie 2008). As nutrient inputs from natural sources are generally low, pristine streams usually show SRP concentrations <10 µg SRP L^{-1}, while average DIN concentrations may amount to 0.1 mg NO_3-N L^{-1}, 0.015 mg NH_4-N L^{-1}, and 0.001 mg NO_2-N L^{-1} (Allan and Castillo 2007). DON can reach proportions of 40–90% of total nitrogen and, thus, often constitutes a major component in pristine systems.

Anthropogenic sources import significant amounts of nutrients into streams and rivers which lead to the eutrophication of the aquatic system and impair its ecological state. The main anthropogenic inputs to riverine systems include increased atmospheric deposition of N_2 due to the burning of fossil fuels, cultivation of N-fixing crops, municipal wastewater and industrial effluents, and agricultural fertilizers (Bernot and Dodds 2005; Burgin and Hamilton 2007). Depending on their temporal and spatial extent, input pathways can be distinguished as (1) point sources, largely in form of municipal wastewater and industrial effluents, (2) nonpoint or diffuse sources from agricultural areas, and (3) intermediate sources, such as runoff from impervious surfaces (Withers and Jarvie 2008). Municipal wastewater is usually dominated by dissolved inorganic and, thus, bioavailable phosphorus and nitrogen (Mainstone and Parr 2002; Withers and Jarvie 2008), the concentrations of which depend on the efficiency of the sewage treatment. Point sources constitute permanent and localized delivery pathways, which are comparatively easy to control. Diffuse inputs from agricultural land use include agricultural fertilizers and increased soil leachate and erosion due to tillage. As SRP and ammonium easily adsorb to charged soil particles, these nutrients enter streams and rivers mainly via soil erosion (Craig et al. 2008; Withers and Jarvie 2008) (Figs. 10.1 and 10.3). By contrast, nitrate is highly soluble and mobile. Excess NO_3-N from agricultural areas is, thus, usually leached to groundwater or drainage waters and transported into river systems via subsurface flow paths (Grizetti et al. 2011) (Figs. 10.1 and 10.3). Intermediate sources include runoff from various urban areas and farmyards and vary greatly in nutrient amounts and composition. Both diffuse and intermediate sources occur mainly during storm events and are difficult to control. Stormwater management structures in urban areas, such as vegetated ponds and wetlands, bio-retention devices, and porous pavements, can help to control both water and nutrient fluxes to urban streams (Bernhardt and Palmer 2007).

10.4 Nutrient Cycling in Streams and Rivers

Once in the aquatic system, phosphorus and nitrogen undergo numerous biogeo-chemical transformations (Fig. 10.2). Biotic transformations include the autotrophic and heterotrophic uptake of nutrients from the water, their assimilation into biomass, and their release by excretion and microbial decomposition (Reddy et al. 1999; Bernot and Dodds 2005; Birgand et al. 2007). In deep and slow-flowing rivers and floodplain channels, nutrient uptake by macrophytes and emergent plants plays a major role in nutrient cycling. Plants can take up SRP and DIN from soil or sediment pore water via roots or directly from the water column (Birgand et al. 2007). While nutrient storage in aboveground plant tissue is usually short-term, resulting in the release of nutrients after the vegetation period, belowground storage in roots and rhizomes may provide long-term storage, depending on the hydrological situation, the vegetation type, the physicochemical properties of the water, and the climate.

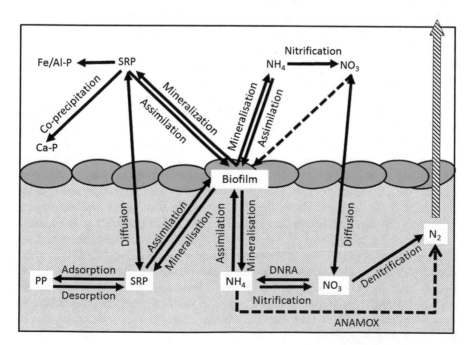

Fig. 10.2 Dominant sources and transformation processes for nitrogen and phosphorus in running waters (after Mainstone and Parr 2002; Revsbech et al. 2006; Withers and Jarvie 2008)

Enhanced nutrient levels and increased water temperatures, for example, accelerate microbial decomposition of plant tissue (Withers and Jarvie 2008). Drying and rewetting of floodplain sediments in the course of water level fluctuations have also been shown to foster nutrient release from organic matter (Schönbrunner et al. 2012). In small streams, benthic processes dominate over water column processes. Benthic algae and microorganisms can assimilate nutrients from both the water column and the pore water of the sediments. In the case of phosphorus surplus, algae are capable of luxury phosphorus uptake, i.e., excess uptake that is not immediately needed but can be used subsequently for later growth, while microorganisms can store phosphorus via the formation of polyphosphate compounds (Reddy et al. 1999).

Bacterial mineralization of organic matter results in the release of SRP and NH_4–N to the water column. Under aerobic conditions, NH_4–N is turned into NO_3–N by nitrifying bacteria via nitrification (Bernot and Dodds 2005; Birgand et al. 2007) (Fig. 10.2). Under anaerobic conditions, denitrifying bacteria use nitrate as an electron acceptor to oxidize organic matter, thus reducing NO_3–N via NO_2–N and N_2O to N_2 (Bernot and Dodds 2005; Birgand et al. 2007; Burgin and Hamilton 2007). This denitrification process is promoted by low oxygen concentrations and high concentrations of organic matter and nitrate, as occur, e.g., in wetland soils, sediments of agricultural streams, and groundwater, and represents a permanent N sink for streams and rivers. Denitrification is often restricted to anoxic microzones in the sediments and can be coupled to nitrification by using nitrate originating from decomposition and subsequent nitrification rather than nitrate imported from external sources (Birgand et al. 2007). In aquatic systems with high organic carbon, but low nitrate concentrations, dissimilatory nitrate reduction to ammonium (DNRA) may become important under anoxic conditions (Birgand et al. 2007; Burgin and Hamilton 2007). However, the significance of DNRA for nitrogen cycling in running waters remains to be investigated yet. Another potential sink for nitrogen is the anaerobic oxidation of ammonium to N_2 using nitrite (Anammox). So far, Anammox has been mainly observed in anoxic environments with high nitrogen, but low carbon concentrations, such as wastewaters and marine systems (Burgin and Hamilton 2007).

In addition to these biological processes, phosphorus availability in running waters is influenced by various physical and chemical transformations. The adsorption of phosphorus to sediment particles plays a key role in phosphorus cycling, especially in streams and rivers impaired by agricultural land use. Adsorption comprises all physical and chemical processes in which phosphorus is bound to the surface of particles, such as ligand exchange, electrostatic attraction, and ion exchange (Reddy et al. 1999; Withers and Jarvie 2008). Sedimentation of particle-bound phosphorus constitutes an important phosphorus sink in retention zones of rivers, such as pools, floodplain lakes, or impounded sections. However, the adsorbed phosphorus can be released again to the water column, depending on the adsorption capacity of the sediments, which is highest in clay and sand, and on the concentration gradient between the pore water and the water column (Reddy et al. 1999; Mainstone and Parr 2002). In general, phosphorus adsorption occurs at high SRP concentrations in the surface water, while desorption is favored by low

SRP concentrations in the surface water. The zero equilibrium phosphorus concentration (EPC_0) is the SRP concentration in the water where no net phosphorus exchange between the water column and the sediments occurs (House 2003). High EPC_0 values are especially evident in streams and rivers in agricultural catchments due to phosphorus overloading of the sediments. As a consequence, sediments function as an internal phosphorus source for the water column during most of the time (Sharpley et al. 2014).

Under aerobic conditions, dissolved inorganic and organic phosphorus may complex with metal oxides and hydroxides to form insoluble precipitates (Reddy et al. 1999; House 2003). This phosphorus is released under anaerobic conditions as may occur in organic-rich sediments of floodplain lakes and agricultural streams and rivers. In addition, phosphorus can coprecipitate with calcite in calcareous waters under high pH conditions resulting from the photosynthetic activity of macrophytes and benthic algae.

To summarize, both biotic and abiotic processes significantly influence phosphorus and nitrogen retention in riverine systems, depending on various factors such as hydrology, climate, the activity of primary producers and decomposers, and the loading of the system by organic matter and nutrients. In particular, the role of fine sediment accumulations as potential sink or source for phosphorus has to be taken into account in nutrient management concepts.

10.5 Human Impacts on Nutrient Cycling

Due to the various biogeochemical transformation processes, nitrogen and phosphorus are continuously recycled between their inorganic and organic forms as well as among the water column, the sediments, and the biota. The continuous downstream movement of water in streams and rivers transforms these nutrient cycles into spirals (nutrient spiraling concept; see review by Ensign and Doyle 2006). The length of the spirals depends on the nutrient uptake capacity of the river relative to the water transport and represents the efficiency of the aquatic system for nutrient retention. This nutrient retention efficiency depends on two factors: (1) the physical (hydrological) retention of the water within the system and (2) the nutrient demand of the aquatic community.

The hydrological retention determines the contact time and the contact area between nutrients and biogeochemically reactive sites in the riverine systems (Ensign and Doyle 2006). It depends on the three-dimensional connectivity of the river channel with adjacent compartments, namely, the longitudinal linkage of headwaters with downstream reaches, the transversal linkage between channel and riparian areas, and the vertical linkage between surface water and the hyporheic zone (Ward 1989). Time, in the form of hydrological dynamics (e.g., flooding of riparian zones), adds a fourth dimension to the complex nutrient exchange processes in riverine systems.

Regarding the longitudinal aspect, nutrient retention is highest in headwater streams, which are important links between the catchment and downstream reaches (Reddy et al. 1999). Due to their diverse channel morphology and their low discharge, pristine headwaters can retain large amounts of nutrients via benthic uptake, thereby controlling nutrient transport into downstream reaches (Craig et al. 2008). Stream regulation due to urbanization or agricultural land use results in a homogenization of the stream channel and an acceleration of water flow and, thus, decreases the physical retention function of headwater streams (Ensign and Doyle 2005).

The vertical dimension of nutrient retention via the hyporheic zone is especially important in small streams (Boulton 2007). The hyporheic water exchange depends on the porosity of the sediments and on pressure imbalances at the sediment surface induced by local obstacles in the channel. Removal of flow obstacles in the channel, coverage of sediment surfaces with concrete or pavement, and the clogging of sediments due to siltation restrict the hyporheic water exchange and heavily impair the nutrient retention of streams (Boulton 2007). Besides, sedimentation of nutrient-loaded soil particles from agricultural landscapes may lead to internal eutrophication.

In larger streams and rivers, the lateral hydrological connectivity with riparian zones and floodplains determines the efficiency of nutrient retention. Here, the dimension of time gains in importance. Natural water level fluctuations of the river lead to the repeated connection and disconnection of floodplain water bodies with the main river, inducing the frequent exchange of chemically different water sources (Weigelhofer et al. 2015). Floodplains have proven to be especially efficient in trapping nutrients associated with particles (Reddy et al. 1999; Fisher and Acreman 2004). The remobilization of sediments is usually low as old sediments are buried by freshly deposited sediments. However, the role of floodplains in retaining dissolved nutrients is less clear and depends on the hydrology and the balance between uptake and release processes. Floodplain soils may constitute hotspots for denitrification in the case of high organic matter contents and high water tables (Forshay and Stanley 2005). However, the repeated drying and rewetting of floodplain soils can also cause the release of substantial amounts of nitrogen and phosphorus during flooding (Schönbrunner et al. 2012; Weigelhofer et al. 2015). Disconnections of floodplains from the main river, resulting from river regulation, impoundments, and channel incision, as well as hydrological alterations due to land use and climate changes have largely reduced the lateral hydrological connectivity of river-floodplain systems in Europe, thereby depriving rivers of these important retention structures (Hein et al. 2016).

The second aspect of nutrient retention in running waters is the nutrient demand of the aquatic primary producers and decomposers, which is controlled by the specific C–N–P (carbon-to-nitrogen-to-phosphorus) ratio of their bodies compared to the C–N–P ratio of their food (ecological stoichiometry; Cross et al. 2005). In pristine streams and rivers, nitrogen and phosphorus are mainly delivered by terrestrial plant material, which has substantially higher C-nutrient ratios than algae or microorganisms, thus limiting production. In eutrophic streams and rivers, nutrient supply from anthropogenic sources can exceed the demand of the community and lead to the saturation of the aquatic system (Bernot and Dodds 2005). Saturation of lotic systems occurs as a result of (1) a limited nutrient demand of aquatic organisms as they become limited by other factors (e.g., light, oxygen), (2) the internal release of nutrients due to mineralization and abiotic release processes, and (3) reduced adsorption capacities of overloaded sediments (Bernot and Dodds 2005; Withers and Jarvie 2008). Sediments enriched with nutrients and organic matter from the catchment may serve as internal eutrophication source for the aquatic system as they continuously provide benthic communities with nutrients from below even though external inputs have been reduced. Organic matter accumulations in sediments occur especially in agricultural streams due to manure application, soil erosion, and the mowing of the riparian vegetation.

Numerous in-field nutrient addition experiments have determined the retention efficiency of streams for dissolved nutrients (Ensign and Doyle 2006). Decreased uptake efficiencies for dissolved N and P have been mainly reported from agricultural and urban streams as these streams are often subject to both degraded stream morphology and increased nutrient loads (Bernhardt and Palmer 2007). While uptake lengths for ammonium or SRP range from less than 100 m to a few 100 m in oligotrophic headwater streams (e.g., Hall et al. 2002; Gibson et al. 2015), agricultural headwater streams often yield uptake lengths of several kilometers (e.g., Gücker and Pusch 2006; Weigelhofer et al. 2013; Sheibley et al. 2014). Such streams have lost their natural retention function and act as mere transport systems, impairing the water quality of downstream reaches and recipient standing waters.

10.6 Potential and Limitations of Mitigation Measures

The following chapter focuses on measures within riverine systems, including riparian zones, for the management of diffuse nutrient inputs to streams and rivers (end-of-pipe measures). Measures for treatment of point sources, especially technical measures for wastewater treatment, are not discussed. For management measures in the catchment, we refer to Chap. 13.

Riparian areas constitute important interaction and buffer zones for river ecosystems as they control the fluxes of material and energy from the terrestrial catchment and the adjacent groundwater to the surface water (Hoffmann et al. 2009). Floodplains and riparian areas have the ability to retain, transform, and release nutrients, thereby influencing the water quality of the recipient water body (Hoffmann et al. 2009; Roberts et al. 2012). The sink and source function of riparian areas depends on the delivery pathway (surface runoff, drainage water, groundwater, or floodwater), the form of the delivered nutrient (particulate or dissolved), the specific biogeochemical conditions in the riparian area (e.g., soil humidity), the riparian vegetation, and the temperature (Fisher and Acreman 2004). Depending on the groundwater table in the riparian area relative to the water table of the surface water, riparian soils may favor oxic or anoxic processes. The use of riparian areas for nutrient management has led to a variety of initiatives, such as the restoration and reconnection of former floodplains, the establishment of vegetated (managed) riparian buffer strips, the reestablishment of wetlands on agricultural land, and the installation of denitrifying bioreactors along streambanks and within channels.

Vegetated buffer strips are narrow, tillage-free, uncultivated border zones between agricultural areas and streams (Hoffmann et al. 2009; Roberts et al. 2012). While natural riparian zones vegetated with floodplain forests can also function as buffer zones, vegetated buffer strips are often optimized for nutrient removal as to vegetation type, width, and location, and they can be managed (Vought et al. 1994; Mayer et al. 2007). Vegetated buffer strips aim at reducing P and N inputs from soil erosion and surface runoff via deposition of soil particles, infiltration of water, and the subsequent geochemical and biological retention through sorption, precipitation, plant, and microbial uptake. Studies show that the retention of total phosphorus (TP) may be fairly efficient, depending on the type of vegetation and the morphology of the buffer strip (e.g., slope, width; Mayer et al. 2007), yielding TP retention between 40% and 95% of the original loads (Hoffmann et al. 2009). However, non-managed buffer strips usually provide no permanent P sink. Part of the deposited TP can be remobilized in the soil, e.g., in the course of mineralization of organic matter or P desorption, and be delivered as SRP to the surface water (Reddy et al. 1999; Fisher and Acreman 2004). The plantation and harvest of fast-growing species on buffer strips removes accumulated P and reduces P saturation, thus decreasing also DRP losses in surface runoff (Vought et al. 1994). So far, there is little evidence for significant N removal in riparian zones via plant uptake (Vought et al. 1994). However, riparian zones may be hotspots for nitrate removal in subsurface water due to denitrification, especially if a high water table is maintained in the biologically active soil (Mayer et al. 2007). Sabater et al. (2003) measured annual nitrate removal rates via denitrification between 5% and 30% m^{-1} in the riparian zones of 14 streams across Europe. The amounts of nitrate removed by denitrification depend more on the hydrology and the soil of the riparian zone than on buffer width and vegetation type (Vought et al. 1994). In general, complex buffer zones combining different vegetation types, such as grassland and forest communities, are the most efficient structures for nutrient removal and provide additional ecosystem services, such as shading, bank stabilization, increased habitat diversity, and improved microclimate (Mander et al. 2005). The effectiveness of riparian buffer strips for nutrient retention is largely reduced if water from terrestrial areas can circumvent the riparian buffer via subsurface flow or preferential surface flow paths, such as ditches which drain road runoff directly into streams (Mainstone and Parr 2002).

In large streams and rivers, riparian zones expand to *floodplains* surrounding the river channels (Fig. 10.3). The retention capacity of these floodplains is largely determined by the lateral hydrological connectivity. Like riparian buffer strips, natural floodplains can show a high retention capacity for particles as well as a high denitrification potential (Fisher and Acreman 2004; Hoffmann et al. 2009). Pinay et al. (2007), for example, measured denitrification rates in floodplain soils of European rivers of up to 30 g N m^{-2} month^{-1}. Floodplains also provide a multitude of ecosystem services apart from nutrient retention, including groundwater replenishment, flood protection, and habitats for a diverse flora and fauna, such as spawning habitats and nurseries for fish (Hein et al. 2016). Therefore, river managers

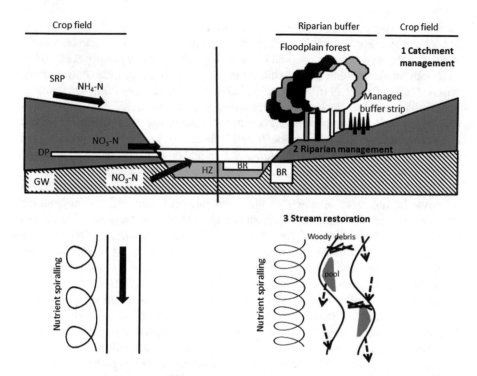

Fig. 10.3 Management options for the mitigation of phosphorus and nitrogen in riverine systems at different scales: 1. the catchment; 2. the riparian zone; 3. the stream/river. *BR* denitrifying bioreactors, *DP* drainage pipe, *GW* groundwater, *HZ* hyporheic zone

have to consider possible detrimental side effects of using floodplains for nutrient mitigation. In the case of reconnecting isolated backwaters with the main river, nutrient loading of the floodplain by inflowing river water can affect the species composition in the wetland negatively (Verhoeven et al. 2006). Besides, flooding can increase the terrestrialization of shallow backwaters through increased sedimentation, and it can enhance greenhouse gas emissions due to the inundation of organic-rich floodplain soils. In the case of reestablishing wetlands on former agricultural land, nutrient legacies in the soils have to be considered, too (Reddy et al. 1999).

Denitrifying bioreactors, also known as denitrification beds, are one of the newest technologies for edge-of-field nitrate reduction (Schipper et al. 2010; Christianson et al. 2012). Denitrifying bioreactors are porous containers which are filled with an organic carbon source, such as wood chips, sawdust, and straw, in order to facilitate denitrification (Schipper et al. 2010; Christianson et al. 2012). For nutrient management in riverine systems, such bioreactors can be installed within groundwater or drainage water flow paths, along streambanks for diffuse inputs, or directly within the stream channel (Fig. 10.3). Due to the increased hydraulic conductivity of the reactors compared to the surrounding soils or sediments, diffuse inputs are concentrated by the reactor, thereby improving nitrate removal. Long-term nitrate removal rates of denitrification beds range between 2 and 22 g N m^{-3} day^{-1} in groundwater, depending on the water residence time, the organic carbon source, nitrate concentrations, water temperature, pH, and the hydrological regime (Schipper et al. 2010). In-stream reactors within an agricultural drainage ditch yielded maximum nitrate removal rates of 160 g N m^{-2} month^{-1} (Robertson and Merkley 2009). In drainage water with fluctuating flow regimes, denitrifying bioreactors may be less effective as alternating high flow rates and intermittency restrict the denitrification process (Christianson et al. 2012; Weigelhofer and Hein 2015). Detrimental side effects of denitrifying bioreactors on adjacent surface waters and the atmosphere are an increased output of DOC, especially during the initial phase, and the production of N$_2$O.

Apart from measures in the riparian zone, *stream restoration* can significantly improve the in-stream retention of dissolved nutrients. Channel reconfiguration, such as channel widening and remeandering, and the restoration of structural complexity via the addition of flow obstructions (e.g., debris dams, side pools, and diversification of bed materials) may enhance nutrient uptake by increasing water residence time, promoting contact between the water and the sediment surface, and enhancing the hyporheic water exchange (Bukaveckas 2007; Craig et al. 2008; Hines and Hershey 2011). Woody material on the stream bed additionally increases the nutrient demand of the decomposing microorganisms due to the high C–N and C–P ratios (Roberts et al. 2007). Besides, debris dams provide organic carbon for in-stream denitrification (Craig et al. 2008). For example, the creation of riffles, cross vanes, and step pools within a restored stream shortened NH$_4$–N uptake lengths from 200 to 70 m (Hines and Hershey 2011). Bukaveckas (2007) observed reductions in P and N uptake lengths from 1370 to 380 m and from 20 km to 620 m, respectively, after channel reconstruction and reconnection with the floodplain, while Roberts et al. (2007) measured reductions in NH$_4$–N uptake lengths of about 50–70% after addition of woody debris in stream channels.

However, the efficiency of stream restoration measures on the in-channel nutrient retention has not yet been evaluated systematically so far, especially in comparison with management measures in the catchment and the riparian zone. Firstly, the majority of stream restoration measures primarily aim at restoring functions other than nutrient retention, such as channel stabilization or habitat diversity. Thus, effects of restoration on nutrient retention are seldom evaluated. Secondly, the tight connection between the various biogeochemical transformations of nutrients in the water and the sediments and the temporal dimension of these processes (e.g., diurnal patterns in primary production) complicates the evaluation of overall nutrient retention. Increased residence time, for example, may increase nutrient uptake by the biota but may also promote sedimentation of fine particles, creating anoxic conditions in the sediments which favor nutrient release (Weigelhofer et al. 2013). Finally, effects of reach-scale restoration measures on the water quality may be distorted by effects of catchment-scale factors, such as the hydrological regime of the catchment, the stream size, and the nutrient loading. For example, the positive effects of stream restoration are usually overshadowed by excessive nutrient loading in agricultural catchments (Weigelhofer et al. 2013) and by altered hydrographs with high stormwater flows in urbanized areas (Bernhardt and Palmer 2007). In those cases, stream restoration concepts should incorporate the establishment of functional units within the stream course, which possess high nutrient uptake capacities, but often need certain maintenance activities. Examples for such functional units are in-stream sediment traps, in-stream wetlands, and slow-flowing stream reaches with planted submerged macrophytes (Filoso and Palmer 2011; Hines and Hershey 2011; Richardson et al. 2011). In the USA, such functional restoration concepts involving the creation of stream-wetland complexes in lowland streams have proven to successfully increase in-stream nutrient uptake (Filoso and Palmer 2011).

10.7   Conclusions and Open Questions

This review shows that efficient mitigation and management of nutrients in riverine systems need measures on both the catchment and the reach scale. On the reach scale, riparian zones are key components for nutrient retention. Stream restoration measures may additionally improve in-stream nutrient retention. However, in catchments with excessive nutrient loading, stream restoration needs a priori reductions of nutrient inputs into the riverine system to avoid detrimental effects on water quality through nutrient release from sediments.

   This review also shows that the efficiency of the various measures in the riparian zone and the stream channel can vary widely depending on the environmental conditions. Thus, for a sustainable and efficient management of nutrients in riverine systems, more investigations are needed which evaluate and compare the nutrient retention efficiency of different management measures under varying conditions, considering especially the temporal variability of nutrient transformation processes. In particular, studies need to concentrate on small headwater streams which have the highest potential for nutrient retention due to their strong linkage with surrounding ecosystems. So far, these systems have been largely neglected in restoration efforts in Austria.

   Future studies need also to address climate change impacts on nutrient cycling. Increased air and water temperatures may accelerate nutrient cycling in riverine systems, while altered hydrology may significantly influence nutrient input and exchange pathways. In addition, an increased variability in water temperature or water levels may change the natural balance of nutrient transformation processes, thereby impacting the nutrient cycling in the aquatic system. Thus, modern stream restoration needs to consider future changes of environmental conditions for a sustainable mitigation of nutrients in riverine systems.

References

Allan JD, Castillo MM (2007) Stream ecology. The structure and function of running waters. Springer, Dordrecht

Bernhardt ES, Palmer MA (2007) Restoring streams in an urbanizing world. Freshw Biol 52:738–751

Bernot MJ, Dodds WK (2005) Nitrogen retention, removal, and saturation in lotic ecosystems. Ecosystems 8:442–453

Birgand F, Skaggs RW, Chescheir GM, Gilliam JW (2007) Nitrogen removal in streams of agricultural catchments—a literature review. Crit Rev Environ Sci Technol 37:381–487

BMLFUW (2017) Nationaler Gewässerbewirtschaftungsplan 2015. BMLFUW, Wien, 356 S

Boulton AJ (2007) Hyporheic rehabilitation in rivers: restoring vertical connectivity. Freshw Biol 52:632–650

Bukaveckas PA (2007) Effects of channel restoration on water velocity, transient storage, and nutrient uptake in a channelized stream. Environ Sci Technol 41:1570–1576

Burgin AJ, Hamilton SK (2007) Have we overemphasized the role of denitrification in aquatic ecosystems? A review of nitrate removal pathways. Front Ecol Environ 5:89–96

Christianson LE, Bhandari A, Helmers MJ (2012) A practice-oriented review of woodchip bioreactors for subsurface agricultural drainage. Appl Eng Agric 28:861–874

Craig LS, Palmer MA, Richardson DC, Filoso S, Bernhardt ES, Bledsoe BP, Doyle MW, Groffman PM, Hassett BA, Kaushal SS, Mayer PM, Smith SM, Wilcock PR (2008) Stream restoration strategies for reducing river nitrogen loads. Front Ecol Environ 6:529–538

Cross WF, Benstead JP, Frost PC, Thomas SA (2005) Ecological stoichiometry in freshwater benthic systems: recent progress and perspectives. Freshw Biol 50:1895–1912

Ensign SH, Doyle MW (2005) In-channel transient storage and associated nutrient retention: evidence from experimental manipulations. Limnol Oceanogr 50:1740–1751

Ensign SH, Doyle MW (2006) Nutrient spiraling in streams and river networks. J Geophys Res Biogeosci 111:G04009. https://doi.org/10.1029/2005JG000114

Filoso S, Palmer MA (2011) Assessing stream restoration effectiveness at reducing nitrogen export to downstream waters. Ecol Appl 21:1989–2006

Fisher J, Acreman MC (2004) Wetland nutrient removal: a review of the evidence. Hydrol Earth Syst Sci 8:673–685

Forshay KJ, Stanley EH (2005) Rapid nitrate loss and denitrification in a temperate river floodplain. Biogeochemistry 75:43–64

Gibson CA, Reilly CMO, Conine AL, Lipshutz SM (2015) Nutrient uptake dynamics across a gradient of nutrient concentrations and ratios at the landscape scale. J Geophys Res Biogeosci 120:326–340

Grizetti B, Bouraoui F, Billen G, Grinsven H, Van Cardoso AC, Thieu V, Garnier J, Curtis C, Howarth R, Johnes P (2011) Nitrogen as a threat to European water quality. In: Sutton MA, Howard CM, Willem EJ, Billen G, Bleeker A, Grennfelt P, van Grinsven H, Grizetti B (eds) The European nitrogen assessment. Cambridge University Press, Cambridge, pp 379–404

Gücker B, Pusch MT (2006) Regulation of nutrient uptake in eutrophic lowland streams. Limnol Oceanogr 51:1443–1453

Hall RO Jr, Bernhardt ES, Likens GE (2002) Relating nutrient uptake with transient storage in forested mountain streams. Limnol Oceanogr 47:255–265

Hein T, Schwarz U, Habersack H, Nichersu I, Preiner S, Willby N, Weigelhofer G (2016) Current status and restoration options for floodplains along the Danube River. Sci Total Environ 543:778–790

Hines SL, Hershey AE (2011) Do channel restoration structures promote ammonium uptake and improve macroinvertebrate-based water quality classification in urban streams? Inland Waters 1:133–145

Hoffmann CC, Kjaergaard C, Uusi-Kämppä J, Hansen HCB, Kronvang B (2009) Phosphorus retention in riparian buffers: review of their efficiency. J Environ Qual 38:1942–1955

House WA (2003) Geochemical cycling of phosphorus in rivers. Appl Geochem 18:739–748

Kroiss H, Lampert C, Zessner M (2005) Nutrient management in the Danube Basin and its impact on the Black Sea. DANUBS EVK1-CT-2000-00051. Final Report. http://iwr.tuwien.ac.at/fileadmin/mediapool-wasserguete/Projekte/daNUbs/daNUbs_Endbericht.pdf

Kronvang B, Jeppesen E, Conley DJ, Søndergaard M, Larsen SE, Ovesen NB, Carstensen J (2005) Nutrient pressures and ecological responses to nutrient loading reductions in Danish streams, lakes and coastal waters. J Hydrol 304:274–288

Mainstone CP, Parr W (2002) Phosphorus in rivers – ecology and management. Sci Total Environ 282–283:25–47

Mander Ü, Kuusemets V, Hayakawa Y (2005) Purification processes, ecological functions, planning and design of riparian buffer zones in agricultural watersheds. Ecol Eng 24:421–432

Mayer PM, Reynolds SK, McCutchen MD, Canfield TJ (2007) Meta-analysis of nitrogen removal in riparian buffers. J Environ Qual 36:1172–1180

Oenema O, van Liere L, Schoumans O (2005) Effects of lowering nitrogen and phosphorus surpluses in agriculture on the quality of groundwater and surface water in the Netherlands. J Hydrol 304:289–301

Pinay G, Gumiero B, Tabacchi E, Gimenez O, Tabacchi-Planty AM, Hefting MM, Burt TP, Black VA, Nilsson C, Iordache V, Bureau F, Vought L, Petts GE, Decamps H (2007) Patterns of denitrification rates in European alluvial soils under various hydrological regimes. Freshw Biol 52:252–266

Reddy KR, Kadlec RH, Flaig E, Gale PM, Kadlec RH, Flaig E, Retention PMGP, Reddy KR, Kadlec RH, Flaig E, Gale PM (1999) Phosphorus retention in streams and wetlands: a review. Crit Rev Environ Sci Technol 29:83–146

Revsbech NP, Risgaard-Petersend N, Schramm A, Nielsen LP (2006) Nitrogen transformations in stratified aquatic microbial ecosystems. Antonie van Leewenhoek 90:361–375

Richardson CJ, Flanagan NE, Ho M, Pahl JW (2011) Integrated stream and wetland restoration: a watershed approach to improved water quality on the landscape. Ecol Eng 37:25–39

Roberts BJ, Mulholland PJ, Houser JN (2007) Effects of upland disturbance and instream restoration on hydrodynamics and ammonium uptake in headwater streams. J N Am Benthol Soc 26:38–53

Roberts WM, Stutter MI, Haygarth PM (2012) Phosphorus retention and remobilization in vegetated buffer strips: a review. J Environ Qual 41:389–399

Robertson WD, Merkley LC (2009) In-stream bioreactor for agricultural nitrate treatment. J Environ Qual 38:230–237

Sabater S, Butturini A, Clement J-C, Burt T, Dowrick D, Hefting M, Maitre V, Pinay G, Postolache C, Rzepecki M, Sabater F (2003) Nitrogen removal by riparian buffers along a European climatic gradient: patterns and factors of variation. Ecosystems 6:20–30

Schipper LA, Robertson WD, Gold AJ, Jaynes DB, Cameron SC (2010) Denitrifying bioreactors—an approach for reducing nitrate loads to receiving waters. Ecol Eng 36:1532–1543

Schönbrunner IM, Preiner S, Hein T (2012) Impact of drying and re-flooding of sediment on phosphorus dynamics of river-floodplain systems. Sci Total Environ 432:329–337

Sharpley A, Jarvie HP, Buda A, May L, Spears B, Kleinman P (2014) Phosphorus legacy: overcoming the effects of past management practices to mitigate future water quality impairment. J Environ Qual 42:1308–1326

Sheibley RW, Duff JH, Tesoriero AJ (2014) Low transient storage and uptake efficiencies in seven agricultural streams: implications for nutrient demand. J Environ Qual 43:1980–1990

Smith VH, Schindler DW (2009) Eutrophication science: where do we go from here? Trends Ecol Evol 24:201–207

Tchobanoglous G, Burton FL, Stensel HD (2003) Wastewater engineering: treatment and reuse, 4th edn. McGraw-Hill Higher Education, Boston, 1819

Verhoeven JTA, Arheimer B, Yin CQ, Hefting MM (2006) Regional and global concerns over wetlands and water quality. Trends Ecol Evol 21:96–103

Vought L, Dahl J, Pedersen CL, Lacoursiére JO (1994) Nutrient retention in riparian ecotones. R Swed Acad Sci 23:342–348

Ward JV (1989) The four-dimensional nature of lotic ecosystems. J N Am Benthol Soc 8:2–8

Weigelhofer G, Hein T (2015) Efficiency and detrimental side effects of denitrifying bioreactors for nitrate reduction in drainage water. Environ Sci Pollut Res 22:13534–13545

Weigelhofer G, Welti N, Hein T (2013) Limitations of stream restoration for nitrogen retention in agricultural headwater streams. Ecol Eng 60:224–234

Weigelhofer G, Preiner S, Funk A, Bondar-Kunze E, Hein T (2015) The hydrochemical response of small and shallow floodplain water bodies to temporary surface water connections with the main river. Freshw Biol 60:781–793

Withers PJA, Jarvie HP (2008) Delivery and cycling of phosphorus in rivers: a review. Sci Total Environ 400:379–395

Chapter 11
Climate Change Impacts in Riverine Ecosystems

Florian Pletterbauer, Andreas Melcher, and Wolfram Graf

11.1 Introduction

The unprecedented rates of warming observed during recent decades exceed natural variability to such an extent that it is widely recognized as a major environmental problem not only among scientists. The role of our economy in driving such change has made it an economic and political issue. There is ample evidence that climate characteristics are changing due to greenhouse gas emissions caused by human activities. As a source of extreme, unpredictable environmental variation, climate change represents one of the most important threats for freshwater biodiversity (Dudgeon et al. 2006; Woodward et al. 2010).

The Intergovernmental Panel on Climate Change (IPCC), a scientific intergovernmental institution, documents knowledge on climate change research since 1988. The last assessment report (Hartmann et al. 2013) noted the following significant trends: The period from 1983 to 2012 was likely the warmest 30-year period of the last 1400 years in the Northern Hemisphere. The observed increase in global average surface temperature from 1951 to 2010 is extremely likely to have been caused by anthropogenically induced greenhouse gas (GHG) emissions. This is underpinned by the fact that the best estimate of the human-induced contribution to warming is

F. Pletterbauer (✉) · W. Graf
Institute of Hydrobiology and Aquatic Ecosystem Management, University of Natural Resources and Life Sciences, Vienna, Austria
e-mail: florian.pletterbauer@boku.ac.at; wolfram.graf@boku.ac.at

A. Melcher
Institute of Hydrobiology and Aquatic Ecosystem Management, University of Natural Resources and Life Sciences, Vienna, Austria

Centre for Development Research, University of Natural Resources and Life Sciences, Vienna, Austria
e-mail: andreas.melcher@boku.ac.at

© The Author(s) 2018
S. Schmutz, J. Sendzimir (eds.), *Riverine Ecosystem Management*, Aquatic Ecology Series 8, https://doi.org/10.1007/978-3-319-73250-3_11

similar to the observed warming. Since 1950, high-temperature extremes (hot days, tropical nights, and heat waves) have become more frequent, while low-temperature extremes (cold spells, frost days) have become less frequent (EEA 2012). Since 1950, annual precipitation has increased in Northern Europe (up to +70 mm/decade) and decreased in parts of Southern Europe (EEA 2012). Hence, climate change is arguably the greatest emerging threat to global biodiversity and the functioning of local ecosystems.

Climate is an extremely important driver of ecosystem processes in general, but especially so in freshwater ecosystems as thermal and hydrological regimes are strongly linked to climate. Atmospheric energy fluxes and heat exchange strongly influence river water temperature, which is one of the most important factors in the chemo-physical environment of aquatic organisms (Caissie 2006). Besides temperature, climate directly affects runoff through the amount and type of precipitation. Increasingly, rising trends of surface runoff have been driven by more frequent episodes of intense rainfall. All river flow derives ultimately from precipitation, although geology, topography, soil type, and vegetation can help to determine the supply of water and the pathways by which precipitation reaches the river channel (Poff et al. 1997).

Riverine ecosystems are particularly vulnerable to climate change because (1) many species within these habitats have limited dispersal abilities as the environment changes, (2) water temperature and availability are climate-dependent, and (3) many systems are already exposed to numerous human-induced pressures (Woodward et al. 2010). Aquatic organisms such as fish and macroinvertebrates are ectothermic. Hence, they are directly and indirectly dependent on the surrounding temperatures. Climate conditions affected species distributions already in the past. Species richness patterns across Europe can still be linked to the Last Glacial Maximum with the highest species richness in Peri-Mediterranean and Ponto-Caspian Europe (Reyjol et al. 2007).

The ecological consequences of future climate change in freshwater ecosystems will largely depend on the rate and magnitude of change related to climate forcing, i.e., changes in temperature and streamflow. These changes not only imply absolute changes (increases or decreases) but also the increasing variation between extremes. The hydrological and thermal regimes of rivers directly and indirectly trigger different ecological processes. In the following section, we discuss water temperature and related processes in more detail. General principles of river hydrology are discussed in Chap. 4.

11.2 Water Temperature

Water temperature is, among others, one of the most important habitat factors in aquatic ecosystems, perhaps even the master variable (Brett 1956). Riverine fish and macroinvertebrates are ectothermic organisms, and thus, all life stages are dependent on their ambient temperatures. Generally, many factors are involved in the formation of water temperature. According to Caissie (2006), the factors, which drive the

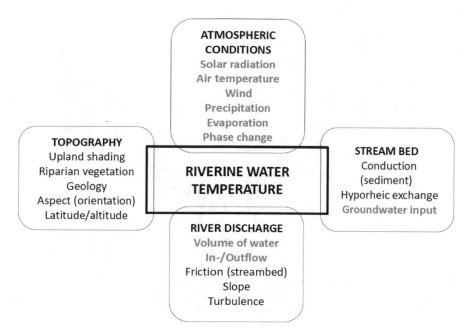

Fig. 11.1 Four groups of factors influencing the water temperature in rivers and factors that are strongly related to the climatic conditions (in blue), thus impacted by climate change (adapted after Caissie 2006)

thermal regime, can be summarized in four groups (Fig. 11.1): (1) atmospheric conditions, (2) stream discharge, (3) topography, and (4) streambed. The atmospheric conditions are highly important and mainly responsible for heat exchange processes occurring at the water surface. Topography covers the geographical setting, which in turn can influence the atmospheric factors. Stream discharge mainly determines the volume of water flow, i.e., affecting the heating capacity. Consequently, smaller rivers exhibit faster and more extreme temperature dynamics because they are more vulnerable to heating and to cooling due to lower thermal capacity. Lastly, streambed factors are related to hyporheic processes. Heat exchange processes, which are highly relevant for water temperature modeling, mainly occur at the interfaces of air and water as well as water and streambed. The former is mainly triggered by solar radiation, long-wave radiation, evaporative heat fluxes, and convective heat transfer. The contribution of other processes, such as precipitation or friction, is small in comparison. Several studies have highlighted the importance of radiation in the thermal regime. This implies the importance of riparian vegetation, which protects a stream against excessive heating (Caissie 2006).

Thermal regimes of rivers show some general trends: Water temperature increases nonlinearly from the river source to its mouth, at which the increase rate is greater for small streams than for large rivers. This general, large-scale pattern is counteracted by small-scale variabilities occurring at confluences with tributaries, in deep pools, or at groundwater inflows. While water temperature is relatively uniform

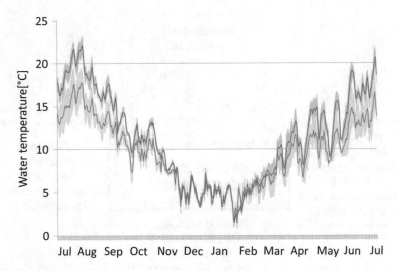

Fig. 11.2 Mean daily temperatures (solid lines) and min/max range (shaded) in the upstream (distance to source 20 km; orange) and downstream (distance to source 70 km; turquoise) section of the river Pielach, a river of the Alpine foothills in Austria

in cross sections, streams and rivers are turbulent systems where stratification is generally not expected. However, groundwater intrusion and hyporheic water exchange in pools can create cold water spots (Caissie 2006).

Besides spatial variations, the thermal regime shows temporal fluctuations of water temperature in diel and annual cycles. Daily minimum temperatures can be observed in the morning hours and maximum temperatures in the late afternoon. The magnitude of daily variations differs on the longitudinal gradient of rivers (Fig. 11.2).

Water temperature is a central feature in the chemo-physical environment of ectotherm aquatic organisms. Temperature controls almost all rate reactions (chemical and biological) and is thus a strong influence on biological systems at all levels of organization directly and indirectly triggering a magnitude of processes in aquatic life (Woodward et al. 2010). The biological dependences of the aquatic fauna and the according responses due to changes in the thermal regime are discussed in more detail below. General impacts of hydrological regimes on freshwater fauna are described in Chaps. 4, 5, and 6 allowing for the inference of potential climate change impacts.

11.3 Impacts

Riverine ecosystems are among the most sensitive to climate change because they are directly linked to the hydrological cycle, closely dependent on atmospheric thermal regimes, and at risk from interactions between climate change and existing,

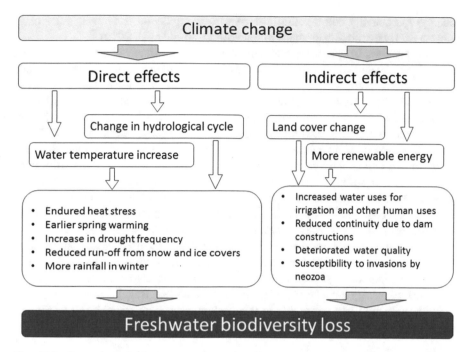

Fig. 11.3 Conceptual description of direct and indirect climate change effects on freshwater biodiversity in rivers (adapted after Fenoglio et al. 2010)

multiple, anthropogenic stressors (Dudgeon et al. 2006; Ormerod 2009). Figure 11.3 conceptually summarizes direct and indirect effects of climate change, combining hydrology and temperature. Water temperature has received much less attention with respect to ecological effects than other facets of water quality, such as eutrophication, suspended sediments, and pollution. The following section highlights climate change impacts on thermal as well as hydrological regimes. Furthermore, the interactions of climate change with other pressures are shortly discussed. Finally, this chapter addresses the ecological implications of climate change.

11.3.1   Climate Change Impacts on Thermal Regimes

An increase in air temperature will directly translate into warmer water temperatures for most streams and rivers. This change in thermal characteristics fundamentally alters ecological processes. Even though over the past 30 years warming in rivers and streams is consistently reported from global to regional scales (e.g., Webb and Nobilis 1995; Kaushal et al. 2010; Orr et al. 2014), climate change is not in all cases the exclusive reason for this warming. Temporal trends in thermal regimes can be also influenced by human-induced pressures such as impoundment, water abstraction, warm-water emissions from cooling and wastewater discharges, land use change (particularly deforestation), or river flow regulation. However, these other causes for river warming are hard to quantify (Kaushal et al. 2010).

At many sites, long-term increases in the water temperatures of streams and rivers typically coincided with increases in annual mean air temperatures. Warming trends also occur in rivers with sparsely settled catchments with intact forest cover. A comprehensive study by Orr et al. (2014) comprising 2773 sites across the United Kingdom showed warming trends (0.03 °C per year) from 1990 to 2006, which are comparable to those reported for air temperature. Similarly, Markovic et al. (2013) showed increasing temperature trends for the Elbe and Danube rivers, which accelerated at the end of the twentieth century. Furthermore, seasonal shifts were indicated by earlier spring warming and an increase in the duration of summer heat phases. During the next century, global air temperatures are projected to increase by 1.5–4.5 °C (Hartmann et al. 2013). This temperature increase will have manifold consequences for aquatic fauna, which are discussed in more detail in Sect. 11.3.4.

Another important, human-induced impact that directly affects water temperature and thermal regimes is deforestation and removal of riparian vegetation. The removal of riparian vegetation can have tremendous effects on water temperatures as increased energy input from radiation induces heating. Small streams with lower heat capacity are quite vulnerable to this impact, especially where a full canopy of riparian vegetation naturally occurs.

11.3.2 Climatic Aspects in Hydrology

Despite the strongly consistent pattern of hydrological change in some regions, e.g., reduced runoff during summer and more runoff during winter due to shifts from snow to rainfall, there is considerable uncertainty in how climate change will impact river hydrology. In Europe, already dry regions such as the Mediterranean area or the Pannonian lowlands will become drier, and already wet regions such as Scandinavia or the Alps will become a bit wetter.

However, streamflow trends must be interpreted with caution because of confounding factors, such as land use changes, irrigation, and urbanization. In regions with seasonal snow storage such as in the Alps, warming since the 1970s has led to earlier spring discharge maxima and has increased winter flows due to more precipitation as rainfall instead of snow. Moreover, where streamflow is lower in summer, decrease in snow storage has exacerbated summer dryness.

The projected impacts in a catchment under future climate conditions depend on the sensitivity of the catchment to change in climatic characteristics and on the projected change of precipitation, temperature, and resulting evaporation. Catchment sensitivity is a function of the ratio between runoff and precipitation. Accordingly, a small ratio indicates a higher importance of precipitation for runoff. Proportional changes in average annual runoff are typically between one and three times as large as proportional changes in average annual precipitation (Tang and Lettenmaier 2012). In turn, the smaller the ratio, the greater the sensitivity. However, the uncertainties in the hydrological models can be substantial. In some regions and especially on medium time scales (up to the 2050s), uncertainties in hydrological models can be greater than climate model uncertainty, i.e., uncertainty in the results of the hydrological model is larger than the predicted change induced by altered climate conditions and thus having no significant meaning.

In alpine regions, glaciers can contribute appreciable amounts of water to the discharge of rivers. All projections for the twenty-first century show continuing net mass loss from glaciers. In glaciered catchments, runoff reaches an annual maximum during summer, which strongly influences river thermal conditions as well. Reduced contributions from glacial runoff induce shifts of peak flows toward spring. Furthermore, the reduced glacial input can lead to more erratic and variable discharge dynamics in response to rain events. The relative importance of high-summer glacier meltwater can be substantial, for example, contributing 25% of August discharge in basins draining the European Alps (Huss 2011). Observations and models suggest that global warming impacts on glacier and snow-fed streams and rivers will pass through two contrasting phases. In the first phase, river discharge increases due to intensified melting. In the second phase, snowfields melt early and glaciers have shrunken to a point that late-summer streamflow is strongly reduced. The turnover between the first and second phase is called "peak meltwater." Peak meltwater dates have been projected between 2010 and 2040 for the European Alps (Huss 2011).

River discharge also influences the response of thermal regimes to increased air temperatures. Simulated discharge decreases of 20 and 40% may result in additional increases of river water temperature of 0.3 and 0.8 °C on average (Van Vliet et al. 2011). Consequently, where drought becomes more frequent, freshwater-dependent biota will suffer directly from changed flow conditions and also from drought-induced river temperature increases. Furthermore, increased temperature will accompany decreased oxygen and increased pollutant concentrations.

Hydrology itself is a driver of aquatic communities, and disturbances, such as floods, have regulatory effects on riverine biota as dominant populations are reduced, pioneers are supported, and free niches are opened. Hydrological dynamics are therefore essential to maintain overall biodiversity in aquatic ecosystems. Riverine species have evolved specific life-cycle adaptations to seasonal differences in hydrological regimes that are specific to different eco- and bioregions. For instance, larval growth rates of benthic invertebrates are high during winter as the hydraulic stress is reduced in low-flow periods in alpine rivers. Disturbances, such as acyclic extreme events, may be linked with severe losses in biomass, with species richness, and with the selection of species-specific traits. Unstable environments favor small, adaptive species with short life cycles, whereby larger organisms with longer life spans are generally handicapped (Townsend and Hildrew 1994; see Chap. 4).

11.3.3 Interactions of Climate Change with Other Stressors

Climate change is not the only source of stressors impacting water resources and aquatic ecosystems. Non-climatic drivers such as population increase, economic development, pollutant emissions, or urbanization challenge the sustainable use of resources and the integrity of aquatic ecosystems (Dudgeon et al. 2006; Nelson et al. 2006). Changing land uses are expected to affect freshwater systems strongly in the future: Increasing urbanization and deforestation may decrease groundwater recharge and increase flood hazards with consequences for hydrology. Furthermore, agricultural practices are strongly related to the climatic conditions (Bates et al. 2008). Thus, agricultural land use will be of particular importance for the integrity of freshwater systems in the future (see Chap. 13). Irrigation accounts for about 90% of global water consumption and severely impacts freshwater availability for humans and ecosystems (Döll 2009).

Climate can induce change in human uses or directly interact with human pressures. Hydropower generation, for example, causes major pressures on riverine ecosystems. Through damming, water abstractions, and hydropeaking, hydropower plants affect habitat quality by, e.g., altering river flow regimes, fragmenting river channels, or disturbing discharge regimes on hourly time scales (Poff and Zimmerman 2010) (for more details, see Chaps. 4–7). However, climate change affects hydropower generation itself through changes in the mean annual streamflow, shifts of seasonal flows, and increases of streamflow variability (including floods and droughts) as well as by increased evaporation from reservoirs and changes in sediment fluxes. Some of these interactions can have negative effects on hydropower generation as well. Especially, run-of-the-river power plants are more susceptible than storage-type power plants to climate change impacts, such as increased flow variability. However, the existing pressures of hydropower generation can be augmented by climate change; e.g., low-flow conditions in river reaches downstream of diversion power plants may be amplified through drought.

Another important field of interacting effects is water quality. On the one hand, increased water temperatures influence many biogeochemical processes such as the self-purification of water. On the other hand, rising temperatures will lead to increasing water demands by socioeconomic systems (e.g., for irrigation or cooling). Water quality aspects are discussed in more detail in Chap. 10.

11.3.4 Ecological Impacts of Thermal Regimes on Aquatic Fauna

As discussed above, climate change will affect several ecosystem processes relevant for aquatic life. The most pervasive impact of climate change will be the change of the thermal regime and mostly a warming of water temperatures. Therein, climate change will affect several characteristics of the thermal regime (e.g., mean, minima, and maxima), which are relevant for aquatic life.

Almost all fishes and macroinvertebrates are obligate poikilotherms or thermal conformers; as such, almost every aspect of the ecology of an individual is influenced by the temperature of the surrounding water from the egg to the adult individual (Brett 1956). Fry (1947) outlined five main categories of temperature effects on fishes that are likely to influence macroinvertebrates too: controlling (metabolic and developmental rates), limiting (affecting activity, movement, and distribution), directing (stimulating an orientation response), masking (blocking or affecting the expression to other environmental factors), and lethal effects that act either directly to kill the organism or indirectly as a stress effect. Thus, the responses of the aquatic fauna to water temperature changes might occur at various levels of organization from the molecular through organismal and population to the community level (McCullough et al. 2009; Woodward et al. 2010). Climate and thus climate change can affect almost every component of an individual fish's life including availability and suitability of habitats, survival, reproduction, and successful hatching, as well as metabolic demands. The temperature thresholds associated with these effects differ not only between species but also between different life stages. Besides the different organizational levels, the responses of aquatic fauna to climate change will be heterogeneous due to regional and taxonomic variations. In the following, the different organizational levels will be discussed and related to climate change impacts with a focus on the population (including species) and community level.

At the molecular level, the thermal tolerance of an organism and its physiological limits are key determinants as to whether the organism is able to adapt to the thermal conditions due to its genetic constitution. Biological reactions to impacts on the molecular level include heat shocks, stress responses, and changes to enzyme function or to genetic structure. However, the physiological response of an organism is also linked to other parameters, such as sex, size, season, and water chemistry. Thus, the thermal preference of a species cannot adequately be described by a single temperature value, such as the mean. Several metrics can be used to quantitatively describe the thermal preference and tolerance of a species and its life stages: optimum growth temperature supporting the highest growth rate, final temperature preference indicating the temperature toward which a fish tends to move when exposed to a temperature range, upper incipient lethal limit, the upper temperature value that 50% of fish survive in an experiment for an extended period, critical thermal maximum that describes the upper temperature in an experiment at which fish loses its ability to maintain the upright swimming position, optimum spawning temperature, and optimum egg development temperature. Actually, lethal temperatures relate not only to a fixed maximum threshold. The maximum temperature a species or a specific life stadium withstands is also strongly related to the acclimation time, i.e., the time over which temperature changes.

According to Magnuson et al. (1997), aquatic organisms can be classified into three thermal guilds: (1) cold-water species with physiological optimums <20 °C, (2) coolwater species having their physiological optimums between 20 and 28 °C, and (3) warm-water species with an optimum temperature > 28 °C. Even though it is possible to delineate thermal niches in the laboratory, evidence from field data is much more heterogeneous (Magnuson et al. 1979) as in complex and dynamic river systems the interplay of several biotic and abiotic factors is relevant for the aquatic organisms.

At the organismal level, fish are able to react behaviorally to stay within the range of their thermal tolerance and to avoid stress effects or sublethal effects. Even though fish, as exotherms, cannot physiologically regulate their body temperature, they are able to select thermally adequate microhabitats by movement within the range of temperatures available in their environment. Movements to avoid stressful thermal conditions and stay within adequate habitat conditions are important behavioral responses to changing spatial and temporal patterns of temperature (McCullough et al. 2009). In contrast to large-scale migrations into new habitats that will be discussed below under the population level, behavioral movements do not change the potential distribution area of a species. Such movements are temporarily limited habitat changes. Thermal stress can lead to reduced disease resistance or changed feeding and foraging, all having negative effects on the fitness and viability of the individual. By contrast, macroinvertebrates do not have the possibility for directional movements within flowing water. Macroinvertebrates have the option to retract into interstitial spaces within the bottom substrate or to drift by passive movement downstream (into warmer river reaches).

At the population level (including the species), factors relevant to responses to thermal variability are spatial distributions on the species level as well as population viability including abundance, productivity, and genetic diversity. If thermal conditions continuously exceed the preferred range of a species and adequate habitats diminish in the current environment, temporal movements into adequate microhabitats, as discussed under the organismal level, become insufficient to secure survival of the population. In this case, temperature drives changes in potential distribution area. Aquatic organisms have two options to stay within a specific thermal niche under warming environments due to climate change: either migrate to northern latitudes or to higher altitudes (see Fig. 11.4).

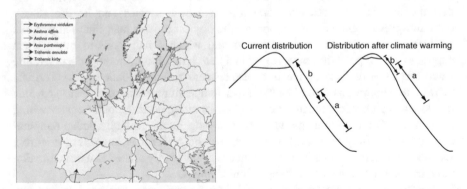

Fig. 11.4 Changes in distribution patterns due to climate change. Left: migration pathways of selected Odonata from southern to northern latitudes (Ott 2010) (reproduced from Ott J. (2010) The big trek northwards: recent changes in the European dragonfly fauna. In: Settele J., Penev L., Georgiev T., Grabaum R., Grobelnik V., Hammen V., Klotz S., Kotarac M. & Kühn I. (Eds) (2009): Atlas of Biodiversity Risk, with permission of Pensoft Publishers). Right: migration of species to higher altitudes and segregation of species groups along elevational and temperature gradients in mountainous regions currently and under climate warming. Elevational ranges of species in group b would be reduced due to displacement by the expanding ranges of species in group a (Rahel et al. 2008) (reproduced from Rahel F.J. et al. (2008) Assessing the effects of climate change on aquatic invasive species. Conservation Biology, 22, 521–33, with the permission of John Wiley & Sons. Ltd., © 2008 Society for Conservation Biology)

However, the possibility to migrate and thus the possibility to follow or to reach thermally suitable habitat depend on two criteria: firstly on the dispersal ability of the species and secondly on the availability of passable migration pathways and corridors to suitable habitats, respectively. Capacity for the former, i.e., dispersal abilities, can be measured in terms of how much time it takes a species to follow the thermal niche or how far species can follow this niche, but these are still not well investigated and largely unknown. In the latter case, migration pathways for endemic species are uncertain. Endemic species have limited distributions for several reasons. Purely aquatic species are expected to be severely challenged by climate change, especially if the river network is not connected to higher latitudes or elevations, and thus to cooler habitats. For example, fish species of the Mediterranean region, where endemism is high, may find no passable route to migrate northward in river systems draining to the Mediterranean Sea.

Another example where migration is impossible is the springs of rivers. Springs, i.e., the real source of the river, are colonized by specific species and assemblages of benthic invertebrates. These assemblages are assumed to be especially vulnerable to any environmental changes in terms of temperature or hydrology, since these habitats have "extratropical" character, i.e., the habitat conditions are and have been extremely constant over time. These assemblages and habitats are especially vulnerable in medium elevation ranges, around 1500 m, where climate-induced temperature increases will raise the source temperature of rivers. These species are among potential losers of climate change effects as they are trapped in sky islands, i.e., mountain refugia, and are not able to shift to suitable thermal or hydrological conditions, either up- or downstream (Bässler et al. 2010; Sauer et al. 2011; Dirnböck et al. 2011; Vitecek et al. 2015a, b; Rabitsch et al. 2016). Another vulnerable stream type is glacier-fed streams with cold and turbid waters inhabited by species specialized for these exceptional conditions. The shrinkage of glaciers will reduce local and regional diversity (Jacobsen et al. 2012).

Generally, the change of distribution patterns is a central topic in climate change impact research in aquatic ecosystems. Climate is a strong determinant in biogeographical distribution patterns (Reyjol et al. 2007), and hence, climate change will have huge impacts on the biogeographical configuration of aquatic communities. Comte et al. (2013) reviewed observed and predicted climate-induced distribution changes for fish. Most evidence was found for cold-water fishes and within cold-water fishes for salmonids (Fig. 11.5). This is not surprising as the different species of salmon and trout are economically highly relevant in angling and fisheries and often represent species with a high cultural value too. Nonetheless, climate change impacts are less well studied in freshwater environments than in the terrestrial or marine realm.

In most cases, climate change-induced distribution shifts of cold-water species lead to shrinking habitat availabilities due to the loss of habitats at the downstream end of the distribution area or to an upstream shift into cooler areas. Filipe et al. (2013) forecasted future distribution of trout across three large basins in Europe covering a wide range of climatic conditions. The predictions clearly showed tremendous losses of habitats. In turn, the Alps represented a stable distribution

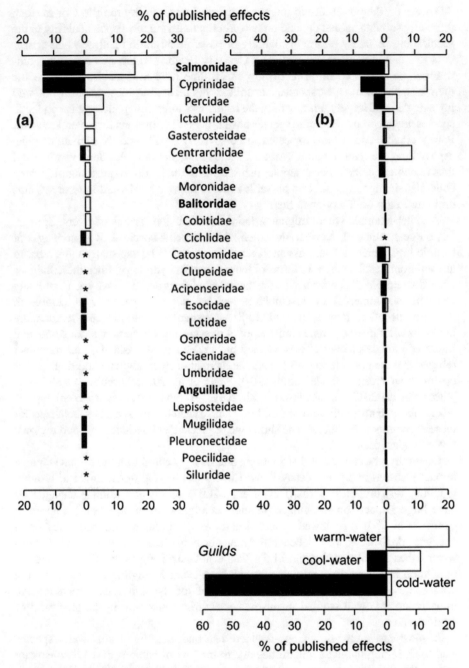

Fig. 11.5 Proportion of negative (black bars) and positive (white bars) effects reported: (**a**) observed effects and (**b**) predicted effects according to the level of biological organization for which predictions have been made (thermal guilds versus species). Asterisks indicate families of which no species has been studied. Bold indicates families for which the proportion of categorical

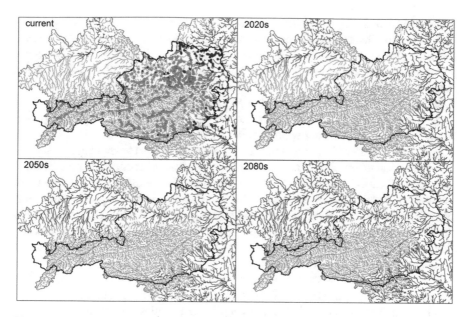

Fig. 11.6 Change of trout (*Salmo trutta*) distribution due to climate change impacts in the upper Danube Basin covering the alpine range of Austria based on the model of Filipe et al. (2013)

area in the models (Fig. 11.6). On a large-scale, continental perspective, the conservation of such habitats is highly important in the face of climate change, since the habitats will dramatically reduce in other areas. If these thermally suitable habitats and their trout populations are impacted by other pressures, the species can be also extirpated in this area. Hari et al. (2006) underlined the relationships of warming rivers in the Swiss Alps and the already occurred decline of trout populations at the end of the twentieth century.

In the case of trout, the species already occupies the upstream sections of upland rivers. Potentially in some areas, trout may extend its distribution further upstream, but in most cases, a further migration may be limited by habitat factors other than temperature or by topographical barriers, respectively. In turn, species that are currently occurring more downstream would have the possibility to track their thermal niche into upstream reaches. However, Comte and Grenouillet (2015) showed that riverine fish species consistently lagged behind the speed at which climate change gains elevation and latitude (Isaak and Rieman 2013) with higher rates for habitat losses than for habitat gains, i.e., the preferred thermal range and the

◀

Fig. 11.5 (continued) effects differed between the observed and predicted effects, according to binomial tests ($P < 0.05$) (Comte et al. 2013) (reproduced from Comte et al. (2013) Climate-induced changes in the distribution of freshwater fish: Observed and predicted trends. Freshwater Biology, 58, 625–639, with the permission of Wiley & Sons Ltd., © 2012 Blackwell Publishing Ltd.)

actually occupied thermal environment drift apart from each other. This lag can be also caused by insufficient connectivity that represents a highly important issue for migration but is impaired by other human-induced impacts such as barriers or also water abstraction. Macroinvertebrates may overcome the problem of migration barriers by overland (aerial) dispersal in their adult life stage, since most aquatic insects have winged adult stages. However, some of these species are poor fliers (e.g., mayflies) and would most likely not be successful to migrate upward, particularly in regions with strong winds or distinct topography.

Temperature effects on communities comprise responses to temperature via food web dynamics, interactions among fish species or biotic interactions among different taxa, as well as the role of diseases and parasites. Furthermore, the emergence of non-native, exotic species is highly relevant in community aspects. Thus, this organizational level is highly relevant with respect to biodiversity that is especially under pressure in freshwater ecosystems (Dudgeon et al. 2006). However, distribution shifts of single species as discussed under the population level are linked to the dynamics of community composition.

The transition of fish species along the river continuum is characterized by two trends: (1) downstream increase of species richness and biomass and (2) turnover in species composition from salmonid to cyprinid communities. In Europe, the species-poor assemblages of the upstream reaches are dominated by cold-water species and the downstream reaches by warm-water-tolerant species. The, in comparison with fish communities, species-rich macroinvertebrate communities change in similar fashion along the river continuum in distinct reaction to temperature and other parameters such as oxygen saturation, substrate composition, flow velocity, and food resources. Temperature increases can thus induce assemblage shifts.

Pletterbauer et al. (2015) investigated fish assemblage shifts based on the Fish Zone Index (FiZI) that considers not only the occurrence of a species but also its abundance (Schmutz et al. 2000). The results showed significant assemblage shifts across major parts of Europe with strongest impacts on fish assemblages in upstream sections of small- and medium-sized rivers as well as in Mediterranean and alpine regions. By comparing distribution shifts for different taxa groups in different regions, Gibson-Reinemer and Rahel (2015) recently found that responses are idiosyncratic for plants, birds, marine invertebrates, and mammals. The authors stated that "inconsistent range shifts seem to be a widespread response to climate change rather than a phenomenon in a single area or taxonomic group." Thus, distribution shifts will not occur for all species at the same time and to the same extent. Accordingly, vulnerabilities have to be addressed on the different levels of organization. Hering et al. (2009) analyzed the vulnerability of the European Trichoptera fauna to climate change and found that parameters such as endemism, preference for springs or for cold water temperatures, short emergence period, and restricted ecological niches in terms of feeding types are responsible for the species-specific sensitivity to climate change impacts. Accordingly, species of the Mediterranean peninsulas and mountainous areas in Central Europe are potentially more threatened than species of Northern Europe (Fig. 11.7).

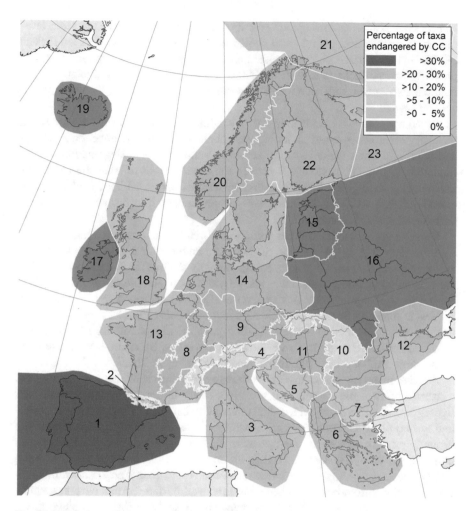

Fig. 11.7 Fraction of Trichoptera taxa potentially endangered by climate change in the European ecoregions (Hering et al. 2009), numbers indicate the Ecoregion number (© Aquatic Sciences, Potential impact of climate change on aquatic insects: A sensitivity analysis for European caddisflies (Trichoptera) based on distribution patterns and ecological preferences, 71, 2009, 3–14, Hering D., Schmidt-Kloiber A., Murphy J., Lücke S., Zamora-Muñoz C., López-Rodríguez M.J., Huber T., Graf W. With permission of Springer)

11.4   Adaptation and Restoration

Successful climate change adaptation requires responses at the appropriate temporal and spatial scales. However, sustaining integral ecosystem processes and functions will need inter- and transdisciplinary approaches to address climate change impacts. The effects of climate change are already visible and measurable in aquatic ecosystems.

Hence, conservation and restoration practitioners and researchers need to share information effectively and with diverse audiences such as policy- and decision-makers, NGOs, and other stakeholders to ensure sharing most recent findings and to enable proactive management (Seavy et al. 2009). Rapid environmental change urgently requires society to be informed about the ongoing and upcoming threats related to climate change.

Broad suggestions for adapting rivers to climate change impacts are similar to those for other ecosystems, including the enhancement of resilience, connectivity, and legal protection while reducing stressors, such as habitat degradation or fragmentation (Palmer et al. 2008). However, the development of adequate and robust management strategies is key to conserve intact, freshwater habitats. With respect to climate change and aquatic ecosystems, water temperature is one of the master variables that requires attention.

Riparian vegetation contributes various important functions in relation to aquatic habitats, including the moderation of water and ambient air temperature via evapotranspiration and reduction of solar energy input by shading. It thus provides a buffer zone that filters sediments and nutrients, provides food, and creates woody debris as habitat for xylobiont species (Richardson et al. 2007). Evapotranspiration rates are highest in forest habitats due to their high leaf area index (Tabacchi et al. 1998). In this context, a major issue is the mitigation potential of riparian vegetation to keep rivers cooler. Recent studies have shown that shading by riparian vegetation can buffer the warming effects of climate change (Bond et al. 2015).

Another important aspect in climate change adaptation is habitat connectivity. As discussed above, species will tend to follow their preferred thermal niche in their river network. Accordingly, the spatial connection between different river reaches is highly important, especially for cold-water taxa, as long-term thermal refugia are located upstream where water temperature is lowest along the longitudinal continuum. As shown by Isaak et al. (2015), thermal habitats in mountain streams seem highly resistant to temperature increases. As a result, many populations of cold-water species currently exist where they are well-buffered from climate change. However, connectivity is not only relevant on the scale of the river network. On shorter time scales, cold-water refugia may occur as patchy distributions along the river course. Deep pools with high groundwater exchange rates or other river sections with groundwater intrusion can provide valuable habitats where species can endure heat waves. Accordingly such refugia must be connected to the surrounding habitats such that they can be accessed and used. However, morphological degradation impedes the availability of such refugia. Thus, habitat heterogeneity and morphological integrity, including natural riverbed and sediment dynamics, are essential to provide adequate habitat patches for different species and their life stages, also from the thermal point of view.

In addition to climate change, the future of freshwater ecosystems will be strongly influenced by other sources of stress: socioeconomic and technological changes as well as demographic developments on the global scale (Dudgeon et al. 2006; Nelson et al. 2006). Ultimately, as climate change impacts start to overwhelm the capacity of society and of ecosystems to cope or adapt, substantial reduction in GHG emissions becomes inevitable. Until some combination of foresight, technological advances, and political will makes such reduction possible, research, monitoring, and experimental advances in practice must be pursued to inform society and to slow the effects of climate change on riverine ecosystems.

11.4.1 Case Study BIO_CLIC: Potential of Riparian Vegetation to Mitigate Effects of Climate Change on Biological Assemblages of Small- and Medium-Sized Running Waters

The transdisciplinary research project BIO_CLIC investigated the impact of riparian vegetation on the water temperature regime as well as on aquatic organisms of small- and medium-sized rivers in southeastern Austria. Its objectives were to identify and understand the potential of riparian vegetation to mitigate climate change impacts on water temperature and, ultimately, on benthic invertebrate and fish species assemblages. Finally, BIO_CLIC aimed to support river managers in implementing integrative management for sustainable river restoration toward climate change adaptation that incorporates ecosystem services and socioeconomic consequences.

The study area in the Austrian lowlands, represented by the rivers Lafnitz and Pinka, was chosen, because in this area an increase of air temperature of ca. 2–2.5 °C is predicted by 2040. Moreover, climate change effects combined with a rising numbers of rivers without or with low levels of riparian vegetation will lead to an increase of water temperature. It can be assumed that climate change effects will exacerbate ecological consequences by impacting water temperature and also hydrology (e.g., increasing the incidence and duration of low-flow periods).

The river Lafnitz amply exhibits hydrologically and morphologically intact river sections with near-natural riparian vegetation. By contrast, the river Pinka is impacted by river straightening and riparian vegetation loss. Due to the spatial proximity of these two rivers, the climatic conditions are comparable, but their different hydro-morphological settings qualify them for analysis to distinguish the effect of riparian vegetation on the thermal regime as well as climate change impacts. Additionally, specific sites along the rivers Lafnitz and Pinka were analyzed according to elements influencing the biological quality of fish and benthic invertebrates, e.g., water temperature, riparian vegetation, and morphological (e.g., channelization, riverbed structure) characteristics.

The results of time series analysis show clearly the difference between the two rivers. In the upper and middle reaches, the mean July water temperature in the Pinka exceeds 15 °C, which sharply contrasts with a more flattened gradient of lower temperatures in the water column of the river Lafnitz. One key reason is the lack of shading effects by the riparian vegetation that is generally missing on the Pinka. For both rivers, water temperature and fish and benthic invertebrate distributions are highly correlated along the longitudinal gradient. This underlines the strong influence of water temperature on the longitudinal distribution of aquatic organisms and highlights the importance of mitigation of global climate change effects by shading. Shifts of their associated species to cold and warm water within the biocenotic (fish) zones will be inevitable with increasing temperatures, forcing the cold-water species to move to higher altitudes, if river connectivity allows.

In more natural river sections with fewer human pressures, in summer months, the water temperature difference between shaded and unshaded biocenotic zones is about 2–3 °C. As temperature increases, other river characteristics such as river dimension, flow, and substrate composition, but also migration barriers, might prove to be limiting factors leading to relatively unpredictable changes in the biotic assemblages. Riparian vegetation and shading could ameliorate such threats by harmonizing and flattening maximum temperature peaks in hot periods by up to 2 °C. This is about the same range of temperature increase that was predicted as an impact of climate change effects in 2050.

Global warming has already shown impacts on European freshwater ecosystems and the services they provide to humans. The main impacts are related to biodiversity, water quality, and health: Environmental parameters specify boundary conditions for habitat availability, and likewise human-induced restraints reduce further opportunities for a dynamic, ever-changing ecosystem. The results clearly demonstrate that efficient river restoration and mitigation requires the reestablishment of riparian vegetation as well as an open river continuum and hydro-morphological improvement of habitats (Melcher et al. 2016).

11.5   Conclusions, Open Questions, and Outlook

Rivers have experienced centuries of human-induced modifications (Hohensinner et al. 2011). While climate change may already impact riverine ecosystems, in the future it is much more likely that human-induced modifications will clearly and unequivocally be accompanied by climate change effects. Consequently, the challenge of how to preserve the status quo or to get back to a more pristine status will become more difficult as fundamental ecosystem processes, such as the thermal regime, will shift. From an applied perspective, climate change has the potential to undermine many existing freshwater biomonitoring schemes, which focus mostly on human pressures like organic pollution or hydro-morphological alterations with little consideration for the increasing influence of climatic effects. Thus, how we currently assess "ecological status" could become increasingly obsolete over time, as the environmental conditions drift away from assumed earlier (and cooler) reference conditions (Woodward et al. 2010) and causal relationships underpinning ecological processes realign. Thus, we may assume that sustaining and restoring habitat heterogeneity and connectivity will continue to enhance ecosystem resilience, but it may be increasingly difficult to know how much or how fast. Long-term monitoring is essential to observe changes induced by climate that are currently lacking for biological quality elements in rivers. However, improving the research focus of monitoring programs to directly address uncertainties raised by climate change should make data available that will better inform future management decisions. Tracking data over the long term will provide the baseline trajectories against which scenarios of simulated management policies can be compared. While surprise from climate change is inevitable, challenging simulation of policies with real data will make it more possible to project the consequences of river policies over longer time periods and to identify and respond to emerging trends in changing conditions.

References

Bässler C, Müller J, Hothorn T, Kneib T, Badeck F, Dziock F (2010) Estimation of the extinction risk for high-montane species as a consequence of global warming and assessment of their suitability as cross-taxon indicators. Ecol Indic 10:341–352

Bates BC, Kundzewicz ZW, Wu S, Palutikof JP (eds) (2008) Climate change and water. Technical paper of the intergovernmental panel on climate change. IPCC Secretariat, Geneva, p 210

Bond RM, Stubblefield AP, Van Kirk RW (2015) Sensitivity of summer stream temperatures to climate variability and riparian reforestation strategies. J Hydrol 4:267–279

Brett JR (1956) Some principles in the thermal requirements of fishes. Q Rev Biol 31:75–87

Caissie D (2006) The thermal regime of rivers: a review. Freshw Biol 51:1389–1406

Comte L, Grenouillet G (2015) Distribution shifts of freshwater fish under a variable climate: comparing climatic, bioclimatic and biotic velocities. Divers Distrib 21:1014–1026

Comte L, Buisson L, Daufresne M, Grenouillet G (2013) Climate-induced changes in the distribution of freshwater fish: observed and predicted trends. Freshw Biol 58:625–639

Dirnböck T, Essl F, Rabitsch W (2011) Disproportional risk for habitat loss of high-altitude endemic species under climate change. Glob Chang Biol 17:990–996

Döll P (2009) Vulnerability to the impact of climate change on renewable groundwater resources: a global-scale assessment. Environ Res Lett 4:1–13

Dudgeon D, Arthington A, Gessner M, Kawabata Z-I, Knowler D, Lévêque C, Naiman R, Prieur-Richard A-H, Soto D, Stiassny M, Sullivan C (2006) Freshwater biodiversity: importance, threats, status and conservation challenges. Biol Rev Camb Philos Soc 81:163–182

EEA (2012) Climate change, impacts and vulnerability in Europe 2012: an indicator-based report

Fenoglio S, Bo T, Cucco M, Mercalli L, Malacarne G (2010) Effects of global climate change on freshwater biota: a review with special emphasis on the Italian situation. Ital J Zool 77:374–383

Filipe AF, Markovic D, Pletterbauer F, Tisseuil C, De Wever A, Schmutz S, Bonada N, Freyhof J (2013) Forecasting fish distribution along stream networks: brown trout (Salmo Trutta) in Europe. Divers Distrib 19:1059–1071

Fry FEJ (1947) Effects of the environment on animal activity. Publications of the Ontario fisheries research. Laboratory 55:1–62

Gibson-Reinemer DK, Rahel FJ (2015) Inconsistent range shifts within species highlight idiosyncratic responses to climate warming. PLoS One 10:1–15

Hari RE, Livingstone DM, Siber R, Burkhardt-Holm P, Guttinger H (2006) Consequences of climatic change for water temperature and brown trout populations in alpine rivers and streams. Glob Chang Biol 12:10–26

Hartmann DL, Tank AMGK, Rusticucci M (2013) IPCC Fifth assessment report, climate change 2013: The physical science basis IPCC, AR5

Hering D, Schmidt-Kloiber A, Murphy J, Lücke S, Zamora-Muñoz C, López-Rodríguez MJ, Huber T, Graf W (2009) Potential impact of climate change on aquatic insects: a sensitivity analysis for European caddisflies (Trichoptera) based on distribution patterns and ecological preferences. Aquat Sci 71:3–14

Hohensinner S, Jungwirth M, Muhar S, Schmutz S (2011) Spatio-temporal habitat dynamics in a changing Danube River landscape 1812-2006. River Res Appl 27:939–955

Huss M (2011) Present and future contribution of glacier storage change to runoff from macroscale drainage basins in Europe. Water Resour Res 47:1–14

Isaak DJ, Rieman BE (2013) Stream isotherm shifts from climate change and implications for distributions of ectothermic organisms. Glob Chang Biol 3:742–751

Isaak DJ, Young MK, Nagel DE, Horan DL, Groce MC (2015) The cold-water climate shield: delineating refugia for preserving salmonid fishes through the 21st century. Glob Chang Biol 21:2540–2553

Jacobsen D, Milner AM, Brown LE, Dangles O (2012) Biodiversity under threat in glacier-fed river systems. Nat Clim Chang 2:361–364

Kaushal SS, Likens GE, Jaworski NA, Pace ML, Sides AM, Seekell D, Belt KT, Secor DH, Wingate RL (2010) Rising stream and river temperatures in the United States. Front Ecol Environ 8(9):461–466

Magnuson JJ, Crowder LB, Medvick PA (1979) Temperature as an ecological resource. Am Nat 19:331–343

Magnuson JJ, Webster KE, Assel RA, Bowser CJ, Dillon PJ, Eaton JG, Evans HE, Fee EJ, Hall RI, Mortsch LR, Schindler DW, Quinn FH (1997) Potential effects of climate changes on aquatic systems: laurentian great lakes and precambrian shield area. Hydrol Process 11:825–871

Markovic D, Scharfenberger U, Schmutz S, Pletterbauer F, Wolter C (2013) Variability and alterations of water temperatures across the Elbe and Danube River basins. Clim Chang 119:375–389

McCullough DA, Bartholow JM, Jager HI, Beschta RL, Cheslak EF, Deas ML, Ebersole JL, Foott JS, Johnson SL, Marine KR, Mesa MG, Petersen JH, Souchon Y, Tiffan KF, Wurtsbaugh WA (2009) Research in thermal biology: burning questions for Coldwater stream fishes. Rev Fish Sci 17:90–115

Melcher A, Dossi F, Graf W, Pletterbauer F, Schaufler K, Kalny G, Rauch HP, Formayer H, Trimmel H, Weihs P (2016) Der Einfluss der Ufervegetation auf die Wassertemperatur unter gewässertypspezifischer Berücksichtigung von Fischen und benthischen Evertebraten am Beispiel von Lafnitz und Pinka. Österreichische Wasser- und Abfallwirtschaft 68:308–323

Nelson GC, Bennett E, Berhe AA, Cassman K, DeFries R, Dietz T, Dobermann A, Dobson A, Janetos A, Levy M, Marco D, Nakicenovic N, O'Neill B, Norgaard R, Petschel-Held G, Ojima D, Pingali P, Watson R, Zurek M (2006) Anthropogenic drivers of ecosystem change: an overview. Ecol Soc 11:29

Ormerod SJ (2009) Climate change, river conservation and the adaptation challenge. Aquatic conservation: marine and freshwater. Ecosystems 19:609–613

Orr HG, Simpson GL, des clers S, Watts G, Hughes M, Hannaford J, Dunbar MJ, Laizé CLR, Wilby RL, Battarbee RW, Evans R (2014) Detecting changing river temperatures in England and Wales. Hydrol Process 766:752–766

Ott J (2010) The big trek northwards: recent changes in the European dragonfly fauna. In: Settele J, Penev L, Georgiev T, Grabaum R, Grobelnik V, Hammen V, Klotz S, Kotarac M, Kühn I (eds) Atlas of biodiversity risk. Pensoft, Sofia, p 280

Palmer MA, Reidy Liermann CA, Nilsson C, Flörke M, Alcamo J, Lake PS, Bond N (2008) Climate change and the world's river basins: anticipating management options. Front Ecol Environ 6:81–89

Pletterbauer F, Melcher AH, Ferreira T, Schmutz S (2015) Impact of climate change on the structure of fish assemblages in European rivers. Hydrobiologia 744:235–254

Poff NL, Zimmerman JKH (2010) Ecological responses to altered flow regimes: a literature review to inform the science and management of environmental flows. Freshw Biol 55:194–205

Poff NL, Allan JD, Bain MB, Karr JR, Prestegaard KL, Richter BD, Sparks RE, Stromberg JC (1997) The natural flow regime: a paradigm for river conservation and restoration. Bioscience 47:769–784

Rabitsch W, Graf W, Huemer P, Kahlen M, Komposch C, Paill W, Reischütz A, Reischütz PL, Moser D, Essl F (2016) Biogeography and ecology of endemic invertebrate species in Austria: a cross-taxa analysis. Basic Appl Ecol 17(2):95–105

Rahel FJ, Bierwagen B, Taniguchi Y (2008) Assessing the effects of climate change on aquatic invasive species. Conserv Biol 22:521–533

Reyjol Y, Hugueny B, Pont D, Bianco PG, Beier U, Caiola N, Casals F, Cowx IG, Economou A, Ferreira MT, Haidvogl G, Noble R, de Sostoa A, Vigneron T, Virbickas T (2007) Patterns in species richness and endemism of European freshwater fish. Glob Ecol Biogeogr 16:65–75

Richardson DM, Holmes PM, Esler KJ, Galatowitsch SM, Stromberg JC, Kirkman SP, Pysek P, Hobbs RJ (2007) Riparian vegetation: degradation, alien plant invasions, and restoration prospects. Divers Distrib 13:126–139

Sauer J, Domisch S, Nowak C, Haase P (2011) Low mountain ranges: summit traps for montane freshwater species under climate change. Biodivers Conserv 20:3133–3146

Schmutz S, Kaufmann M, Vogel B, Jungwirth M (2000) Methodische Grundlagen und Beispiele zur Bewertung der fischökologischen Funktionsfähigkeit Österreichischer Fließgewässer

Seavy NE, Gardali T, Golet GH, Griggs FT, Howell CA, Kelsey R, Small SL, Viers JH, Weigand JF (2009) Why climate change makes riparian restoration more important than ever: recommendations for practice and research. Ecol Restor 27:330–338

Tabacchi E, Correll DL, Hauer R, Pinay G, Planty-Tabacchi AM, Wissmar RC (1998) Development, maintenance and role of riparian vegetation in the river landscape. Freshw Biol 40:497–516

Tang Q, Lettenmaier DP (2012) 21st century runoff sensitivies of major Global River basins. Geophys Res Lett 39:1–5

Townsend CR, Hildrew AG (1994) Species traits in relation to a habitat template for river systems. Freshw Biol 31:265–275

Van Vliet MTH, Ludwig F, Zwolsman JJG, Weedon GP, Kabat P (2011) Global river temperatures and sensitivity to atmospheric warming and changes in river flow. Water Resour Res 47:W02544

Vitecek S, Graf W, Previšić A, Kučinić M, Oláh J, Bálint M, Keresztes L, Pauls SU, Waringer J (2015a) A hairy case: the evolution of filtering carnivorous Drusinae (Limnephilidae, Trichoptera). Mol Phylogenet Evol 93:249–260

Vitecek S, Kučinić M, Oláh J, Previšić A, Bálint M, Keresztes L, Waringer J, Pauls SU, Graf W (2015b) Description of two new filtering carnivore Drusus species (Limnephilidae, Drusinae) from the Western Balkans. ZooKeys 513:79–104

Webb BWW, Nobilis F (1995) Long term water temperature trends in Austrian rivers/Tendance a long terme de la temperature des cours d'eau autrichiens. Hydrol Sci J 40:83–96

Woodward G, Perkins DM, Brown LE (2010) Climate change and freshwater ecosystems: impacts across multiple levels of organization. Philos Trans R Soc Lond B Biol Sci 365:2093–2106

Chapter 12
Ecotoxicology

Ralf B. Schäfer and Mirco Bundschuh

12.1 Introduction

Chemicals are used widely in all spheres of modern society (Table 12.1), for example, in industrial production (e.g., solvents, coolants), medicine (e.g., pharmaceuticals), agriculture (e.g., pesticides), and consumer products (e.g., sunscreens). In 2015, more than 33 million chemicals were commercially available (Chemical Abstracts Service 2015) with approximately 100,000 estimated to be in current use and 30,000 produced with more than 1 ton per year in the European Union (EU) (Breithaupt 2006). The production of chemicals with harmful impacts to the aquatic environment in the EU totaled more than 130,000,000 tons in 2013 (EUROSTAT 2015). The widespread use of such chemicals comes with their intentional (e.g., pesticide spraying) and unintentional (e.g., leaching or gassing out) release into the environment, where they may enter river ecosystems via different paths depending on geological, hydrological, climatic and use patterns, often associated with land use, in the river's catchment (see Chap. 13). These paths include (1) direct discharge, such as from industrial facilities, mining, or wastewater treatment plants (WWTPs), but also accidental spills; (2) runoff from the land surface or subsurface flows, often after precipitation; (3) erosion or disposal of waste, which can lead to the re-suspension, desorption, or diffusion of chemicals into the water phase; and (4) atmospheric deposition of chemicals.

R. B. Schäfer (✉)
Institute for Environmental Sciences, University Koblenz-Landau, Landau, Germany
e-mail: schaefer-ralf@uni-landau.de

M. Bundschuh
Department of Aquatic Sciences and Assessment, Swedish University of Agricultural Sciences,
Uppsala, Sweden
e-mail: mirco.bundschuh@slu.se

Table 12.1 Effects from widely occurring chemicals in river systems

Chemical source or use	Chemical or chemical group	Effects on structure or functions	References
Pyrolysis (incomplete combustion) of organic material	Polycyclic aromatic hydrocarbons (PAH)	Genotoxicity and embryotoxicity and changes in metabolism of fish in Chinese rivers	Floehr et al. (2015)
Pharmaceuticals	Antibiotics	Higher antibiotic resistance genes in US river sediments with high concentrations of antibiotics	Pei et al. (2006)
Synthetic hormones and additives	17a-Ethynylestradiol, nonylphenol	Highest feminization of fish in surface waters in the Netherlands with highest potential exposure to endocrine-disrupting chemicals	Vethaak et al. (2005)
Sunscreen UV filters	3-Benzylidene camphor	Environmental concentrations related to histological and reproductive effects in fish under laboratory conditions	Fent et al. (2008)
Agricultural pesticides	Insecticides and fungicides	Pesticide toxicity associated with changes in microbial and invertebrate communities and reduction in organic matter processing in Australian streams	Schäfer (2012)
Biocides in consumer products	Triclosan	Changes in microbial communities in US streams	Drury et al. (2013)
Mining and geogenic background	Metals	Changes in invertebrate communities in US mountain streams	Clements et al. (2000)
Salts related to mining and road de-icing	Chloride	Changes in mortality and reproduction of aquatic plants and animals	Cañedo-Argüelles et al. (2013)
Fuel combustion and geogenic background	Acids	Reduction in organic matter processing and loss of sensitive invertebrate species in acidic streams	Pye et al. (2012)
Coolants, plasticizers, and insulating fluids	Polychlorinated biphenyls and other organochlorines	Reduced gonad size, decreased plasma levels of 11-ketotestosterone, EROD and vitellogenin induction, and histopathologies of male gonads	Randak et al. (2009)
Municipal wastewater	Complex mixture of micropollutants	Reduced feeding of invertebrates and leaf litter decomposition	Englert et al. (2013)
Road runoff	Metals and polyaromatic hydrocarbons	Changes in invertebrate communities and organic matter processing	Maltby et al. (1995)

The entry paths can be classified as point sources (e.g., path 1) and nonpoint (also diffuse, e.g., paths 2 and 4) sources, a distinction that is important with respect to management actions. The relevance of the entry paths is compound specific (i.e., influenced by physicochemical properties such as solubility, lipophilicity, vapor pressure) and depends on the environmental context of the river in terms of pH, temperature, ultraviolet radiation, and soil type. For example, compounds with high water solubility and a high vapor pressure (e.g., urea herbicides) enter aquatic systems predominantly via discharge, runoff, or subsurface flows, whereas compounds with a low water solubility and low vapor pressure (e.g., polychlorinated biphenyls) primarily enter aquatic systems via atmospheric deposition. Furthermore, compounds with a low water solubility and high vapor pressure (e.g., pyrethroid insecticides) enter aquatic systems primarily adsorbed or bound to particulate matter, often during runoff events. Moreover, the environmental context influences the movement and, in the case of organic compounds, degradation and bioavailability of chemicals. Similarly, several environmental variables (e.g., pH, temperature, ionic composition) govern the speciation of metals and consequently their ecotoxicological effects.

Catchment hydrology is another important driver of the chemical exposure regime. The magnitude, timing, frequency, and duration of spates as well as the lateral and longitudinal connectivity influence transport and concentration levels of chemicals in rivers (see Chap. 3, 5, 9). For example, the river discharge determines the dilution potential for micropollutants (i.e., organic and inorganic chemical stressors detected at concentrations of μg/L or lower) released from WWTPs (Englert et al. 2013). Thus, discharge minima may be associated with the highest chemical concentrations from point sources. Moreover, discharge maxima are typically associated with heavy precipitation events that can induce surface runoff or subsurface drainage leading to the diffuse entry of agrochemicals (e.g., pesticides and nutrients) and of chemicals from roads and urban environments [e.g., biocides, metals, nanomaterials, and polycyclic aromatic hydrocarbons (PAHs)]. Again this can lead to transient peak concentrations, particularly in streams, and in rivers with a strong variability in discharge.

Depending on the compound, peak concentrations can be a more important driver of ecological effects than the average concentration (see next section). Consequently, the hydrological profile should also be considered when designing monitoring programs for chemicals in a river. An adequate exposure characterization relies on sampling methods that allow for the sampling at discharge minima and maxima. However, current governmental monitoring is largely independent from such considerations and is conducted at fixed temporal intervals. For example, an analysis of European river monitoring data showed that the sampling frequency is monthly or lower in 80% of sampling sites (Malaj et al. 2014). Such a monitoring strategy is likely to miss short-term exposure events such as runoff, which result in peak concentrations in streams and in rivers with a strong variability in discharge. For streams, a modeling study revealed that monthly monitoring would miss almost all peak exposure events of pesticides, which lasted from several hours to a few days (Stehle et al. 2013).

Multiple sampling methods that could be used in river systems have been developed to capture peak exposure events. These methods include automated sampling devices that can take event-triggered or flow-dependent samples (Mortimer et al. 2007). Passive sampling, i.e., the deployment of devices with receiving phases into which chemicals passively diffuse or adsorb to (for an overview, see Vrana et al. 2005), can also be tailored to obtain estimates of peak exposures. However, the abovementioned methods need adaptation to the spectrum of chemicals. In the case of passive sampling, for example, a suitable receiving phase for the chemicals of interest needs to be identified (Vrana et al. 2005). In this context, the analysis of land use in the catchment can aid in selecting target chemicals for the chemical monitoring (see Chap. 13): Agricultural land use is the dominant driver of pesticide input in water bodies, whereas biocides rather enter surface waters through urban runoff or via WWTPs (Wittmer et al. 2010). Nonetheless, a complete characterization of chemical exposure remains a formidable challenge, especially in catchments with mixed land use, where several hundred or thousand compounds may be present in the different environmental media, while most monitoring programs are limited to a few tens to hundreds of chemicals. Biomonitoring in concert with stressor-specific biotic metrics, bioassays that indicate toxic exposure, and effect-directed analysis and related approaches therefore represent important complementary tools for a more holistic characterization of the chemical exposure regime (Brack et al. 2015).

12.2 Impacts

12.2.1 Propagation of Impacts Across Levels of Biological Organization

Chemical exposure can impact individual freshwater organisms and higher levels of biological organization (e.g., populations, species). Several examples of effects from compounds representative for different chemical groups on ecosystem properties are provided in Table 12.1. In this context, a toxic chemical acts initially on the physiological level of an organism (Fig. 12.1). For example, polychlorinated biphenyls and other organochlorine chemicals have been associated with changes in enzyme activity and chemical signaling such as induction of vitellogenin or reduced plasma levels of 11-ketotestosterone (Table 12.1). Such physiological effects represent the basis for effects on higher biological levels (Fig. 12.1): Changes in the metabolism and chemical signaling can lead to individual-level effects such as reduced fitness or mortality. This in turn, depending on the magnitude of effects on individuals, translates to a reduction in the population size or, in the case of endocrine-disrupting chemicals, to a change in the gender ratio (males to females). On the next hierarchical level, impacts on populations may change the composition of communities, as reported for microorganisms in response to antibiotics, to the biocide triclosan, and to metals such as copper (Table 12.1). Community changes

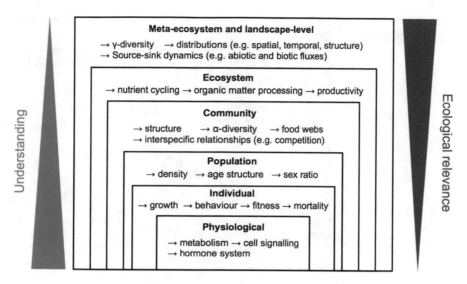

Fig. 12.1 Examples of end points impacted by chemicals on different levels of biological organization

can result in a reduction of ecosystem functioning, for example, of organic matter processing (Table 12.1), if community-level impacts are not buffered by functional redundancy. In other words, a reduction in functioning is induced by the loss of sensitive organisms, which is not compensated by an increase in tolerant organisms. Finally, ecosystem-level impacts can influence meta-ecosystem dynamics and the spatial and temporal distribution of species as well as macroecological characteristics, such as species-abundance distributions (Fig. 12.2). The ecological relevance of an impact increases from the physiological level toward the meta-ecosystem and landscape level, whereas understanding of impacts is highest on the lower levels of biological organization (e.g., physiological and individual). This discrepancy between relevance and understanding is partly due to the fact that most ecotoxicological studies focus on these lower levels of organization (Beketov and Liess 2012).

12.2.2 Relevance of Chemical Input into River Ecosystems

The relevance of chemical impacts on streams and rivers has been highlighted in several large-scale studies. Toxic chemicals, organic and inorganic, were listed among the most important pollutants in the Millennium Ecosystem Assessment for Rivers (Fynlayson and D'Cruz 2005). Similarly, a global analysis identified water pollution and water resource development as major ecological threats (Vörösmarty et al. 2010), with pesticides, nutrients, and metals comprising the dominant drivers of water pollution. Recent studies on insecticides on the global scale and on organic chemicals on the European scale estimated that approximately 70 and 42% of water

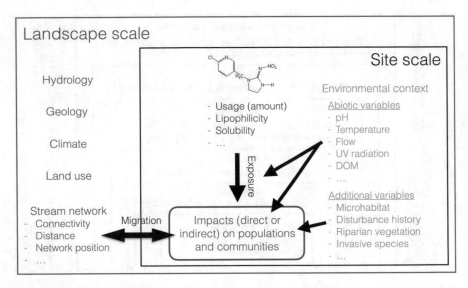

Fig. 12.2 Scales and contexts that influence the impact of chemicals on organisms in river ecosystems

bodies are at risk of adverse effects (Stehle and Schulz 2015; Malaj et al. 2014). However, all of the abovementioned studies relied on simplified risk assessment approaches, which involve the comparison of a measured or predicted environmental concentration of chemicals to an effect metric (e.g., effect concentration, threshold concentration), which is often related to ecotoxicological laboratory or mesocosm (e.g., artificial outdoor ponds) experiments. Such experiments mainly capture acute toxic effects of chemicals, e.g., direct mortality. Thus, the majority of experimental data refer to acute toxic effects, whereas data coverage is considerably lower for long-term chronic effects (Malaj et al. 2014) or indirect effects within food webs, which are further discussed below.

Combining this widespread availability of reliable acute toxicity data for many organic chemicals and metals with the fact that their toxicity in aquatic ecosystems varies strongly, the risk and exposure are often standardized based on the compounds' acute toxicity, especially in scientific studies. However, this biases risk assessments, such as those mentioned above, toward chemicals that are likely to cause short-term acute toxicity, such as pesticides and metals. Notwithstanding, several other chemicals that occur widely in the environment can also have strong impacts on populations and communities (Table 12.1) through other modes of action than those linked to direct acute toxicity. Endocrine disruption, i.e., effects on the hormone system, may compromise long-term population viability (Kidd et al. 2007), and long-term chronic exposure from hydrophobic chemicals can have dramatic effects similar to acute short-term toxicity (see discussion in Malaj et al. 2014). For instance, exposure to low levels of estrogens, which can occur downstream from WWTPs, induced a collapse of a fish population after 3 years of chronic exposure

(Kidd et al. 2007). Thus, such chemicals and their potential to affect the integrity of aquatic ecosystems over the long term may have been overlooked in these larger-scale risk assessment studies. Moreover, chemical monitoring has been tailored toward pesticides and metals that are long-known culprits of chemical impacts, and exposure data on other micropollutants (e.g., pharmaceuticals, biocides, nanoparticles) is rather scarce. Nevertheless, chemical monitoring programs have started to include pharmaceuticals and biocides over the last decade. To sum up, although metal and pesticide toxicity are important drivers of chemical effects and related risks are widespread, several other chemical groups that occur widely and can affect riverine organisms have not received sufficient recognition due to a lack of exposure and effect data, especially related to real-world conditions. As outlined above, effect-directed analysis and related approaches represent important tools to unravel emerging toxicants (Brack et al. 2015), but they are rather eligible for application in case studies than for large-scale routine monitoring.

The relevance of chemical groups varies with land use (see above and Chap. 13) and, more broadly, with the management of chemical inputs. For example, many European countries strongly regulate certain chemicals, and wastewater is usually treated before discharge into surface water bodies. Consequently, in most European regions, chemical risks from nonpoint sources are presumably highest for invertebrates and primary producers (see discussion in Malaj et al. 2014), whereas fish are at low risk, albeit the efficiency of wastewater treatment varies with the compound group and related discharges may entail ecotoxicological risks (see below). Contrasting examples occur in several South Asian countries, where water regulation, and enforcement of the latter, is weaker, leading to higher chemical pollution levels, which can pose serious risks for all aquatic organisms including fish but also for human health, if the contaminated river water is used for human consumption (Vörösmarty et al. 2010). In addition, the catchment hydrology, climate, and geology influence the relevance of chemicals. For example, particularly in arid and semiarid regions, irrigation agriculture can result in increasing ion concentrations in water bodies and changes in ion ratios with associated adverse ecological effects (Cañedo-Argüelles et al. 2013). Furthermore, acid rain is particularly relevant in regions with hard-rock geology, such as granite, because of the lack of buffer capacity. The influence of the environmental context on impacts of chemicals is discussed in more detail in the next section.

12.2.3   Assessing and Predicting Impacts of Chemicals in River Systems

The assessment and prediction of chemical impacts in river systems are complicated by a range of context-dependent factors that act on different spatial scales (Clements et al. 2012). Here, we focus particularly on factors that influence the impacts of chemicals in river systems on the site and landscape scales (Fig. 12.2). Several of the factors listed in Fig. 12.2 can influence the impacts of chemicals on freshwater organisms twofold: firstly, through their influence on chemical concentrations, bio-availability, and speciation in the case of metals or ionizable organic compounds and, secondly, as filters that shape the composition and sensitivity of populations and communities on which chemical toxicants act. Above, we have already outlined that catchment hydrology, climate, land use, and geology influence the exposure patterns of chemicals. However, they also influence the impact on aquatic organisms by determining which species are present and potentially modulating their physiological fitness and ultimately sensitivity. For example, hydrology, in terms of water level fluctuations, resulted in a tenfold increase in the sensitivity of invertebrate communities to an insecticide in a study with artificial ponds (Stampfli et al. 2013). In fact, a wide range of additional stressors, including the abiotic variables such as pH and temperature, have been identified as modulators of chemical impacts (Laskowski et al. 2010). This is hardly surprising given that other pairs of stressors in rivers such as habitat degradation (see Chap. 3) and eutrophication (Chap. 10) are also known to interact in their impact on freshwater organisms (Jackson et al. 2015). Consequently, prediction of the impacts of a chemical in a river would require understanding of the interactions of chemicals with potentially co-occurring stressors or the environmental context in general (Clements et al. 2012).

This is highly relevant for risk assessment as it implies that the same chemical concentration can have different impacts depending on the context (Fig. 12.2). Indeed, the multiple-stressor situation is rather the rule than the exception. An analysis of four stressor classes in German rivers, namely, habitat degradation, eutrophication, organic chemicals, and invasive species, found that almost all river sampling sites (97%) were simultaneously subjected to two or more stressor classes (Schäfer et al. 2016). Thus, the multiple-stressor situation has stimulated large research efforts, such as the EU project MARS involving 24 research institutions (http://www.mars-project.eu), though chemicals play a minor role in this project. However, chemicals typically occur as mixtures and estimation of mixture effects of chemicals in real-world ecosystems still represents a major challenge (Brack et al. 2015).

Besides co-occurring stressors and abiotic variables, a wide range of context-dependent variables govern the impacts on populations and communities. These include, for instance, the history of disturbance, where communities with a long history of chemical exposure may embody a history of adaptations (Johnson and Sumpter 2015). Furthermore, the capacity to cope with or adapt to stress is related to the energetic state of the community. Riparian vegetation contributes to the energetic state of a river through provisioning of organic matter. This energetic state depends also on the position in the river network as conceptualized by the river continuum concept (Vannote et al. 1980). At the same time, the local populations and communities are linked to meta-population and meta-community dynamics, which also depend on river network connectivity and overland distances to other network branches as well as in-stream distances to recolonization pools (Kärnä et al. 2015). These aspects influence the resilience of river ecosystems to chemical effects, which is further discussed below in the context of mitigation.

An additional source of complexity when aiming to predict impacts based on chemical exposure results from species interactions inside communities: If a chemical affects a species A that interacts with another species B, the chemical impact can propagate onto species B, although this species was not directly affected by the toxicant. Analogous to general ecology, this phenomenon is called an indirect effect, though the distinction between trait-mediated and density-mediated indirect effects is rarely made in ecotoxicological studies due to their strong laboratory orientation (see Schäfer et al. 2016). An example for an indirect (density-mediated) effect is represented by an herbicide input that eradicates periphyton leading to a reduction of invertebrate grazers that had not been directly impacted by the herbicide.

Overall, we have outlined that several factors influence the effects of a chemical under real-world conditions. Rivers often integrate, depending on the size of the upstream catchment area, the outputs of different land use types leading to exposure from point sources such as industrial or wastewater effluents with micropollutants, pesticides from agriculture, and potentially biocides and salts from roads and urban areas. These mixtures of chemicals can have a variety of direct and indirect effects on the different trophic levels in a river, which, as highlighted above, also depend on the general environmental context. This complexity may explain the mismatch between effect thresholds based on laboratory or semi-field studies under simplified conditions and those from field studies that have been observed for pesticides and metals (Schäfer 2014). However, the assessment and prediction of chemical effects are crucial for river management, taking also into account that affected river sections are interconnected longitudinally with other sections and rivers as well as with the surrounding landscape.

Given that both chemical exposure and effects in rivers are currently incompletely captured, the true risk is difficult to predict. Therefore, the sheer amount of chemicals in use that can potentially occur as mixtures in rivers requires intelligent prioritization based on adapted monitoring strategies that enable a realistic characterization of exposure and effects. The establishment of exposure scenarios, based on land use and point sources, through joint modeling and monitoring approaches might aid in defining priority substances and priority mixtures that are likely to occur and promote a more realistic exposure characterization. Such scenarios could also help to assess potential mixture effects as well as the interactions with co-occurring stressors, which currently hamper a realistic evaluation of the contribution of chemicals to the ecological degradation of rivers. The actual situation is partly due to the too narrow focus of current ecotoxicological risk assessment approaches (Beketov and Liess 2012). More holistic and field-realistic ecotoxicological approaches are required and may include tools that detect effects in rivers on different biological levels such as biomarkers, bioassays, and trait-based community, meta-population, and meta-community approaches. Moreover, tools that allow for the detection of effects on ecosystem processes and in turn for the evaluation of the consequences of such effects for ecosystem services are relevant for river managers to balance management measures and to highlight the relevance of measures to the public and stakeholders potentially involved in decision-making. Additional research in the abovementioned areas is needed to deliver such tools, though several promising avenues exist (Brack et al. 2015; Segner et al. 2014).

12.3   Mitigation

With the goal to reduce the load of chemical stressors in ecosystems, various mitigation measures of different categories have been proposed that range from substitution and reduction of chemicals to effluent treatment and landscape design. Ideally, chemicals would only be active at their place of intended action (e.g., pharmaceuticals inside the organism, pesticides on the crop area) and would be harmless by design when entering nontarget ecosystems. This could, for example, be achieved through rapid and complete degradation in the case of organic chemicals, which has been characterized as "benign by design" (Kümmerer 2007). However, implementation of this concept seems challenging for chemicals such as biocides and pesticides that are used in products with an intended high longevity (e.g., paint) or that are released to crop areas to harm specific target organisms but subsequently often enter other environments, where nontarget organisms that are physiologically similar to the target organisms are exposed.

Use reduction and substitution with compounds that have a lower environmental risk are mitigation measures to decrease exposure and impacts in ecosystems such as rivers. Apart from measures that aim at reducing the amount of a chemical released into a receiving ecosystem at its source (e.g., discharge from industrial facilities and mines), many "end-of-pipe" technologies and measures are currently under discussion to decrease the chemical load of rivers. These can be separated into measures targeting point sources and nonpoint sources. Regarding point sources, the most widely used technique is the treatment of the effluent before it enters a water body, for example, in municipal WWTPs. However, such treatments currently are incomplete for several chemicals including metals and organic micropollutants. Consequently, advanced treatment technologies have been developed to reduce the concentrations of metals or organic micropollutants, the latter for which WWTP effluents are considered the major source (Schwarzenbach et al. 2006). For metals, chemical precipitation has widely been employed to reduce metal concentrations in effluents. This process, however, produces sludge that needs to be disposed. Alternative methods include adsorption to new adsorbents (bio-sorption), membrane filtration, and electrodialysis that have a higher removal efficiency than conventional chemical precipitation but, depending on the method, also lead to toxic waste or incur high operational costs (Barakat 2011). For organic micropollutants, activated carbon filtration and ozonation can reduce the concentrations on average by approximately 80%, where the efficiency depends on the chemical properties of the individual chemical (Margot et al. 2013). Such reductions in the chemical concentrations are also reflected by alleviation of various ecotoxicological measures at subcellular, organism, and population levels, as documented by a meta-analysis (Bundschuh et al. 2011).

While they do offer increases in treatment efficacy, both advanced treatment technologies increase treatment costs and can also impact aquatic organisms. Activated carbon, for instance, may adsorb, besides organic micropollutants, essential trace elements. This results in negative, sublethal effects on aquatic life. The impacts of these effects in the receiving ecosystem may be limited if the incoming wastewater is sufficiently enriched with nutrients from the receiving ecosystem (Bundschuh et al. 2011). Moreover, ozonation may induce the formation of toxic oxidation by-products such as aldehydes and carboxylic acids that can adversely affect fish and other organisms. These by-products can eventually be removed by means of sand filtration, thus detoxifying the wastewater prior to its release into the receiving stream or river ecosystem (Bundschuh et al. 2011). Furthermore, treatment technologies have been developed to remove salt concentrations in effluents originating from resource extraction activities such as mining or hydraulic fracturing (Cañedo-Argüelles et al. 2013).

Although empirical evidence suggests that advanced treatment technologies improve the chemical quality of the wastewater, it has been argued that aquatic communities downstream of such point sources are adapted to the environmental conditions including pollution after several decades of chronic exposure. Consequently, Johnson and Sumpter (2015) cautioned that a modification of water quality, albeit an improvement from the human perspective, might have negative consequences for communities downstream from WWTP potentially forcing an ecosystem to leave its steady state by undergoing a regime shift. These considerations call for a careful monitoring of the responses of wildlife within the receiving aquatic ecosystem to understand the long-term consequences of implementing advanced treatment technologies.

In contrast to these point sources, nonpoint sources of pesticides as well as metals and salts require different approaches to achieve a reasonable level of mitigation. In particular pesticides are intentionally released to the environment when applied on crops to control for weeds (i.e., herbicides) as well as for fungal and for insect pests (i.e., fungicides and insecticides). During or following their application, they non-intentionally enter off-crop areas, such as aquatic ecosystems, via spray drift and runoff (Schulz 2004). Similarly, salts are intentionally released into the environment as road de-icers and can run off from roads and urban areas and in turn contribute to the salinization of freshwater ecosystems (Cañedo-Argüelles et al. 2013). Another nonpoint source of salts is agriculture, where, especially in arid and semiarid regions, irrigation dissolves soil-borne salts and the clearing of natural vegetation, especially trees, can lead to rising groundwater tables of saline water (see Jolly et al. 2001 for examples in the Murray-Darling basin in Australia). Both processes potentially cause diffuse inputs of saline effluents into water bodies. Similar to salt input from de-icing, nonpoint sources of metals are primarily runoff from roads and urban areas following heavy precipitation. Vegetated buffer strips are one mitigation measure that can reduce the concentrations of toxicants (pesticides, metals, and salts from agricultural fields, roads, and urban areas) in runoff flowing into aquatic ecosystems (Schulz 2004), though erosion rills within these buffer strips can jeopardize their retention efficiency. Moreover, such buffer strips can substantially decrease the input of pesticides into rivers via spray drift, depending on the density of the vegetation, with dense shrubbery being most efficient (Schulz 2004).

Within the aquatic environment, natural and constructed wetlands are decentralized mitigation measures that can strongly reduce the concentrations and associated effects of hydrophobic organic substances and metals [partly up to 90% (Schulz 2004)]. Within these wetland systems, various processes contribute to the reduced load of pesticides and metals in the water phase. The primary processes are adsorption to sediments and macrophytes as well as trapping of the suspended particles carrying the insecticides. The secondary contributions of photolysis, hydrolysis, and microbial degradation to mitigation efficacy of wetlands are lower (Stehle et al. 2011). Recent studies showed that wetlands can also mitigate the peak concentrations of rather hydrophilic chemicals, albeit to a lower degree compared to hydrophobic substances (Stehle et al. 2011). Hence, wetlands but also buffer strips seem suitable low-tech solutions for environmental managers to mitigate loads and effects of nonpoint source inputs of chemicals, at least for compounds that are subjected to degradation processes (Fenner et al. 2013).

Other chemicals such as PAHs and phenomena such as acidification are linked to the wet and dry deposition of chemicals from the atmosphere. In such cases, the reduction of chemical emissions, for example, in the case of acidification of sulfur dioxide and nitrogen oxides, and treatment before emission are the most feasible mitigation measure. Finally, given that chemical effects such as diffuse inputs from extreme weather events cannot always be avoided and, in fact, are partly acceptable within the framework of current pesticide regulation, as long as recovery occurs, fostering the resilience of the river ecosystem represents an important measure. This can be achieved through improving the connectivity of river systems (Chap. 9) as well as through preservation and creation of recolonization pools on the landscape level. Indeed, several studies in lentic and lotic water bodies emphasized the importance of external recolonization from such pools for recovery from toxicant effects (Schäfer 2012; Trekels et al. 2011). The importance of healthy and intact pools and wetlands as nurseries and refugia on the floodplain was amply demonstrated following a massive cyanide spill on the Tisza river in 2000. Whereas all life in the river channel was extinguished by the cyanide plume, recolonization from the neighboring floodplain habitats was swift (weeks and months) and efficient (Sáyli et al. 2000).

12.4 Conclusions

Toxicants still represent a relevant stressor in river ecosystems, despite major improvements of the situation over the last decades, at least in regions with a strong governmental regulation. Intelligent strategies are required to deal with the complex exposure situation in terms of mixtures, multiple stressors, and other features of the environmental context that all influence the magnitude of potential negative effects. The identification of land use-specific exposure paths, the subsequent prioritization of compounds and adaptation of monitoring strategies, as well as the development of more holistic and field-realistic ecotoxicological tools would enable a reliable characterization of exposure and effects, which is of paramount importance. Notwithstanding, a variety of mitigation measures is available to address point and nonpoint source pollution that range from local effluent treatment to the enhancement of resilience of river ecosystems through landscape approaches. Several of these approaches have the advantage that they would not only alleviate the problem of chemical toxicants in rivers but also, for instance, reduce excessive nutrient inputs and enhance riverine connectivity.

References

Barakat MA (2011) New trends in removing heavy metals from industrial wastewater. Arab J Chem 4:361–377

Beketov MA, Liess M (2012) Ecotoxicology and macroecology – time for integration. Environ Pollut 162:247–254

Brack W, Altenburger R, Schüürmann G, Krauss M, López HD, van Gils J, Slobodnik J, Munthe J, Gawlik BM, van Wezel A, Schriks M, Hollender J, Tollefsen KE, Mekenyan O, Dimitrov S, Bunke D, Cousins I, Posthuma L, van den Brink PJ, López de Alda M, Barceló D, Faust M, Kortenkamp A, Scrimshaw M, Ignatova S, Engelen G, Massmann G, Lemkine G, Teodorovic I, Walz K-H, Dulio V, Jonker MTO, Jäger F, Chipman K, Falciani F, Liska I, Rooke D, Zhang X, Hollert H, Vrana B, Hilscherova K, Kramer K, Neumann S, Hammerbacher R, Backhaus T, Mack J, Segner H, Escher B, de Aragão Umbuzeiro G (2015) The SOLUTIONS project: challenges and responses for present and future emerging pollutants in land and water resources management. Sci Total Environ 503–504:22–31

Breithaupt H (2006) The costs of REACH. EMBO Rep 7:968–971

Bundschuh M, Gessner MO, Ternes TA, Sögding C, Schulz R (2011) Ecotoxicologial evaluation of wastewater ozonation based on detritus-detritivore interactions. Chemosphere 82:355–361

Cañedo-Argüelles M, Kefford B, Piscart C, Prat N, Schäfer RB, Schulz C-J (2013) Salinisation of rivers: an urgent ecological issue. Environ Pollut 173:157–167

Chemical Abstracts Service (2015) CHEMCATS – chemical suppliers database. http://www.cas.org/content/chemical-suppliers

Clements WH, Carlisle DM, Lazorchak JM, Johnson PC (2000) Heavy metals structure benthic communities in Colorado mountain streams. Ecol Appl 10:626–638

Clements WH, Hickey CW, Kidd KA (2012) How do aquatic communities respond to contaminants? It depends on the ecological context. Environ Toxicol Chem 31:1932–1940

Drury B, Scott J, Rosi-Marshall EJ, Kelly JJ (2013) Triclosan exposure increases triclosan resistance and influences taxonomic composition of benthic bacterial communities. Environ Sci Technol 47:8923–8930

Englert D, Zubrod JP, Schulz R, Bundschuh M (2013) Effects of municipal wastewater on aquatic ecosystem structure and function in the receiving stream. Sci Total Environ 454–455:401–410

EUROSTAT (2015) Production of environmentally harmful chemicals. http://ec.europa.eu/eurostat/tgm/table.do?tab=table&init=1&language=de&pcode=ten00011&plugin=1

Fenner K, Canonica S, Wackett LP, Elsner M (2013) Evaluating pesticide degradation in the environment: blind spots and emerging opportunities. Science 341:752–758

Fent K, Kunz PY, Gomez E (2008) UV filters in the aquatic environment induce hormonal effects and affect fertility and reproduction in Fish. CHIMIA Int J Chem 62:368–375

Floehr T, Scholz-Starke B, Xiao H, Hercht H, Wu L, Hou J, Schmidt-Posthaus H, Segner H, Kammann U, Yuan X, Roß-Nickoll M, Schäffer A, Hollert H (2015) Linking Ah receptor mediated effects of sediments and impacts on fish to key pollutants in the Yangtze Three Gorges Reservoir, China — a comprehensive perspective. Sci Total Environ 538:191–211

Fynlayson C, D'Cruz R (2005) Inland water systems. In: Millenium ecosystem assessment. Island Press, Washington, DC, pp 551–583

Jackson MC, Loewen CJG, Vinebrooke RD, Chimimba CT (2015) Net effects of multiple stressors in freshwater ecosystems: a meta-analysis. Glob Chang Biol 22:180–189

Johnson AC, Sumpter JP (2015) Improving the quality of wastewater to tackle trace organic contaminants: think before you act! Environ Sci Technol 49:3999–4000

Jolly ID et al (2001) Historical stream salinity trends and catchment salt balances in the Murray–Darling Basin, Australia. Mar Freshw Res 52(1):53–63

Kärnä O-M, Grönroos M, Antikainen H, Hjort J, Ilmonen J, Paasivirta L, Heino J (2015) Inferring the effects of potential dispersal routes on the metacommunity structure of stream insects: as the crow flies, as the fish swims or as the fox runs? J Anim Ecol 84:1342–1353

Kidd KA, Blanchfield PJ, Mills KH, Palace VP, Evans RE, Lazorchak JM, Flick RW (2007) Collapse of a fish population after exposure to a synthetic estrogen. Proc Natl Acad Sci USA 104:8897–8901

Kümmerer K (2007) Sustainable from the very beginning: rational design of molecules by life cycle engineering as an important approach for green pharmacy and green chemistry. Green Chem 9:899–907

Laskowski R, Bednarska AJ, Kramarz PE, Loureiro S, Scheil V, Kudlek J, Holmstrup M (2010) Interactions between toxic chemicals and natural environmental factors – a meta-analysis and case studies. Sci Total Environ 408:3763–3774

Malaj E, von der Ohe PC, Grote M, Kühne R, Mondy CP, Usseglio-Polatera P, Brack W, Schäfer RB (2014) Organic chemicals jeopardize the health of freshwater ecosystems on the continental scale. Proc Natl Acad Sci USA 111:9549–9554

Maltby L, Forrow DM, Boxall ABA, Calow P, Betton CI (1995) The effects of motorway runoff on freshwater ecosystems: 1. Field study. Environ Toxicol Chem 14:1079–1092

Margot J, Kienle C, Magnet A, Weil M, Rossi L, de Alencastro LF, Abegglen C, Thonney D, Chèvre N, Schärer M, Barry DA (2013) Treatment of micropollutants in municipal wastewater: ozone or powdered activated carbon? Sci Total Environ 461–462:480–498

Mortimer M, Mueller JM, Liess M (2007) Sampling methods in surface waters. In: Nollet LML (ed) Handbook of water analysis. CRC Press/Taylor & Francis Group, Boca Raton, FL, pp 1–45

Pei R, Kim S-C, Carlson KH, Pruden A (2006) Effect of river landscape on the sediment concentrations of antibiotics and corresponding antibiotic resistance genes (ARG). Water Res 40:2427–2435

Pye MC, Vaughan IP, Ormerod SJ (2012) Episodic acidification affects the breakdown and invertebrate colonisation of oak litter. Freshw Biol 57:2318–2329

Randak T, Zlabek V, Pulkrabova J, Kolarova J, Kroupova H, Siroka Z, Velisek J, Svobodova Z, Hajslova J (2009) Effects of pollution on chub in the River Elbe, Czech Republic. Ecotoxicol Environ Saf 72:737–746

Sáyli G, Csaba G, Gaálné-Darin E, Orosz E, Láng M, Majoros G, Kunsági Z, Niklesz C (2000) Effect of the cyanide and heavy metal pollution of the Szamos and Tisza rivers on the aquatic flora and fauna with special attention to fish. Magyar Állatorvosok Lapja 122(8):493–500

Schäfer RB (2014) In response: why we need landscape ecotoxicology and how it could be advanced – an academic perspective. Environ Toxicol Chem 33:1193–1194

Schäfer RB, von der Ohe P, Rasmussen J, Kefford JB, Beketov M, Schulz R, Liess M (2012) Thresholds for the effects of pesticides on invertebrate communities and leaf breakdown in stream ecosystems. Environ Sci Technol 46:5134–5142

Schäfer RB, Kühn B, Malaj E, König A, Gergs R (2016) Contribution of organic toxicants to multiple stress in river ecosystems. Freshw Biol. https://doi.org/10.1111/fwb.12811

Schulz R (2004) Field studies on exposure, effects, and risk mitigation of aquatic nonpoint-source insecticide pollution: a review. J Environ Qual 33:419–448

Schwarzenbach RP, Escher BI, Fenner K, Hofstetter TB, Johnson CA, von Gunten U, Wehrli B (2006) The challenge of micropollutants in aquatic systems. Science 313:1072–1077

Segner H, Schmitt-Jansen M, Sabater S (2014) Assessing the impact of multiple stressors on aquatic biota: the receptor's side matters. Environ Sci Technol 48:7690–7696

Stampfli NC, Knillmann S, Liess M, Noskov YA, Schäfer RB, Beketov MA (2013) Two stressors and a community – effects of hydrological disturbance and a toxicant on freshwater zooplankton. Aquat Toxicol 127:9–20

Stehle S, Schulz R (2015) Agricultural insecticides threaten surface waters at the global scale. Proc Natl Acad Sci USA 112:5750–5755

Stehle S, Elsaesser D, Gregoire C, Imfeld G, Passeport E, Payraudeau S, Schäfer RB, Tournebize J, Schulz R (2011) Pesticide risk mitigation by vegetated treatment systems: a meta-analysis. J Environ Qual 40:1068–1080

Stehle S, Knabel A, Schulz R (2013) Probabilistic risk assessment of insecticide concentrations in agricultural surface waters: a critical appraisal. Environ Monit Assess 185:6295–6310

Trekels H, Van de Meutter F, Stoks R (2011) Habitat isolation shapes the recovery of aquatic insect communities from a pesticide pulse. J Appl Ecol 48:1480–1489

Vannote RL, Minnshall WG, Cummins KW, Sedell JR, Cushing CE (1980) The river continuum concept. Can J Fish Aquat Sci 37:130–137

Vethaak AD, Lahr J, Schrap SM, Belfroid AC, Rijs GBJ, Gerritsen A, de Boer J, Bulder AS, Grinwis GCM, Kuiper RV, Legler J, Murk TAJ, Peijnenburg W, Verhaar HJM, de Voogt P (2005) An integrated assessment of estrogenic contamination and biological effects in the aquatic environment of the Netherlands. Chemosphere 59:511–524

Vörösmarty CJ, McIntyre PB, Gessner MO, Dudgeon D, Prusevich A, Green P, Glidden S, Bunn SE, Sullivan CA, Liermann CR, Davies PM (2010) Global threats to human water security and river biodiversity. Nature 467:555–561

Vrana B, Allan IJ, Greenwood R, Mills GA, Dominiak E, Svensson K, Knutsson J, Morrison G (2005) Passive sampling techniques for monitoring pollutants in water. TrAC Trends Anal Chem 24:845–868

Wittmer IK, Bader HP, Scheidegger R, Singer H, Luck A, Hanke I, Carlsson C, Stamm C (2010) Significance of urban and agricultural land use for biocide and pesticide dynamics in surface waters. Water Res 44:2850–2862

Chapter 13
Land Use

Clemens Trautwein and Florian Pletterbauer

> ...In every respect the valley rules the stream. [. . .] We must,
> in fact, not divorce the stream from its valley in our thoughts
> at any time. (Hynes 1975)

13.1 Introduction

Dendritic stream-river networks are the backbone of most landscapes on earth's surface and determine linkages between terrestrial and aquatic ecosystems. These networks are hierarchically organized from microhabitats to the scale of whole catchments (Frissell et al. 1986). Accordingly, many processes of lotic freshwater ecosystems are determined by this hierarchically nested structure of river networks and their interlinkages. Hynes (1975) emphasized the importance of the linkage between the conditions within a river catchment and the flows of energy, materials, and organisms in the river as a dynamic ecosystem. These flows are interwoven in complex, cross-linked relationships of ecosystem functioning. Up to now, it is a prime challenge to understand these functions in detail and to robustly manage riverine landscapes in a sustainable manner. Starting with the work of Hynes (1975), the scale of riverine management was understood to occur over entire river catchments or even river basins with several concepts that integrated this perception (Vannote et al. 1980; Frissell et al. 1986; Ward 1989; Poff 1997; Fausch et al. 2002; Ward et al. 2002; Burcher et al. 2007). These theoretical frameworks seek to understand and to quantify interactions between landscape conditions over large spatial extents and instream responses. Ultimately, the catchment approach was even implemented into legal frameworks, such as the Water Framework Directive (WFD), which recognizes the river basin as relevant management scale (European Parliament 2000).

C. Trautwein (✉) · F. Pletterbauer
Institute of Hydrobiology and Aquatic Ecosystem Management, University of Natural
Resources and Life Sciences, Vienna, Austria
e-mail: florian.pletterbauer@boku.ac.at

However, the bases of these concepts originate in landscape ecology and its inherent landscape-scale thinking, which was traditionally focused on terrestrial ecosystems. In turn, relationships between landscape patterns and their consequences are dependent on the characteristics of the mosaic of the surrounding landscape at multiple temporal and spatial scales. Moreover, river networks are highly effective in linking different landscape elements even over large distances. Hence, Wiens (2002) suggested to take the "land" out of landscape ecology to emphasize the importance of integrating landscape ecological approach into river research and to deepen the understanding on the interplay between terrestrial and aquatic ecosystems which should enable to perceive riverine ecosystems as "riverscapes" as proposed by Ward (1998). After Fausch et al. (2002) and Wiens (2002), encouraging researchers and managers to consider the entire river environment, it was Burcher et al. (2007) who introduced the *land cover cascade* concept in which disturbance stimuli are propagated through a series of hierarchical entities until they ultimately affect biota in their habitat. Subsequently, researchers expanded their thinking to integrate this "bigger picture" into pressure-impact analyses. Several studies statistically linked landscape characteristics, that describe the conditions over large spatial extents, to biological and environmental indicators of ecosystem states.

This chapter emphasizes the importance of the catchment scale in river ecology and how human actions in the landscape with related multiple alterations result in pressures on riverine ecosystems impacting habitat quality and aquatic biota. At the beginning, land use terminology will be explained, followed by methods and data that are used to describe, detect, and investigate landscape-scale patterns in relation to lotic ecosystems. Subsequently, the linkage of land uses and pressures will be highlighted followed by the characterization of possible management actions to mitigate negative human-induced land use effects.

13.2 Land Use and Land Cover Definitions

Land use in river catchments is widely understood to have an effect on terrestrial but also aquatic ecosystems and especially on their interlinkage. Humans utilize land area with the intention to obtain products and/or benefits by using resources and—in most cases—changing the properties of land patches. Any human action and usually a series of operations on land define a type of land use. Hence, the term *land use* refers to the purpose the land serves, e.g., agriculture or urban areas, and it does not necessarily describe the (bio)physical materials at the earth's surface. The description of vegetation, materials, objects, and bare surface, either natural or man-made, commonly defines *land cover* types (Fisher et al. 2005).

These two terms, land use and land cover, are often erroneously used interchangeably, although each term has a very specific meaning as mentioned above. They are fundamentally distinct in their genesis, purpose, and application. Land cover is the direct observation, mostly from top view imagery, while land use requires additional socio-economic interpretation of the activities that take place on that surface (Comber 2008a).

Land use gives answers to the "why?" people need, use, and change the land, whereas land cover is important to the "how?" are ecosystems affected when inferring process-based fluxes of energy and material. Therefore, land use and land cover are distinct concepts in earth surface modeling approaches. The research community requires land cover for environmental models and land use for policy making (Comber 2008b).

There is no one-to-one relationship between land use and land cover types. The land cover "grass" can be found in multiple land uses like urban parks, residential land, pasture, etc. On the other hand, many homogeneous land use types encompass more than one land cover; residential land may contain trees, grass, buildings, and paved surfaces. Land use for recreation describes possible activities but can be applicable to different land cover types: the surface for recreational activities can, for instance, be a sandy beach, a built-up area like a pleasure park, woodlands, or even a pond. Hierarchical classification of land use is one approach to integrate cover information and human purpose in one dataset. Lower levels of land use classes are likely to have a one-to-one land cover equivalent.

In planning and decision-making processes, both the land use and land cover of a particular area (e.g., a river catchment) provide a comprehensive picture of the situation. Depending on the processes intended to be studied, and depending on the available data, it is crucial to select and prepare land use datasets with thematic and spatial resolution that is appropriate for the spatial extent of the study.

In communication between disciplines, research and practice, management, and conservation, this distinction between land use and land cover should be made clear. Such data on land use and land cover are often used indifferently in practice, leading to thematic inconsistencies in mapping and classification that can cause problems of further interpretation (Loveland et al. 2005).

In riverine ecosystem research and management, land use is seen as integrator of human actions in the landscape (Allan and Castillo 2007). For certain purposes, land use evidentially affects the ecological integrity of river ecosystems by altering spatially nested controlling factors of different scales (Allan 2004). Many, although not all, impacts on streams are entirely or partly linked to human actions in the landscape and thus can be quantified from data on land use.

Datasets on land use are often preferred over land cover datasets because the latter, emphasizing vegetation cover, is subject to seasonal changes and geographic particularities (latitude, biogeographical regions, etc.), making a universal classification scheme difficult to realize.

13.3 Methods and Data in Land Use Analysis

Current land use and land cover data are most often obtained from analysis of either satellite or aerial images. New technologies both in computer science and remote sensing provided the data and imagery that accelerated the incorporation of landscape thinking into riverine research (Johnson and Gage 1997). Before large-scale datasets became available, habitat mapping was field intensive and necessarily limited to discrete reaches. Research methods in freshwater ecology have historically been applied on individual stream reaches that were placed in their landscape context by conceptual models.

Since the 1940s, aerial photography has provided a landscape-scale perspective, but a suite of recent new technologies has dramatically improved our ability to conduct research over large areas (Steel et al. 2010). For decades, remote sensing has enabled synoptic views of entire rivers and their catchments at increasingly fine resolutions. It can now provide data at resolutions that capture reach-scale riparian and instream habitat structuring (Hall et al. 2009).

The era of satellite images for environmental monitoring started in 1972 with the launch of the Landsat program in a joint mission of the US Geological Survey (USGS) and the National Aeronautics and Space Administration (NASA). Since then satellites have continuously acquired multispectral (visible and infrared light) images of the entire earth's land surface (US Geological Survey 2016). This extensive data archive provides valuable resources for land use classification and allows for tracking land use changes through time.

There are an increasing number of sensing technologies used for mapping both landscape and water properties including optical sensors, light detection and ranging (LiDAR), forward-looking infrared (FLIR), and radio detection and ranging (RADAR) (Mertes 2002). These high-tech devices are deployed from multiple platforms that range from low-altitude tethered balloons to drones, to helicopters and fixed-wing aircraft up to outer space satellites, thereby producing data with sub-centimeter to multimeter grain resolution across scales from microhabitats to channel units to valleys to catchments.

Spatial land use data finally is the product of interpretation, separation, and classification of raw data based on four linked conditions: (1) *map scale* (e.g., degree of generalization, 1:50,000) and *spatial grain resolution* (e.g., 30 × 30 m pixel size), (2) minimum mapping unit, (3) the basic information used (e.g., satellite image or field mapping survey), and (4) the structure and number of items of the nomenclature.

One prominent example of a land use dataset that is often used for catchment analysis is the CORINE Land Cover dataset. The program runs since 1985 and was initiated by the European Union. The four characteristics of this product are map scale 1:100,000; based on Landsat 7 satellite images with 30 m pixel size; minimum mapping area for patches = 25 ha; and nomenclature with 3 hierarchical levels and 44 categories.

At present, most current studies rely on static maps that may represent land use with some years' asynchrony to the field-based stream data. However, remotely sensed data are likely to become more widely used in the future. Availability and accessibility of such data are greatly increasing in quantity and quality as opportunities grow to better synchronize time frames between data sources (remotely sensed, field based) (Allan 2004).

Basically, geographic information systems (GIS) enabled the analyses of landscape patterns. The most common practice is to calculate landscape composition metrics, i.e., expressing the total amount of various land use categories as percentage of the area under investigation (e.g., total catchment area or riparian corridor). Landscape composition metrics quantify the amounts of various land use/cover types without taking into account the spatial arrangement.

Besides a catchment's land use composition, a multitude of other landscape metrics can express the geometric properties of landscape elements and their relative positions and distributions (Botequilha Leitao et al. 2006). Landscape configuration describes arrangement, position, orientation, and neighborhoods of elements. Configurational metrics include patch size, patch shape, and compactness, as well as distance between patches. In this respect, composition metrics tell an incomplete story (Fahrig 2003). A watershed with 25% forest consisting of one large patch of contiguous forest provides significantly different ecological functions than a watershed that also contains 25% forest but is fragmented by small, scattered clear cuts.

Relatively fewer studies used landscape configuration to evaluate its divergent effects on landscape processes and ultimately on aquatic communities. Gergel et al. (2002) reviewed the advantages of landscape indicators for monitoring human impacts and finally concluded that simple metrics are useful as complements to other aquatic indicators. Simple metrics like proportion of the catchment or a buffer zone in different land covers and a measure of the arrangement or connectivity of natural and human-modified cover types in the riparian zone appear straightforward to interpret. In a later study, Gergel (2005) tested spatial and nonspatial landscape metrics within simulated landscapes of different percentages and arrangements of nutrient source and sink areas to understand watershed nutrient loadings. They found that a wide variety of different spatial configurations for watersheds with intermediate relative abundances of sources and sinks can lead to either very high or very low loading.

13.4 Land Use as Human Pressure and Its Impacts on Rivers

The pathways of cause-effect chains that are induced by land use act via hydromorphological and physicochemical controls in the ecosystem. Accordingly, natural factors of the landscape have also to be considered in analyses of land use effects. Geological bedrock material (e.g., siliceous vs. calcareous), soil types (e.g., soil fertility), and topography (steep vs. smooth slopes) are features of the natural terrain driving and underlying anthropogenic land use. Thus, anthropogenic and natural factors covary, because the latter influences the suitability of locations for the former such as agricultural and urban development (Feld et al. 2016). Land use can also be interlinked with natural gradients of geo-climatic factors, i.e., land use patterns over large extents, such as Europe, are strongly dependent on climatic factors (Lucero et al. 2011). Similarly, Feld et al. (2016) showed significant interactions of land use and geo-climatic factors in a comprehensive study covering eleven organism groups across several freshwater systems such as rivers, floodplains, lakes, ponds, and groundwater. In this study, different biodiversity indices were linked to land use descriptors, geo-climatic factors, as well as their interactions. The results showed consistently strong shared effects of land use and geo-climatic descriptors. Potentially, geo-climatic factors not only dominate but act in concert with land use. Hence, broad-scale studies on the role of environmental factors driving biodiversity must not overlook the shared effects of natural and anthropogenic descriptors. In turn, whenever anthropogenic and natural gradients covary, and only anthropogenic land use is assessed, the influence attributed to land use can be overestimated (Allan 2004).

The emphasis of landscape ecology on scale and their hierarchies has led riverine landscape research to use multi-scale studies in order to gain a better understanding of processes acting in a stream network (Lammert and Allan 1999; Fausch et al. 2002; Lowe et al. 2006). Many studies reported that a high proportion of forest cover in a catchment is normally associated with good stream conditions. Conversely, agricultural or urban areas in the catchment have been documented to have a negative influence on downstream river conditions. Agricultural land use degrades streams by increased diffuse inputs of fine sediments, nutrients, and pesticides, impacting riparian and stream channel habitats and altering flows. Increased urban land area can change the amounts and variety of pollutants in runoff, cause more erratic hydrology owing to increased impervious surface area, increase water temperatures owing to loss of riparian vegetation and warming of surface runoff on exposed surfaces, and degrade instream habitat structure owing to sediment inputs and channelization (Allan 2004).

Many studies have attempted to identify and measure land use effects and quantify the changes in land cover and the related effects on stream characteristics and associated biotic assemblages (Hughes et al. 2006). A schematic overview of the manifold effects of land use and the mechanisms disturbing aquatic biota are given in Fig. 13.1 by putting the relationships in the DPSIR framework. In most cases, the local, instream pressures and impacts related to land uses comprise sedimentation, nutrient enrichment, toxics, and hydromorphologic alterations (Allan 2004). Sedimentation clogs the interstitial of the river bed substrate, thus impairing spawning habitat for fish and habitat for benthic invertebrates. Furthermore, increased nutrient load can decrease levels of dissolved oxygen. Pesticides and other toxic substance like heavy metals impact vitality of biota. Finally, drainages and sealed surfaces change the natural characteristics of flood events (e.g., storm water runoff).

Most of these studies have investigated landscape-scale effects of single land use categories. For example, the proportion of agriculture or of urbanization in a river catchment is well-studied, and both are understood to have detrimental effects on instream conditions. From a global perspective, urban land use may appear insignificant, as urban areas only occupy less than 2% of the earth's land surface. Besides the extreme land cover changes induced by urbanization (e.g., soil sealing, canalization of creeks), the true significance of urban land use becomes obvious when considering urban-rural linkages (Lambin et al. 2001). Cities depend on land-intensive systems for food, energy, and other natural resources intensifying land use change in the surrounding rural environment. This link might become spatially uncoupled in a more-and-more globalized world and can put even higher land use intensification pressure on basins with favorable conditions for agriculture or forestry.

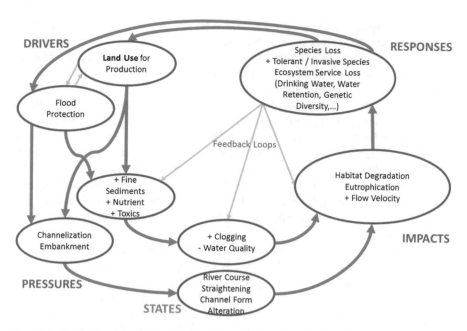

Fig. 13.1 DPSIR framework for land use and human alteration with impact on aquatic ecosystems

In particular, agriculture triggers several processes with negative effects for ecological integrity, not only for river stretches directly adjacent to farmed fields. Fertilizers are deployed over large areas. They are washed out by precipitation and transported into the water column of rivers, raising nutrient levels. This effect accumulates, increasing nutrient concentrations over larger spatial extents. In topographically pronounced, mountainous areas, arable land is situated on the valley floors of rivers. Typically, the floodplains were cut off and drained to make this area accessible for agricultural use. Accordingly, agriculture is often found in the direct vicinity of rivers, reducing riparian vegetation and, hence, the buffering function that reduces surface flows of fine sediments. Combining this lost buffer function with shorter distances between fertilizer and pesticides emissions and recipient aquatic ecosystems downstream increases environmental degradation.

Trautwein et al. (2011) showed that the causal links between land uses and impacts on river fish communities in Austrian rivers are more pronounced than mere correlation implies. The study revealed cumulative effects of several land use categories and quantified the thresholds at which land use's impact on fish communities could be predicted. Agriculture and urbanization were the best predictors for low ecological status based on fish community metrics. Poor and moderate ecological status was observed when >23.3% was classified as agriculture. Urbanization effects could be detected when >2% of the upstream catchment was urbanized. This detrimental effect of urbanization is a disproportionately large effect on instream conditions as compared to other human land uses (agriculture, pasture). Furthermore, effects of agriculture and urbanization were amplified when both categories occurred above their threshold in the catchment.

Multiple impacts from land use are most often coupled to human impacts on the local and reach scale. To develop robust management strategies for riverine ecosystems, the cause-effect chains have to be understood through multiple system components at multiple spatial (and temporal) scales. However, land use characteristics often represent an "umbrella pressure indicator," i.e., if urbanization or intense agriculture occurs, then also other human-induced pressures can be found. Besides nutrient and pesticide pollution, agriculture is often linked to increased water use, fine sediment input, and habitat loss due to physical modifications (see Fig. 13.1). Agricultural intensification along rivers induces channelization to increase the area of arable land (e.g., a so-called "10th federal state of Austria" was created by draining wet meadows after World War II) and to prevent flooding of the crops. However, channelization itself affects the functioning of the riverine ecosystem in different ways; on the one hand, it degrades the morphology; on the other hand, hydrology is also affected.

The synergistic and antagonistic effects of multiple pressures are still not fully untangled (Schinegger et al. 2012) despite much ongoing research (e.g., Nõges et al. 2016; Piggott et al. 2015; Schinegger et al. 2016). As many catchments face pressures related to more than one land use category (e.g., agricultural and urban land) as well as multiple human pressures, it is essential to understand in detail the different impact pathways and the interrelationships between the different pressures to develop robust management strategies.

(Best) Management Practice for Mitigating Land Use Effects
Managing land use effects can take place at different scales. At local scales (i.e., small spatial extent), it is more feasible to plan and implement land use with less detrimental effects than on the catchment scale. For example, near natural vegetation can be sustained by a variety of extensive land uses in riparian and floodplain area. Riparian zone management was one of the four most applied goals in 37,000 river restoration projects in the USA (Bernhardt et al. 2005).

In fact, riparian zones are critical transition areas between streams and their catchments. Riparian management can effectively influence stream conditions directly and buffer detrimental effects of land use practices further away from the river. The benefits promise to be highly disproportionate to the land area required, especially at the local and reach scale. For example, the protective effect of riparian forests against mixed agricultural and urban pressures was demonstrated in three regions in France. Riparian corridors appear to be manageable areas, and these results strongly support the idea of including their restoration in priority actions for achieving good ecological status (Wasson et al. 2010). However, in a nested hierarchy of landscape types, large-scale uses can constrain and even overwhelm smaller-scale processes. For example, the cumulative impacts of intensive land use over entire catchments are likely to override riparian zone mitigation (Hughes et al. 2006; Wang et al. 2006).

Land use management across entire catchments is much more challenging. Reversal of land use to a less-developed state over vast areas is usually economically and politically infeasible. However, mitigation of land use effects can be accomplished by promoting best management practices (BMPs) and improvements in landscape management. Modernization of agriculture can reduce and change fertilizer applications in the catchment and reduce soil erosion through conservation tillage and cover crops. Mitigation measures in urban areas address stormwater runoff and input of toxics from point sources.

One example of runoff management in small rural catchments is provided by Wilkinson et al. (2014). They tested runoff attenuation features (RAFs) based on the concept of the storage, slowing, filtering, and infiltration of runoff already at its source. RAFs are various constructions like bunds, drain barriers, runoff storage features within the main channel, and adjacent and large woody debris dams. By placing multiple measures all over the catchment, both flood risk and fine sediment input to streams could be reduced and improvements for water quality and stream ecological status could be achieved. While the high performance of RAFs can be achieved at relatively low cost for installation, to effectively manage runoff at larger scales would require a distributed network of RAFs, and the transactional overhead of negotiating with multiple land owners and government agencies can be quite challenging.

The previous example of RAFs well illustrates that restoration actions are expected to be most effective when they follow process-based principles by addressing causes of degradation in line with physical and biological potential at the appropriate scale (Beechie et al. 2008, 2010). Still, they do not solve root causes of the detrimental effects, because surface runoff and erosion from arable or pasture land still occur and the features need maintenance following floods regularly.

13.5 Research Outlook

Much research is needed to identify and test specific mechanisms that link human practices with aquatic ecosystems over multiple scales. This is especially difficult over vast areas, e.g., large river catchments. Well-defined hypotheses of cause-and-effect networks have to be formulated for testing models with empirical data and improve predictions. The lack of such mechanistic hypotheses was already outlined by Johnson and Gage (1997) and reviewed again by Allan (2004) as the majority of studies report correlative relationships between landscape and instream conditions but do not permit specific predictions of instream responses. Current knowledge is therefore limited to apply prescriptive management.

Two areas should be exploited to significantly advance the research frontier. First, we should exploit the increasingly available remotely sensed data to develop synoptic evidence at multiple scales, especially macro-scales. Large-scale, long-term data of multiple investigation sites should be assembled in large (global) datasets and linked to global environmental observation data that are produced in increasing resolution and rate at relatively less cost compared to field investigations.

Second, such macro-data should be used to challenge a new generation of hypotheses that are better formulated to explore causal mechanisms at larger scales and across scales, e.g., micro up to macro and vice versa. This should help uncover mechanisms about how streams interact with their surrounding landscapes. Better hypotheses will grow out of improved theoretical ecological models of dynamic landscapes and species distribution and evolution, developed to predict ecosystem dynamics. With this move forward, also management actions will better ensure rehabilitation success. Nevertheless, the scientific challenge will persist, because more complex models with increased number of factors are more difficult to interpret. Solving models with higher complexity is achieved on higher computational cost and at expense of tractability.

Large-scale, long-term experiments would be ideal to test mechanistic relationships between land and stream but they are costly and extremely difficult to design and manage. Instead, a promising way to go is still via analyzing "natural" experiments (i.e., space for time substitution, natural disturbance comparison, before and after policy change) (Paulsen et al. 2008).

From the management perspective, it is of considerable interest to distinguish the effects of various pressures within the catchment. Very often multiple pressures occur simultaneously, but to identify which one is most important is a difficult task. Von der Ohe and Goedkoop (2013) could distinguish the effects of co-occurring pressures on benthic invertebrates by stressor-specific metrics. But when effects of pressure interaction are not additive but antagonistic or synergistic, the untangling and identification of such specific metrics become very challenging (Schinegger et al. 2016). Development of ecological assessment methods that are able to indicate most important pressures and even pressure combinations is still highly desired by water managers (Hering et al. 2014).

References

Allan JD (2004) Landscapes and riverscapes: the influence of land use on stream ecosystems. Ann Rev Ecol Evol Syst 35:257–284

Allan JD, Castillo MM (2007) Stream ecology: structure and function of running waters. Springer, Dordrecht

Beechie TJ, Pess G, Roni P, Giannico G (2008) Setting river restoration priorities: a review of approaches and a general protocol for identifying and prioritizing actions. N Am J Fish Manag 28:891–905

Beechie TJ, Sear DA, Olden JD, Pess GR, Buffington JM, Moir H, Roni P, Pollock MM (2010) Process-based principles for restoring river ecosystems. BioScience 60:209–222

Bernhardt ES, Palmer MA, Allan JD, Alexander G, Barnas K, Brooks S, Carr J, Clayton S, Dahm C, Follstad-Shah J, Galat D, Gloss S, Goodwin P, Hart D, Hassett B, Jenkinson R, Katz S, Kondolf GM, Lake PS, Lave R, Meyer JL, O'donnell TK, Pagano L, Powell B, Sudduth E, Gleick P, Sand-Jensen K, Ricciardi A, Rasmussen JB, Baron JS, Palmer M, Postel S, Richter B, Buijse AD, Whalen PJ, Lavendel B, Malakoff D, Koebel JW, Cohn JP, Gillilan S, Palmer MA (2005) Ecology. Synthesizing U.S. river restoration efforts. Science 308:636–637

Botequilha LA, Miller J, Ahern J, McGarigal K (2006) Measuring landscapes. Island Press, Washington, DC

Burcher CL, Valett HM, Benfield EF (2007) The land-cover cascade: relationships coupling land and water. Ecology 88:228–242

Comber AJ (2008a) The separation of land cover from land use using data primitives. J Land Use Sci 3:215–229

Comber AJ (2008b) Land use or land cover? J Land Use Sci 3:199–201

European Parliament (2000) Water framework directive – establishing a framework for community action in the field of water policy. Off J Eur Communities 21:72

Fahrig L (2003) Effects of habitat fragmentation on biodiversity. Ann Rev Ecol Evol Syst 34:487–515

Fausch KD, Torgersen CE, Baxter CV, Li HW (2002) Landscapes to riverscapes: bridging the gap between research and conservation of stream fishes. BioScience 52:483

Feld CK, Birk S, Eme D, Gerisch M, Hering D, Kernan M, Maileht K, Mischke U, Ott I, Pletterbauer F, Poikane S, Salgado J, Sayer CD, van Wichelen J, Malard F (2016) Disentangling the effects of land use and geo-climatic factors on diversity in European freshwater ecosystems. Ecol Indic 60:71–83

Fisher PF, Comber AJ, Wadsworth R (2005) Land use and land cover: contradiction or complement. In: Fisher PF (ed) Re-Presenting GIS. Wiley, Chichester

Frissell CA, Liss WJ, Warren CE, Hurley MD (1986) A hierarchical framework for stream habitat classification: viewing streams in a watershed context. Environ Manag 10:199–214

Gergel SE (2005) Spatial and non-spatial factors: when do they affect landscape indicators of watershed loading? Landsc Ecol 20:177–189

Gergel SE, Turner MG, Miller JR, Melack JM, Stanley EH (2002) Landscape indicators of human impacts to riverine systems. Aquat Sci 64:118–128

Hall RK, Watkins RL, Heggem DT, Jones KB, Kaufmann PR, Moore SB, Gregory SJ (2009) Quantifying structural physical habitat attributes using LIDAR and hyperspectral imagery. Environ Monit Assess 159:63–83

Hering D, Carvalho L, Argillier C, Beklioglu M, Borja A, Cardoso AC, Duel H, Ferreira T, Globevnik L, Hanganu J, Hellsten S, Jeppesen E, Kodeš V, Solheim AL, Nõges T, Ormerod S, Panagopoulos Y, Schmutz S, Venohr M, Birk S (2014) Managing aquatic ecosystems and water resources under multiple stress – an introduction to the MARS project. Sci Total Environ 504:10–21

Hughes RM, Wang L, Seelbach PW (2006) Landscape influences on stream habitats and biological assemblages. American Fisheries Society, Bethesda

Hynes HBN (1975) The stream and its valley. Verhandlungen Internationaler Vereinigung Theoretischer und Angewandter Limnologie 19:1–15

Johnson LB, Gage SH (1997) Landscape approaches to the analysis of aquatic ecosystems. Freshw Biol 37:113–132

Lambin EF, Turner BL, Geist HJ, Agbola SB, Angelsen A, Bruce JW, Coomes OT, Dirzo R, Fischer G, Folke C, George PS, Homewood K, Imbernon J, Leemans R, Li X, Moran EF, Mortimore M, Ramakrishnan PS, Richards JF, Skånes H, Steffen W, Stone GD, Svedin U, Veldkamp TA, Vogel C, Xu J (2001) The causes of land-use and land-cover change: moving beyond the myths. Glob Environ Chang 11:261–269

Lammert M, Allan JD (1999) Environmental auditing: assessing biotic integrity of streams: effects of scale in measuring the influence of land use/cover and habitat structure on fish and macroinvertebrates. Environ Manag 23:257–270

Loveland TR, Vogelmann JE, Gallant AL (2005) Perspectives on the use of land-cover data for ecological investigations. In: Wiens JA, Moss MR (eds) Issues and perspectives in landscape ecology. Cambridge University Press, Cambridge

Lowe WH, Likens GE, Power ME (2006) Linking scales in stream ecology. BioScience 56:591

Lucero Y, Steel EA, Burnett KM, Christiansen K (2011) Untangling human development and natural gradients: implications of underlying correlation structure for linking landscapes and riverine ecosystems. River Syst 19:207–224

Mertes LAK (2002) Remote sensing of riverine landscapes. Freshw Biol 47:799–816

Nõges P, Argillier C, Borja Á, Garmendia JM, Hanganu J, Kodeš V, Pletterbauer F, Sagouis A, Birk S (2016) Quantified biotic and abiotic responses to multiple stress in freshwater, marine and ground waters. Sci Total Environ 540:43–52

Paulsen SG, Mayio A, Peck DV, Stoddard JL, Tarquinio E, Holdsworth SM, Sickle J, Van YLL, Hawkins CP, Herlihy AT, Kaufmann PR, Barbour MT, Larsen DP, Olsen AR (2008) Condition of stream ecosystems in the US: an overview of the first national assessment. J N Am Benthol Soc 27:812–821

Piggott JJ, Townsend CR, Matthaei CD (2015) Reconceptualizing synergism and antagonism among multiple stressors. Ecol Evol 5(7):1538–1547

Poff NL (1997) Landscape filters and species traits: towards mechanistic understanding and prediction in stream ecology. J N Am Benthol Soc 16:391

Schinegger R, Trautwein C, Melcher A, Schmutz S (2012) Multiple human pressures and their spatial patterns in European running waters. Water Environ J 26:261–273

Schinegger R, Palt M, Segurado P, Schmutz S (2016) Untangling the effects of multiple human stressors and their impacts on fish assemblages in European running waters. Sci Total Environ 573:1079–1088

Steel EA, Hughes RM, Fullerton AH, Schmutz S, Young J, Fukushima M, Muhar S, Poppe M, Feist BE, Trautwein C (2010) Are we meeting the challenges of landscape-scale riverine research? A review. Living Rev Landsc Res 4:1. https://doi.org/10.12942/lrlr-2010-1

Trautwein C, Schinegger R, Schmutz S (2011) Cumulative effects of land use on fish metrics in different types of running waters in Austria. Aquat Sci 74:329–341

U.S. Geological Survey (2016) Landsat—Earth observation satellites: U.S. geological survey fact sheet 2015–3081, 4 p. https://doi.org/10.3133/fs20153081

Vannote RL, Minshall GW, Cummins KW, Sedell JR, Cushing CE (1980) The river continuum concept. Can J Fish Aquat Sci 37:130–137

von der Ohe PC, Goedkoop W (2013) Distinguishing the effects of habitat degradation and pesticide stress on benthic invertebrates using stressor-specific metrics. Sci Total Environ 444:480–490

Wang L, Seelbach PW, Lyons J (2006) Effects of levels of human disturbance on the influence of catchment, riparian, and reach-scale factors on fish assemblages. American Fisheries Society Symposium, p 199–219

Ward JV (1989) The 4-dimensional nature of lotic ecosystems. J N Am Benthol Soc 8:2–8

Ward JV (1998) Riverine landscapes: Biodiversity patterns, disturbance regimes, and aquatic conservation. Biol Conserv 83:269–278

Ward JV, Tockner K, Arscott DB, Claret C (2002) Riverine landscape diversity. Freshw Biol 47:517–539

Wasson J-G, Villeneuve B, Iital A, Murray-Bligh J, Dobiasova M, Blacikova S, Timm H, Pella H, Mengin N, Chandresis A (2010) Large-scale relationships between basin and riparian land cover and the ecological status of European rivers. Freshw Biol 55:1465–1482

Wiens JA (2002) Riverine landscapes: taking landscape ecology into the water. Freshw Biol 47:501–515

Wilkinson ME, Quinn PF, Barber NJ, Jonczyk J (2014) A framework for managing runoff and pollution in the rural landscape using a Catchment Systems Engineering approach. Sci Total Environ 468-469:1245–1254

Chapter 14
Recreational Fisheries: The Need for Sustainability in Fisheries Management of Alpine Rivers

Günther Unfer and Kurt Pinter

14.1 Introduction

Fishing is an ancient practice in the acquisition of natural resources dating back to the Middle Stone Age. The principal reasons why humans visit waters to catch fish underwent a substantial transition in many countries throughout the preceding decades. While fishing to gain food still is an important factor in tropical areas of the world, especially in Africa and Asia, it is mostly for sport in inland waters of economically higher developed countries, as in major parts of Europe and North America (Welcomme 2016). There, the majority of fishermen nowadays fish solely to obtain recreation or to experience the aesthetics of nature.

However, many people still like to fish, and recreational fishing has developed into a notable economic sector in European countries (Arlinghaus 2004). Besides the economic values related to recreational fishing, social and ethical components are of increasing importance. Along with the growing common perception that fishing is a reasonable pastime, animal welfare and nature conservation issues are raised that, in extreme cases, deem fishing morally reprehensible (Arlinghaus et al. 2012).

Aside from social perceptions of fishing, it is significant that anglers represent the most prominent stakeholder group for aquatic ecosystem concerns in many areas of the world. Fishermen represent a very valuable source of experience and knowledge that can be explicitly valuable whenever nature conservationists are in need of support from a larger group of people (see Chap. 16). Often they are the "memory of a river," recalling fish sizes, catch rates, and ecological conditions. Therefore, many ideas or campaigns to protect or restore freshwater ecosystems are driven by people who enjoy fishing and thus have developed a closer tie to aquatic ecosystems.

G. Unfer (✉) · K. Pinter
Institute of Hydrobiology and Aquatic Ecosystem Management, University of Natural Resources and Life Sciences, Vienna, Austria
e-mail: guenther.unfer@boku.ac.at; kurt.pinter@boku.ac.at

© The Author(s) 2018
S. Schmutz, J. Sendzimir (eds.), *Riverine Ecosystem Management*, Aquatic Ecology Series 8, https://doi.org/10.1007/978-3-319-73250-3_14

As testament to their reliable support for nature conservation and their serious interest in "fish welfare," recreational fisheries/fishermen have to work on strategies as to how (1) to use fish stocks in a sustainable way, (2) to protect healthy or to restore impaired habitats, and (3) to practice fishing in a morally/ethically defensible way.

In general, management is required wherever human activities negatively impact fish habitats or where commercial and/or recreational fisheries use stocks in an unsustainable way. Next to those impacts the reestablishment of formerly endangered piscivorous predators (e.g., cormorants, otters), the spreading of invasive fish species (some of them introduced by fishermen), or the consequences of global change comprise further interferences that can cause substantial problems for natural fish stocks. The probability that such problems combine or overlap is very likely in Europe, severely complicating the challenge of managing fisheries sustainably. In the end, all potential problems of wild fish stocks are related to human activities. Therefore, the need to manage fish, ultimately, is always associated with human influences, attitudes, behavior, and expectations. Modern fisheries management in waters dedicated primarily to recreational fishing must try to merge nature conservation needs and the satisfaction of a still-growing number of anglers.

The majority of Austrian running water bodies can be assigned to the rhithral and are predominantly colonized by brown trout (*Salmo trutta*) and other salmonid species, like the European grayling (*Thymallus thymallus*) and the Danube salmon (*Hucho hucho*). Similar to other Central European countries, the rainbow trout (*Oncorhynchus mykiss*), which was first brought to Austria in 1886 (MacCrimmon 1971) and successfully established self-sustaining populations, evolved rapidly as a target species for recreational fishing. Additionally, many anglers fish in natural lakes as well as other stagnant water bodies like artificial ponds, reservoirs, and floodplain oxbows, targeting a broader diversity of fish species.

In this chapter we focus on the management of salmonid rivers and streams and present an example of a highly valuable trout fishing beat in a pre-alpine river. On that basis we discuss the cornerstones of our understanding of sustainable fisheries management, the tools fisheries management can use and the restrictions or limits sound management has to address.

14.2   The Ybbs Case Study

One fishing beat that is an example of modern fisheries management is situated in Lower Austria in the upper reaches of the River Ybbs. There, the so-called River Ois drains from the foothills of the Northern Limestone Alps. Its constrained and largely preserved natural riverbed can be characterized as pool-riffle channel type (Frissell et al. 1986; Montgomery and Buffington 1997) with a mean discharge of 4.5 m^3/s. The river stretch features a high variance of structural diversity, water depth, and heterogeneous substrate conditions. This fishing beat comprises a wetted area of about 5.2 ha over a length of 4 km and an average width of about 12 m (Fig. 14.1).

Fig. 14.1 Pool-riffle sequences define the character of the Ois River, a trout stream located in the foothills of the Northern Limestone Alps in Lower Austria (Source: C. Ratschan)

An important prerequisite for the sustainable harvest of fish is the analysis of key parameters related to the population size of the extant, exploited species. As a first step we distinguish two population aspects. The first one estimates stock density, which is typically described as the sum of individuals or the sum of weight for a given area and for the existing size classes. The second one estimates total stock size and then calculates the number of harvestable fish. To gain these data, regular fish censuses are required. In this context, it is critical for fisheries managers to recognize the key relationships that link fish abundance and biomass with density-independent factors, such as the carrying capacity of a river, which, over the long term, can change due to a variety of natural or anthropogenic influences (Fig. 14.2). In other words, changing environmental conditions entail changing population densities. However, historically overexploitation has repeatedly occurred when management practices are tied to habits or routines rather than regular environmental updates (e.g., Sánchez-Hernández et al. 2016). Consequently, exploitation of natural resources such as wild fish stocks requires a constant reconsideration of what an ecosystem under current conditions is able to yield (Fig. 14.2a, see also Chap. 16 regarding *path dependence*).

By means of regular stock assessments deeper insights into the magnitude of short-term population dynamics can be gained, which describe a further parameter to be considered (Fig. 14.2b). Rapidly changing stocks are predominantly regulated by density-dependent factors as well as seasonal environmental influences and could be relevant for determining harvesting quotas.

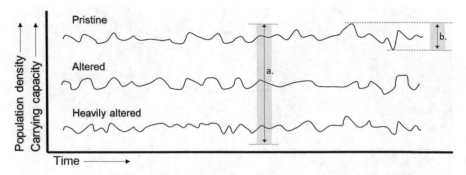

Fig. 14.2 Natural and human-induced levels and variations of the population densities. a: The size of a population is a matter of the carrying capacity of the respective water body. The carrying capacity is related to natural conditions and/or anthropogenic influences. In this example three different levels of population density are illustrated, assuming a heavily altered situation of low density up to a pristine situation of high population density. b: Small-scale dynamics of the population density as a consequence of varying density-dependent and natural environmental factors

In the respective river section of the River Ois, a quantitative fish sampling campaign has been carried out yearly (1997–2016) to assess the named parameters of the salmonid species. In terms of species distribution, brown trout dominates, holding an average share of 66%. It is followed by the nonnative rainbow trout (29%) and European grayling (5%). Further species to be found are bullhead (*Cottus gobio*), occasionally arctic char (*Salvelinus umbla*), and nonnative brook trout (*Salvelinus fontinalis*). To further illustrate the management approach, the as yet unexploited brown trout (catch-and-release management) is taken as example.

The first step to sustainably harvesting brown trout is to capture the demographics of the population. In so doing we can see that recruitment is subject to extensive natural fluctuations (Fig. 14.3). According to data from almost two decades of semiquantitative sampling (cf. Unfer et al. 2011), high reproduction success of brown trout occurs every 2.8 years on average. The observed population dynamics can mostly be attributed to hydrological conditions during the incubation period (Unfer et al. 2011). The latter are seasonally differing, flow conditions and further density-independent factors responsible for short-term fluctuations in fish populations (Fig. 14.2b). Differences in reproductive success are further manifested in the density of the total stock, with fluctuations of up to 200% of the total biomass in any respective time period (see also Table 14.1).

Following the determination of stock densities, the yearly production of brown trout has to be considered in order to identify regions in the recruitment curve where harvest becomes possible. Production is defined as the amount of tissue elaborated per unit time per unit area (Clarke et al. 1946; Waters 1977). Our monitoring data over successive years allows us to illustrate the production balance for single age or size categories. For example, reading cells of the same color on a diagonal from upper left to lower right, the net production (production minus loss) of the 0+ cohort (age class 1 in Table 14.1) from 2010 to 2011 totals 10 kg/ha. The same cohort gains a further increment of 22 kg/ha by the year 2012 before the net production becomes

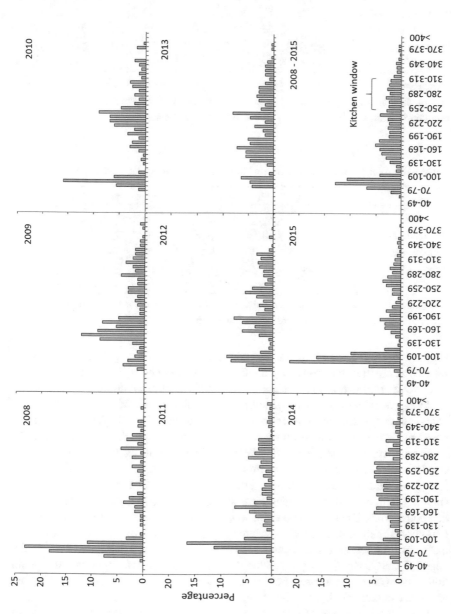

Fig. 14.3 Relative length frequency plots of the brown trout population in the years 2008–2015 and cumulated for the 8-year period 2008–2015 also showing the suggested "kitchen window" in the size class 250–320 mm

Table 14.1 Biomass and net production (kg/ha) of brown trout between 2008 and 2015

Size class	Age class	2008	2009	2010	2011	2012	2013	2014	2015
<120 mm	1	1^{+1}	$0.2^{+0.2}$	1^{+1}	2^{+2}	1^{+1}	$0.5^{+0.5}$	$1.4^{+1.4}$	3^{+3}
120–220 mm	2	2	9^{+8}	6^{+6}	11^{+10}	12^{+10}	8^{+7}	10^{+10}	8^{+7}
220–320 mm	3	12	20^{+18}	27^{+17}	39^{+33}	32^{+22}	23^{+11}	29^{+21}	30^{+20}
>320 mm	4	8	$12^{-0.6}$	15^{-5}	$27^{+0.2}$	17^{-22}	12^{-20}	9^{-14}	11^{-18}
Total biomass$^{\text{Total net production}}$		24	41^{+26}	48^{+19}	78^{+45}	63^{+12}	44^{-2}	50^{+18}	53^{+12}

The production (superscript figures) is calculated on the yearly increment of biomass of the respective cohort. Delimitation of size classes follows the age classes

Table 14.2 Abundance and production (Ind/ha) of brown trout between 2008 and 2015.

Size class	Age class	2008	2009	2010	2011	2012	2013	2014	2015
<120 mm	1	199^{+199}	39^{+39}	114^{+114}	282^{+282}	226^{+226}	85^{+85}	265^{+265}	452^{+452}
120–220 mm	2	48	251^{+52}	109^{+70}	199^{+85}	262^{-20}	203^{-23}	189^{+104}	163^{-102}
220–320 mm	3	69	107^{+59}	190^{-61}	172^{+63}	201^{+3}	146^{-116}	196^{-7}	157^{-32}
>320 mm	4	25	29^{-41}	37^{-70}	59^{-131}	53^{-119}	34^{-167}	25^{-121}	30^{-166}
Total abundance $^{\text{Total net production}}$		341	426^{+110}	450^{+52}	712^{+299}	743^{+90}	468^{-222}	675^{+241}	802^{+153}

The production (superscript figures) is calculated on a yearly increment of individuals for the respective cohort. Delimitation of size classes follows the age classes

negative (-20 kg/ha) in the subsequent year. Over the long run, the described scheme of positive and negative production turns out to be typical, with culminating positive net production in age class 3 and negative net production for older classes.

Along with the increment of biomass, an increase of fish abundance can be documented for the transition from age class 1 to age class 2 (Table 14.2), unless fish densities have already been very high in the first year. In the following year, fish abundance typically stabilizes, apparently by interacting again with the previous year's level, before it decreases again in age class 4 when fish grow older. General life cycle characteristics of brown trout become evident in any respective river stretch when one considers long-term stock developments of both biomass and fish abundance. Especially in case of low reproductive success, it becomes evident that downstream movement of juvenile stages from the headwaters and tributaries increases production (higher abundances of age class 2 and 3 compared to preceding years). Finally, when fish grow older, natural mortality, out-migration, and potentially otter (*Lutra lutra*) predation explain decreasing fish abundance and biomass, hence the negative net production in age class 4.

The natural decrease of fish abundance in size class 4 further supports recommendation that the harvest of fish needs to focus on the most productive, i.e., the third class. Therefore, instead of applying the usually prescribed minimum size of harvestable fish, we recommend a harvest slot (*kitchen window*) with a minimum fish length of 250 mm and a maximum length of 320 mm (Fig. 14.2) that corresponds to an average weight of 200 g per fish. By the application of harvest slots within a realm of high productivity, fishing mortality becomes a sustainable expansion of natural mortality that leaves enough excess for future generations to persist. The harvest slot furthermore leads to benefits such that fish outgrowing the kitchen window remain in the ecosystem, further developing and releasing their high value for the reproductive success of the population. Also, from a genetic point of view advantages arise, as the removal of intermediate-sized fish potentially decreases the risks of reducing fish genetic heterogeneity (Birkeland and Dayton 2005).

Finally, for the determination of harvest quotas, Mertz and Myers (1998) assume that, if fishing mortality is equal to the natural mortality, at least one half of the production of the stock may be harvested. Based on the available data (Table 14.1, total biomass), the average total net annual production ($\bar{x}_p = 19$ kg/ha) of the whole river Ois fishing beat (5.2 ha) can be calculated as 96 kg per year (2008–2015). Half of the yearly net production divided by an average weight of a harvested fish of 200 g results in an average possible sustainable harvest of almost 250 brown trout per year. In comparison, the current stock of brown trout in a similar size/age class, e.g., between 22 and 32 cm, averages more than 800 available individuals in the total fishing beat. On average the proposed harvest quota would therefore range between one third to one fourth of the respective stock, which means a sufficient amount of fish remaining to continue and grow bigger, even in years of very low abundance. Additionally, the exploitation of fish within the limits of the "kitchen window" would reduce the total biomass below the river's carrying capacity, i.e., below unsustainable mortality levels. That increases the chances of survival for smaller fish, and this extra production again can result in surplus or sustainable production (Wallace and Fletcher 2001). In summary, the example of the River Ois illustrates the necessity to develop fisheries management approaches on the basis of careful consideration of (changing) stock quantities. The analysis of quantitative fish data reveals the size and the dynamics that are inherent in the stock and therefore form the basis for management decisions.

Note that the observed dynamics are specific to habitat characteristics at several scales nested inside of each other, e.g., to the local characteristics of the fishing beat, as well as to the location of a beat within the distribution boundaries of a species and to the characteristics and the quality of the surrounding catchment. In this context, the abovementioned fishing beat is located at the upper distribution boundaries of grayling and rainbow trout. An impassable migration barrier at the lower section of the beat (Fig. 14.4) proves to be responsible for the decreasing numbers of grayling, since adult returners are not able to recolonize their nursery river reaches. Rainbow trout, by comparison, are able to maintain stable stocks within the reach of the beat. However, despite the fact that brown trout abundance and biomass are twice that of rainbow trout, only rainbow trout are currently harvested. The continuing monitoring of the stocks provides information on how rainbow trout are affected by fisheries exploitation and provides valuable knowledge for a future exploitation of brown trout. Grayling, however, are generally not harvested, which is due to the location of the beat in the upper most distribution area of this species, the small size of the stock, and the aforementioned deficits in the life cycle of grayling.

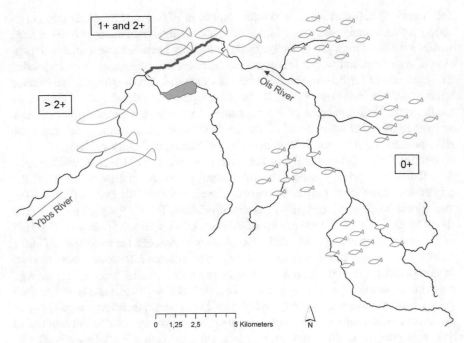

Fig. 14.4 Simplified life cycle scheme of brown trout in the catchment of the River Ybbs (Ois). Juvenile (0+) brown trout out-migrate from nursery headwaters and tributaries to lower river sections. The respective fishing beat (bold line) benefits from immigrating trout. Older fish (>2+) partly out-migrate to lower river sections. Boxes indicate highest abundance of respective age class within the catchment. A migration barrier (dashed line) at the lower end of the fishing beat prevents upstream migration

14.3 Managing Impacted Habitats

The high habitat quality of the fishing beat on the Ois is an exception in the Austrian river landscape. Most of Austria's waters are impacted to a varying extent by river channelization, impoundments, water abstraction, and hydropeaking (see Chaps. 2–13). The degree of deterioration and the interplay and severity of different impact combinations can be manifold. The consequences of impaired habitat quality in any case are the reduction of the habitats' carrying capacity and bottlenecks in the life cycles of affected fish populations. To assess the consequences and to derive proper management measures, a process called "deficit analysis" (Holzer et al. 2004) is carried out in three steps:

1. Analysis of the habitat quality to isolate and specify potential bottlenecks
2. Analysis of the stock (development)
3. Analysis of preceding management activities

14.3.1 Analyzing Habitat Quality

The first step of deficit analysis aims to detect the occurring habitat deficits, which typically are related to hydrological (water abstraction, hydropeaking, thermal alterations, etc.) as well as morphological impacts (bank stabilization measures, longitudinal/lateral barriers, etc.). Both types of interventions, but also the retention of bedload in upper reaches of the catchment, can have further negative consequences for the quality of bed sediments and the availability of food. The analysis of habitat quality is intended to serve fisheries management purposes. The main focus is on the habitat requirements of all different life stages of the river-type-specific fish species, clearly highlighting the species relevant for angling and relevant prey fish. This is in many cases congruent or at least in line with processes of river restoration projects, but the scope of river restoration is generally broader, and fish fauna are just one out of many important aspects related to the ecological integrity of running waters (see Chaps. 15 and 19).

The starting point for the habitat analysis is set by the life cycle of the fish species of interest. A fish life begins at the spawning ground—therefore the quality of spawning habitats is a major issue and has to be thoroughly analyzed. The most prominent fish species of alpine rivers, such as brown trout, grayling, and Danube salmon, rely on loose gravel for spawning and successful recruitment. Potential spawning habitat deficits are typically related to increased accumulation rates of fine sediments, on the one hand, or to an artificial coarsening of bed sediments on the other hand (see Chap. 8). Increased input of fines leads to clogging of the interstitial pores and consequently degraded gravel beds, hindering redd excavation or the successful development of incubated eggs. Retention of sediments through torrent control structures and impoundments as well as flushing out gravel due to reservoir management practices lead to coarsening of bed sediments, impeding redd construction for interstitial spawning salmonid species. Overall, alterations of natural sediment regimes are a severe and widespread problem in alpine rivers, and deficits of suitable spawning habitats are consequently among the major bottlenecks in many rivers and streams (e.g., Hauer et al. 2013; Pulg et al. 2013). For example, trout fry develops inside interstitial pores for a period of up to 6 months. Thus, not only the spawning itself but also the early development after hatching is affected by deteriorated riverbeds. Even though eggs are able to develop, high losses can occur in the alevin or early fry stage when sealed river beds prevent juveniles from successfully emerging.

Further on in the life cycle, early juvenile stages are threatened by a variety of human-induced, hydrological impacts, such as stranding due to hydropeaking surges (see Chap. 5), reservoir flushing (see Chap. 6), thermal changes (see Chap. 11), etc. Morphological alterations due to damming and other river control measures (see Chap. 3) can lower habitat quality for all different life stages. While residual flow stretches mainly reduce the amount of adult fish habitat as the amount of flow and consequently habitats are reduced (see Chap. 4), many regulated channels lose important habitats for juvenile fish, such as shallow gravel banks or adjacent side arms and backwaters. Another major problem is the disruption of migration pathways, both laterally and longitudinally (see Chap. 9). Beside weirs and ramps, riverbed degradation and, consequently, disrupted connections between main stem and tributaries hamper spawning migrations and decrease the original longitudinal range of populations. In addition to hindering upstream migration, specifically hydropower weirs and the associated turbines can cause high mortality of downstream migrating fish. Even this incomplete list of possible habitat perturbations reveals how analyzing habitat quality constitutes a major task of fisheries management. As the life cycle stages of any fish species are related to distinct habitat features, the quality of these features has to be assessed and contrasted with data on fish demographics and distributions. In many cases, fish population structures specifically reflect the habitat situation and help identify potential quality shortcomings.

The most powerful management action to sustainably support healthy fish populations is habitat restoration. Consequently, especially for the conservation of wild fish stocks, the primary task of fisheries managers is to pursue all options to restore habitat conditions to as close to a pristine situation as possible. Small-scaled mitigation measures, such as the maintenance or improvements of spawning grounds (e.g., Pulg et al. 2013), can be carried out and financed relatively easily by associations responsible for fisheries. However, mitigation or restoration at large- or even catchment-scales needs broader efforts (see Chap. 15) that should nonetheless be supported by fisheries managers. To attain objectives on a larger scale, fisheries managers dealing with common issues (e.g., along the same river) have the chance to gain greater influence when they form coalitions and join with local communities to speak with a common voice (see Chap. 16). Combining a critical mass of expert and public opinion is a vital necessity in Austria, where riverine water bodies are characterized by a small-scale segmentation of management units.

However, when the habitat quality analyses are completed, the results have to be contrasted and merged with the results of step 2, the survey of the current fish stocks. Quantitative electrofishing data must first be generated to enable a comprehensive assessment of the actual population status, e.g., abundances, biomass, and population structures. As highlighted in the Ois example, yearly surveys and long-term data series create the most desirable basis for analysis. In many cases fish stock data are missing or collected only sporadically. However, it is an important management task to gather stock data. Although the financial expenditure for fish stock surveys is substantial, it will pay off, since in combination with the habitat quality survey potential bottlenecks become detectable. Furthermore, the elimination of habitat deficits will sustainably improve the stocks, as opposed to stocking as a continuous management measure, which generates costs without solving the underlying problems (see below).

Further, to have sufficient knowledge on the actual fish populations and habitat quality provides the opportunity to reflect and evaluate preceding management actions, specifically success or failure of stocking campaigns (see below). The results of all the three steps of deficit analyses provide the basis for the elaboration of management strategies and to derive management actions.

14.3.2 Stocking Fish: Restrictions and Possibilities

When to Consider Stocking?

If bottlenecks remain after all possibilities to improve the habitat are exhausted, then a fisheries manager has to consider other options to improve the fish stock (cf. Fig. 14.5). Not only in Austria but also in other regions of the world, stocking of artificially propagated fish has been seen as the major (often the only) tool and duty of fisheries management in recreational fisheries. While large amounts of fish of various species and age classes are still stocked, stocking lost its status as panacea and is nowadays more and more questioned. The majority of recent scientific literature dealing with stocking fish in riverine environments stresses potential ecological problems and threats deriving from stocking activities (e.g., Christie et al. 2014). On the one hand, stocked fish suffer from high mortality and emigration rates after release, so economic success is increasingly in doubt. On the other hand, there is clear proof that propagated fish have negative consequences for wild stocks and populations due to genetic admixture or homogenization as well as to increasing competition for habitat and food (e.g., Fraser 2008; Olden et al. 2004).

However, if habitat problems remain and essential environmental prerequisites for different life stages are lacking, stocking might be the only option to sustain recreational fishing. There are different motives to stock (Laikre 1999; Welcomme and Bartley 1998) whereby stocking to mitigate/compensate environmental impacts is the most common reason but stocking for conservation purposes is becoming more and more popular. While compensatory stocking can help to sustain recreational fishing, and can be seen as ecologically reasonable, stocking solely to attract or satisfy anglers can hardly be justified in the context of sound or ecologically orientated fisheries management. Whenever fishes are stocked, one should account for possible negative consequences for the receiving ecosystems and contrast them with potential benefits (mainly socio-ecological). Economic issues or the benefit of

Fig. 14.5 Fisheries management actions have to be adopted according to the ecological status of a river

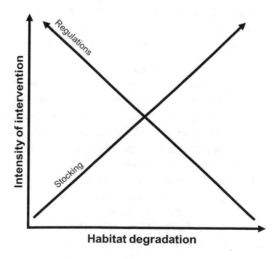

the owners of fishing rights play an important role in Austria, where the right to fish is a private law and profits are part of the income of the respective fishing right holder. This fact hampers the elaboration and implementation of large-scale, e.g., catchment wide, fisheries management strategies or plans and stands in contrast to countries where the right to fish is public and ecological or nature conservation issues typically have priority over commercial aspects.

Quality of Stocked Fish

Besides the negative impacts that stocked fish might have on wild stocks due to resource competition processes, two further major issues have to be considered: aspects of genetic descent as well as deficits deriving from the artificial propagation in hatchery environments. The latter can result in both behavioral and/or phenotypical deficiencies that, in many cases, are already genetically fixed. Typically, hatchery fish suffer from various abnormalities as they are kept under hatchery conditions for a period of time. As the time spent in the hatchery environment increases, the process of adapting wild animals to human-controlled environments leads to a wide range of behavioral as well as physiological alterations. In general, domestication results in increased fitness under hatchery conditions but decreased fitness under natural conditions (Berejikian et al. 2005) leading to high mortality rates after release (e.g., Weiss and Schmutz 1999). Among the reasons for weak performance of stocked hatchery fish are reduced ability to identify and take natural food, failure to adaptively react to variable food availability, increased boldness in responding to novel objects, reduced flight response, etc. (e.g., Järvi 2002). About 10–15 years ago, researchers postulated that post-release survival has to be enhanced (Maynard et al. 2004). However, if large hatchery fish (catchable sizes) are stocked, it might be more favorable that mid- to long-term survival rates are low, to avoid crossbreeding with wild conspecifics that spreads negative impacts on the genetic integrity of wild populations.

Domestication effects are more likely to be avoided if fish for stocking purposes originate from wild breeders and the duration of their stay in hatcheries is as short as possible. However, as shall be explained below, stocking of juvenile size classes does not guarantee high survival rates. (Unfer et al. 2009; Pinter et al. 2018) nor can undesired genetic consequences be fully prevented (Christie et al. 2016).

Stocking for Conservation

Stocking to achieve conservational aims is the only ecologic reason to release propagated fish. However, fish deriving from artificial propagation will always create repercussions, e.g., reduced reproductive fitness (Araki et al. 2007; Christie et al. 2014). This is so even if the aim of stocking is the reestablishment of reproducing populations in cases where a habitat bottleneck has been restored or a viable river (reach) should be recolonized (e.g., after a fish kill). As stated above, effort to (re) establish a population or stock is only meaningful and promising if the habitat prerequisites for self-sustainability are given or habitat quality has been restored. The foremost aim of any conservational stocking campaign should be the initiation of primary (re)colonization, which implies stocking over a restricted period of time, not continuously. Although the aims of conservational stocking are respectable, the task is far from being simple. The most important precondition is to select or identify a donor population of suitable genetic origin. In Austria, decades of stocking fish of foreign origin and trans-basin spreading of populations of both brown trout and European grayling substantially altered the genetic integrity of wild fish populations, particularly those of salmonid species (Duftner et al. 2005; Pinter 2008; Meraner et al. 2013b; Schenekar et al. 2014). Nowadays, fish species, predominantly cyprinids, e.g., nase (*Chondrostoma nasus*) and barbel (*Barbus barbus*), are stocked with increasing regularity and in increasing quantities. This is so, not least because of increasing problems related to successful protection and subsequent recolonization of riverine habitats by fish-eating predators (cormorant, otter, merganser). The distribution of further species across different river basins increases the risk of modifying the up-to-now widely unimpaired genetic integrity of riverine fish species, which were of minor importance for recreational fishing. Exemplarily, in the Po River catchment (Italy), the native barbel (*Barbus plebejus*) already got virtually replaced by the non-native European barbel (Meraner et al. 2013a).

Which Size/Age Classes Can Be Stocked?

In case a suitable donor population (of local origin and adequate size) is available, the next question is how to carry out the restocking? As it is mandatory to avoid domestication effects as far as possible, the first choice will be to stock fertilized eggs. Eggs can be stocked in "artificial nests" and/or breeding boxes. There are different types of boxes, while the application of eggs using breeding boxes in Austria is known by the term "Cocooning" (Holzer et al. 2011). A major advantage of stocking eggs to avoid domestication is that hatchlings can adapt to the natural environment from the earliest stage on. If potential spawning sites for egg deposition are selected, fish emerging from the nests can home back to the site after they are themselves ready to spawn, and a regularly used spawning ground/site can be established.

While the optimum life stage for conservational stocking is evident, various problems remain. As already mentioned the first task is to spot a suitable donor population, which has to be large enough and genetically appropriate. Furthermore, the number of eggs needed to be successful is unclear but in any case high. The European grayling provides a good example: A female grayling spawns about 5000 eggs over a period of 4 years, corresponding to a lifetime amount of 20,000 spawned eggs. For a population to remain stable, each couple has to produce two adult fish during their life spans. This means that, out of 20,000 eggs on an average, two adult fish will develop. The minimum number of adults for self-sustaining populations is 50, following the 50/500 rule (Franklin 1980). Consequently, to receive these 50 adult fish a few years later the planting of half a million eggs would be required; in other words, the eggs of 100 females (5000 eggs each).

Populations where it is ecologically acceptable to take such a high amount of spawners are scarce. Admittedly, fertilization rates of artificial spawning will be higher compared to natural redds as well as to survival rates to the eye-point stage. On the other hand, large egg numbers are needed so as to establish a founder population and to compensate for losses of reproductive fitness and further imperfections due to, e.g., the inhibition of natural partner selection and modification of other processes of natural selection. We suggest continuing yearly plantings for the duration of a full life cycle to stabilize the initial stock by more than just one cohort. Survival rates among the cohorts can, of course, vary markedly, as the juveniles are exposed to natural hazards (see the Ois example).

A study of the success of stocking juvenile age classes of brown trout carried out in different Austrian streams between 2005 and 2008 clearly demonstrated that stocking of 0+ and 1+ trout had very low success. For this study, juveniles of a hatchery strain as well as fish derived from wild spawners of local origin were used, and the performance (survival) was contrasted with the resident population (Pinter et al. 2018). In all streams native wild trout outcompeted the stocked 0+ strains. Similar results were obtained for 1+ trout, which were stocked into further natural streams after being reared for 1 year under three different rearing conditions (a natural stream, a structured flow channel, conventional hatchery round tank). Also in these experiments, independent of their rearing history, survival rates of stocked trout after one and a half years were below 10%. Survival was far below that of the resident fish, which again outcompeted the stocked 1+ trout (Unfer et al. 2009).

Monitoring the Success of Stocking

Whenever fisheries managers release fish, they watch them swimming away, convinced that they did a good deed, following the agricultural maxim "who will reap must sow." But as already noted, managing fish stocks in riverine environments is a complex challenge. Following a stocking campaign that releases catchable sizes, subsequent fishing often satisfies anglers. This is because hatchery fish entering natural waters are easy to catch since they soon begin to starve and are therefore prone to take all kinds of bait. Inside the hatchery they received artificial food in great quantities, but in the wilderness they are in many cases neither adept at recognizing natural food items nor able to react to varying food availability (Järvi 2002). As high rates of these fish will die or move away soon after release, they are economically helpful only for a limited time period after stocking. If it is the overall management aim to satisfy anglers, who like to easily catch naive hatchery fish, the aim might be reached best by regularly releasing hatchery fish, e.g., every second week. If stocking is aimed to support the natural populations, then the targeted purpose would definitely not be achieved. But failure or success can easily be monitored. Nowadays, different tagging methods for all size classes of fish are established. Even eyed eggs, e.g., of brown trout, can be marked using chemical dyes (e.g., Unfer and Pinter 2013), or their origin can be classified through molecular-biological methods (e.g., Meraner et al. 2013b). It is surprising that the majority of fisheries managers spend huge amount of money for stocking but monitoring studies on the success are scarce and often judged as too expensive. Recent studies on fish stocking in rivers show either failure (e.g., Persat et al. 2016; Vonlanthen and Schlunke 2015; Mielach et al. 2015) or at least limited success (e.g., Caudron et al. 2011).

How to Regulate Fishing

All management perspectives, conservational, economic, or fish-ecological, hold that fishing regulations should primarily aim to preserve viable populations. Waters supporting healthy stocks need to be managed without stocking interventions but nevertheless can be harvested in a sustainable way following the necessary regulations (cf. Fig. 14.5): Fisheries regulations can protect the long-term productivity of river ecosystems and fish populations by taking fish following guidelines and catch limits set by natural production and, likewise, by releasing fish in size and age classes of limited availability. This follows investment principles of withdrawing only the interest while leaving the fund intact. However, if fish shall be released, it would be counterproductive to harm these fish. Therefore, it is mandatory to restrict the fishing gear if management strategies aim to release certain species or size classes. The closer to a pristine habitat situation, i.e., a very good status according to the WFD, the stricter must be the formulation and implementation of regulations regarding gear restrictions and angling pressure (see below). Examples of good practices to minimize hooking mortality include the use of barbless hooks, bait that can't be swallowed, or the minimization of handling procedures of fishes dedicated to release. Regarding fish handling ethical sound practices have to be mandatory in any case.

The example of sustainable harvest in the river Ois represents more than a concept or slogan. It is a realistic management option. On the other hand, at least from an ethical point of view, pure "catch-and-release" (releasing all caught individuals) is questionable. While it can make sense to release all individuals of a threatened species or to preserve a small stock or population, pure "catch-and-release" regulations are hard to explain to people who generally conceive of angling as cruel. If people go fishing with the intention to release their entire catch, they are indeed playing with creatures, which is hardly acceptable for animal welfare proponents, irrespective of the debate as to whether fish feel pain or not (Braithwaite 2010; Rose et al. 2014).

Our view is that fishing can and should be a reasonable pastime as long as we aim at finessing, catching, and taking home healthy and tasty food, as the human race has done for millennia, provided that modesty finds its way into the understanding of the way natural resources are used. In this context, a further regulatory lever comes into play: *angling pressure*. Angling pressure can be expressed by days or hours of angling per river length or water surface area. As it can be quantified, so can it be restricted. Limiting angling pressure means that fish are caught less frequently. This helps to avoid learning effects and reduces timidity, which supports angler satisfaction as it will be easier to hook a fish compared to intensively fished beats. Furthermore, limited angling pressure reduces insurance rates and, consequently, hooking mortality. According to Fig. 14.5, specifically near-natural habitats have to be protected from overfishing to meet conservational requirements, while altered or artificial water bodies, where in many cases stocking will be a frequently used management tool, can also be burdened with higher pressure. The general scheme of adjusting fishing regulations to fit the ecological status of water bodies can be used to guide anglers and therefore also angling pressure (Fig. 14.5). Near-natural streams have to be managed and fished appropriate to conservational requirements, while heavily altered or artificial water bodies require a broader range of management opportunities. One cannot forget however that flowing waters remain open ecological systems. Therefore, management actions should always be considered thoroughly in advance, as their effects may reach far beyond the boundaries of a management unit. Finally, as people, specifically children or urban societies, should get the chance to experience angling and to develop a closer relationship to fish and aquatic systems, proper strategies as how to guide as well as foster recreational fishing must be developed, safeguarding the future of this leisure activity and of aquatic ecosystems.

14.4 Conclusions

Contemporary management of recreational fisheries needs to balance between the poles of anglers' desire and the sociopolitical and moral obligation to conserve nature. Therefore, management goals should be defined by involving all relevant parties, i.e., authorities, legislators, fishing right owners, or fishing associations. If we subscribe to adaptive management, then the authorities would work with local practitioners and scientists to establish a vision, define what is known and not known, set goals, develop and implement policies, monitor results, and periodically repeat the entire process. Otherwise, we are stuck in the rut of conventional, top-down management (see Chaps. 15 and 16). As soon as the goals for a water body are defined, the different tools a fisheries manager has can be used. It is our conviction that recreational fishing and environmental conservation can and should be merged, whereby the fisheries have to accept their subordinate role to nature conservation in near-natural waters. Subordination, however, does not mean a loss of rights or benefit, but can resemble a successful strategy provided that modest and sustainable harvest schemes are elaborated and angling is carried out in an ethical acceptable way. The example of the River Ois illustrates that if the management of fishing beats is done thoughtfully, sustainable harvest and maintenance of vital stocks can be guaranteed.

References

Araki H, Cooper B, Blouin MS (2007) Genetic effects of captive breeding cause a rapid, cumulative fitness decline in the wild. Science 318:100–103

Arlinghaus R (2004) Angelfischerei in Deutschland – eine soziale und ökonomische Analyse. Leibniz-Institut für Gewässerökologie und Binnenfischerei, Berlin, 160pp

Arlinghaus R, Schwab A, Riepe C, Teel T (2012) A primer on anti-angling philosophy and its relevance for recreational fisheries in urbanized societies. Fisheries 37(4):153–164

Berejikian BA, Kline P, Flagg TA (2005) Release of captively reared adult anadromous salmonids for population maintenance and recovery: biological trade-offs and management considerations. In: Nickum M, Mazik P, Nickum J, MacKinlay D (eds) Propagated fish in resource management. American Fisheries Society Symposium 44. American Fisheries Society, Bethesda, MD, pp 233–245

Birkeland C, Dayton PK (2005) The importance in fishery management of leaving the big ones. Trends Ecol Evol 20(7):356–358

Braithwaite V (2010) Do fish feel pain? OUP, Oxford

Caudron A, Champigneulle A, Guyomard R, Largiader CR (2011) Assessment of three strategies practiced by fishery managers for restoring native brown trout (Salmo trutta) populations in Northern French Alpine Streams. Ecol Freshw Fish 20:478–491

Christie MR, Ford MJ, Blouin MS (2014) On the reproductive success of early-generation hatchery fish in the wild. Evol Appl 7(8):883–896

Christie MR, Marine ML, Fox SE, French RA, Blouin MS (2016) A single generation of domestication heritably alters the expression of hundreds of genes. Nat Commun 7:10676

Clarke GL, Edmondson WT, Ricker WE (1946) Dynamics of production in a marine area. Ecol Monogr 16(4):321–337

Duftner N, Koblmueller S, Weiss S, Medgyesy N, Sturmbauer C (2005) The impact of stocking on the genetic structure of European grayling (Thymallus thymallus, Salmonidae) in two alpine rivers. Hydobiologia 542:121–129

Franklin IR (1980) Evolutionary change in small populations. In: Soule' ME, Wilcox BA (eds) Conservation biology: an evolutionary–ecological perspective. Sinauer Associates, Sunderland, pp 135–150

Fraser DJ (2008) How well can captive breeding programs conserve biodiversity? A review of salmonids. Evol Appl 1:535–586

Frissell CA, Liss WJ, Warren CE, Hurley MD (1986) A hierarchical framework for stream habitat classification: viewing streams in a watershed context. Environ Manag 10(2):199–214

Hauer C, Unfer G, Habersack H, Pulg U, Schnell J (2013) Bedeutung von Flussmorphologie und Sedimenttransport in Bezug auf die Qualität und Nachhaltigkeit von Kieslaichplätzen. Korrespondenz Wasserwirtschaft 4(13):189–197

Holzer G, Unfer G, Hinterhofer M (2004) Gedanken und Vorschläge zu einer Neuorientierung der fischereilichen Bewirtschaftung österreichischer Salmonidengewässer. Österreichs Fisch 57(10):232–248 ISSN 0029–9987

Holzer G, Unfer G, Hinterhofer M (2011) "Coocooning" eine alternative Methode zur fischereilichen Bewirtschaftung. Österreichs Fisch 64:16–27

Järvi T (2002) Performance and ecological impacts of introduced and escaped fish: physiological and behavioural mechanisms–AQUAWILD. Final report to: European Commission EC Contract No. FAIR CT, 97–1957

Laikre L (1999) Conservation genetic management of brown trout (Salmo trutta) in Europe. Report by the concerted action on identification, management and exploitation of genetic resources in the brown trout (Salmo trutta) ("TROUTCONCERT"; EU FAIR CT97–3882)

MacCrimmon H (1971) World distribution of rainbow trout (Salmo gairdneri). J Fish Res Board Can 28:663–704

Maynard DJ, Flagg TA, Iwamoto R, Mahnken CV (2004) A review of recent studies investigating seminatural rearing strategies as a tool for increasing Pacific salmon postrelease survival. Development of a Natural Rearing System to Improve Supplemental Fish Quality, 24

Meraner A, Venturi A, Ficetola GF, Rossi S, Candiotto A, Gandolfi A (2013a) Massive invasion of exotic Barbus barbus and introgressive hybridization with endemic Barbus plebejus in Northern Italy: where, how and why? Mol Ecol 22:5295–5312

Meraner A, Unfer G, Gandolfi A (2013b) Good news for conservation: mitochondrial and microsatellite DNA data detect limited genetic signatures of inter–basin fish transfer in Thymallus thymallus (Salmonidae) from the Upper Drava River. Knowl Manag Aquat Ecosyst 409:01

Mertz G, Myers RA (1998) A simplified formulation for fish production. Can J Fish Aquat Sci 55(2):478–484

Mielach C, Pinter K, Unterberger A, Unfer G (2015) AlpÄsch – Genotypisierung, nachhaltige Sicherung und Bewirtschaftung regionaler Äschenbestände in anthropogen veränderten Gewässersystemen. Analyse der Lebensraumqualität und der Äschenbestände und Erarbeitung von Managementkonzepten. Studie im Auftrag des Land- und Forstwirtschaftlichen Versuchszentrums Laimburg und des Tiroler Fischereiverbands, 118pp

Montgomery DR, Buffington JM (1997) Channel-reach morphology in mountain drainage basins. Geol Soc Am Bull 109(5):596–611

Olden JD, LeRoy Poff N, Douglas MR, Douglas ME, Fausch KD (2004) Ecological and evolutionary consequences of biotic homogenization. Trends Ecol Evol 19:18–23

Persat H, Mattersdorfer K, Charlat S, Schenekar T, Weiss S (2016) Genetic integrity of the European grayling (Thymallus thymallus) populations within the Vienne River drainage basin after five decades of stockings. CYBIUM 40(1):7–20

Pinter K (2008) Rearing and stocking of brown trout, Salmo trutta L.: Literature review and survey of Austrian fish farmers within the frame of the project-initiative TROUTCHECK. Diploma thesis. Vienna

Pinter K, Weiss S, Lautsch E, Unfer G (2018) Survival and growth of hatchery and wild brown trout (Salmo trutta) parr in three Austrian headwater streams. Ecol Freshw Fish 27(1):146–157

Pulg U, Barlaup BT, Sternecker K, Trepl L, Unfer G (2013) Restoration of spawning habitats of brown trout (Salmo trutta) in a regulated chalk stream. River Res Appl 29:172–182

Rose JD, Arlinghaus R, Cooke SJ, Diggles BK, Sawynok W, Stevens ED, Wynne CDL (2014) Can fish really feel pain? Fish and Fisheries 15(1):97–133

Sánchez-Hernández J, Shaw SL, Cobo F, Allen MS (2016) Influence of a minimum-length limit regulation on Wild Brown trout: an example of recruitment and growth overfishing. N Am J Fish Manag 36(5):1024–1035

Schenekar T, Lerceteau-Kohler E, Weiss S (2014) Fine-scale phylogeographic contact zone in Austrian brown trout Salmo trutta reveals multiple waves of post-glacial colonization and a pre-dominance of natural versus anthropogenic admixture. Conserv Genet 15:561–572

Unfer G, Pinter K (2013) Marking otoliths of brown trout (Salmo trutta L.) embryos with alizarin red S. J Appl Ichthyol 29(2):470–473

Unfer G, Pinter K, Weiss S, Lercetau-Köhler E, Sturmbauer C (2009) Projektinitiative Troutcheck Niederösterreich. Abschluss-Kurzbericht, 81p

Unfer G, Hauer C, Lautsch E (2011) The influence of hydrology on the recruitment of brown trout in an Alpine river, the Ybbs River, Austria. Ecol Freshw Fish 20(3):438–448

Vonlanthen P, Schlunke D (2015) Erfolgskontrolle Besatzmassnahmen und Populationsgenetische Untersuchung der Äschen im Kanton Aargau. Aquabios GmbH, Auftraggeber: Departement Bau, Verkehr und Umwelt, Sektion Jagd und Fischerei, Kanton Aargau

Wallace RK, Fletcher KM (2001) Understanding fisheries management. Mississippi-Alabama Sea Grant Consortium

Waters TF (1977) Secondary production in inland waters. Adv Ecol Res 10:91–164

Weiss S, Schmutz S (1999) Performance of hatchery-reared brown trout and their effects on wild fish in two small Austrian streams. Trans Am Fish Soc 128:302–316

Welcomme RL (2016) Fisheries governance and management, Freshwater Fisheries Ecology. Wiley, Oxford, pp 467–482

Welcomme RL, Bartley DM (1998) An evaluation of present techniques for the enhancement of fisheries. FAO Fisheries Technical Paper (FAO)

Part II
Management, Methodologies, Governance

Chapter 15
Restoration in Integrated River Basin Management

Susanne Muhar, Jan Sendzimir, Mathias Jungwirth, and Severin Hohensinner

15.1 Introduction

The European Water Framework Directive (WFD; European Commission 2000) introduced a new focus in river management by putting the protection and restoration of the aquatic environment as a key issue on the water policy agenda. This expanded emphasis on restoration activities reflects global efforts to make river management more sustainable by better integrating policy and science to harmonize engineering, ecological, and social concerns in governing river basins. Over the last 20–30 years, several management frameworks such as Integrated River Basin Management (IRBM) or adaptive management (AM) have been developed in a series of separate, parallel experiments to achieve these goals. While specific details may vary, most of these management lineages converged on broadly common ways to sustainably manage natural resources and human activities in river basins in an integrated, interdisciplinary approach. The need to put restoration and conservation activities in a social context is increasingly considered mandatory in recent management programs (see Chap. 16).

What Is Meant by Restoration?
Various definitions illustrate how diverse restoration activities are perceived and implemented as "restoration," based on a different, partly contrasting understanding of the general objectives and the methods of restoring river ecosystems. In numerous cases, so-called restoration projects are merely attempts to convert selected sections of riverine systems to some predetermined structure and function, e.g., as spawning areas of a target fish species. Here, as in attempts limited to partial

S. Muhar (✉) · J. Sendzimir · M. Jungwirth · S. Hohensinner
Institute of Hydrobiology and Aquatic Ecosystem Management, University of Natural Resources and Life Sciences, Vienna, Austria
e-mail: susanne.muhar@boku.ac.at; jan.sendzimir@boku.ac.at; mathias.jungwirth@boku.ac.at; severin.hohensinner@boku.ac.at

© The Author(s) 2018
S. Schmutz, J. Sendzimir (eds.), *Riverine Ecosystem Management*, Aquatic Ecology Series 8, https://doi.org/10.1007/978-3-319-73250-3_15

restoration (Roni and Beechie 2013) or to artificial stimulation of natural processes or structures, the term *rehabilitation* rather than restoration should be used. Furthermore, creation, reclamation, or reallocation refers to the conversion of an ecosystem into a different one and so likewise should not be confounded with restoration (Jungwirth et al. 2002).

Briefly, restoration initializes the reestablishment of specific river-type conditions. These processes work to reinforce each other in achieving a self-sustaining status. As such they reflect characteristic structures, processes, and functions of a comparable river/river type with only minor human impacts that corresponds to at least "good ecological and chemical status" of rivers required by the WFD. Thus, river restoration refers to a large variety of measures addressing as key components the morphology and hydrology of rivers as well as measures linked to land use practices and spatial planning. Often such approaches are packaged as bundles of measures directed toward a self-sustaining status by promoting multiple functions and services of river systems *in support of biodiversity, recreation, flood management, and landscape development* (ECRR 2016).

In this chapter, the term "restoration" will be used to refer to any of the above addressed activities.

15.2 Guiding Principles for River Restoration

A diversity of concepts and methods is used in restoration science and practice, described in a variety of applications (Kondolf 2011; Wohl 2005; Bernhardt et al. 2007; Palmer et al. 2005; Palmer 2009; Roni and Beechie 2013). The following section summarizes a set of core guiding principles for sustainable river restoration.

15.2.1 The Riverine Landscape Perspective: Restoration Strategies Across Spatial Scales

The beneficial services of river basins emerge from a functional space far wider than river channels. Riverine landscapes can be viewed as expansive systems, whose functional and structural elements are determined by the river and its flow and sediment regime, yielding an intricately connected system consisting of the river and its surroundings (Wiens 2002). Rivers and their adjacent floodplains act as functionally interrelated systems, depending on processes and structures which affect each other mutually, e.g., by floods, habitat-forming and habitat-providing processes, nutrient retention and provision, etc. (see Chap. 3). Beyond the lateral dimension of interaction within a given river site/landscape, those landscapes mostly depend on natural driving forces (e.g., hydrological, bedload, vegetation regime) in its longitudinal dimension of up- and downstream interactions, at the entire river system, catchment, and even the higher ecoregion/biogeographical region scale. Superimposed on these patterns of "natural" interactions are alterations, e.g., barriers, water diversions and impoundments, and land-use shifts, which apply human-induced pressures at the river site or in upstream and even downstream parts of the catchment.

Conventional strategies of reestablishing endangered species or iconic images of river landscapes have given way to a focus on *process-oriented* restoration (Roni and Beechie 2013; see also Sect. 15.2.2). Initial rehabilitation efforts achieved success but only at the local-to-reach scale (Frissell and Ralph 1998). Successful basin level restoration requires a conceptual framework that accounts for interactions within and between all spatial levels. Consequently, restoration concepts for alluvial rivers have to consider the longitudinal linkages between processes at catchment down to reach scale, lateral exchange processes across the floodplain, vertical interactions (riverbed/aquifer), and the effects on biota. This is especially important, as multiple human-induced alterations with cumulative effects most often occur at the larger spatial scales of stream reaches, valley segments, or entire drainage basins. Because these large-scale changes can seriously limit the recovery potential, e.g., of anadromous fishes like salmon, restoration efforts have to be scaled up to address these problems (Frissell and Ralph 1998).

The design of restoration concepts must be informed by a deep understanding of the time/space scales of processes that support the functions and features (biotic and abiotic) of river landscapes. As with the restoration of the Kissimmee River (Toth et al. 1995; Wetzel et al. 2001; Whalen et al. 2002), reestablishing natural flood retention functions may be as important as the recovery of indigenous species. A focus on processes (see Sect. 15.2.2) can guide all restoration phases from initiation through long-term monitoring, accumulating success as short- and eventually long-term processes are recovered. In this way, restoration practices can be fine-tuned, strongly revised, or replaced as experience provides the knowledge to improve our understanding and our management policies (Roni and Beechie 2013) (see Chap. 16).

15.2.2 Process-Orientated Versus Static Approaches

As summarized in Chap. 3, alpine rivers are in general characterized as highly dynamic fluvial systems. Their morphology shifts frequently, reflecting "...cumulative responses to recent events and deferred responses to previous events" (Brierley and Fryirs 2005). They can undergo irreversible changes at two levels: internal, e.g., when one channel adjusts to input from a confluent channel, and external, e.g., altered land cover and/or sediment load. This sense of dynamism must inform restoration policy such that it can work with and steer the natural forces that alter the geomorphological configurations of the riverbed, the riparian corridor, and the floodplain. Policies that properly use such natural forces can exploit the inherent capacity of rivers to adapt to such dynamism and "passively restore" them, regenerating and maintaining their ecological integrity and status (Middleton 1999; Jungwirth et al. 2002).

Such process-oriented strategies reflect initiatives to make national and international river restoration efforts more holistic (Palmer and Allan 2006) and to better address primary causes of ecosystem degradation through restoration actions (Kondolf et al. 2006; Roni et al. 2008; Beechie et al. 2010). Underlying these recommendations is a history of failed conventional policies that narrowly focused on the recreation of specific habitat characteristics to meet certain uniform habitat standards (Wohl et al. 2005; Newson and Large 2006). Such restoration actions favor engineered solutions that create artificial and unnaturally static habitats. These approaches therefore attempt to control processes and dynamics rather than restore them (Beechie and Bolton 1999).

What Is Process-Based Restoration?

The dynamism of rivers and their associated habitats is evident in how they periodically reshape themselves. Policy can be designed to exploit their natural tendency to "evolve in response to geomorphic processes." Process-based restoration (PBR) is designed to reestablish the rhythm and magnitude of the processes, e.g., physical, chemical, and biological, which support the functions of river and floodplain ecosystems. Processes often are measured as rates of change or movement of mass or biota in ecosystems (Beechie and Bolton 1999). Process examples include plant succession and growth, sediment transport and erosion, water and routing, inputs of thermal energy and nutrients, and nutrient cycling in the aquatic food web. PBR is increasingly applied in most European countries, especially to renew or recreate type-specific habitats (Kondolf 2011).

PBR aims to counter anthropogenic disturbances by setting an ecosystem on a trajectory toward fully functioning processes that require minimal further correction (Sear 1994; Wohl et al. 2005). Restoring key processes increases the resilience of the system to future sources of variation and disturbance, e.g., climate, by increasing the capacity to make physical, chemical, and biological adjustments. Success of PBR can be measured in terms of the recovery of habitat, biodiversity, or patterns of undisturbed river dynamics, e.g., flooding buildup and recession, channel migration, and erosion. By restoring the key functions of undisturbed states, PBR strategies avoid common failures of conventional policies, e.g., piecemeal stabilization of newly established habitats or creation of habitats that are beyond a site's potential to maintain and will be eventually undone by system drivers that have not been addressed (Beechie et al. 2010).

15.2.3 Setting Goals and Benchmarks for River Restoration: The "Leitbild Concept"

Restoration ecology and practice revolve around the definition of overall goals for restoration as well as the "vision" of what should be achieved in terms of abiotic and biotic conditions. Initially, this idea—called Leitbild concept—was applied primarily in restoration projects in Germany and Austria (Kern 1992; Muhar 1996). The Leitbild—as benchmark for a river status' assessment as well as for restoration planning—relates to the "natural potential" of a river ecosystem in the absence of "human disturbance," e.g., unprecedented variation in ecological, economic, or political factors. The general idea underlying this concept has become widely accepted (Kern 1992; Hughes 1995; Hughes et al. 2000). In 2000, this approach was adopted by the WFD for the definition of "reference conditions." Accordingly, reference conditions correspond to the high ecological status of rivers and should reflect *totally, or nearly totally, undisturbed conditions for hydromorphological elements, general physical and chemical elements*, and *biological quality elements* for the surface water body type from those normally associated with that type under undisturbed conditions (European Commission 2000; CIS 2003).

In a world where anthropogenic impacts are evident in almost every ecosystem, attaining such a reference state is highly unlikely. This challenge prompted suggestions to replace this reference-based strategy with an objective-based one (Dufourd and Piegay 2009; Bouleau and Pont 2015). This means a strategy that responds realistically to the challenges of establishing sustainable ecosystems by refocusing on achievable and desirable outcomes: the ecosystem services generated by a functioning ecosystem. In a methodological framework of river restoration and management (see Fig. 15.1), this is taken into account by a second step of defining another "benchmark" in terms of "operable targets for the future development of rivers" ("operational Leitbild"). This means to consider framework conditions/limitations, comparable to the "good ecological status," which, at the very least, should be achieved for surface water bodies, according to the obligations of the WFD.

This "two-step approach" moves the Leitbild process from visionary (high ecological status/reference conditions) to operational (good status) stages and at the same time provides an objective and comparable benchmark for the evaluation of the ecological status of water bodies by classifying the deviation from this benchmark according to WFD.

In application, this concept serves as a template to define the overall perspective for restoration measures as precisely as possible, and it will help to assess both the current deficits prior to restoration and success once restoration measures have been implemented. It is important to note that the reference condition approach does not attempt to recreate the past or return to a specific, spatially defined previous state, as argued by several authors, e.g., Dufour and Piegay (2009). The advantages of this concept are that (1) definition of the natural, intact system in the "visionary" first step provides an objective benchmark; (2) the natural reference remains valid even under modified frame conditions (e.g., land use, infrastructure, legal situation; Muhar 1994; Muhar et al. 2000); (3) the type-specific visionary Leitbild is based on a holistic approach that accommodates large-scale aspects; and (4) the integrative operational Leitbild encompassing the requirements of all disciplines readily identifies the most feasible solution (Jungwirth et al. 2002).

Methodological Approaches to Develop a "Leitbild"/Define Reference Conditions

A long-standing and highly reliable method for defining reference conditions is the field assessment of close-to-natural reference sites for the related river type. This is ideally done within the respective river system or the geographical region. If appropriate sites are not available, comparable sites on other rivers should be investigated. In practice, it is often time-consuming to investigate such relatively undisturbed habitats. However, such meticulous approaches will provide the most reliable information, even quantitative data, which is often missing because of its high cost. Collected over several time slots, such "long-term data" can help to define natural variability.

Severe hydromorphological modifications of most large rivers in the industrialized world have all but eliminated near-natural reference sites (Dynesius and Nilsson 1994; Nilsson et al. 2005). Efforts to rigorously identify type-specific conditions increasingly rely on historical data (maps and records) from preindustrial eras. These help describe facets of "reference ecosystems," e.g., communities of aquatic species or geomorphological characteristics prior to large-scale river engineering for transport (channels) and energy (dams) (Muhar et al. 2000, 2008). Most such data was collected for larger river systems, e.g., for the Mississippi and Illinois rivers (Sparks 1995) and also for many large European rivers (Petts 1989; Hohensinner et al. 2008). Surveys have preserved centuries-old data of geography (land surveys), botany, and zoology (plant and animal communities) or decades-long records of hydrology (daily water-level records) and plant and animal surveys extending more than 100 years back (Jungwirth et al. 2002). Spatially based data as well as history-based definitions of reference conditions can be complemented by modeling approaches (Petts and Amoros 1996) as long as a reliable database of sufficient size is available or can be established.

The development pathway of a riverine landscape can then be simulated on computer and used as a reference scenario to compare with current conditions. Several reference scenario approaches have been developed in the Netherlands (Harms and Wolfert 1998) where drastic changes to the environmental context of lowland rivers make it difficult to extrapolate from past conditions to the present. GIS data has been used to simulate trajectories of vegetation and fauna in space and time for a landscape ecological decision support system (LEDESS). Alternatively, with current hydrodynamical and hydromorphological characteristics as a "starting point," one can use the "intrinsic ecological potential" of vegetation and fauna to explore what development paths are possible and thereby define rehabilitation goals (Pedroli et al. 2002; Egger et al. 2015). Nijboer et al. (2004) recommended not to rely only on a single method for defining reference conditions. Rather, complementary methodological approaches should help to "get the full picture." Finally, all qualitative and quantitative data gained should undergo an expert appraisal in order to review the data with regard to plausibility and potentially complete the reference model by expert knowledge.

The processes described above emphasize the core approach for defining reference conditions from a river ecosystem perspective. Equally important as part of an integrative river management process is to define multidisciplinary goals—further specified by setting qualitative or quantitative benchmarks—for other thematic fields. As such, a Leitbild concept was applied as an initially separate initiative that gradually united the "whole restoration community" around one methodological approach, which was then developed into a comprehensive, inter-, and transdisciplinary river management program for the river Kamp (Preis et al. 2006; Muhar et al. 2006; Renner et al. 2013). For example, the definition of problems, goals, and objectives and identification of solutions was done in participatory processes by scientists together with representatives (administration, NGOs, general public) in the field of river engineering/flood protection, sewage treatment, infrastructure, hydropower development, nature conservation, agriculture and silviculture, fishing, and tourism.

15.2.4 Socio-political Forces That Restore River Basins

Development projects framed by only one or two perspectives, e.g., engineering and/or economics, may deliver some short-term profit for intensive investments in a few sectors, e.g., transport, industry, and large-scale agriculture. But the long-term decline of riverine communities highlights how such narrow policies often fail to support all the factors whose functioning sustains ecosystems and society over the long run (Pahl-Wostl et al. 2011; Sendzimir et al. 2007; Gleick 2003). Reversing this decline requires making management more comprehensive by expanding our view from economy far beyond ecology to include society (see above). Efforts to restore streams and river reaches increasingly do so within the broader context of a socio-ecological system (SES) by accounting for the human social and cultural requirements of riverine SES (Wohl 2005).

Expanding the frame within which we assess and then manage riverine systems requires more than simply increasing the diversity of expert opinion among academic disciplines. Often citizens active in NGOs or in practicing their livelihoods on or around rivers have experience and insights that are unique and otherwise unavailable to decision-makers. To make such nonprofessional perspectives available, stakeholder participation in river governance is recognized universally (Renner et al. 2013) as an integral component of river restoration practices, especially in design, funding, and authorization of such projects (Bennett et al. 2011). The need for more flexible, adaptive, and integrative approaches to water management (e.g., Gleick 2003; Pahl-Wostl 2007; Milly et al. 2008; Viviroli et al. 2011) has driven experiments in making governance more adaptive through reflective, participatory, and deliberative dialogue (Pahl-Wostl 2002; Rist et al. 2007; Wiek and Larson 2012). These processes and the lessons gained from these experiments are more thoroughly covered in Chap. 16.

15.3   Comprehensive Restoration Planning

Restoration must address the cumulative impacts of history while increasing the system's resilience to the impacts of multiple sources of environmental variation in the present and in the future. Even if human impacts are mitigated, systems will always have to adapt to varying and unpredictable forces. This means that even the best restoration design eventually encounters novelty that could not be anticipated, so learning must inform efforts to adjust policies and practices as the restoration process proceeds. Therefore, the logical core of restoration design is a stepwise process linking a series of logical working phases: river and catchment assessments, definition of restoration goals, selection and prioritization of actions, design of projects, and development of a monitoring program (see Fig. 15.1 step (1)–(8) and also Jungwirth et al. 2002; Roni and Beechie 2013). But, as detailed in Chap. 16, such a thread has to be coiled in a loop that links design with updated information from monitoring as well as changing frame conditions (e.g., advances in restoration methods and practices, changing legal regulations, etc.) so as to promote learning and revision over time. Additionally, in the face of uncertainty arising from, e.g., climate change, learning and adaptation must periodically be ramped up to reset the course of river management.

The design process moves forward as each step's purposes and output informs the next. Examining the environmental conditions, problems, and needs at the macroscale of the catchment is the initial phase of the process. Based on standardized monitoring programs according to the WFD/national legal regulations (Chap. 17), the ecological status of rivers will be identified (see Fig. 15.1) (1). Depending on the individual restoration case/program, this assessment phase will go beyond mere freshwater ecological subjects and will pool data and information obtained across disciplines (water management, flood protection, agriculture, recreation, etc.). This assessment phase should evaluate the environmental status and the "values" and "deficits" of the river and its catchment (3). These analyses help to specify the restoration as well as the conservation needs for the subsequent planning process and, at the same time, to determine whether human impacts have caused changes that are irreversible within any reasonable time frame. For example, no restoration, rather mitigation or rehabilitation, is possible when factors controlling the shape and profile of channels, such as sediment supply and flow, have been fundamentally changed. According to the WFD, such river stretches are classified as "heavily modified water bodies," requiring deviating goals and mitigation procedures (Chap. 17).

Summarizing, steps (1)–(3) identify the critical restoration goals and needs at different spatial scales (river basin down to river reach), which are adjusted and further specified by additional investigations and definitions of reference conditions (3a) as well as of frame conditions and restrictions (3b). Those combined analyses are crucial to finally determine what status can be achieved (4), which then is transformed in an "operational guiding view" (equating to the good ecological status or potential, according to WFD obligations), leading to step (5)—the design of a comprehensive restoration program including the prioritization of actions.

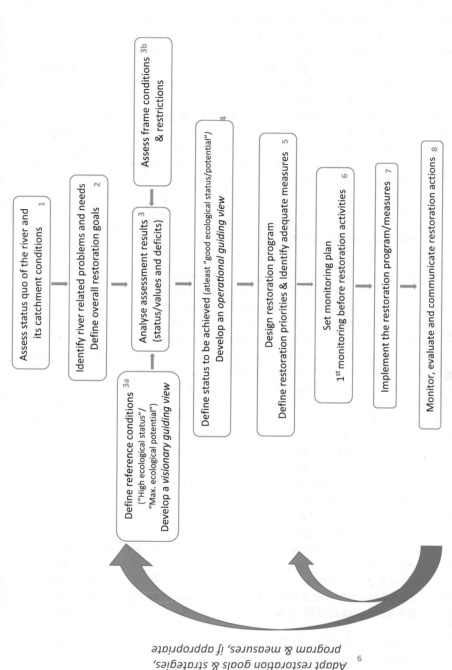

Fig. 15.1 Main working steps and phases within a restoration process: in all working steps, scientists and experts are cooperating with stakeholder groups at different extents. One exception is the definition of reference conditions and benchmark setting for all river-related professional (ecological, technical, and social) disciplines

A coherent assessment (steps 1–3) and monitoring program (steps 6 and 8) are needed to serve as the basis to determine what potential adaptive strategies are needed within this management cycle than to design the restoration program adequately according to new findings and experiences about the success or failure of individual restoration activities (9). Nilsson et al. (2016) underline the high importance of monitoring and evaluation to avoid problems arising from inaccurate design and implementation of restoration.

In sum, to be comprehensive, a program must include the complete sequence of steps from river basin assessment to monitoring, wherein each step is clearly linked to the next by its information output. In the frame of the whole procedure, stakeholder groups will be involved at different extents considering and including their experience, their goals, and their current concerns.

15.4 Restoration Measures

All over Europe, river stretches are being restored to achieve the targets of the EU Water Framework Directive (WFD; European Commission 2000), e.g., "good ecological status" or "good ecological potential," respectively, and to fulfill the obligations of the Habitats and Birds Directive (European Commission 1992, 2009)—using also synergies with natural water retention measures promoted by the Floods Directive (European Commission 2007). However, beside the legal obligations, restoration efforts are more and more targeted to improve other functions or services, e.g., cultural ecosystem services, providing access to and enhancing the attractiveness of rivers and their surroundings for recreational use and tourism (see Chap. 21). An increasing number of restoration projects to date are designed to modify processes that will finally restore desirable, e.g., pre-disturbance, morphology. This is often done by widening to initiate the development of a braiding or meandering river course and by creating new, sinuous river channels and reconnecting the river and floodplain. As such, certain "close-to-natural" morphologies are used as iconic targets to choose the appropriate type of measure, whose achievement suggests significant improvement of a variety of ecological functions. Beside the importance of type-specific restoration approaches, the (length and width) dimension seems to be a crucial criterion for successful restoration. In general, higher restoration effects are expected in larger-scale projects, since longer reaches might better provide a minimum area for hydromorphological processes to act and viable populations to establish (Muhar et al. 2016). Nevertheless, there are only few studies up to now that document higher effects in larger-scale projects (e.g., Schmutz et al. 2014).

There are relatively few ambitious restoration programs that address all the multiple pressures that are typical for the majority of alpine river landscapes, e.g., altered flow and sediment regimes (Schinegger et al. 2012; Muhar et al. 2013).

Below we provide an overview of common restoration strategies and applications (see Table 15.1) and list several examples of interventions, directed toward

Table 15.1 Summary of common restoration strategies and applications

Restoration category	Restoration type	Example of intervention	Human impacts
Reestablish morphological river type	Initiate a river-type-specific morphology	Remove bed and bank stabilization	1,2,4
	Set initial measures for type-specific self-development of the river/the river floodplain system (e.g., braiding, meandering) and instream structures	Reconnect or create side arms	1,2,4
		Reconnect oxbows/meanders	1,2,4
		Restructure riparian zone (wood structure, bays)	1,2,4
		Initiate dynamic aquatic/terrestrial transition zone	1,2,4
		Small-scale river widening	1,2,4
		Excavate/reestablish natural riverbed	1,2,4
Reestablish lateral connectivity/ floodplain habitat restoration	Increase/reconnect floodplain/ natural retention areas	Remove/replace/lower dams	1,2,4
		Lower the floodplain area	1,2,4
	Initiate/create floodplain habitats	Initiate/create aquatic floodplain habitats	1,2,4
		Initiate/plant floodplain vegetation	1,2,4
Flow management	Increase residual flow	Increase and adapt residual (environmental) flow	1,2,3,4,5
	Increase dynamic flow	Increase and adapt dynamic (environmental) flow	1,2,3,4,6
Mitigate hydropeaking	Modify power plants	Produce hydropower without hydropeaking	5
	Alter the mode of hydropower production	Mitigate hydropeaks	5
	Create compensating reservoirs	Mitigate hydropeaks through compensatory reservoirs	5
	Divert hydropeak flow	Divert hydropeak flow to a larger receiving river or to a dammed river section	5
	Coordinate hydropeaks	Avoid overlapping hydropeaks	5
Temperature management	Modify turbine intakes	Create multiple entry devices and operate water release according to environmental criteria	4,5
Sediment management	Reopen sediment sources	Remove, lower, reconstruct torrent controls, weirs, ramps	1,5,6,7
	Active sediment input	Open riparian zones and floodplain areas	1,5,6,7
		Donate sediment to the river	1,5,6,7

(continued)

Table 15.1 (continued)

Restoration category	Restoration type	Example of intervention	Human impacts
Flushing management	Flush reservoirs	Targeted flushing with defined environmental thresholds	1,5,6,7
	Dredge or suck sediment	Dredge or suck sediment from reservoirs, and release sediments downstream in a controlled way	1,5,6,7
Reestablish longitudinal continuum	Remove migration barriers	Deconstruct barrier (e.g., weir, bed sill, ramp, etc.)	4,6,7
	Modify migration barriers	Rebuild a passable construction (ramp)	4,6,7
	Implement fish bypass(es)	Construct a close-to-nature fish bypass channel, side arm	4,6,7
		Technical fish bypass	4,6,7
Land use	Modify land use	Encourage extensive sustainable agriculture	1,2
		Create buffer zones	1,2

*Human impacts: 1. regulations/engineering, 2. land-use change, 3. water withdrawal, 4. damming, 5. hydropeaking, 6. altered sediment regime, 7. migration barriers

improving the morphology and hydrology of river-floodplain systems, such as the interaction of land-use practices with longitudinal and lateral connectivity. The latter will be addressed through detailed case studies in Sects. 15.5.1 and 15.5.2, whereas "morphological restoration" or rehabilitation approaches of river-floodplain systems will be summarized below as the most common and applied measures and exemplified by two good practice cases in Sect. 15.5.

15.4.1 Common Restoration Measures Improving the Morphological Character of the River-Floodplain Systems

Depending on the physical configuration of the individual fluvial system and the diverse forms of human impairments, river restoration may comprise a variety of measures at different spatial scales. Improving instream habitats, in particular those of key species, e.g., salmonids, was one of the principal goals, at least in the first period of restoration activities starting in the 1990s, both in Europe, particularly in the Alpine region, and in the Pacific Northwest of the United States and other areas.

Those human interventions were often made at local-to-reach scale. They comprise measures, e.g., restructuring the water-land transition zone by small-scale river widening or enhancing/improving instream habitats through riffles, pools, sediment bars, or wood accumulations. At the same time, such interventions lead to a pattern of diverse flow, substrate, and depth conditions, providing shelter or refuges or feeding sources for aquatic organisms (Allan and Castillo 2007). Boulders, however, should only be used for river restoration where they might function as an integral part of the physical river environment, e.g., in high gradient rivers with coarse bed material. Woody debris are often added directly in the channel, and new woody plants have been reintroduced along the river course to address the lack of any riparian vegetation on riverbanks in cultural landscapes (Reich et al. 2003). Both instream and riparian wood enhancement may improve the diversity of fish and macroinvertebrates, increase storage of organic material and sediment, and strengthen bank stability (Lester and Boulton 2008). Compared to other restoration techniques, such measures have logistical advantages and generally require low costs and low maintenance. Instream woody debris may be used either for improving channel and/or bank stability or facilitating dynamic processes that, in turn, enhance aquatic habitat complexity (Gurnell et al. 2005).

In many cases, the success of wood reintroduction as a crucial restoration type that can either re- or demobilize the sediment within the river basin can be limited by other human impairments, such as remote impacts (e.g., dams and reservoirs) or sectional channel narrowing, causing riverbed degradation due to high shear stress and increased sediment output. Over recent decades, widening of river channels has become a major focus of ecological restoration, in particular, in former braided or anabranched reaches (see Chap. 3). Riverbed widening aims to reduce shear stress (specific stream power) and to enhance instream habitat complexity and, consequently, riverine biodiversity. Potentially, it may initiate braiding within a limited area (see River Drava, Sect. 15.5.1). Such restoration measures offer new opportunities for establishing riparian habitats and, in particular, initiate pioneer succession stages (Rhode et al. 2004). New shallow shore zones generally favor fish reproduction and provide habitats for juveniles (Muhar et al. 2008). Removal of bank protections increases lateral erosion processes and instream sediment turnover. Both reduced transport capacity (specific stream power) and increased sedimentation generally lead to aggradation of the riverbed (Habersack and Piégay 2008). As a consequence, water tables may be heightened, which may improve hydrological connectivity to lateral water bodies or to the aquifer. Such effects are common goals of restoration projects from the ecological point of view but might amplify the risk of floods in nearby settlement areas. In regions that are not specifically sensitive to altered flood levels, such side effects of channel widening may be an option for enhancing the flood retention capacity of the respective area (Leyer et al. 2012). Comprehensive restoration of formerly island-braided or wandering gravel-bed rivers (see Chap. 3) involves a doubling or even tripling of current channel widths. This may significantly boost aggradation in the widened river section. In case of degrading channel sections, sediment deposition induced by substantial channel widening may intentionally help to stabilize the bed level (Habersack and Piégay 2008). However, in the longitudinal view, such reaches function as sediment traps, causing increased bedload deficits in downstream river sections as long as a new equilibrium slope has not yet been established. Besides locally implemented measures, the success of channel widenings is closely tied to the available sediment supply from upstream and, in general, depends on a process-based restoration approach using a catchment perspective in river restoration (see above). Laboratory experiments clearly proved that braided channels transform to sinuous-meandering single channels when bedload input is reduced (Marti and Bezzola 2004). This highlights the fact that the consequence of a restoration project does not necessarily mean the restoration of historical channel patterns (see Sect. 15.2.3). Under significantly changed boundary conditions, channel widening may result in alternative states of fluvial systems (e.g., sinuous or meandering), which may potentially also be the goal of such measures, in particular in heavily modified rivers.

In multichannel systems, e.g., wandering gravel-bed rivers and anabranching sand-bed rivers, channel widening is also an option to reduce bed degradation or to enhance habitat complexity. Because such river systems commonly suffer severely truncated lateral hydrological connectivity, however, the reconnection of the cutoff side arms channel is the main concern (Buijse et al. 2002). Reestablishing hydrological connectivity between the main channel and water bodies in the floodplains, e.g., side arms and oxbows of the floodplains, which are often remnants abandoned by human interventions, generally amplifies fluvial dynamics and counteracts the predominating terrestrialization processes in the floodplain. Other positive effects are the improvement of aquatic habitat diversity (e.g., when formerly lentic water bodies then become lotic ones), the stimulation of erosional and depositional processes within sustainable limits, and the reestablishment of migration pathways for aquatic species (Jungwirth et al. 2000). Enhanced exchange of surface water between the river and the diverse floodplain water bodies also stimulates the various exchange processes in the aquifer (Amoros and Bornette 2002). Both surface and subsurface flows are fundamental for the improvement of the ecological effects related to the "flow pulse" and the "flood pulse" (Junk et al. 1989; see Chap. 3.4). Nevertheless, reconnection of cutoff floodplain water bodies may also induce new problems from the hydraulic and ecological point of view. Dividing the flow into several river arms reduces the transport capacity of individual channels and may lead to bed aggradation with similar consequences as already described for channel widenings. Though man-made, cutoff water bodies in many cases provide ecologically valuable habitats for endangered, stagnophilic species. Today, such biotopes and their coenoses are often designated as protected natural capital (e.g., an ecological good) according to national or international legislation (e.g., EU Habitats and Birds Directives). Accordingly, "dynamization" measures may lead to difficult-to-solve conflicts involving ecological rebalancing between the reestablishment of system-inherent fluvial processes and of nature conservation levels, such as promotion of endangered animals, plants, and habitats (Muhar et al. 2011). Other potential conflicts to be addressed in restoration programs are the undesirable increases of non-native species or of nutrient influx in biotopes that have been hydrologically recoupled, especially to the river channel (Hobbs et al. 2009; Paillex et al. 2009). As restoration proceeds, fluvial disturbances can trigger the evolution of new habitats from pioneer succession stages. However, though partly restored, the regeneration potential of new habitats can be constrained by vertical decoupling between rivers and their floodplains as a consequence of channel incisions and floodplain depositions (see Chap. 3). If the restoration of former pioneer and softwood forest sites that are characterized by high groundwater levels and frequent inundations is a major goal, the artificial lowering of the floodplain terrain may provide a promising option (compare "Cyclical Floodplain Rejuvenation" in Duel et al. 2001; see also case study Traisen, Sect. 15.5).

Mitigation of the ecological consequences related to progressive terrestrialization is also a critical ecological objective along meandering or anastomosing rivers in lowlands (Buijse et al. 2002). Along meandering rivers, improvement of flood conveyance capacity has been achieved by cutting off meander bends, a wholesale straightening of the course and in cases by a widening of the channel (resectioning). Accordingly, restoration measures on channelized meander rivers today comprise the enhancement of lateral hydrological connectivity by reconnecting former meander loops and promotion of the natural inundation regime. Since straightening of the main channel is often accompanied by a distinct incision that lowers the bed, leaving the former meander loops (oxbow lakes) stranded up to several meters above the current elevation of the river (Muhar et al. 2011), reconnecting such oxbow lakes would result in a largely drainage of the water body. Therefore, to adapt to this rebalancing as it occurs, management should experimentally probe for ways to restore lateral connectivity (e.g., reconnection only at higher flows or evolution of new, deeper lying oxbows). A promising solution from the ecological point of view includes the removal of bank protections and the initiation of channel migration. Depending on the external channel controls, self-adjustment processes may lead to the evolution of new meander bends over the mid to long term (Larsen 2008). Because such measures may counteract the general objectives of flood protection, channel shifts and, consequently, sinuosity may be improved only within a limited channel belt that functions as a flood channel at higher flows.

The described measures for restoration of channelized river sections highlight typical management options related to specific morphological river types (see Chap. 3). Apart from these a multitude of potential small and large restoration measures exists owing to the complex nature of fluvial systems and different forms of human impacts (e.g., Calow and Petts 1994; Rutherfurd et al. 2000; Brierley and Fryirs 2008). All the schemes implemented at the local or reach scale have in common that they are closely interlinked to the sectional and basin-wide boundary conditions (Gurnell et al. 2016). Depending on the extent of the individual measure, nonlinear channel adjustments that may affect longer river sections up- and downstream have to be considered.

As Frissell and Ralph (1998) conclude, decisions as to which restoration measures are appropriate depend on "(1) the type and degree of human pressures on river ecosystems as well as on the (2) characteristics of river, where restoration should take place (primarily channel type and its geomorphological and hydrological setting)."

15.5 Good Practice Examples of Morphological River Restoration

15.5.1 River Restoration Drava

The River Drava (*Drau* in German) is part of the Danube catchment, with its source in Italy, close to the boarder to Austria and its mouth in Osijek (Croatia), where the river joins the Danube River after 748 km. The Austrian section of the River Drava flows 264 km along the border between the Central Alps and the Southern Alps, characterized by a catchment area of 12,058 km^2, respectively, out of a total of 41,000 km^2.

Until approximately 140 years ago, the Upper Drava was a free flowing, meandering mountain river with numerous braiding stretches due to alluvial cones where its tributaries entered. In this dynamic river system with its annual floods and high bedload transport, the river course frequently changed. A braiding river-floodplain system with large gravel banks and extensive softwood forests as well as wetland meadows characterized the valley bottom.

Human Impacts

The first substantial human changes began with the building of the railroad line through the Upper Drava valley in 1868. In the following years, river engineering channelized the river to reduce flood risk as well as to allow intensive agricultural land use and the expansion of settlements. When the river was forced into a single main channel, the river dynamics were restricted, and the number of side arms, gravel banks, water bodies and vegetation populations in the floodplains decreased. Riverbed incision occurred due to the regulation processes, e.g., torrent control structures, and to reduced sediment supply by the tributaries. This incision reduced groundwater levels, causing desiccation of the remaining adjacent wetlands.

Restoration Measures

In 1998, the river and its riparian zones became protected by designation as a Natura 2000 area. Between 1999 and 2011, two consecutive EU LIFE projects were implemented in the Upper Drava River (across a total length of 80 km) under the title "Restoration of the wetland and riparian area on the Upper Drau River" and "Life vein-Upper Drau river". These projects defined goals such as species and habitat regeneration and protection as well as water management interests. The overall and most fundamental objective was halting further riverbed incision and recoupling the main river and its side arms and floodplains and even the stepwise rising of the river bottom. This has been the essential prerequisite for all follow-up activities aimed at an increased lateral connectivity and habitat improvement of the river/floodplain corridor.

Approximately 30 km of bank protection structures were removed to enhance lateral erosion, which leads to increased sediment input and contributes to the development of gravel/sandbars and islands (see Fig. 15.2a and b). Additionally, numerous local instream measures and several large-scale riverbed widenings

Fig. 15.2 (**a**) "Self-development" of the Drava river course (at Obergottesfeld/Carinthia) after removing bank protection and initial channel creation (LIFE project: Life vein-Upper Drau River) (photographs: Herbert Mandler, © Amt der Kärntner Landesregierung). (**b**) Morpho-dynamic processes supported the development of key aquatic habitat types (sediment bars and islands; woody structures along the riparian corridor of the Upper River Drava) (photographs: Susanne Muhar)

including new side arms were initialized. Furthermore, in the adjacent areas, the development of new water bodies and floodplain forests began, providing adequate habitats for animal and plant species that were typical prior to river engineering impacts. In total, about 1/3 (about 25 km) of the former straightened river was restored, addressing the river-floodplain system as a functional unit and offering room for dynamic processes and self-development of the river course. Approximately 25 ha of river habitats were created or initialized in the floodplain area. The restored river stretches comprise habitat types according to the EU Habitats Directive [e.g., dynamic gravel banks (3220) and tamarisk and willow pioneer communities (3230, 3240)], which will further develop to alluvial forests (91 E0).

Summary
The Upper River Drava (Carinthia) is one of the most comprehensive and well-monitored flagship restoration cases in the Alps and even in Europe. It is one of the large-scale good practice examples investigated within a standard monitoring program financially supported by the LIFE instrument as well as a recently finished EU

project (REFORM 2011–2015). Results demonstrate that spawning habitats for amphibians and fish, in particular the Danube salmon (*Hucho hucho*), grayling (*Thymallus thymallus*) and Souffia (*Leuciscus souffia*), European bullhead (*Cottus gobio*) and the Ukrainian brook lamprey (*Eudontomyzon mariae*), successfully increased that, e.g., two plants which were both nearly extinct in Austria, could be reestablished again: German tamarisk (*Myricaria germanica*) and the dwarf bulrush (*Typha minima*). See also (Unfer et al. 2011; Schmutz et al. 2016; Muhar et al. 2016; Poppe et al. 2016).

The projects aimed to increase natural flood retention, to achieve good ecological status, and to provide an appropriate river landscape for additional ecosystem services, e.g., recreation (Chiari 2010; Böck et al. 2015), by a synergistic restoration approach. In general, the projects were designed to reduce human intervention as much as possible and to promote dynamic, self-sustaining river processes.

Specific characteristics of the restoration case study Drava:

- Very long period and intensive process of implementation of restoration measures (slowly built up, starting with ecologically orientated flood protection measures in 1990s, ended up in two EU LIFE projects).
- Adaptive management approach—stepwise process of implementation—monitoring and practical experiences and adaption of the restoration approaches.
- Intensive process of stakeholder involvement (from local to national level: general public, different actor groups, administration).
- Broad communication and documentation of the project (gained wide acknowledgement at regional up to international level; excursions, documented as best practice example in international literature).
- Open-minded and engaged partners at different administration level/responsible persons (administration, planning offices, researchers) have their specific interest in this project.
- Mutual confidence (no decisions over the head of the people; flood protection continues to be guaranteed).
- Stakeholders recognized the synergistic of the projects, e.g., in terms of ecosystem services (flood retention/protection, restoration/ecological functions, land use, recreation/tourism).
- Restoration planning:

    Interdisciplinary team (different sectors, administration, practitioners, researchers)
    Profound restoration program

- Restoration measures (technical point):

    Well supported by EU financing instruments (life nature)
    Stepwise, more process-based approach (River Drava is developing its own course).
    Dynamic processes are allowed and supported until predefined benchmarks.

15.5.2 River Restoration "Traisen"

General Characteristic

Until the late nineteenth century, prior to channelization, the lower course of the Traisen River featured typical alpine bar- and island-braided river patterns. Almost 1000 km^2 large catchment is located in Lower Austria, stretching from the Northern Limestone Alps in the south over the Alpine Foreland to the north, where the river discharges into the Danube River 55 km upstream from Vienna. While mean flow at the mouth only amounts to 14 m^3/s, the river still shows a torrential behavior during floods, with approximately 800 m^3/s discharge during 100-year floods. Owing to its alpine character, gravel constituted the dominant substrate type, and softwood communities dominated by willows, alders, and poplars developed in the riparian zone. The mouth of the river is located in the extensive floodplain of the Danube River that was designated as a Nature 2000 reserve according to the EU Habitats (FFH) and Birds Directives.

Human Impacts

In the course of the construction of the hydropower plant, Altenwörth at the Danube River between 1973 and 1976, the estuary of Traisen River was relocated 8.5 km further downstream, just below the weir of the new power plant. The reason therefore was the uplift of the water table in the impoundment, which otherwise would have required a costly pump station for pumping the discharge of the Traisen into the Danube reservoir. The relocation of the river mouth involved the lengthening of the river channel through the Danube floodplain in the form of a straight canal (bypass) with a double trapezoidal profile for mean flow and floods. In addition, check dams to back up the water at mean flow were installed in combination with flood protection levees along the riverbanks (Fig. 15.3). The artificial channel lacked longitudinal and lateral hydrological connectivity, which inhibited fish migration, and dredging prevented bed aggradation as well as the formation of new instream habitats. Over recent decades, ongoing deposition of fine sediments (silt and sand) during Danube floods has increased the terrain level in the floodplain. Both the artificially incised bypass canal and the elevated floodplain terrain resulted in lower groundwater levels and, consequently, drier site conditions for riparian vegetation.

Restoration Measures

During the EU "LIFE+ Project Traisen" between 2012 and 2016, a new, significantly wider river channel more than 9 km long was excavated parallel to the bypass canal for reasons of ecological restoration. Restoration to its pre-engineered state was prevented by current physical conditions, e.g., a much lower channel slope due to the lengthening of the river in the 1970s and missing bedload transport. Consequently, instead of the originally braided channel pattern, a sinuous, partly meandering channel and several secondary water bodies were excavated (Fig. 15.4). Largely devoid of bank protections, the river can adjust and shift its new bed within a limited area. Introduction of large wood structures and future mobilization of riparian vegetation due to lateral bank erosion will additionally contribute to habitat

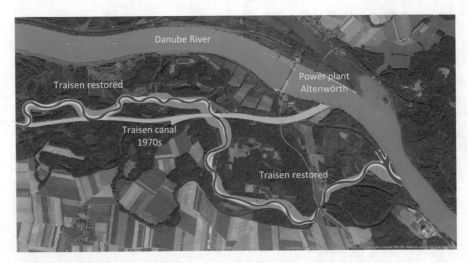

Fig. 15.3 "LIFE+ Project Traisen" in the plan view—Traisen canal from the 1970s (orange) and restored river section (blue) with sediment bars (grey) and both-sided lowered floodplain terrain for river-type-specific riparian vegetation (green) (Source: modified after Technisches Büro Eberstaller GmbH (2010); orthophoto: Copyright © 2014 Esri, DigitalGlobe, Earthstar Geographics, CNES/ Airbus DS, GeoEye, USDA FSA, USGS, Getmapping, Aerogrid, IGN, IGP, swisstopo, and the GIS User Community; all rights reserved)

Fig. 15.4 Bifurcation of the newly excavated riverbed with the adjoining lowered floodplain terrain (left) from the old straight Traisen canal (right) prior to its flooding (in the background: Danube River; source: M. Haslinger and Verbund AG 2014)

diversity. In order to initiate new pioneer and softwood communities, 54 ha of the adjacent floodplain terrain were artificially lowered by several meters (Fig. 15.4). This measure shall guarantee an intensified interaction with the aquifer. The old canal from the 1970s will remain as a backwater system connected to the new river channel and will additionally function as a flood bypass.

Summary

Both the type of measures and their dimensions make this an outstanding example of river restoration. To date, in Europe, only a few restoration projects exist that are comparable with the "LIFE+ Project Traisen." Consequently, there is an EU-wide shortage in distinct experiences in such projects. In addition, the approximately 9.5 km long restored river section replaces an artificial canal rather than a former natural and later channelized river reach. Considering these framework conditions, the project can be seen as a large field observatory that helps to investigate pending research questions. How intensively will the channel change in future? What are the annual channel migration rates? How does the river cope with missing bed load from upstream sections? Will neophytes dominate the colonization of the new large gravel bars? Which fish species will adopt the new river course as habitat? A comprehensive monitoring project will help to answer such open questions in the following years. It is assumed that the wide gravel channel will become narrower within a few years as a consequence of missing bedload transport and deposition of fine sediments on the new bars. Initial monitoring results show that native willow communities will successfully establish on the new vegetation sites and rheophilic fish species are using the new possibilities for migration from the Danube River to new habitats and spawning sites in the Traisen River.

Specifics of this restoration case study:

- One of the largest river restoration projects in Europe that was implemented in a relatively short period of time.
- Besides the river channel, the project also includes various floodplain biotopes, such as abandoned arms, alluvial ponds, and 54 ha of floodplain terrain, which were artificially lowered in order to improve the establishment of natural vegetation communities.
- The restored river section replaces an artificial bypass canal from the 1970s and not a former natural river section. Accordingly, no direct comparison with the natural state of that river section can be made.
- The given physical framework conditions, such as missing bed load, predominance of fine sediments, reduced fluvial dynamics, etc., will lead to a new morphological state of the restored section. Here, the newly constructed channel functions as a starting situation for future hydromorphological changes.
- The project was implemented in an outstandingly integrative manner. Initiated by the Institute of Hydrobiology and Aquatic Ecosystem Management (BOKU Vienna) and by the landowners of the project area, the restoration was also strongly supported by the federal and provincial administrations and politicians, by the Danube waterway authority (via donau), and most of all by the operator of the Danube hydropower plant Altenwörth (Verbund AG) that functioned as the lead partner of the project.

15.6 Conclusions

In the last decade, considerable progress has been made in river restoration activities—in terms of (1) more integrative approaches to planning, implementation, monitoring, and adaptive revision of policies and practices, giving higher attention particularly to stakeholder participation; (2) type and dimension of restoration measures, which have been implemented; as well as (3) increasing awareness of the importance of monitoring programs.

Nevertheless, the success of habitat restoration in channelized rivers is constrained by several factors. Locally, along the channelized rivers, the availability of space lateral to the river channel is often limited due to the intensification of land uses and the enlargement of settlement areas. Remote impacts, such as severely truncated sediment supply and altered flow regimes, inhibit the recovery of the original river-type-specific fluvial processes and dynamics. Though the importance of basin-wide sediment management to prevent further bed degradation and to improve habitat diversity has been internationally recognized by water authorities, an integrative solution of that question is still pending. Long-term legacies of channelization, such as vertical decoupling of the river and floodplain levels up to several meters, have significantly changed the hydromorphological configuration of alluvial river systems. Ongoing challenges to successful restoration are currently demonstrated by the persistence of large-scale impacts, e.g., extensive areas of intensive land use in the catchment or numerous migration barriers (weirs, impoundments, ramps) limiting the recolonization through interrupted longitudinal connectivity. Moreover, current fluvial systems are confronted with multiple pressures that cannot be solved solely by morphological habitat restoration (Schinegger et al. 2012; Tockner et al. 2010; Hering et al. 2015). This calls for further improvements to an integrative and basin-wide approach that goes beyond fluvial morphology to account for all functional aspects of river systems.

> *The most fundamental challenge facing successful restoration of aquatic systems is to establish a clear understanding of the cause and effect relationships between the physical processes at work within a watershed, how the expression of these processes have been altered by human activities, and what short and long term restoration strategies best address such factors. (Frissell and Ralph 1998)*

References

Allan JD, Castillo MM (2007) Stream ecology. Structure and function of running waters, 2nd edn. Springer, Dordrecht

Amoros C, Bornette G (2002) Connectivity and biocomplexity in waterbodies of riverine floodplains. Freshw Biol 47:761–776

Beechie TJ, Bolton S (1999) An approach to restoring salmonid habitat-forming processes in Pacific northwest watersheds. Fisheries 24:6–15

Beechie TJ, Sear DA, Olden JD, Pess GR, Buffington JM, Moir H, Roni P, Pollock MM (2010) Process-based principles for restoring river ecosystems. BioScience 60:209–222. http://www.bioone.org/doi/full/10.1525/bio.2010.60.3.7

Bennett SJ, Simon A, Castro JM, Atkinson JF, Bronner CE, Blersch SS, Rabideau AJ (2011) The evolving science of stream restoration. In: Simon A, Bennett SJ, Castro JM (eds) Stream restoration in dynamic fluvial systems: scientific approaches, analyses, and tools. American Geophysical Union, Washington, DC

Bernhardt ES, Sudduth EB, Palmer MA, Allan JD, Meyer JL, Alexander G, Follastad-Shah J, Hassett B, Jenkinson R, Lave R, Rumps J, Pagano L (2007) Restoring rivers one reach at a time: results from a survey of U.S. river restoration practitioners. Restor Ecol 15:482–493

Böck K, Muhar S, Muhar A, Polt R (2015) The ecosystem services concept: gaps between science and practice in river landscape management. GAIA – Ecol Perspect Sci Soc 24:32–40

Bouleau G, Pont D (2015) "Did you say reference conditions?" ecological and socioeconomic perspectives on the European water framework directive. Environ Sci Pol 47:32–41

Brierley GJ, Fryirs KA (2005) Geomorphology and river management. Applications of the river styles framework. Wiley, Chichester

Brierley GJ, Fryirs KA (2008) River futures. An integrative scientific approach to river repair. Society for Ecological Restoration International, Island Press, Washington

Buijse AD, Coops H, Staras M, Jans LH, Van Geest GJ, Grift RE, Ibelings BW, Oosterberg W, Roozen FCJM (2002) Restoration strategies for river floodplains along large lowland rivers in Europe. Freshw Biol 47:889–907

Calow P, Petts GE (1994) The rivers handbook. Hydrological and ecological principles, vol 2. Blackwell Science, Oxford

Chiari S (2010) Raumbedarf für multifunktionale Flusslandschaften – Potentielle Synergien zwischen ökologischen Erfordernissen und den Bedürfnissen der Freizeit- und Erholungsnutzung. Universität für Bodenkultur, Wien

CIS (2003) Common implementation strategy for the water framework directive (2000/60/EC)

Duel H, Baptist MJ, Penning WE (2001) Cyclical floodplain rejuvenation. A new strategy based on floodplain measures for both flood risk management and enhancement of the biodiversity of the river Rhine. Executive summary CFR Project, Delft, Netherlands

Dufour S, Piégay H (2009) From the myth of a lost paradise to targeted river restoration: forget natural references and focus on human benefits. River Res Appl 25:568–581

Dynesius M, Nilsson C (1994) Fragmentation and flow regulation of river systems in the northern third of the world. Science 266:753–762

ECRR (2016) European centre for river restoration. The network for best practices of river restoration in Greater Europe. Retrieved from http://www.ecrr.org/Home/tabid/2535/Default.aspx

Egger G, Politti E, Lautsch E, Benjankar R, Gill KM, Rood SB (2015) Floodplain forest succession reveals fluvial processes: a hydrogeomorphic model for temperate riparian woodlands. J Environ Manag 161:72–82

European Commission (1992) Council Directive 92/43/EEC of 21 May 1992 on the conservation of natural habitats and of wild fauna and flora

European Commission (2000) Directive 2000/60/EC of the European parliament and the council of 23 October 2000. Establishing a framework for community action in the field of water policy

European Commission (2007) Directive 2007/60/EC on the assessment and management of flood risks

European Commission (2009) Directive 2009/147/EC of the European parliament and of the council of november 2009 on the conservation of wild birds (the codified version of Council Directive 79/409/EEC as amended)

Frissell CA, Ralph SC (1998) Stream and watershed restoration. In: Naiman RJ, Bilby RE (eds) River ecology and management: lessons from the Pacific coastal ecoregion. Springer, New York

Gleick PH (2003) Global freshwater resources: soft-path solutions for the 21st century. Science 302:1524–1528

Gurnell A, Tockner K, Edwards P, Petts G (2005) Effects of deposited wood on biocomplexity of river corridors. Front Ecol Environ 3:377–382

Gurnell AM, Rinaldi M, Belletti B, Bizzi S, Blamauer B, Braca G, Buijse AD, Bussettini M, Camenen B, Comiti F, Demarchi L, García de Jalón D, González del Tánago M, Grabowski RC, Gunn IDM, Habersack H, Hendriks D, Henshaw AJ, Klösch M, Lastoria B, Latapie A, Marcinkowski P, Martínez-Fernández V, Mosselman E, Mountford JO, Nardi L, Okruszko T, O'Hare MT, Palma M, Percopo C, Surian N, van de Bund W, Weissteiner C, Ziliani L (2016) A multi-scale hierarchical framework for developing understanding of river behaviour to support river management. Aquat Sci 78:1–16

Habersack H, Piègay H (2008) River restoration in the alps and their surroundings: past experience and future challenges. In: Habersack H, Piégay H, Rinaldi M (eds) Gravel-bed rivers VI. Elsevier, Philadelphia, pp 703–735

Harms WB, Wolfert HP (1998) Nature rehabilitation for the river Rhine: a scenario approach in different scales. In: Nienhuis P, Leuven RSEW, Ragas AMJ (eds) New concepts for sustainable management of river basins. Backhuys, Leiden, pp 95–113

Hering D, Carvalho L, Argillier C, Beklioglu M, Borja A, Cardoso AC, Duel H, Ferreira T, Globevnik L, Hanganu J, Hellsten S, Jeppesen E, Kodeš V, Solheim AL, Nõges T, Ormerod S, Panagopoulos Y, Schmutz S, Venohr M, Birk S (2015) Managing aquatic ecosystems and water resources under multiple stress – an introduction to the MARS project. Sci Total Environ 503-504:10–21

Hobbs RJ, Higgs E, Harris JA (2009) Novel ecosystems: implications for conservation and restoration. Trends Ecol Evol 24:599–605

Hohensinner S, Herrnegger M, Blaschke AP, Habereder C, Haidvogl G, Hein T, Jungwirth M, Weiß M (2008) Type-specific reference conditions of fluvial landscapes: a search in the past by 3D-reconstruction. Catena 75:200–215

Hughes RM (1995) Defining acceptable biological status by comparing with reference conditions. In: Davis WS, Simon RF (eds) Biological assessment and criteria; tools for water resource planning and decision making. Lewis, Boca Raton, FL, pp 31–47

Hughes RM, Paulsen SG, Stoddard JL (2000) EMAP-surface waters: a multiassemblage, probability survey of ecological integrity in the USA. In: Jungwirth M, Muhar S, Schmutz S (eds) Assessing the ecological integrity of running waters. Kluwer, Dordrecht, pp 429–433

Jungwirth M, Muhar S, Schmutz S (2000) Fundamentals of fish ecological integrity and their relation to the extended serial discontinuity concept. Hydrobiologia 422-423:85–97

Jungwirth M, Muhar S, Schmutz S (2002) Re-establishing and assessing ecological integrity in riverine landscapes. Freshw Biol 47:867–887

Junk WJ, Baylay PB, Sparks RE (1989) The flood pulse concept in the river flood-plain systems. Canadian Special Publications of Fisheries and Aquatic Sciences 106:110–127

Kern K (1992) Rehabilitation of streams in south-West Germany. In: Boon PJ, Calow P, Petts GE (eds) River conservation and management. Wiley, Chichester, pp 321–335

Kondolf GM, Boulton AJ, O'Daniel S, Poole GC, Rahel FJ, Stanley EH, Wohl E, Bång A, Carlstrom J, Cristoni C, Huber H, Koljonen S, Louhi P, Nakamura K (2006) Process-based ecological river restoration: visualizing three-dimensional connectivity and dynamic vectors to recover lost linkages. Ecol Soc 11:5

Kondolf GM (2011) Setting goals in river restoration: when and where can the river "heal itself"? In: Simon A, Bennett SJ, Castro JM (eds) Stream restoration in dynamic fluvial systems: scientific approaches, analyses, and tools. American Geophysical Union, Washington, DC

Larsen EW (2008) Modeling revetment removal and implications for meander migration of selected bends – river miles 222 to 179 of the Sacramento River. Technical report. Ducks Unlimited for CALFED Ecosystem Restoration Program, Sacramento, California

Lester RE, Boulton AJ (2008) Rehabilitating agricultural streams in Australia with wood: a review. Environ Manag 42:310–326

Leyer I, Mosner E, Lehmann B (2012) Managing floodplain-forest restoration in European river landscapes combining ecological and flood-protection issues. Ecol Appl 22:240–249

Marti C, Bezzola GR (2004) Sohlenmorphologie in Flussaufweitungen. Mitteilungen der Versuchsanstalt fur Wasserbau, Hydrologie und Glaziologie an der Eidgenossischen Technischen Hochschule Zurich, 184:173–188

Middleton B (1999) Wetland restoration flood pulsing and disturbance dynamics. Wiley, New York

Milly PCD, Betancourt J, Falkenmark M, Hirsch RM, Kundzewicz ZW, Lettenmaier DP, Stouffer RJ (2008) Climate change – Stationarity is dead: whither water management? Science 319:573–574

Muhar S (1994) Stellung und Funktion des Leitbildes im Rahmen von Gewässerbetreuungskonzepten. Wiener Mitteilungen 120:136–158

Muhar S (1996) Habitat improvement of Austrian rivers with regard to different scales. Regul Rivers Res Manag 12:471–482

Muhar S, Schwarz M, Schmutz S, Jungwirth M (2000) Identification of rivers with high and good habitat quality: methodological approach and applications in Austria. Hydrobiologia 422:343–358

Muhar S, Preis S, Hinterhofer M, Jungwirth M, Habersack H, Hauer C, Hofbauer S, Hittinger H (2006) Partizipationsprozesse im Rahmen des Projektes "Nachhaltige Entwicklung der Kamptal-Flusslandschaft". Österreichische Wasser- und Abfallwirtschaft 11–12:169–173

Muhar S, Jungwirth M, Unfer G, Wiesner C, Poppe M, Schmutz S, Hohensinner S, Habersack H (2008) Restoring riverine landscapes at the Drau River: successes and deficits in the context of ecological integrity. In: Habersack H, Piégay H, Rinaldi M (eds) Gravel-bed rivers VI. Elsevier, Philadelphia, pp 779–803

Muhar S, Pohl G, Stelzhammer M, Jungwirth M, Hornich R, Hohensinner S (2011) Integrated river basin management: bringing the requirements of water management, aquatic ecology and nature conservation into agreement on the basis of various EU directives (as practised on the river Enns in Styria). Österreichische Wasser- und Abfallwirtschaft 9-10:167–173

Muhar S, Schinegger R, Fleck S, Schülting L, Preis S, Trautwein C, Schmutz S (2013) Scientific foundations for identifying ecologically sensitive river stretches in the Alpine Arc. Study funded by MAVA and WWF

Muhar S, Januschke K, Kail J, Poppe M, Schmutz S, Hering D, Buijse AD (2016) Evaluating good-practice cases for river restoration across Europe: context, methodological framework, selected results and recommendations. Hydrobiologia 769:3–19

Newson MD, Large ARG (2006) 'Natural' rivers, 'hydromorphological quality' and river restoration: a challenging new agenda for applied fluvial geomorphology. Earth Surf Process Landf 31:1606–1624

Nijboer RC, Johnson RK, Verdonschot PFM, Sommerhauser M, Buffagni A (2004) Establishing reference conditions for European streams. Hydrobiologia 516:91–105

Nilsson C, Reidy CA, Dynesius M, Revenga C (2005) Fragmentation and flow regulation of the world's large river systems. Science 308:405–409

Nilsson C, Aradottir AL, Hagen D, Halldórsson G, Høegh K, Mitchell RJ, Raulund-Rasmussen K, Svavarsdóttir K, Tolvanen A, Wilson SD (2016) Evaluating the process of ecological restoration. Ecol Soc 21:41

Pahl-Wostl C (2002) Towards sustainability in the water sector: the importance of human actors and processes of social learning. Aquat Sci 64:394–411

Pahl-Wostl C (2007) The implications of complexity for integrated resources management. Environmental Modeling and Software 22:561–569

Pahl-Wostl C, Jeffrey P, Isendahl N, Brugnach M (2011) Maturing the new water management paradigm: progressing from aspiration to practice. Water Resour Manag 25:837–856

Paillex A, Dolédec S, Castella E, Mérigoux S (2009) Large river floodplain restoration: predicting species richness and trait responses to the restoration of hydrological connectivity. J Appl Ecol 46:250–258

Palmer MA, Bernhardt ES, Allan JD, Lake PS, Alexander G, Brooks S, Carr J, Clayton S, Dahm CN, Follstad SJ, Galat DL, Loss SG, Goodwin P, Hart DD, Hassett B, Jenkinson R, Kondolf GM, Lave R, Meyer JL, O'Donnell TK, Pagano L, Sudduth E (2005) Standards for ecologically successful river restoration. J Appl Ecol 42:208–217

Palmer MA, Allan JD (2006) Restoring rivers. Issues Sci Technol 22(2):40–48

Palmer MA (2009) Reforming watershed restoration: science in need of application and applications in need of science. Estuar Coasts 32:1–17

Pedroli B, De Blust G, Van Looy K, Van Rooij S (2002) Setting targets in strategies for river restoration. Landsc Ecol 17:5–18

Petts GE (1989) Historical analysis of fluvial hydrosystems. In: Petts GE, Möller H, Roux AL (eds) Historical change of large alluvial rivers: Western Europe. Wiley, Chichester, pp 1–18

Petts GE, Amoros C (1996) Fluvial hydrosystems. Chapman & Hall, London

Poppe M, Kail J, Aroviita J, Stelmaszczyk M, Giełczewski M, Muhar S (2016) Assessing restoration effects on hydromorphology in European mid-sized rivers by key hydromorphological parameters. Hydrobiologia 769:21–40

Preis S, Muhar S, Habersack H, Hauer C, Hofbauer S, Jungwirth M (2006) Nachhaltige Entwicklung der Flusslandschaft Kamp: Darstellung eines Managementprozesses in Hinblick auf die Vorgaben der EU Wasserrahmenrichtlinie (EU-WRRL). Österreichische Wasser- und Abfallwirtschaft 11–12:159–167

Reich M, Kershner JL, Wildman RC (2003) Restoring streams with large wood: a synthesis. In: Gregory SV, Boyer KL, Gurnell AM (eds) The ecology and management of wood in world rivers. American Fisheries Society, Cornvallis

Renner R, Schneider F, Hohenwallner D, Kopeinig C, Kruse S, Lienert J, Link S, Muhar S (2013) Meeting the challenges of transdisciplinary knowledge production for sustainable water governance. Mt Res Dev 33:234–247

Rhode S, Kienast F, Bürgi M (2004) Assessing the restoration success of river widenings: a landscape approach. Environ Manag 34:574–589

Rist S, Chiddambaranathan M, Escobar C, Wiesmann U, Zimmermann A (2007) Moving from sustainable management to sustainable governance of natural resources: the role of social learning processes in rural India, Bolivia and Mali. J Rural Stud 23:23–37

Roni P, Hanson K, Beechie T (2008) Global review of the physical and biological effectiveness of stream habitat rehabilitation techniques. N Am J Fish Manag 28:856–890

Roni P, Beechie T (2013) Stream and watershed restoration: a guide to restoring riverine processes and habitats. Wiley, Chichester, pp 189–214

Rutherfurd ID, Jerie K, Marsh N (2000) A rehabilitation manual for Australian streams. Vols. 1 and 2. Cooperative Research Centre for Catchment Hydrology, and the Land and Water Resources Research and Development Corporation, Canberra

Schinegger R, Trautwein C, Melcher A, Schmutz S (2012) Multiple human pressures and their spatial patterns in European running waters. Water Environ J 26:261–273

Schmutz S, Kremser H, Melcher A, Jungwirth M, Muhar S, Waidbacher H, Zauner G (2014) Ecological effects of rehabilitation measures at the Austrian Danube: a metaanalysis of fish assemblages. Hydrobiologia 729:49–60

Schmutz S, Jurajda P, Kaufmann S, Lorenz AW, Muhar S, Paillex A, Poppe M, Wolter C (2016) Response of fish assemblages to hydromorphological restoration in central and northern European rivers. Hydrobiologia 769:67–78

Sear DA (1994) River restoration and geomorphology. Aquat Conserv Mar Freshw Ecosyst 4:169–177

Sendzimir J, Magnuszewski P, Flachner Z, Balogh P, Molnar G, Sarvari A, Nagy Z (2007) Assessing the resilience of a river management regime: informal learning in a shadow network in the Tisza River basin. Ecol Soc 13:11

Sparks RE (1995) Need for ecosystem management of large rivers and floodplains. Bioscience 45:168–182

Technisches Büro Eberstaller GmbH (2010) Übersicht Planungskorridor LIFE+ Projekt Traisen. In: Verbund AG, DonauConsult Ingenieurbüro GmbH, ezb, freiwasser & IHG/BOKU Wien (eds), LIFE+ Lebensraum im Mündungsabschnitt des Flusses Traisen – Umweltverträglichkeitserklärung – UVE – Kurzzusammenfassung. Technical Report, Vienna

Tockner K, Pusch M, Borchardt D, Lorang MS (2010) Multiple stressors in coupled river-floodplain ecosystems. Freshw Biol 55:135–151

Toth LA, Arrington DA, Brady MA, Muszick DA (1995) Conceptual evaluation of factors potentially affecting restoration of habitat structure within the channelized Kissimmee River ecosystem. Restor Ecol 3:160–180

Unfer G, Haslauer M, Wiesner C, Jungwirth M (2011) Lebensader Obere Drau – Fischökologisches Monitoring. Amt der Kärntner Landesregierung, Abt. 18 - Wasserwirtschaft, 113

Wetzel PR, van der Valk AG, Toth LA (2001) Restoration of wetland vegetation on the Kissimmee River floodplain: potential role of seed banks. Wetlands 21:189–198

Whalen PJ, Toth LA, Koebel JW, Strayer PK (2002) Kissimmee river restoration: a case study. Water Sci Technol 45:55–62

Viviroli D, Archer DR, Buytaert W, Fowler HJ, Greenwood GB, Hamlet AF, Huang Y, Koboltschnig G, Litaor MI, Lopez-Moreno JI, Lorentz S, Schädler B, Schreier H, Schwaiger K, Vuille M, Woods R (2011) Climate change and mountain water resources: overview and recommendations for research, management and policy. Hydrol Earth Syst Sci 15:471–504

Wiek A, Larson KL (2012) Water, people, and sustainability – a systems framework for analyzing and assessing water governance regimes. Water Resour Manag 26:3153–3171

Wiens JA (2002) Riverine landscapes: taking landscape ecology into the water. Freshw Biol 47:501–515

Wohl E (2005) Compromised rivers: understanding historical human impacts on rivers in the context of restoration. Ecol Soc 10:2

Wohl E, Angermeier PL, Bledsoe B, Kondolf GM, MacDonnell L, Merritt DM, Palmer MA, Poff NL, Tarboton D (2005) River restoration. Water Resour Res 41:W10301

Chapter 16
Adaptive Management of Riverine Socio-ecological Systems

Jan Sendzimir, Piotr Magnuszewski, and Lance Gunderson

16.1 Becoming Adaptive in an Increasingly Variable World

Understanding and managing rivers demands much more than it did in the past. For centuries conventional engineering and economics oversimplified the challenge with assumptions of stability and stationarity (Milly et al. 2007). Despite vast increases in technical capacity, current management strategies cannot provide durable solutions to crises like more intense and frequent floods and droughts (Gleick 2004; Pahl-Wostl 2007; Huntjens et al. 2011; World Water Development Report 2009). Efforts to address intensifying trends of such crises revealed that our perspectives must expand along at least two dimensions.

First, we must broaden our vision horizontally with perspectives from across society. No single lens can reveal all the causes and possible cures for such problems. Management should become more inclusive and integrate multiple perspectives from academic disciplines with the shifting expectations from society. Our vision must extend out beyond engineering and economics and even the natural sciences to include insights from the social sciences and bridge the disciplines (*interdisciplinary*). This expanding perspective now sees riverine communities as

J. Sendzimir (✉)
Institute of Hydrobiology and Aquatic Ecosystem Management, University of Natural Resources and Life Sciences, Vienna, Austria
e-mail: jan.sendzimir@boku.ac.at

P. Magnuszewski
International Institute for Applied Systems Analysis, Laxenburg, Austria

Centre for Systems Solutions, Wrocław, Poland
e-mail: magnus@iiasa.ac.at

L. Gunderson
Department of Environmental Sciences, Emory University, Atlanta, GA, USA
e-mail: lgunder@emory.edu

© The Author(s) 2018
S. Schmutz, J. Sendzimir (eds.), *Riverine Ecosystem Management*, Aquatic Ecology
Series 8, https://doi.org/10.1007/978-3-319-73250-3_16

social-ecological systems (SES) to reflect a broader focus on critical interactions within and between the domains of engineering, ecology, economics, and social sciences (Sendzimir et al. 2008). However, decision-making needs to expand further and become *transdisciplinary*, embracing experience from all across society, including business, and governmental and nongovernmental organizations. Drawing from such a wide horizon, we can more completely define problems, probe for solutions, and be confident that our policies are acceptable to all sectors of society.

Second, we must extend our vision vertically to include all the levels at which management issues occur. Part of our uncertainty arises because the scales at which problems emerge and solutions might be found shift and expand. What had been local problems that responded to simply switching populations or resources now appear at broader scales, e.g. catchments, basins, and continents, globally (Allen et al. 2011). This scale challenge demands a perspective that can account for all the levels as well as how some problems can jump levels under extreme circumstances, and these influences can move either up or down the chain.

Some managers frame all problems within the boundaries of their administration. Problems that spring from over the horizon and cross those boundaries can prove to be catastrophic surprises. Surprises can emerge "from below," when the effects of local processes accumulate into something that encompasses the whole landscape or the region. For example, increases in local surface runoff from overdevelopment of fields and forests can aggregate up to massive erosion and flooding problems at larger scales. On the other hand, natural and human communities along rivers are increasingly vulnerable "from above" to global variability in climate, politics, or finance. Addressing such uncertainty with authority from larger (EU, national) scales can often increase uncertainty in governance. As the scale of accountability expands from individual river valleys to basins and beyond, managers must navigate increasing institutional complexity arising from overlapping administrations at different levels across each basin. Managers increasingly must adapt to this institutional complexity since almost half the earth's land surface is covered by some 260 transboundary river basins (Wolf et al. 1999), and that complexity is amplified when global processes cross those boundaries.

Systems like riverine SES are complex in structure, and that complexity is compounded by their dynamism as unforeseen consequences of policies and management are becoming apparent. Initial success at restoring ecosystem integrity often cannot be sustained (Scheffer 2004), and the system reverts to an undesirable state. So often have initial policy successes collapsed and remained so, despite all efforts at restoration, that the dysfunctional inertia following these surprising reversals has come to be known as *policy resistance* (Sterman 1994, 2000). Attempts to control disturbances (flood, fire, and pests) have often led to larger and more profound disruptions (Holling 1978). In riverine SES, for example, policies to constrain flood volumes within channels bounded by dikes result either in increased crop damages due to water stagnation (see Chap. 28) or in larger flood volumes moving faster when dikes do fail. As a result, the trend of increasing flood damages continues to rise (Sendzimir et al. 2007; Gleick 2004; Pahl-Wostl et al. 2007).

Conventional management policies have been aimed to "control" river dynamism (Poff et al. 1997) to achieve *water security* (*sensu* Cook and Bakker 2012) but at the expense of losing ecosystem services (Nilsson et al. 2005; Palmer et al. 2008; Vörösmarty et al. 2010). Most rivers are already over-allocated and stressed by the impacts of control (dams, diversions, etc.) to meet rising social demands for services (Chen et al. 2015). The paradox is that reducing river dynamism yields short-term economic gain and security from extreme events (flood, drought), but it shrinks the range of options for managing all the economic and ecological functions that make a river basin resilient to uncertainty over the long run (see Chap. 28).

Rising variability of large-scale drivers such as climate, population, or finance may overwhelm such narrow, exploitative management approaches, raising a mandate for more flexible and adaptive policies (Medema et al. 2008). However, transition to a more flexible management regime may be blocked by previous decisions that locked management onto a path of defensive and inflexible policies and technologies. When the scope of current decisions is confined to a narrow path defined by previous ones, this is known as *path dependence*. For example, massive investments in static strategies based on dams and dikes have established a concrete development path bounded by mental and technical walls that choke off innovation. First, they set a historic precedent that makes it difficult to even imagine more flexible ways to manage water. Second, they have monopolized the financial and policy resources of government, denying support for future management experiments to explore new ways to adapt to change. Experimentation is further blocked by a sense of hopelessness when path dependence on short-term control gives rise to policy resistance in the long term.

These possibilities have driven experimentation to integrate science and policy in one decision-making process. Policy in this sense refers to the policy formulation process. Policy is formulated as management interventions in the short to midterm to explore and refine different strategies with an eye to developing long-term policies (administrative) and even legislation. If ongoing change in ecosystems and society can render inflexible policies obsolete, then management must dynamically adapt as a counter to perennial uncertainty. This chapter describes a general synthesis of how to make decision-making more adaptive and then explores the barriers to learning in management. We then describe how one such process, known as adaptive management (AM), has been applied in different river basins, on which basis we discuss AM's strengths and limitations in various resource management contexts.

16.2 Management as an Adaptive Learning Process

Adapting to and managing change requires the sustained capacity to learn and to flexibly manage. Over the past century, these challenges have been independently recognized and addressed in a variety of disciplines and contexts. Similar but separate experimental lineages have tried to make decision-making more adaptive in, e.g., business (Vennix 1999; Senge 1990; Checkland 2000), policy (Toth 1988;

Sabatier 2006), and natural resource management (Holling 1978; Walters 1986; Gunderson et al. 1995), to name but a few. This chapter describes principles common to many approaches and focuses on adaptive management of natural resources for its application to river management.

The move to make management adaptive starts with the recognition that uncertainty is inevitable in an evolving world. If "nature" and "society" never stop changing, then reality is always several steps ahead of what we know, and certainty is a dangerous trap. Uncertainty from unprecedented scales, e.g., global climate, is daunting, but inaction is not an option, and ignorance is not an excuse. We cannot postpone management actions until "enough" information is available (Richter et al. 2006). We must manage even as we learn how, which means "learning while doing." Four kinds of uncertainty confront a manager: *environmental variability*, *partial observability* (e.g., ambiguity about resource status), *partial controllability* (e.g., ambiguity about how policy is implemented), and *structural or process uncertainty* (e.g., ambiguity about what causal relations produce the resource trends of concern) (Williams 2011). Given that change never ceases, the older goal of control that eliminates uncertainty has been replaced by the adaptive vision of reducing uncertainty as much as possible through sustained learning.

One can argue that all forms of management involve learning. However, the history of management experiments that failed because learning was too haphazard (e.g., unstructured trial and error) has driven generations of experiments in how to formally structure that learning such that management policy, science, and local practice advance knowledge together. Starting after World War II a series of sudden, irreversible collapses in fisheries, agriculture, rangeland grazing, and forestry forced natural scientists to recognize how strategies to eliminate uncertainty may produce deeper crises (Beverton and Holt 1957; Holling 1978). Failures to reverse or even explain these collapses may provoke even more profound uncertainty and highlighted the need to learn even as restoration efforts continued. This hard experience suggested that learning to reduce uncertainty was a much more achievable and practical goal than eliminating uncertainty, though other goals can be pursued at the same time, e.g., stabilizing and restoring ecosystem functions and services, to name but one.

16.2.1 Fundamentals of Adaptive Management

> For the things we have to learn before we can do them, we learn by doing them. – Aristotle, The Nicomachean Ethics

Adaptive management (AM) emerged from these resource collapse crises as one lineage of experiments started by Holling (1978) to integrate policy and science in a cycle of learning (Fig. 16.1). This process structures learning into four phases and iteratively integrates a series of linked processes in which we modify assessment, policy formulation, implementation, and monitoring in order to track and manage

Fig. 16.1 Adaptive
management—a decision-
learning process driven by
cyclic, experimental policy
implementation and
monitoring (after
Magnuszewski et al. 2005)

change in the world (Walters 1986; Magnuszewski et al. 2005). The assessment
process highlights uncertainties in societal goals. Policies are formulated as hypotheses that test those uncertainties. Management actions can be used to test uncertainties, along with achieving other societal objectives. Monitoring the consequences
of actions leads to an evaluative effort, which can be used to adjust or reaffirm
policies. AM has burgeoned into many forms, some of them fully institutionalized,
such as the implementation of the Water Framework Directive in Europe. However,
we confine our description of AM to one of its most fundamental forms to highlight
the most essential elements in its basic structure (for a more detailed survey, see
Allen et al. 2011).

The AM process starts with the assembly of a team of a size range (e.g., not more
than 25, 20 is better) that is big enough to encompass a diversity of perspectives but
not so large that discussion becomes unmanageable. Efforts to increase diversity
have invited participation from natural and social scientists, as well as key actors in
the business, policy, and administration sectors. Special benefit has been gained by
including local practitioners (e.g., fishermen, hunters, environmental activists)
whose unique experience (often over time spans much longer than most science
research projects) can raise vital questions and add practical insights to the discussion. The challenge is to achieve the right balance between power and perspective.
That means including the key actors that have real power to affect the outcome of
any decision as well as a range of practitioners with the sufficient experience and
training to provide a healthy diversity of perspectives. Insufficient diversity means
that no breakthrough innovation may occur due to lack of key alternative analysis or
policies. Insufficient power means that any gains in understanding may not be
realized in practice.

In its simplest form, the AM learning cycle starts with an *assessment* phase
wherein stakeholders explore a range of assumptions in order to formulate a suite
of hypotheses that provide separate predictions of why the problem in question
occurs (Walters 1986; Gunderson et al. 1995). Within this phase participants can

articulate different objectives, identify both alternative strategies or policies for management and their consequences, and recognize the key uncertainties that remain even as, in later phases, we monitor the consequences of our policies (Williams 2011). Modeling can serve as a useful exercise for participating stakeholders to define and bound the problem and examine the key variables and interactions they consider crucial to the dynamics of resilience and vulnerability in the system. Modeling can be done to explore assumptions about the structure of causal relations (conceptual models) or the dynamic implications of those assumptions (mathematical simulation), in explaining how the problem trend occurs. Further, stakeholders can collaborate in role-playing games to simulate the social relationships that may be key to the management problem and/or its solution.

In the assessment phase, stakeholders winnow down their list of important hypotheses to one or two key questions to investigate for causes and solutions of the problem of concern. The following phases use those questions as a base on which to formulate (*policy formulation*) and then implement policies (*management actions*) that test those questions. While conventional management focused policy on solutions, AM first designs policy that tests the most useful questions with the expectation that the results will eventually produce more durable solutions. The *monitoring and evaluation* phase uses those questions and the indicators and thresholds derived therefrom, to survey and document how the system responds to these policies. Eventually the lessons gained from that survey data will be used when management recognizes that the objectives were not achieved or the world, e.g., boundary conditions, has changed sufficiently to merit starting the whole process again. A new learning cycle will begin with an assessment phase to reexamine and perhaps redefine the problem, the management goals, the underlying assumptions about causal relations, etc. Ideally AM would transcend the boundaries on learning placed by budgets (project timelines), and politics (election cycles), and periodically revive the process to keep our science and policy in step with how the world is changing. The European Union implements such a strategy by reviewing river management plans every 6 years under the aegis of the Water Framework Directive.

16.2.2   Challenges to the Adoption of Adaptive Management

Management may serve to ameliorate threatening change, but because of path dependence innovations in management can be blocked as another kind of unwelcome change. There are many reasons why the AM process has not succeeded either in being attempted, sustained, or adopted by communities and/or governance agencies. Understanding these barriers may catalyze better experiments to improve AM in concept and practice. We first describe these barriers to AM in general and then consider refinements to integrate AM better in the policy world and conclude by examining specific barriers to different phases of the AM cycle.

AM can appear as a complex and intimidating challenge to any manager or agency considering to apply it. To start with, its very flexibility has engendered so many interpretations that no single model appears as a tested example (Fontaine 2011). Secondly, previous experiments on North American rivers, e.g., the Columbia River basin, the Colorado River in the Grand Canyon, the Everglades, and the Sacramento-San Joaquin river basin and delta in California (Johnson 1999), have proven very costly in terms of time and money invested as well as the institutional complexities that had to be harmonized (Richter et al. 2006). In these cases the process entailed multiple layers (executive, manager, and technical) of decision and review (Walters et al. 2000) and was agitated by the levels of sociopolitical controversy surrounding the significant economic trade-offs at hand (Richter et al. 2006). If success requires years of sustaining such effort, many might opt for less costly approaches. Finally, despite some exceptions (Allan and Stankey 2009), there are few, convincing success stories in AM's history of risky and costly experiments (Lee 1999; Medema et al. 2008). One reason may be that often success is hard to detect when clear thresholds are difficult to establish and measure (van Wilgen and Biggs 2011), though adaptive learning successfully developed flow thresholds on the Letaba (Pollard et al. 2009) and Kissimmee (Toth et al. 1998) rivers. Another reason is that no history of successfully repeated experiments can be built on the basis of contradictory definitions or applications of AM (Allen et al. 2011).

The slow adoption of AM can broadly be explained by path dependence on conventional, control-centered management strategies. A variety of factors can contribute to this: funding and administrative priorities based on reactive rather than proactive management paradigms (Walters 1997), failure to identify or address the potential to shift goals to acknowledge the increased risk of surprise because of social sources of uncertainty (Tyre and Michaels 2011), defaulting to extant institutional norms rather than reflecting, box-ticking as opposed to learning and favoring competition over cooperation (Allan and Curtis 2005), and resource limitations in terms of funding and human capital.

The novel capacities that AM requires of any river manager would be a major barrier to adoption. These capacities include developing new ways for managers and stakeholders to collaborate under high uncertainty, especially in integrating science and policy in participatory processes, considering a wide range of practices over long time periods projected in future scenarios (Pahl-Wostl et al. 2007). As opposed to direct administration of agency employees, managers might have to patiently attend to repeated phases of social learning by the community in implementing and then sustaining new and innovative management approaches (Pahl-Wostl et al. 2007).

16.2.3 Advances in Adaptive Management

AM began as an initiative by natural scientists to integrate policy and science by inviting policy makers, managers, and local practitioners to share the risk of raising good hypotheses by blending their experience to confront the multiple issues in these complex systems. In turn, the AM process encourages scientists to share the risk of formulating credible policy that really sustains SES. Many perspectives from both sides have motivated and informed experiments to improve AM. Policy makers in general found the core AM cycle (Fig. 16.1) far too simple in its appreciation of the diverse influences that create the policy world in which they operate (Walters 1997). For example, Weick (1995) claims that a key challenge to adopting and implementing AM is that the basic concept may aim for the key management goal, but it does not give enough attention to one of the founding tenets of AM: the diversity of perspectives and goals among stakeholders. Many of them are driven by very different goals, which are tied to politics, business lobbying, or ideological stance and which are in turn linked with broader organizational and social contexts. Understanding these interactions is critical to a successful application of AM.

Innovators in business management reached the same conclusion as government policy makers: the science of trends and causal mechanisms is critical to identify potential paths and related causes, but it must be complemented by appreciation of the most powerful factors that drive those causes and trends, e.g., the goals, paradigms, and mental models that drive the stakeholders (Senge 1990). Paradigms and mental models are like archetypes or basic templates of how we see the world, and we use them to filter information into categories like true or false and believable or incredible. For example, some see the world as a collection of individuals, whereas others see a hierarchical structure of power relations that connects everyone (see Kahan 2008; Yazdanpanah et al. 2014). The structure of relationships and the trends they generate arise out of these goals and mental models of society. How we interact and how those relations generate trends depend on how we think. How we think and what goals we use can be changed as part of the decision-making process. AM has been applied to foster this by creating a space that's safe to reveal our underlying thoughts and open the door for innovative compromises in thinking, policy, and action.

Efforts to make business organizations more adaptive revealed a hierarchy of relationships between the factors that influence how stakeholders consider evidence (Fig. 16.2). Moving from the top to the bottom one encounters factors of increasing influence on how we perceive how our world works. Shallow analysis never goes beyond the tip of the iceberg, the events that periodically appear with no pattern that connects, and makes sense of them. Historians and statisticians might look deeper for those patterns that show us trends that brought us here or that suggest where we are going. Looking deeper one searches for the ways that we interact, the patterns of behavior, that create those trends. The network of such causal mechanisms is also referred to as the *system structure*. At the base are the most powerful forces that determine all the patterns above: the myths and paradigms we nurture or the goals that we as individuals or as members or an organization develop and follow. To summarize a vast and rapidly expanding field in applying cognitive psychology to management, one still must pursue rigorous science and technical implementation, but achieving mutual understanding of the science and policy based on very different goals and underlying paradigms is essential. Science only becomes useful as the different perspectives in any dispute come to agree on how to define the means to interpret and then implement science.

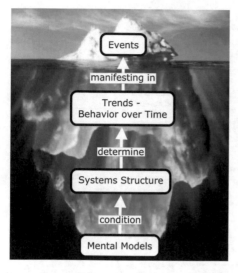

What has just happened?

What has been happening? What will happen?
Have we been here or some
place similar before?

What are the forces at play
contributing to these patterns?

What in our thinking allows
this situation to persist?

Fig. 16.2 System structure as a determinant of its behavior (adapted from Senge 1990)

The importance of goals and mental models to decision-making becomes clear if the AM cycle is expanded to try to account for the complexity of the policy world (Fig. 16.3). In the assessment phase, all factors in Fig. 16.2 are used to define the problem (usually as a trend) and the pattern of likely causes. From the outset this is complicated by the fact that the nature of the problem itself is in question. Problems may have multiple interpretations, based on conflicting values and goals (Weick 1995). Even if the basis of a problem is relatively simple and clear, the goals and mental models of individuals and of the surrounding culture and politics will dominate how it is defined and interpreted. This is because goals and mental models act like a filter that determines what information is selected and how it is measured in evaluating policy performance or in decisions as to how policy is implemented.

One example of this is the physics of increasing the concentration of CO_2 in the atmosphere, which is not complex. It can only cause more heat to be retained, and this reliably predicts why the years since 2000 have exhibited the hottest atmospheric temperatures ever measured. Yet many deny the evidence (that climate change exists at all) or the theory that society is driving it by generating more greenhouse gases. Political conservatives, who often ascribe to hierarchist or individualist paradigms, tend to deny climate change theory or evidence. Political liberals, who tend to subscribe to paradigms centered on community (communitarian) and social equality (egalitarian), tend to accept the science supporting climate change theory (Kahan 2008, 2013). However, both conservatives and liberals are equally likely to interpret data ideologically. There is a third, social and political, aspect involved. As Kahan (2013) concludes: "...ideologically motivated cognition ...[is] a form of information processing that promotes individuals' interests in forming and maintaining beliefs that signify their loyalty to important affinity groups." Thus, while the perceived political and economic consequences of climate

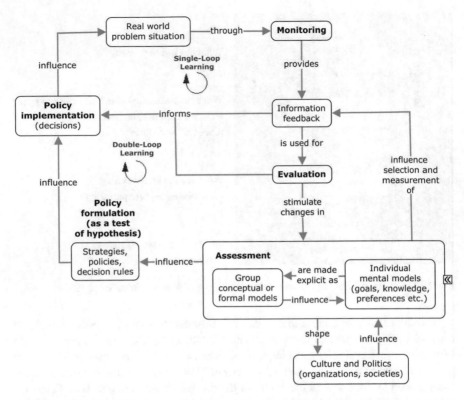

Fig. 16.3 Adaptive management supported by cycles of learning to improve policy performance (single-loop) and clarify goals and mental models (double-loop)

change are so unacceptable to political conservatives that they deny climate change and the science behind it, their public stance has much to do with being "in social solidarity" (*sensu* Thompson et al. 1990) with those subscribing to the same view.

As previously stated, the assessment phase serves to define the management challenge and identify which factors may be involved in its genesis or in any solution(s). How factors are assessed depends both on mental models and goals held by individual stakeholders as well as the model(s) of the problem developed by the group. The process of building a model of the problem that is understandable and acceptable to the group offers a chance to clarify what goals and mental models are being used by individuals. If those are ignored, there is a good chance that efforts to reform policy or encourage the adoption of new innovative policies may be rejected by stakeholders. This highlights the importance of sustaining the learning process long enough to achieve such clarification. This deeper level of learning, e.g., *double-loop learning*, involves probing the decision-making process itself and how social and institutional relations and stakeholder perspectives influence it (Torbert et al. 2004; Argyris 1976, 2005; Argyris and Schön 1978; Williams 2011). Double-loop learning can enable shifts in underlying beliefs, values, and mental models. This stands in contrast to *single-loop learning* that improves policy performance through

Table 16.1 Phases of cyclic learning cycles in natural resource management (AM), business (Deming/Shewart), and education (Kolb)

Learning cycle	Learning phases			
AM	Assess	Policy	Implement	Monitor
Deming/ Shewart	Act	Plan	Do	Study
Kolb	Reflective observation	Abstract conceptualization	Active experimentation	Concrete experience

trial and error in implementation without reflecting on the whole system of feedbacks that influence decisions, policies, and performance. The expansion from single- to double-loop learning represents a major challenge of management modernization in education, business, and natural resource management over the past 50 years. This is reflected in similarities between AM and other learning cycles tested in the fields of education (Kolb) and business [Deming/Shewhart—(Deming 1986; Best and Neuhauser 2006)] (Table 16.1).

The structure of the group model, e.g., the pattern of interactions, can be used in the assessment phase to identify feedbacks and other interaction patterns that are critical to the emergence of the problem. Changing, e.g., stopping, slowing, or speeding up, such patterns could therefore be part of a solution, so such group models can be used to identify critical points of intervention for management policies. Such models can also help identify what is not known about critical factors or patterns and thus help set a research agenda for long-term monitoring. Thus, the policy formulation phase uses these data and insights to devise strategies and policies that are used to guide the next two concurrent phases, e.g., policy implementation and monitoring. The latter is guided by a research agenda aimed to answer the critical questions and unknowns identified in the assessment.

16.2.4 Specific Barriers to Different Phases of the Adaptive Management Cycle

AM experiments over 40 years have revealed the diversity of factors that must be integrated to successfully manage adaptively. This has thrown a sharper light on what barriers can block it. Table 16.2 groups these barriers into the phases of the AM cycle which they impact.

As discussed above, barriers in the assessment phase mostly revolve around failure to clarify conflicting values, goals, and perceptions. Two barriers emerge from rigidity of organizational structure, e.g., bias toward authority that only passes downward in a decision hierarchy (*verticality*) or bias toward one's own working group that overrules the functioning of the whole (*departmentalism*). Barriers in the policy phase are mostly due to procedural failures in formulating policy, e.g., lack of

Table 16.2 Barriers to adaptive learning and decision-making. Reference numbers are enclosed in parentheses and the author/dates listed at the bottom of the table

AM phase: Assessment

Learning factor: Individual mental models

Barriers to learning:

- Nature of problem is itself in question (1); multiple conflicting interpretations (1); different value orientations (1); unclear and/or conflicting goals (1); manager's understanding of the goals or the problem of concern is confounded by vague, competing definitions, especially if the roots of one problem are messily entangled with the roots of other problems (1)
- Misperceptions of feedbacks and delays, unscientific reasoning, defensive routines, judgmental biases (2)

Learning factor: Group mental models

Barriers to learning:

- Inadequate stakeholder participation (3)
- Lack of a "safe and authorized space" where scientists and decision-makers could meet together and develop a constructive dialogue and build mutual trust (4,5)

Learning factor: Politics and culture

Barriers to learning:

- No long-term perspective, shallow analysis, contradictory perspectives, political pressures, verticality and departmentalism in structure (2)
- Deeply entrenched view of science as objective and neutral (6)
- The lack of rigorous science—and policy that uses that science—to regain its social context; the lack of "reorientation of fundamental values regarding human relationships with the biosphere, whether through political, ethical, or religious movements" (4,5,6,7,8,9)

AM phase: Policy

Learning factor: Policy formulation

Barriers to learning:

- Inability to conduct controlled experiments, preference for quick "technical fixes," lack of relevant science, neglect of uncertainties (2)
- Lack of transparency in decision-making (10); success measures are lacking (1)

AM phase: Actions

Learning factor: Policy implementation

Barriers to learning:

- Implementation failure, policy resistance, high cost of error, gaming the system, inconsistency (2)
- Time, money, or attention is lacking (1)

Learning factor: Real world

Barriers to learning:

- Dynamic complexity arising because social-environmental systems are dynamic, tightly coupled, governed by feedback, nonlinear, history-dependent, self-organizing, adaptive, counterintuitive, policy resistant with many trade-offs (2)
- "Sisyphus situation": using science to solve an environmental issue is an endless task, as immediately new issues subsequently emerge to be solved (8)

(continued)

Table 16.2 (continued)

AM phase: Monitor and evaluate
Learning factor: Information feedback
Barriers to learning:
• Limited information, delayed feedback, selective perception, ambiguity, bias, distortion, measurement error (2)

References: (1) Weick 1995, (2) Sterman 1994, (3) Pahl-Wostl et al. 2011, (4) Michaels 2009, (5) Perreira et al. 2009, (6) Surridge and Harris 2013, (7) Litfin 1994, (8) Sverrisson 2001, (9) Konijnedijk 2004, (10) Holmes and Savgård 2009

transparency, failures to adequately address uncertainties or the science (lack of controlled experiments), often related to preferences for quick answers and fixes to the problem rather than profound learning and better understanding.

Barriers in the action phase revolve around execution, either overall (implementation failure) or due to confounding feedbacks that reverse initial success and scuttle any attempts to improve or revise a strategy (policy resistance). One particular barrier that may be common to more than one phase is "gaming the system," i.e., gaining personal advantage by manipulating or exploiting the governing rules that were originally designed to protect everyone. This can block mutual efforts to assess, formulate policy, act, or monitor. Finally, the monitoring and evaluation phase is blocked by barriers similar to those that block perception and understanding in the assessment phase, e.g., selective perception, ambiguity, bias, etc.

16.3 Diverse Approaches to Adaptive Water Management

Adaptive management is not the only alternative to conventional, "command-and-control" decision-making in natural resource management. Alternative or "soft" approaches (*sensu* Gleick 2004) to water management have been tested in parallel for decades, mostly in the global North. In addition to AM, two prominent approaches are integrated water resources management (IWRM) and ecosystem-based approaches (EBAs). All three share similar goals concerning equity, human well-being, and sustainability, though with slight differences in emphasis: sustainability (IWRM), conservation (EBAs), and a combination of the first two through the lens of learning (AM) (Schoeman et al. 2014). IWRM promotes sustainability through a governance framework that allows actors to negotiate integrated land and water management at the scale of the river basin, at which point it is called integrated river basin management (IRBM) (Grigg 2008). EBAs foster conservation by incorporating in decision-making the valuation of ecosystem services (De Groot et al. 2002) and Ramsar's "wise use of wetlands" (Finlayson et al. 2011). All three approaches increasingly converge in their application in water management as a result of debate among policy scientists (Pahl-Wostl et al. 2011).

Schoeman et al. (2014) find that the strengths of all three can be combined to assist water managers in the following ways:

- Broad stakeholder participation across scales, disciplines, and sectors (IWRM and AM) promotes adaptive capacity by diversifying the knowledge base (skills and experience) and sharing it within and across networks.
- Navigating better through institutional (policy, law) complexity by fostering better information sharing through networks.
- Improving the cohesion between policy formulation and implementation by basing monitoring and policy application on hydrological boundaries (IWRM).
- Supporting water security in the face of climate variability by offering platforms to share values and complex information so as to resolve conflicts and encourage innovative experiments.

16.4 Adaptive Management: The Law and Governance

16.4.1 Law

The resilience of a social-ecological system (SES) in the face of uncertainty arises not only from flexibility and adaptability but also from reliable stability during stress or turbulence. Environmental law is a legal buttress that provides a measure of certainty in the institutions that it formalizes in legislative and administrative code. But that investment in stability can encourage an institutional rigidity that hinders efforts to flex, innovate, and develop novel management strategies to probe a shifting environment (Benson and Garmestani 2011; Allen et al. 2011). This rigidity can be reinforced if the information gathered in the legislative and lobbying process loses its diversity as it is filtered and synthesized through a small minority of individual politicians. The possibility for innovative experimentation is cut off if the legislative process is only informed by small networks of technocrats who exclusively share a vocabulary narrowed by their common experience. This path dependence on technocratic elites was one impetus for the search to expand the range of perspectives, experience, and training that informed the policy debate (Gunderson et al. 1995) driving the move toward interdisciplinary and transdisciplinary dialogue in decision-making. Therefore, though in the European Union AM for water resources has been institutionalized within the WFD with a 6-year management cycles (EU 2016), each new cycle should begin by recognizing the danger of path dependence on a technocratic elite and search for a sufficiently wide range of perspectives across society. That search seeks to increase our capacity to adapt to change by expanding beyond management to governance in general.

16.4.2 Governance

Managing aquatic SES proceeds over time scales (decades) that far exceed those of individual science or development projects or individual management campaigns within political administrations. To make SES sustainable, the adaptive potential raised by AM must be sustained over periods long enough to institutionalize adaptive and sustainable practices. This drive to build long-term SES sustainability proposed *adaptive governance* as a framework that would foster AM while addressing social aspects neglected in initial AM experiments (Allen and Gunderson 2011). Specifically, it should create a workspace where formal and informal institutions can collaborate to understand and manage complex issues in SES (Schultz et al. 2015).

Adaptive governance is distinguished by its capacity to increase the importance of learning at the policy level and to bridge previously separate levels: formal/informal groups and networks as well as scales of administration (*polycentricity*), in ways that embrace cross-scale interactions in ecosystems and society (Chaffin et al. 2014). By encouraging *social learning*, e.g., learning that occurs in a group as a whole when collaboration involves developing shared understanding of meanings and practices, it is easier to resolve conflicts dealing with differences in perspective and community acceptance to implement and sustain innovative management approaches (Huntjens et al. 2011). Learning can also be enhanced among those who devise and implement policy through *policy learning*, e.g., a discursive process that challenges assumptions and goals of policies and thereby develops cognitive frameworks to revise the goals, techniques, and policies (Pahl-Wostl et al. 2007). The degree to which a riverine SES becomes more adaptive to change depends on what level of policy learning it attains (Argyris 2005; Hargrove 2002; Sanderson 2002). In sum, learning should be encouraged for all actors in all phases of the adaptive cycle.

Efforts to make governance more adaptive must be concerted enough to enjoin administrative agencies, which are prone to make small, slow, incremental changes (Allen et al. 2011), to contribute to much more comprehensive policy reform. Two deficiencies often make agencies default to an *organizational inertia* (Allen et al. 2011) or *path dependence* that does not stray from standard operating procedures: a lack of sufficient information to radically reform policy and a lack of institutions that fit the scale at which problems occur (Dietz et al. 2003).

Adaptive governance can address these deficiencies by incorporating the perspectives of different stakeholders working at multiple scales, e.g., local, river reach, basin, and regional (Hughes et al. 2005). Such multi-scale collaboration, enhanced by bridging organizations, improves environmental management by facilitating the creation (governance) and actualization (management) of visions that are ambitious and innovative (Folke et al. 2005). Bridging organizations, often NGOs, enhance policy acceptance by improving stakeholder perceptions of data collected and decisions reached through establishing communication channels and negotiating the meaning of the information as well as the multiple positions and interpretations with stakeholders. This builds trust that increases chances for collaboration and reduces transactions costs while providing means to enforce adherence to policies even in the absence of a regulatory authority (see Chap. 22).

Such informal collaboration builds trust and understanding through sharing management power as well as its responsibilities. Trust is an essential ingredient to making governance more cost-effective (Hahn et al. 2006). It stimulates and consolidates coordination and interaction between different actors from different domains and organizations in the water governance networks. With higher trust actors invest their resources, e.g., money and knowledge, in cross-disciplinary collaborative processes. Therefore, it stimulates experimentation and learning (Edelenbos and van Meerkerk 2015).

These benefits are best realized in two ways. First is with the help of *boundary spanners* to cut across divides that fragment institutions. Such individuals create informal spaces for innovation that connects people operating at other domains, levels, scales, and organizations. Secondly, trust and collaboration can be gained by integrating support from the top (visionary leadership) with inputs from each scale or level. The chances of successful transition to adaptive governance are increased by AM-friendly legislation, sufficient funding, and a dependable stream of useful information from a diversity of sources, e.g., both ecosystem monitoring and participatory discussion (Olsson et al. 2004). Adaptive governance recognizes the unique contributions to all these functions played by innovative social networks. For example, networks act as a bridge, communicating between the variable and unpredictable realms of ecosystems and the more rigid, formal realms of institutions and policy (Folke et al. 2005). Such social networks can be more successful than conventional institutions in government and business at generating political, financial, and legal support for innovative policies (Folke et al. 2005). Success often hinges on visionary leadership to direct the collaborative common of the network (see Allen et al. 2011 for more detailed discussion).

16.5 Putting Adaptive Management in Action

Two challenges confront anyone leading an AM process: the threat of harm from the ongoing problem and the chance that the entire process is halted for any of range of reasons why society often defaults to conventional practices rather than innovates (Sect. 16.3). As a result, even river managers who attempt adaptive governance are cautious. For this reason, most AM experiments have been "passive" in that they implement a single preferred course of action based on the best available modeling and planning, which is then modified as experience grows (Benson and Garmestani 2011). Passive AM relies on nature to provide the variability within which different policy options are tested.

In some cases passive monitoring of how the world changes may not generate clear and reliable knowledge within a time frame useful to make or implement policy. Under those circumstances, decision-makers within an AM process may seek to intervene with a policy that stimulates the system and thereby learn from how the system varies in response. Here policy options are seen not as solutions but as hypotheses to be tested under a regime where formalized learning and management are objectives for which experimentation is the key (Rist et al. 2013).

Interventions can be risky, especially in riverine SES, which are "...open systems that cannot be isolated from their social context" (Konrad et al. 2011). However, some river managers do choose to actively intervene in response to some emerging problem. Often a lack of data challenges managers trying to understand and manage how environmental flows contribute to the serviceable functioning of a river. A history of water control structure construction has almost eliminated naturally flowing rivers. In North America construction of more than 2.5 million control structures have left less than 2% of all rivers naturally flowing (Lytle and Poff 2004), with a similarly small percentage of naturally flowing rivers in Europe (see Chap. 6). This denies most managers naturally occurring reference flow conditions to use as performance criteria.

Social awareness of this extreme state has been heightened by the growing recognition of society's dependence on increasingly fragile aquatic resources. Many rivers are stressed and over-allocated in order to serve the demands of growing populations through hydro-engineering (Chen et al. 2015). In response, societal values have shifted to question this history, and managers have been compelled to modify operations of these control structures to mitigate physical and biological impacts on aquatic ecosystems (Rood et al. 2003).

16.5.1   Case Study: Active Adaptive Governance in Colorado

One celebrated example of active adaptive intervention has been tested for several decades on the Colorado River. Installation of the Glen Canyon Dam in 1963 had cut spring flood volumes by 65% (Collier et al. 1997; Poff et al. 1997), and this deficit was raised as a possible factor in the decline of an indigenous fish species, the humpback chub (*Gila cypha*). This was only one of a multitude of other challenges, but we focus on it to highlight an example of active, adaptive intervention. However, we caution that managers do not have the luxury of managing only one issue to the exclusion of all others. Management interventions must balance a great diversity of trade-offs, which AM should assist in identifying (for full details, see Walters et al. 2000)

As previously noted, the versions of AM briefly outlined above are basic templates that describe an idealized sequence of events. While no case of AM has ever strictly followed such a template, one can use them to design a new project or to roughly follow the "progress" of each AM experiment. For example, in Colorado a number of experimental interventions in dam water releases to test different river flow volumes and rates were attempted *before* an AM process was officially started. After an environmental Impact Statement in 1995, the river managing agencies elected to formally adopt an AM program in order to better harmonize conflicting goals related to water use, recreation, and protection of native species (Walters et al. 2000). This means that the AM assessment phase began *after* a period of management interventions and monitoring.

Following initial assessment, policies to experimentally manipulate river flows were formulated and implemented. The most celebrated was the release of a large "beach/habitat-building flow" (BHBF) in 1996 to assess whether episodic flows could move sand from the main river channel onto lateral deposits on riverine "beaches" used for camping and thereby "reverse successional impacts on the productivity of backwater/slough habitats" (Walters et al. 2000).

By 1997, management elected to boost the assessment process through development of explicit dynamic simulation models of the Colorado River ecosystem. This was done to support future adaptive planning, not through "detailed quantitative predictions about policy options, but rather... to expose broad gaps in data and understanding that are easily overlooked in verbal and qualitative assessments" (Walters et al. 2000). In 1998 the assessment phase proceeded through a series of workshops dedicated to develop an ecosystem model and a family of submodels with participation of up to 40 scientists and managers working within the Colorado River system. This phase concluded with a final meeting to critically review the model to identify key gaps and weaknesses both in the model and the data (Walters et al. 2000).

One conclusion reached by the group modeling exercise was that extreme policy options, e.g., restoring seasonal flows and increasing water temperatures, could produce highly uncertain, potentially deleterious effects with no unquestionable benefits. This may be one reason why more extreme experiments, e.g., BHBF, have rarely been repeated. But flow experiments have continued as part of long-term monitoring over the past 20 years. Complete return to natural flow dynamics was impossible, but an array of simultaneous experiments in flow modification were implemented to address a range of problems: conserve shoreline sandbars and fish, improve navigation (modified low fluctuating flows), and rebuild sandbars with tributary sand inputs below dam (high flow experiments), to warm the river, to benefit juvenile humpback chub (high steady flows and low summer steady flows and fall steady flows), and to limit rainbow trout egg viability (trout management flows) (Melis et al. 2015).

Such strong interventions can produce dramatic results that reveal threshold effects. But they only start a learning process that must be sustained over years of monitoring to discover policies robust to system dynamics over both the short and long terms. While the initial high-flow experiments on the Colorado River generated a surprising amount of riparian beach building (53% of all beaches increased in size (Poff et al. 1997), these beaches and sandbars were eventually eroded by fluctuating flows from hydro-peaking (Schmidt and Grams 2011). Furthermore, if other aspects of serial discontinuity (lotic habitat fragmentation, thermal regime changes, sediment supply reduction, biota migration disruptions) are involved, then experimental manipulations of flow may not be sufficient for learning or treatment (Lytle and Poff 2004).

Experiments in natural flow simulation occur now worldwide to reestablish the ecological integrity of regulated rivers (Lytle and Poff 2004). Robinson and Uehlinger (2008) note that "As a large-scale disturbance, the long-term sequential use of floods provides an excellent empirical approach to examine ecosystem regime shifts in rivers." The Sustainable Rivers Project is a North American example of a program of experimental interventions to reestablish environmental flows by "re-operationalizing" dams at some six sites covering over 1000 km of river (Warner et al. 2014).

16.6 Comparing Adaptive Management with Other Management Approaches

Challenges to sustainability occur under a multitude of circumstances, and no single approach to adaptability fits every circumstance. Experience since 1975 suggests a number of circumstances where AM is not likely to succeed. These include low uncertainty about what policies to apply and what outcomes are likely and low probability of carrying out an effective monitoring program or of feeding monitoring data and its analysis back into the reformulation of management strategy (Williams 2011). Unfortunately, many political regimes have election cycles so short (less than 4 years) that long-term monitoring is hard to secure as a policy standard before a new regime is installed. For those situations where resource variation does not respond to management actions, e.g., climate change, then *controllability* is low, and AM is not appropriate (Fig. 16.4). Challenges occurring at very large, e.g., global, scales with very slow response times are examples of low controllability. However, for such situations AM might be useful for managers if it can help explore and mitigate some

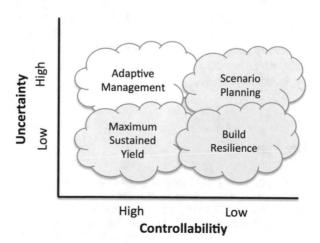

Fig. 16.4 Decision options for aquatic ecosystems based on balancing the degrees of uncertainty and controllability (after Peterson et al. 2003)

of the impacts that occur at scales responsive to management actions. Otherwise, situations of low controllability but high uncertainty might respond best to scenario development (Peterson et al. 2003). Scenarios can be developed which simply outline a sequence of actions and events required to attain some goal. While generally more qualitative than quantitative, under high uncertainty, they can help improve policy implementation if management teams can use them to agree on what they expect and how they will respond given different outcomes.

On the other hand, if there is little uncertainty about the response of resources to management policy, then AM's capacity to experimentally compare different policies is wasted (Allen et al. 2011). For example, in fisheries where rigorous monitoring yields very reliable fish population data, managers might aim to achieve maximum sustainable yield, i.e., the largest yield or catch that can sustainably be taken from a population over an indefinite period. Historic applications of MSY have generally set it around 30% of unexploited population size (Thorpe et al. 2015). Alternatively, with low uncertainty but also low controllability, there is not much chance to intervene precisely. Therefore, the optimum strategy might be to implement policies that increase the resilience of aquatic species to disturbance. For example, with little possibility to influence drought cycles, general restoration of floodplain habitat and reconnection hydraulically to the river channel can increase opportunities to find shelter, food, and chances to reproduce. In this way aquatic species have more capacity to adapt to extremes in rainfall patterns.

In summary, adaptive management will not reassure political regimes that demand only easy answers. It pushes in the opposite direction by exploring exactly what the uncertainties are and how to probe them. This push lowers the chance of falling into traps where dependence on certainty makes policies inflexible to change. AM has higher chances of succeeding where uncertainty may be high but so is controllability, so the chances for experimental learning are high. That means that AM is more appropriate where uncertainty is high, e.g., little is known about the scales at which problems emerge, but where management actions can produce clear and definitive impacts. Chances of AM success improve in situations with lower uncertainty, e.g., simple systems with data clearly indicating causal relationships, and policies can be implemented at small scales with clear effect, e.g., controllability, to test hypothetical causal relationships (Gregory et al. 2006).

References

Allan C, Curtis A (2005) Nipped in the bud: why regional scale adaptive management is not blooming. Environ Manag 36:414–425

Allan C, Stankey GH (2009) Adaptive environmental management, vol 351. Springer, New York

Allen CR, Gunderson LH (2011) Pathology and failure in the design and implementation of adaptive management. J Environ Manag 92(5):1379–1384

Allen CR, Fontaine JJ, Pope KL, Garmestani AS (2011) Adaptive management for a turbulent future. J Environ Manag 92(5):1339–1345

Argyris C (1976) Single-loop and double-loop models in research on decision making. Adm Sci Q 21(3):363–375

Argyris C (2005) Double-loop learning in organizations: a theory of action perspective. In: Smith KG, Hitt MA (eds) Great minds in management: the process of theory development. Oxford University Press, Oxford, pp 261–279

Argyris C, Schon D (1978) Organizational learning: a theory of action approach. Addision Wesley, Reading, MA

Benson MH, Garmestani AS (2011) Embracing panarchy, building resilience and integrating adaptive management through a rebirth of the National Environmental Policy Act (2011). J Environ Manag 92:1420–1427

Best M, Neuhauser D (2006) Walter A Shewhart, 1924, and the Hawthorne factory. Qual Saf Health Care 15(2):142–143

Beverton RJH, Holt SJ (1957) On the dynamics of exploited fish populations. Her Majesty's Stationery Office, London

Chaffin BC, Gosnell H, Cosens BA (2014) A decade of adaptive governance scholarship: synthesis and future directions. Ecol Soc 19(3):56

Checkland P (2000) Soft systems methodology: a thirty year retrospective. Syst Res Behav Sci 17 (S1):S11

Chen A, Abramson A, Becker N, Megdal SB (2015) A tale of two rivers: pathways for improving water management in the Jordan and Colorado River basins. J Arid Environ 112:109–123

Collier MP, Webb RH, Andrews EA (1997) Experimental flooding in the Grand Canyon. Sci Am 276:82–89

Cook C, Bakker K (2012) Water security: debating an emerging paradigm. Glob Environ Chang 22 (1):94–102. https://doi.org/10.1016/j.gloenvcha.2011.10.011

De Groot RS, Wilson MA, Boumans RM (2002) A typology for the classification, description and valuation of ecosystem functions, goods and services. Ecol Econ 41:393–408

Deming WE (1986) Out of the crisis. MIT Center for Advanced Engineering Study

Dietz T, Ostrom E, Stern PC (2003) The struggle to govern the commons. Science 302:1907–1912

Edelenbos J, van Meerkerk I (2015) Connective capacity in water governance practices: the meaning of trust and boundary spanning for integrated performance. Curr Opin Environ Sustain 12:25–29

EU (2016) Introduction to the new EU water framework directive. http://ec.europa.eu/environment/water/water-framework/info/intro_en.htm

Finlayson CM, Davidson N, Pritchard D, Randy M, MacKacy H (2011) The Ramsar convention and ecosystem-based approaches to the wise use and sustainable development of wetlands. J Int Wildl Law Policy 14:176–198

Folke C, Hahn T, Olsson P, Norberg J (2005) Adaptive governance of social- ecological systems. Annu Rev Environ Resour 30:441–473

Fontaine JJ (2011) Improving our legacy: incorporation of adaptive management into state wildlife action plans. J Environ Manag 92(5):1393–1398

Gleick P (2004) Global freshwater resources: soft-path solutions for the 21st century. Science 302:1524–1528

Gregory R, Ohlson D, Arvai J (2006) Deconstructing adaptive management: criteria for applications to environmental management. Ecol Appl 16:2411–2425

Grigg NS (2008) Integrated water resources management: balancing views and improving practice. Water Int 33:279–292

Gunderson LH, Holling CS, Light SS (1995) Barriers and bridges to the renewal of ecosystems and institutions. Columbia University Press, New York

Hahn T, Olsson P, Folke C, Johansson K (2006) Trust-building, knowledge generation and organizational innovations: the role of a bridging organization for adaptive co-management of a wetland landscape around Kristianstad, Sweden. Hum Ecol 34:573–592

Hargrove R (2002) Masterful coaching. Revised Edition. Jossey-Bass, San Francisco, CA

Holling CS (1978) Adaptive environmental assessment and management. International Institute for Applied Systems Analysis, Laxenburg

Holmes J, Savgård J (2009) The planning, management and communication of research to inform environmental policy making and regulation: an empirical study of current practices in Europe. Sci Public Policy 36(9):709–721

Hughes TP, Bellwood DR, Folke C, Steneck RS, Wilson J (2005) New paradigms for supporting the resilience of marine ecosystems. Trends Ecol Evol 20:380–386

Huntjens P, Pahl-Wostl C, Rihoux B, Schlüter M, Flachner Z, Neto S, Koskova R, Dickens C, Nabide Kiti I (2011) Adaptive water management and policy learning in a changing climate: a formal comparative analysis of eight water management regimes in Europe, Africa and Asia. Environ Policy Govern 21(3):145–163

Johnson BL (1999) The role of adaptive management as an operational approach for resource management agencies. Conserv Ecol 3(2) (online at http://www.consecol.org/vol3/iss2/art8)

Kahan DM (2008) Cultural cognition as a conception of the cultural theory of risk. In: Handbook of risk theory, Roeser S (ed) Harvard Law School Program on Risk Regulation Research Paper No. 08-20; Yale Law School, Public Law Working Paper No. 222. Available at SSRN http://ssrn.com/abstract=1123807

Kahan DM (2013) Making climate-science communication evidence-based—all the way down. In: Boykoff M, Crow D (eds) Culture, politics and climate change. Routledge, London

Konijnendijk CC (2004) Enhancing the forest science-policy interface in Europe: urban forestry showing the way. Scand J For Res 19(S4):123–128

Konrad CP, Olden JD, Lytle DA, Melis TS, Schmidt JC, Bray EN, Freeman MC, Gido KB, Hemphill NP, Kennard MJ, McMullen LE, Mims MC, Pyron M, Robinson CT, Williams JG (2011) Large-scale flow experiments for managing river systems. Bioscience 61(12):948–959

Lee KN (1999) Appraising adaptive management. Conserv Ecol 3(2):3. http://www.consecol.org/vol3/iss2/art3/

Litfin KT (1994) Ozone discourses. Columbia University Press, New York

Lytle DA, Poff NL (2004) Adaptation to natural flow regimes. Trends Ecol Evol 19(2):94–100

Magnuszewski P, Sendzimir J, Kronenberg J (2005) Conceptual modeling for adaptive environmental assessment and management in the Barycz Valley, Lower Silesia, Poland. Int J Environ Res Public Health 2(2):194–203

Medema W, McIntosh BS, Jeffrey PJ (2008) From premise to practice: a critical assessment of integrated water resources management and adaptive management approaches in the water sector. Ecol Soc 13(2):29

Melis TS, Walters CJ, Korman J (2015) Surprise and opportunity for learning in Grand Canyon: the Glen Canyon dam adaptive management program. Ecol Soc 20(3):22. https://doi.org/10.5751/ES-07621-200322

Michaels S (2009) Matching knowledge brokering strategies to environmental policy problems and settings. Environ Sci Policy 12(7):994–1011

Milly PCD, Julio B, Malin F, Robert M, Zbigniew W, Dennis P, Ronald J (2007) Stationarity is dead. Ground Water News Views 4(1):6–8

Nilsson C, Reidy CA, Dynesius M, Revenga C (2005) Fragmentation and flow regulation of the world's large river systems. Science 308:405–408. https://doi.org/10.1126/science.1107887

Olsson P, Folke C, Berkes F (2004) Adaptive co-management for building resil ience in social-ecological systems. Environ Manag 34:75–90

Pahl-Wostl C (2007) Transition towards adaptive management of water facing climate and global change. Water Resour Manag 21(1):49–62

Pahl-Wostl C, Craps M, Dewulf A, Mostert E, Tabara D, Taillieu T (2007) Social learning and water resources management. Ecol Soc 12:5

Pahl-Wostl C, Jeffrey P, Isendahl N, Brugnach M (2011) Maturing the new water management paradigm: progressing from aspiration to practice. Water Resour Manag 25:837–856

Palmer MA, Reidy-Liermann C, Nilsson C, Florke M, Alcamo J, Lake PS, Bond N (2008) Climate change and world's river basins: anticipating management options. Front Ecol Environ 6 (2):81–89. https://doi.org/10.1890/060148

Pereira ÂG, Raes F, Pedrosa TDS, Rosa P, Brodersen S, Jørgense MS, Ferreira F, Querol X, Rea J (2009) Atmospheric composition change research: time to go post-normal? Atmos Environ 43:5423–5432

Peterson GD, Cumming GS, Carpenter SR (2003) Scenario planning: a tool for conservation in an uncertain world. Conserv Biol 17:358–366

Poff NL, Allan JD, Bain MB, Karr JR, Prestegaard KL, Richter BD, Sparks RE, Stromberg JC (1997) The natural flow regime: A paradigm for river conservation and restoration. Bioscience 47(11):769–784

Pollard SR, du Toit D, Mallory S (2009) Drawing environmental water allocations into the world of realpolitik: emerging experiences on achieving compliance with policy in the lowveld rivers, South Africa. In: Paper Presented at the International Conference on Implementing Environmental Water Allocations, Port Elizabeth, South Africa, 23–26 February 2009

Richter BD, Warner AT, Meyer JL, Lutz K (2006) A collaborative and adaptive process for developing environmental flow recommendations. River Res Appl 22:297–318

Rist L, Campbell BM, Frost P (2013) Adaptive management: where are we now? Environ Conserv 40(1):5–18

Robinson CT, Uehlinger U (2008) Experimental floods cause ecosystem regime shift in a regulated river. Ecol Appl 18(2):511–526

Rood SB, Gourley CR, Ammon EM, Heki LG, Klotz JR, Morrison ML, Mos-ley D, Scoppettone GG, Swanson S, Wagner PL (2003) Flows for flood- plain forests: a successful riparian restoration. Bioscience 53:647–656

Sabatier PA (2006) Policy change and learning: An advocacy coalition approach (theoretical lenses on public policy)

Sanderson I (2002) Evaluation, policy learning and evidence-based policy making. Public Adm 80:1–22

Scheffer M (2004) Ecology of shallow lakes. Springer

Schmidt JC, Grams PE (2011) The high flows—Physical science results. Effects of Three High-Flow Experiments on the Colorado River Ecosystem Downstream from Glen Canyon Dam, Arizona. US Department of the Interior, US Geological Survey. Circular 1366:53–91

Schoeman J, Allan C, Max Finlayson C (2014) A new paradigm for water? A comparative review of integrated, adaptive and ecosystem-based water management in the Anthropocene. Int J Water Res Dev 30(3):377–390. https://doi.org/10.1080/07900627.2014.907087

Schultz L, Folke C, Österblom H, Olsson P (2015) Adaptive governance, ecosystem management, and natural capital. Proc Natl Acad Sci USA 112:7369–7374

Sendzimir J, Magnuszewski P, Balogh P, Vari A (2007) Anticipatory modelling of biocomplexity in the Tisza river basin: first steps to establish a participatory adaptive framework. Environ Model Softw 22(5):599–609

Sendzimir J, Magnuszewski P, Flachner Z, Balogh P, Molnar G, Sarvari A, Nagy Z (2008) Assessing the resilience of a river management regime: informal learning in a shadow network in the Tisza River basin. Ecol Soc 13(1): 11. http://www.ecologyandsociety.org/vol13/iss1/art11/

Senge PM (1990) The fifth discipline: the art and practice of the learning organization. Currency Doubleday, New York

Sterman J (1994) Learning in and about complex systems. Syst Dyn Rev 10(2-3):291–330

Sterman J (2000) Business dynamics: systems thinking and modeling for a complex world. Irwin/McGraw-Hill, New York

Surridge B, Harris B (2013) Science-driven integrated river basin management: a mirage? Interdiscip Sci Rev 32(3):298–312

Sverrisson A (2001) Translation networks, knowledge brokers and novelty construction: pragmatic environmentalism in Sweden. Acta Sociol 44(4):312–327

Thompson M, Ellis R, Wildavsky A (1990) Cultural theory. Westview, Boulder, CO, p 296

Thorpe RB, LeQuesne WJF, Luxford F, Collie JS, Jennings S (2015) Evaluation and management implications of uncertainty in a multispecies size-structured model of population and community responses to fishing. Methods Ecol Evol 6:49–58

Torbert WR, Cook-Greuter SR, Fisher D, Foldy E, Gauthier A, Keeley J, Rooke D, Ross SN, Royce C, Rudolph J (2004) Action inquiry: the secret of timely and transforming leadership. Berrett-Koehler, San Francisco

Toth FL (1988) Policy exercises objectives and design elements. Simul Gaming 19(3):235–255

Toth L, Melvin SL, Arrington DA, Chamberlain J (1998) Hydrologic manipulations of the channelized Kissimmee river. Bioscience 48(9):757–764

Tyre AJ, Michaels S (2011) Confronting socially generated uncertainty in adaptive management. J Environ Manag 92(5):1365–1370

van Wilgen BW, Biggs HC (2011) A critical assessment of adaptive ecosystem management in a large savanna protected area in South Africa. Biol Conserv 144(4):1179–1187

Vennix J (1999) Group model-building: tackling messy problems. Syst Dyn Rev 15:379–401

Vörösmarty CJ, McIntyre PB, Gessner MO, Dudgeon D, Prusevich A, Green P, Glidden S, Bunn SE, Sullivan CA, Reidy LC, Davies PM (2010) Global threats to human water security and river biodiversity. Nature 467:555–561

Walters C (1986) Adaptive management of renewable resources. Wiley, Hoboken, NJ

Walters C (1997) Challenges in adaptive management of riparian and coastal ecosystems. Conserv Ecol 1 (online at http://www.consecol.org/vol1/iss2/art1)

Walters C Korman J Stevens LE Gold B (2000) Ecosystem modeling for evaluation of adaptive management policies in the Grand Canyon. Conserv Ecol 4 (online at http://www.consecol.org/vol4/iss2/art1)

Warner AT, Bach LB, Hickey JT (2014) Restoring environmental flows through adaptive reservoir management: planning, science, and implementation through the sustainable rivers project. Hydrol Sci J 59(3-4):770–785

Weick KE (1995) Sensemaking in organizations, vol 3. Sage, London

Williams BK (2011) Adaptive management of natural resources-framework and issues. J Environ Manag 92(5):1346–1353

Wolf AT, Natharius JA, Danielson JJ, Ward BS, Pender JK (1999) International river basins of the world. Water Resour Dev 15(4):387–427

World Water Development Report (2009) The 3rd United Nations World Water Development Report: Water in a Changing World (WWDR-3)

Yazdanpanah M, Hayati D, Thompson M, Zamani GH, Monfared N (2014) Policy and plural responsiveness: taking constructive account of the ways in which Iranian farmers think about and behave in relation to water. J Hydrol 514:347–357

Chapter 17
Legislative Framework for River Ecosystem Management on International and European Level

Edith Hödl

17.1 Introduction

Modern water legislation strives to build the institutional foundation for sustainable water resource management and protection of important habitats and species. This chapter provides an overview of the most important legislative framework for river ecosystem management on the international and European levels, profoundly influencing and guiding national water legislation.

At international levels, the most prominent legal acts for improving freshwater governance and fostering the equitable and sustainable sharing of transboundary watercourses are the UNECE (United Nations Economic Commission for Europe) Convention on the Protection and Use of Transboundary Watercourses and International Lakes as well as the UN (United Nations) Convention on the Law of the Non-Navigational Uses of International Watercourses. Due to its importance for the conservation and use of wetlands, especially as waterfowl habitat, the Convention on Wetlands of International Importance is also described.

The most relevant legislation for river ecosystem management in the European Union is the Water Framework Directive (WFD). It establishes a European Union-wide basis for integrated water resources management based on a river basin management approach. This chapter explains how the Directive—legally binding for all EU Member States—serves as an umbrella for measures related to drinking and bathing water, urban wastewater treatment, groundwater protection, floods, and protection of waters against the impacts of agricultural pressures such as from nitrates in organic and chemical fertilizers and/or industrial emissions. Additionally, the WFD ensures a strong linkage with the provisions of the Birds and Habitats

E. Hödl (✉)
Institute of Hydrobiology and Aquatic Ecosystem Management, University of Natural Resources and Life Sciences, Vienna, Austria
e-mail: edith.hoedl@boku.ac.at

© The Author(s) 2018
S. Schmutz, J. Sendzimir (eds.), *Riverine Ecosystem Management*, Aquatic Ecology Series 8, https://doi.org/10.1007/978-3-319-73250-3_17

Directives. The latter aims to protect Europe's most valuable species and aquatic ecosystems by finding a good balance between water and nature protection and the sustainable use of water resources.

Since the adoption of the international water conventions as well as the WFD, many positive achievements have occurred. However, several challenges and short-comings remain in implementing the relevant legal provisions for the sustainable management of rivers and aquatic ecosystems. Selected issues are highlighted in this chapter by describing how legislation can address and adapt to water management issues in the twenty-first century.

17.2   International Law

International law is defined as the the set of principles and rules generally regarded and accepted as binding for those entities subject to it, usually states, but also the United Nations and the European Union, in their conduct with each other (Article 38 Statute of the International Court of Justice). The development of international law is one of the primary goals of the United Nations (United Nations 2016a). International water conventions and multilateral and bilateral agreements are the major sources of international water law, a subset of international environmental law.

The legal form of conventions and agreements is of binding nature on the contracting parties establishing respective rights and obligations governing their relations. As a general rule, a treaty applies only to those states that have expressed their consent to be bound by it. This is an issue of state sovereignty. Depending on the number of parties involved, treaties may be bilateral (two state parties) or multilateral (more than two state parties), either with limited (open for signature by a restricted number of countries) or with universal (open for participation by all states) participation. To become party to an international convention, a state must express, through a concrete act, its willingness to undertake the legal rights and obligations contained in the treaty. This is usually accomplished through signature and ratification of the treaty, or if it is already in force, by accession to it (United Nations 2016a).

More than 3600 international agreements, both bilateral and multilateral, exist which address water-related issues in transboundary rivers, lakes, and seas, focusing on a particular region, a river basin or a part of one (Vinogradov et al. 2003; United Nations 2016b). Although the vast majority of these agreements signed from AD 805 to 1984 relate to navigational issues, boundary delineation, or fisheries-related matters, a growing number focus on water management issues addressing water as a limited and consumable resource (United Nations Food and Agriculture Organization 1978, 1984; United Nations Environment Programme 2002).

The most important international general framework water conventions are the UNECE Convention on the Protection and Use of Transboundary Watercourses and International Lakes (United Nations 1992) and the UN Convention on the Law of the Non-Navigational Uses of International Watercourses (United Nations 1997). Both conventions include, on the one side, substantive elements determining rules for the use of waters and, on the other side, procedural provisions defining certain requirements for the management of transboundary watercourses.

17.2.1   International Water Conventions

The UNECE Convention on the Protection and Use of Transboundary Watercourses and International Lakes, adopted in Helsinki in 1992 and entered into force in 1996, aims at protecting and ensuring the quantity, quality, and sustainable use of transboundary water resources by facilitating cooperation among the riparian countries. It strengthens measures for the ecologically sound management and protection of transboundary surface waters and groundwater. The three central obligations of the Convention are the prevention, control, and reduction of transboundary impacts, the reasonable and equitable use of waters, and the cooperation of riparian countries through agreements and joint bodies (United Nations Economic Commission for Europe 2011). The Convention counts 41 contracting parties as of May 2016 (United Nations 2016c). Almost all countries sharing transboundary waters in the region of the UNECE are Parties to the Convention (Fig. 17.1).

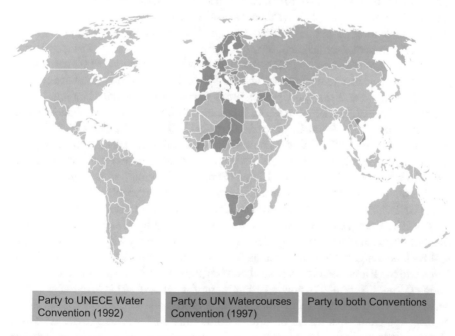

Party to UNECE Water Convention (1992)   Party to UN Watercourses Convention (1997)   Party to both Conventions

Fig. 17.1 Contracting parties to UNECE Convention on the Protection and Use of Transboundary Watercourses and International Lakes and UN Convention on the Law of the Non-Navigational Uses of International Watercourses (as of May 2016, United Nations 2016d)

An amendment of the UNECE Water Convention from the year 2003, which entered into force in 2013, allows all UN member states to become party to it (Trombitcaia and Koeppel 2015; United Nations Economic Commission for Europe 2016b). By the end of 2014, more than 50 states from outside the UNECE region have participated in meetings and activities of the UNECE Water Convention (Trombitcaia and Koeppel 2015).

The UN Convention on the Law of the Non-Navigational Uses of International Watercourses, adopted in New York in 1997 and entered into force in 2014, forms a global legal framework establishing basic standards and rules for cooperation on the use, management, and protection of international watercourses. The Convention counts 36 contracting parties as of May 2016 (United Nations 2016d).

Both Conventions play a role in improving freshwater governance and support interstate cooperation by fostering the equitable and sustainable sharing of transboundary watercourses across the world. The two instruments vary in their scope, substantive and procedural rules, as well as institutional and dispute settlement mechanisms (Wouters and Vinogradov 2003; Rahaman 2009; McCaffrey 2015; Tanzi 2015; United Nations 2016e). While the UN Water Convention is primarily providing some basic principles and is addressed to the global community of watercourse states, the UNECE Water Convention is stricter and includes more detailed provisions with respect to the prevention of transboundary impacts and the mandatory character of institutional cooperation between riparian states (Wouters and Vinogradov 2003; Tanzi 2015). As the differences between the two conventions are not of a conflicting legal character, but an issue of stringency and detail, the UNECE Water Convention is not invalidated through derogation by the later UN Water Convention under the principle of *lex posterior derogat priori* (Tanzi 2013, 2015).

It is questionable as to whether the UN Water Convention will ever find its prominent place in international water law. While the UN Water Convention is in force since the year 2014, the UNECE Water Convention is legally binding, institutionally developed, and supported with funding as leading instrument of international water law since 1996 (Rieu-Clarke and López 2013; Stoa 2014). The UNECE secretariat servicing the UNECE Water Convention is located in Geneva (Switzerland), and working groups on Integrated Water Resources Management and on Monitoring and Assessment are established (United Nations Economic Commission for Europe 2016a). The importance for its global relevance was confirmed in the year 2013 by the opening of the UNECE Water Convention for any UN member state to accede to it.

The UNECE Water Convention has already significantly influenced the management of transboundary waters by forming the basis for many bilateral and multilateral basin-specific agreements such as for the Danube, the Meuse and Scheldt, the Oder, and the Rhine river basin as well as many transboundary waters in the UNECE region (United Nations Economic Commission for Europe 2012; Moynihan 2015). Most of these agreements either refer specifically to the UNECE Water Convention as their "parental" instrument or adopt terminology and various provisions of the Convention's general rules as well as legal principles guiding actions of national public authorities and individuals (Wouters and Vinogradov 2003; Keessen et al. 2008; Tanzi et al. 2015).

The primary *substantive rules* are the governing rules of "equitable and reasonable utilization" of water resources and the "obligation to take all measures necessary not to cause significant harm" to other basin states or to their environment.

Principle of Equitable and Reasonable Utilization

This principle entitles each basin state to a reasonable and equitable share of water resources for the beneficial uses within its own territory. It rests on the principle of shared sovereignty and equality of rights. "Equitable" utilization does not necessarily mean an equal portion of the resource or equal share of uses and benefits. The UNECE Water Convention includes this principle in Article 2.2c; the UN Water Convention stipulates the equitable and reasonable utilization of an international watercourse in Articles 5 and 6. Compared to the UNECE Water Convention, the UN Water Convention provides more guidance on the factors for assessing the equitable and reasonable utilization of an international watercourse (Tanzi 2015).

Obligation Not to Cause Significant Harm

No state in an international drainage basin is allowed to use the watercourses in their territory in a way that would cause significant harm to other basin states or to their environment, including harm to human health or safety, to the use of the waters for beneficial purposes, or to the living organisms of the watercourse systems. This principle is specified in Articles 2 and 3 of the UNECE Water Convention, and included in many Articles of the UN Water Convention, but is particularly addressed in Article 7. More precise and coordinated guidelines and advanced standards of conduct on the prevention of transboundary impact are included in the UNECE Water Convention (Tanzi 2015).

To properly realize the substantive rule of equitable and reasonable utilization and the obligation not to cause significant harm, certain *procedural rules* are necessary, including the prior notification of planned measures, consultation and negotiation, and the principles of cooperation and exchange of information.

Principles of Notification, Consultation, and Negotiation

Every riparian state along an international watercourse is entitled to prior notice, consultation, and negotiation in cases where the proposed use by another riparian of a shared watercourse may cause serious harm to its rights or interest. The UNECE Water Convention regulates the obligation for consultation among riparian countries in Article 10; the UN Water Convention includes provisions as regards notification concerning planned measures, consultation, and negotiation predominantly in Articles 11 to 19.

Principles of Cooperation and Information Exchange

It is a responsibility for each riparian state along an international watercourse to cooperate and exchange data and information regarding the state of the watercourse as well as present and future planned uses along the watercourse. The need for information exchange is specifically addressed in Articles 6 and 13 and the principle of cooperation in Article 9 of the UNECE Water Convention; the general obligation to cooperate and the regular exchange of data and information are specifically enshrined in Articles 8 and 9 of the UN Water Convention.

Institutional mechanisms, including provisions for the *institutional management* of transboundary waters and *dispute settlement procedures*, complete the general provisions of international water legislation. Transboundary water controversies and disputes are often resolved under the auspices of various international organizations and bodies, such as river basin commissions established by multilateral or bilateral river basin agreements (Wouters and Vinogradov 2003; Tanzi and Contartese 2015).

Institutional Management Procedures
Under the UN Water Convention, states are encouraged to create institutional management mechanisms, but are not obliged to do so (Article 24). On the contrary, Article 9 of the UNECE Water Convention stipulates that institutional cooperation between riparian states is mandatory. The riparian parties shall enter into bilateral or multilateral agreements in order to define their mutual relations and conduct regarding the prevention, control, and reduction of transboundary impacts.

Dispute Settlement Procedures
All states in an international watercourse should seek a settlement of any disputes by peaceful means, in case states concerned cannot reach agreement by negotiation. The UNECE Water Convention regulates the settlement of disputes in Article 22 and Annex IV. Article 33 of the UN Water Convention specifies the procedures in the event of a dispute between two or more riparian parties of the Convention.

17.2.2   Convention on Wetlands of International Importance

The conservation and use of wetlands is regulated in the Convention on Wetlands of International Importance, especially as waterfowl habitat (United Nations 1971). The Convention was adopted in Ramsar in 1971 and entered into force in 1975; as of May 2016, the Convention counts 169 contracting parties (Ramsar Convention Secretariat 2016).

The Secretariat coordinating the Convention's activities is based at the headquarters of the International Union for Conservation of Nature in Gland (Switzerland). The Convention's mission is the conservation and wise use of wetlands through local and national actions and international cooperation. The contracting parties commit to work toward the wise use of all their wetlands, designate wetlands of international importance, and ensure their effective management as well as cooperate internationally on transboundary wetlands (Ramsar Convention Secretariat 2016).

Wetlands are defined as any land area that is saturated or flooded with water, either seasonally or permanently (Article 1 of the Convention; Ramsar Convention Secretariat 2016). Inland wetlands include aquifers, lakes, rivers, streams, marshes, peatlands, ponds, floodplains, and swamps. Coastal wetlands include all coastlines, mangroves, saltmarshes, estuaries, lagoons, sea grass meadows, and coral reefs. As of 2015, there are over 2100 designated Ramsar sites covering more than 208 million hectares worldwide (Ramsar Convention Secretariat 2016).

Through its recognition of the fundamental ecological functions of wetlands as regulators of water regimes, the Convention plays a crucial role in integrated water resources management (De Chazournes et al. 2015). The protection of wetland ecosystems and their benefits is therefore essential to ensuring the sustainable utilization of water resources (Ramsar Convention Secretariat 2010).

17.3 European Union Law

The three sources of European Union law are primary law (treaties establishing the European Union), secondary law (such as regulations and directives based on the treaties), and supplementary law, including case law by the European Court of Justice, international treaties, and general principles of European Union law. The entirety of European law is called *acquis communautaire* which has been agreed upon in the community of the European Union (European Commission 2016a). The collection of all European environmental law is called environmental acquis communautaire. European water law is a subset of the environmental acquis communautaire.

While the European Commission has the monopoly on proposing legislation, the European Parliament and the Council of the European Union representing all EU-28 Member States have the role of adopting European law. All EU Member States must obey the acquis, and all candidate countries must accept the acquis to become a member of the European Union; it is translated into the official languages of the European Union. The European Commission is the "guardian of the treaties" having the power to oversee a proper compliance and implementation review system in case EU Member States do not properly apply European law (European Commission 2016b).

The Directive is the most commonly used legal act for environmental law and the major source of European water legislation. It regulates legally binding objectives, but leaves it to the discretion of EU Member States as to how to implement the objectives in national legislation. When an EU Member State fails to implement a Directive, the European Commission may take legal action by launching an infringement procedure and referring the case to the European Court of Justice for a ruling.

European legislation in water management has substantially changed since the early 1970s. From the protection of specific uses of waters (fish and shellfish waters) and sectors (drinking and bathing water) to addressing water pollution at the source (urban wastewater, nitrates from agriculture), it has developed into a system of integrated water resources management in Europe. The most important Directive with regard to river ecosystem management is the WFD, fostering a holistic approach for the protection of all waters and serving as an umbrella for measures related to drinking and bathing water, urban wastewater treatment, groundwater protection, floods, and protection of waters against agricultural pressures or industrial emissions (European Commission 2016c).

17.3.1   Water Framework Directive

The WFD (European Parliament and Council 2000), adopted in the year 2000, establishes a governance framework for all waters in the European Union, which prevents further deterioration and protects and enhances the status of aquatic ecosystems, promotes sustainable water use, aims at enhanced protection and improvement of the aquatic environment, ensures the progressive reduction of pollution of groundwater and prevents its further pollution, and contributes to mitigating the effects of floods and droughts by applying a river basin approach. The Directive is substantially influenced by the guiding principles of European environmental law, namely, the precautionary principle as well as the principles that preventive action should be taken, that environmental damage should as a priority be rectified at the source and that the polluter should pay (Article 191(2) Treaty of the Functioning of the European Union). The protection of all waters in one single piece of framework legislation, the need for achievement of a "good status" for all waters as a legally binding rule, and water management based on a river basin approach have a strong impact on the management of European waters and thereby paved the way for modern water legislation.

The following paragraphs outline the *main legislative provisions* of the WFD by describing the scope of the Directive, the river basin management approach, environmental objectives for all waters and protected areas by a set deadline, the use of economic instruments, the need to establish a Program of Measures and a River Basin Management Plan coordinated with the Flood Risk Management Plan, the implementation cycle and regular reporting, as well as public information and consultation measures.

Expanding the Scope of Water Protection to All Waters
Article 1 defines the purpose of the Directive by expanding the scope to inland surface waters, transitional waters, coastal waters, and groundwater.

Water Management Based on River Basins
Coordination of administrative arrangements shall be made within districts whose outer boundaries are defined by the perimeters of catchments within the river basin in question, i.e., river basin districts. These are natural geographical and hydrological units as opposed to previous definitions based on political boundaries. Article 3 includes detailed provisions as to how EU Member States need to ensure coordination and cooperation in shared river basins across administrative and political borders.

Environmental Objectives Including "Good Status" for All Waters by a Set Deadline
The environmental objectives for surface waters include the prevention of deterioration of status, the need to achieve "good status" by a set deadline and to progressively reduce pollution from priority substances, and the need to cease or phase out emissions, discharges, and losses of priority hazardous substances. As for groundwater, the environmental objectives comprise the prevention or limitation of pollutants leaching into groundwater, the prevention of deterioration of the status, the need

to achieve "good status" by a set deadline, as well as measures to reverse any significant and sustained upward trend in the concentration of any pollutant.

The need to achieve "good status" for all surface waters and groundwaters by set deadlines is one of the most essential rules enshrined in Article 4 of the Directive. The class "good status" is defined in terms of ecology (biology and morphology) and chemistry for surface waters as well as chemistry and quantity for groundwater. By using an intercalibration method to classify results for biological quality elements (five classes), type-specific reference conditions for the boundary between the classes of "high" and "good" status and for the boundary between "good" and "moderate" status, which are consistent with the normative definitions of those class boundaries given in Annex V of the Directive, were established (European Commission 2011a). This process aimed to ensure consistency and comparability in the assessment of good ecological status, representing the same level of ecological quality for aquatic ecosystems everywhere in Europe. The "one out, all out" approach defines the final status of water bodies, i.e. if an individual quality element is not achieving good status for a particular watercourse, then the entire water body is classified as failing. In other words, the worst status of the decisive elements (ecology and chemistry for surface waters, chemistry and quantity for groundwater) used in the assessment of a water body determines its final status.

In cases of surface water bodies that have been heavily modified in their physical structure to serve various uses, including navigation, flood protection, hydropower, and agriculture, a less stringent objective must be reached, as it might not be viable or desirable from a socioeconomic perspective to abandon such uses and to remove the physical modifications that affect the water bodies in order to achieve "good status."

Achieving Compliance with Objectives in Protected Areas by a Set Deadline
According to Article 4 and Annex IV of the Directive, EU Member States shall achieve compliance with any standard and objectives in protected areas by the year 2015, unless otherwise specified in Community legislation. This provision refers to areas designated for the abstraction of drinking water and for the protection of economically significant aquatic species, bathing waters, nutrient-sensitive areas [vulnerable zones under the Nitrates Directive (European Council 1991a) and sensitive areas under the Urban Waste Water Treatment Directive (European Council 1991b)], as well as areas designated for the protection of habitats or species under the Birds and Habitats Directives (European Council 1992; European Parliament and Council 2009).

Economic Instruments and Adequate Water Pricing for Water Services
Article 9 integrates economic instruments and the promotion for a prudent use of water resources into European water management. The introduction of water pricing is certainly one of the Directives' most important and controversial innovations. Adequate water pricing acts as an incentive for the sustainable use of water resources and thus helps to achieve the environmental objectives under the Directive. EU Member States are required to ensure that the price charged to water consumers for water services—such as for the abstraction and distribution of freshwater and the

collection and treatment of urban wastewater—reflects the true costs by including environmental and resource costs, taking the "polluter pays" principle into account.

According to the interpretation of the European Commission, the concept of water services does not only refer to drinking water and wastewater treatment but also to water abstraction for the cooling of industrial installations and for irrigation in agriculture, the use of surface waters for navigation purposes, flood protection or hydropower production, and wells drilled for agricultural, industrial, or private consumption. However, a recent decision of the Court of Justice of the European Union (judgment of 11 September 2014, case C-525/12, Commission v Germany) decided that EU Member States may, subject to certain conditions, opt not to proceed with the recovery of costs for a given water-use activity, where this does not compromise the purposes and the achievement of the objectives of Article 9 of the Directive (Court of Justice of the European Union 2014).

Program of Measures

In order to achieve the environmental objectives of Article 4, EU Member States shall ensure the establishment of a Program of Measures according to Article 11. Each Program of Measures shall include basic measures (Annex VI Part A) and, where necessary, supplementary measures (Annex VI Part B) going beyond legally required basic measures. Basic measures refer to European water legislation as a whole and relate to drinking and bathing water, urban wastewater treatment, groundwater protection, floods, and protection of waters against agricultural pressures or industrial emissions. Measures follow the combined approach (Article 10) for point and diffuse sources by including emission limit values and quality standard measures to achieve the quality objectives. Supplementary measures may, inter alia, include national legislative, administrative, or economic instruments, codes of good practice, and educational or research projects going beyond the legal requirements of basic measures.

River Basin Management Plans Coordinated with Flood Risk Management Plans and Strategic Environmental Assessment

Due to its strong interlinkages, Article 9 of the Floods Directive (European Parliament and Council 2007) explicitly states that EU Member States shall take appropriate steps to coordinate the implementation of the Floods Directive and WFD focusing on opportunities for improving efficiency and information exchange and for achieving common synergies and benefits with respect to the environmental objectives of the WFD (European Commission 2014). Consequently, according to the WFD, River Basin Management Plans need to be closely coordinated with Flood Risk Management Plans required by the Floods Directive. Flood Risk Management Plans highlight the hazards and risks of flooding from rivers, the sea, surface water, groundwater, and reservoirs and address all aspects of flood risk management. They focus on prevention, protection, and preparedness, including flood forecasts and early warning systems and taking into account the characteristics of the particular river basin. Information about how Flood Risk Management Plans will take into account environmental objectives according to the WFD and how measures under the River Basin Management Plan and the Flood Risk Management Plan will be coordinated shall be provided in the plans.

According to the European Strategic Environmental Assessment Directive (European Parliament and Council 2001), plans and programs for water management according to the WFD and the Floods Directive setting the framework for future development consent of projects listed in the European Environmental Impact Assessment Directive (European Parliament and Council 2011) are subject to a mandatory environmental assessment. This includes an environmental report in which the likely significant effects on the environment and the reasonable alternatives of the proposed plan or program are identified. The public and the environmental authorities need to be informed and consulted; in case of plans and programs which are likely to have significant effects on the environment in another EU Member State, the plan or program must be implemented in consultation with the other EU Member State(s).

Implementation Cycle and Regular Reporting
The WFD requires a clearly set 6-year implementation cycle, starting with a report about the analysis of the characteristics of the river basin district(s), including a pressure and impact analysis (Article 5), defining a Program of Measures to achieve the environmental objectives (Article 11), and a reported summary of all elements in the River Basin Management Plan (Article 13). The latter sets the scene for river basin management planning in the 6-year implementation period.

The milestones in the implementation of the WFD since the year 2000 included a first analysis of the river basin district(s) by 2004, the establishment of monitoring programs by 2006, a Program of Measures based on the analysis and the findings of the monitoring measures, and a first River Basin Management Plan by 2009. A review and update was made in the River Basin Management Plan 2015 and will be made every 6 years thereafter. Plans and programs need to be reported to the European Commission, which has the objective of regularly publishing a report on the implementation of this Directive in EU-28 Member States (Article 18).

The first and second implementation report published by the European Commission included information about the first stage of implementation in the year 2007 and the monitoring networks in 2009. The third implementation report providing results of the first River Basin Management Plans was issued by the European Commission in 2012 (European Commission 2012a) and will be published every 6 years thereafter. Additionally, within 3 years of the publication of the implementation report, the European Commission regularly publishes an interim report describing progress in implementation of the Directive; the fourth implementation report includes a review of progress in the implementation of the Programs of Measures planned by EU Member States in their second River Basin Management Plans (European Commission 2015a).

Public Information and Consultation
In order to better involve citizens in water management issues in Europe, Article 14 stipulates an active involvement of all interested parties in the implementation of this Directive. In particular, the production, review, and updating of the River Basin Management Plans have to be subject to public participation; a timetable and work program for the production of the plan need to be made available

for comments 3 years, an overview of the significant water management issues 2 years, and a draft of the River Basin Management Plan 1 year before the final publication of the plan; interested parties have 6 months to comment on these documents (Article 14). This process aims at ensuring that the balancing of diverging interests in the different stages of implementing the Directive is fully taken into consideration and, furthermore, to ensure that the different plans, programs, and measures are subsequently effectively put into operation.

17.3.2 Birds and Habitats Directives

The Birds Directive (European Parliament and Council 2009), adopted in 1979 and amended in 2009, relates to the conservation of all species of naturally occurring birds in the wild state and covers the protection, management, and control of these species. The Habitats Directive (European Council 1992), adopted in 1992, contributes toward ensuring biodiversity through the conservation of natural habitats and of wild fauna and flora in the European territory, taking account of economic, social, cultural, and regional requirements. Both Directives form the cornerstone of Europe's nature conservation policy and establishes the EU-wide Natura 2000 ecological network of protected areas, safeguarded against potentially damaging developments (European Commission 2016d).

The Birds and Habitats Directives and the WFD aim to protect healthy aquatic ecosystems while at the same time ensuring a balance between water/nature protection and the sustainable use of nature's natural resources. Many synergies are possible, since the implementation of measures under the WFD will generally benefit the objectives of the Birds and Habitats Directives (European Commission 2011b). A strong coordination between water and nature conservation authorities is necessary in order to coordinate and streamline implementing measures (Janauer et al. 2015).

A direct linkage between the WFD and the Birds and Habitats Directive is established in Articles 1, 4, and 11 (accompanied by Annex IV and Annex VI) of the WFD. The purpose of the Directive is, inter alia, to prevent further deterioration and protect and enhance the status of aquatic ecosystems and, with regard to their water needs, terrestrial ecosystems and wetlands directly depending on the aquatic ecosystems. The environmental objectives enshrined in Article 4 and Annex IV require the protection of habitats or species where the maintenance or improvement of the status of water is an important factor in their protection, including relevant Natura 2000 sites, the network of areas designated according to Birds and Habitats Directives. In order to achieve the environmental objectives, the Program of Measures according to Article 11 and Annex VI Part A stipulates the need to apply measures required under the Birds and Habitats Directive.

17.4   Challenges for the Future: How Can Modern Water Legislation Address and Adapt to It?

The multidimensional regime of modern water legislation is composed of international water law, water law of the European Union, and national water legislation guided by the concept of integrated water resources management (Reichert 2016). Awareness is increasing at the international and European levels that freshwater resources are limited and need to be legally protected both in terms of quantity and quality.

European water law has introduced clear and legally binding objectives, a common terminology, and managerial framework as well as provides for robust enforcement and compliance mechanisms under judicial review, including financial sanctions for non-compliance. This is the reason why European water law has, to a growing extent, substantial influence on transboundary water cooperation that extends beyond the European territory (Reichert 2016). The adoption of the WFD was considered as a major development affecting the implementation of the UNECE Water Convention (Wouters and Vinogradov 2003).

However, aquatic environments continue to face pressure, often suffering from pollution, over-abstraction, morphological alterations, loss of biodiversity, floods, and droughts. The main challenges on European level, analyzed on the basis of the results of the first WFD cycle covering the implementation period from 2009 to 2015, showed deficits in the identified measures to achieve the environmental objectives, in water pricing and cost recovery of water uses, over-abstraction of water, the change in flow and physical shape of water bodies, and ongoing pollution caused by agriculture, industry, and households (European Commission 2015a). The good ecological status of water bodies improved from 43% in the year 2009 to 53% in the year 2015 (European Commission 2015b). Further progress is needed to implement and achieve the objectives of the WFD and related water-specific legislation.

17.4.1   Is European Water Law Fit for Future Challenges?

In recent years, European freshwater policy and legislation has undergone comprehensive evaluations designed to ascertain whether the regulatory framework is fit for purpose. The so-called Fitness Check looked at relevance, effectiveness, efficiency, and coherence of water policy and legislation, in order to identify any excessive regulatory burden, overlaps, inconsistencies, and/or obsolete measures in place or gaps in the legislative framework (European Commission 2012b, c, 2016e). The results of the assessment formed a substantial basis and present a building block of the Blueprint to Safeguard Europe's Water Resources, thereby defining a new EU strategy on the use of water resources by outlining clear actions for the upcoming years (European Commission 2012b).

The WFD (European Parliament and Council 2000), the Groundwater Directive (European Parliament and Council 2006), the Directive on environmental quality standards in the field of water policy (European Parliament and Council 2008), the Nitrates Directive (European Council 1991a), the Urban Waste Water Treatment Directive (European Council 1991b), and the Floods Directive (European Parliament and Council 2007) were examined in the frame of the Fitness Check. Additionally, quantitative and adaptive water management issues—water scarcity and drought as well as climate change adaption—for which there is currently no legislation at European Union level were included in the evaluation. In the same vein, the Birds and Habitats Directives are currently being evaluated to ensure that they are "fit for purpose" (European Commission 2016f).

Relevance
The Fitness Check of European freshwater policy and legislation has confirmed the relevance of the current water policy framework to address the key challenges faced by European freshwaters, namely, water quality, water scarcity, droughts, and floods as well as significant pressures from the discharge of pollutants, hydro-morphological alterations, and water abstraction. A need has been identified to focus more on water efficiency (water demand management and water availability) and on integrating quantity and quality aspects of water management (European Commission 2012c).

Effectiveness
The effectiveness of most of the legislation is slowed down by the fact that there have been considerable delays in implementation at the national level. However, strong enforcement and compliance mechanisms, including infringement procedures, have prompted action and sped up implementation. An efficient use of EU funding could substantially support implementation, in particular, as regards implementation measures in the field of urban wastewater and by preventing nitrates from agricultural sources polluting groundwater and surface waters. A significant number of EU Member States have relied on exemptions from the environmental objective to reach "good status" by a certain deadline. A thorough check must be made, to ensure that the conditions under which the exemptions were granted are being fulfilled (European Commission 2012c).

Efficiency
The WFD has brought effective improvements in coordination between administrations within and between EU Member States and also with third countries. Due to the current economic situation in most of EU Member States, there is uncertainty about possible gaps in financing water policy. The efficient implementation of EU water policy requires both financial and administrative resources. Assessments showed that many EU Member States lack a consistent methodology for assessing and monetizing the costs and associated benefits of implementation measures that support cost-effectiveness (European Commission 2012c). Thus, it is important to measure and monitor the balance between the costs and benefits of such policies, and more effort needs to be made to develop and apply methodologies that produce reliable and comparable data to underpin the cost-effectiveness of measures.

Coherence

By streamlining and simplifying EU water legislation, the WFD has eliminated potential double requirements in the field of water legislation and considerably reduced the risk of contradiction between different legal instruments. The Fitness Check has identified areas where synergies between water and other environment (related) policies water should be better aligned (European Commission 2012c). It has also underlined the need for improvements in relation to integration of water policy objectives and other policy areas (agriculture, energy, hydropower, navigation, nature protection). In this respect the potential of the River Basin Management Plans as an integration tool needs to be further strengthened. Conflicting interests, such as between water protection and hydropower development, might be addressed in the plans by setting the scene for a cycle of 6 years and defining guiding principles and a framework of how to weigh environmental improvements up against the social benefit of hydropower production.

17.4.2 Review of European Water Legislation

The WFD will be reviewed by the year 2019 to see whether proposals for amendments are necessary to facilitate the achievement of its legally binding objectives; the legislative basis for review is set in Article 19.2 of the WFD. Additionally, legislation regarding urban wastewater treatment and drinking water is currently being looked at. Based on the findings and conclusions as well as defined actions included in the Blueprint to Safeguard Europe's Water Resources (European Commission 2012b) and the Fitness Check of European freshwater policy and legislation (European Commission 2012c), evaluations in the frame of the required review will further assess the adequacy of current legislation and its implementation.

The issues of the different level and quality of implementation, the significant use of exemptions, as well as economic and social impacts of the implementation are topics to be discussed in the reviewing process and are described in the following paragraphs.

Different Level and Quality of Implementation

Due to the flexibility in how to achieve the obligations and objectives set in the Directive, the level and quality of national implementation of the legal framework is often different in EU Member States. Although the need for flexibility due to the regional variations and environmental differences in circumstances is recognized, the risk that unambitious national practices will lead to a lack of practical effectiveness needs to be avoided (Keessen et al. 2010). This is particularly relevant as such divergences in implementation between EU Member States can hamper cooperation in transboundary river basins (European Commission 2012c). Considerable differences were found in the implementation of water-related legislation (nitrates, urban wastewater), in governance structures, in quantitative aspects of water management, and in putting in place measures that target abstractions and flow regulations as well as water pricing (European Commission 2012c; European Commission 2015a).

Environmental Objectives and Significant Use of Exemptions

Based on the findings of the first River Basin Management cycle, it becomes clear that the water environment is still under great pressure from economic activities, urban and demographic developments, as well as climate change (European Commission 2015a). The significant number of exemptions indicates that the good status may not be completed for many years. None of the legislation under consideration has achieved all the objectives it was expected to have achieved at this stage (European Commission 2012c, 2015a). A range of exemptions without appropriate justification occurred because many EU Member States have often only estimated how far existing measures will contribute to the achievement of environmental objectives (European Commission 2015a). More efforts and actions are certainly needed in the implementation process.

On the other side, it is also necessary to look at issues that might hamper better implementation results due to policy reasons. The "one out, all out" approach in defining the status of water bodies is criticized by several EU Member States (European Commission 2015b). Although being consistent with the precautionary principle, it led to the fact that improvements in the status (of one quality element) are somehow hidden in a negative picture that is insensitive in showing progress in achieving the objectives of the Directive. Therefore, additional indicators (or a separate presentation of results to achieve the environmental objectives) are needed to support the communication with stakeholders in order to better acknowledge the measures that have already been implemented and to promote the best and most efficient use of investment for environmental outcomes (European Commission 2015b).

Socioeconomic Components of Implementation

Economic circumstances and limited available budget often impact implementation, especially where costs might be unfeasible for several European States. In this regard, discussion continues as to how to modify the process beyond the year 2027 (end of third implementation cycle) to better account for different economic circumstances and technological developments in the future. There is certainly a need for innovative and fortified solutions in terms of technologies, management approaches, governance, and funding to better support successful implementation of the legislative framework.

17.5 Conclusions

The legislative framework for river ecosystem management on international and European level is guided by an integrated water resources management approach that substantially influences national water legislation. The main principles enshrined in the international water conventions and relevant legally binding provisions of the European WFD have triggered new thinking in water management that crosses administrative boundaries, establishes a cyclical planning process, and paves the way for modern water legislation. Evaluations in the past years concluded that the current water policy and legislative framework addresses the key challenges faced by European freshwaters. However, there might be a need for review and adaption in order to better address water management issues in the twenty-first century. And finally it is a matter of political will and stakeholder motivation and commitment to further foster successful implementation of the legislative framework for river ecosystem management in the years to come.

References

Sources of Law

Court of Justice of the European Union (2014) Judgment of the Court (Second Chamber) of 11 September 2014, case C-525/12, Commission v Germany. Available at http://curia.europa. eu/juris/document/document.jsf?text=&docid=157518&pageIndex=0&doclang=EN& mode=lst&dir=&occ=first&part=1&cid=40121. Accessed 15 June 2016

European Council (1991a) Council Directive of 12 December 1991 concerning the protection of waters against pollution caused by nitrates from agricultural sources (91/676/EEC). Available at http://eur-lex.europa.eu/legal-content/EN/TXT/PDF/?uri=CELEX:31991L0676&from=EN. Accessed 15 June 2016

European Council (1991b) Council Directive of 21 May 1991 concerning urban waste water treatment (91/271/EEC). Available at http://eur-lex.europa.eu/legal-content/EN/TXT/PDF/? uri=CELEX:31991L0271&from=EN. Accessed 15 June 2016

European Council (1992) Council Directive 92/43/EEC of 21 May 1992 on the conservation of natural habitats and of wild fauna and flora. Available at http://eur-lex.europa.eu/legal-content/ EN/TXT/?uri=CELEX:31992L0043. Accessed 15 June 2016

European Parliament and Council (2000) Directive 2000/60/EC of the European Parliament and of the Council of 23 October 2000 establishing a framework for Community action in the field of water policy. Available at http://eur-lex.europa.eu/legal-content/EN/TXT/?uri=CELEX:32000L0060. Accessed 15 June 2016

European Parliament and Council (2001) Directive 2001/42/EC of the European Parliament and of the Council of 27 June 2001 on the assessment of the effects of certain plans and programmes on the environment. Available at http://eur-lex.europa.eu/legal-content/EN/TXT/?uri=CELEX:32001L0042. Accessed 15 June 2016

European Parliament and Council (2006) Directive 2006/118/EC of the European Parliament and of the Council of 12 December 2006 on the protection of groundwater against pollution and deterioration. Available at http://eur-lex.europa.eu/legal-content/EN/TXT/PDF/?uri=CELEX: 02006L0118-20140711&from=EN. Accessed 15 June 2016

European Parliament and Council (2007) Directive 2007/60/EC of the European Parliament and of the Council of 23 October 2007 on the assessment and management of flood risks. Available at http://eur-lex.europa.eu/legal-content/EN/TXT/PDF/?uri=CELEX:32007L0060&from=EN. Accessed 15 June 2016

European Parliament and Council (2008) Directive 2008/105/EC of the European Parliament and of the Council of 16 December 2008 on environmental quality standards in the field of water policy. Available at http://eur-lex.europa.eu/legal-content/EN/TXT/PDF/?uri=CELEX:32008L0105& from=EN. Accessed 15 June 2016

European Parliament and Council (2009) Directive 2009/147/EC of the European Parliament and of the Council of 30 November 2009 on the conservation of wild birds. Available at http://eur-lex. europa.eu/legal-content/EN/TXT/?uri=CELEX:32009L0147. Accessed 15 June 2016

European Parliament and Council (2011) Directive 2011/92/EU of the European Parliament and the Council of 13 December 2011 on the assessment of the effects of certain public and private projects on the environment. Available at http://eur-lex.europa.eu/legal-content/EN/TXT/? uri=CELEX:32011L0092. Accessed 15 June 2016

United Nations (1971) Convention on wetlands of international importance especially as waterfowl habitat. Available at http://www.ramsar.org/sites/default/files/documents/library/scan_certified_ e.pdf. Accessed 15 June 2016

United Nations (1992) United nations treaty collection, convention on the protection and use of transboundary watercourses and international lakes, Helsinki, 17 March 1992. Available at https://treaties.un.org/Pages/ViewDetails.aspx?src=TREATY&mtdsg_no=XXVII-5& chapter=27&lang=en. Accessed 15 June 2016

United Nations (1997) United nations treaty collection, convention on the law of the non-navigational uses of international watercourses, New York, 21 May 1997. Available at https://treaties.un.org/Pages/ViewDetails.aspx?src=TREATY&mtdsg_no=XXVII-12& chapter=27&lang=en. Accessed 15 June 2016

Bibliography

De Chazournes LB, Leb C, Tignino M (2015) The UNECE water convention and multilateral environmental agreements. In: Tanzi A, McIntyre O, Kolliopoulos A, Rieu-Clarke A, Kinna R (eds) The UNECE convention on the protection and use of transboundary watercourses and international lakes, its contribution to international water cooperation. Koninklijke Brill NV, Leiden, pp 60–72

European Commission (2011a) Common Implementation Strategy for the Water Framework Directive (2000/60/EC) Guidance document on the intercalibration process 2008–2011. Available at https://circabc.europa.eu/sd/a/61fbcb5b-eb52-44fd-810a-63735d5e4775/IC_GUID ANCE_FINAL_16Dec2010.pdf. Accessed 15 June 2016

European Commission (2011b) Links between the Water Framework Directive and Nature Directives – Frequently Asked Questions. Available at http://ec.europa.eu/environment/nature/ natura2000/management/docs/FAQ-WFD%20final.pdf. Accessed 15 June 2016

European Commission (2012a) Report from the Commission to the European Parliament and the Council on the Implementation of the Water Framework Directive (2000/60/EC) River Basin Management Plans, COM(2012) 670 final. Available at http://eur-lex.europa.eu/legal-content/ EN/TXT/PDF/?uri=CELEX:52012DC0670&from=EN. Accessed 15 June 2016

European Commission (2012b) Communication from the Commission to the European Parliament, the Council, the European Economic and Social Committee and the Committee of the Regions. A blueprint to safeguard Europe's water resources COM(2012) 673 final. Available at http://eur-lex.europa.eu/legal-content/EN/TXT/PDF/?uri=CELEX:52012DC0673&from=EN. Accessed 15 June 2016

European Commission (2012c) Commission staff working document. The fitness check of EU freshwater policy SWD(2012) 393 final. Available at http://ec.europa.eu/environment/water/ blueprint/pdf/SWD-2012-393.pdf. Accessed 15 June 2016

European Commission (2014) Links between the Floods Directive (FD 2007/60/EC) and Water Framework Directive (WFD 2000/60/EC). Resource document. Technical report – 2014 – 078. Available at https://circabc.europa.eu/. Accessed 15 June 2016

European Commission (2015a) Communication from the Commission to the European Parliament and the Council. The Water Framework Directive and the Floods Directive: actions towards the 'good status' of EU water and to reduce flood risks, COM(2015) 120 final. Available at http://ec.europa.eu/environment/water/water-framework/pdf/4th_report/COM_2015_120_en.pdf. Accessed 15 June 2016

European Commission (2015b) 4th European water conference. Conference Report. Brussels, 23–24 March 2015. Available at http://ec.europa.eu/environment/water/2015conference/pdf/report.pdf. Accessed 15 June 2016

European Commission (2016a) European Commission – enlargement – acquis. Available at http://ec.europa.eu/enlargement/policy/glossary/terms/acquis_en.htm. Accessed 15 June 2016

European Commission (2016b) EUR-Lex access to European Union law. Available at http://eur-lex.europa.eu/. Accessed 15 June 2016

European Commission (2016c) The EU Water Framework Directive – integrated river basin management for Europe. Available at http://ec.europa.eu/environment/water/water-framework/index_en.html. Accessed 15 June 2016

European Commission (2016d) The Habitats Directive. Available at http://ec.europa.eu/environment/nature/legislation/habitatsdirective/index_en.htm. Accessed 15 June 2016

European Commission (2016e) What is REFIT – the European Commission's regulatory fitness and performance programme. Available at http://ec.europa.eu/smart-regulation/refit/index_en.htm. Accessed 15 June 2016

European Commission (2016f) Fitness check of the Birds and Habitats Directives. Available at http://ec.europa.eu/environment/nature/legislation/fitness_check/index_en.htm. Accessed 15 June 2016

Janauer GA, Albrecht J, Stratmann L (2015) Synergies and conflicts between water framework directive and natura 2000: legal requirements, technical guidance and experiences from practice. In: Ignar S, Grygoruk M (eds) Wetlands and water framework directive, GeoPlanet: earth and planetary sciences. Springer, Cham. https://doi.org/10.1007/978-3-319-13764-3_2

Keessen AM, van Kempen JJH, van Rijswick HFMW (2008) Transboundary river basin management in Europe. Legal instruments to comply with European water management obligations in case of transboundary water pollution and floods. Utrecht Law Rev 4(3):35–56

Keessen AM, van Kempen JJH, van Rijswick HFMW, Robbe J, Backes CW (2010) European river basin districts: are they swimming in the same implementation pool? J Environ Law 22(2):197–222. Available at https://doi.org/10.1093/jel/eqq003. Accessed 15 June 2016

McCaffrey S (2015) The 1997 UN convention: compatibility and complementarity. In: Tanzi A, McIntyre O, Kolliopoulos A, Rieu-Clarke A, Kinna R (eds) The UNECE convention on the protection and use of transboundary watercourses and international lakes, its contribution to international water cooperation. Koninklijke Brill NV, Leiden, pp 51–59

Moynihan R (2015) The contribution of the UNECE water regime to transboundary cooperation in the Danube river basin. In: Tanzi A, McIntyre O, Kolliopoulos A, Rieu-Clarke A, Kinna R (eds) The UNECE convention on the protection and use of transboundary watercourses and international lakes, its contribution to international water cooperation. Koninklijke Brill NV, Leiden, pp 296–309

Rahaman MM (2009) Principles of international water law: creating effective transboundary water resources management. Int J Sustain Soc 1(3):207–223

Ramsar Convention Secretariat (2010) Water-related guidance: an integrated framework for the convention's water-related guidance, Ramsar handbooks for the wise use of wetlands, vol 8, 4th edn. Ramsar Convention Secretariat, Gland. Available at http://www.ramsar.org/sites/default/files/documents/pdf/lib/hbk4-08.pdf. Accessed 15 June 2016

Ramsar Convention Secretariat (2016) About Ramsar. Available at http://www.ramsar.org/. Accessed 15 June 2016

Reichert G (2016) Transboundary water cooperation in Europe a successful multidimensional regime? International water law. Brill Res Perspect Int Water Law 1(1):1–111

Rieu-Clarke A, López A (2013) Factors that could limit the effectiveness of the UN watercourses convention upon its entry into force. In: Loures FR, Rieu-Clarke A (eds) The UN watercourses convention in force: strengthening international law for transboundary water management. Routledge, London, pp 77–94

Stoa RB (2014) The United Nations watercourses convention on the dawn of entry into force. Vanderbilt J Transnatl Law 47(5):1321–1370

Tanzi A (2013) UN economic commission for Europe water convention. In: Loures FR, Rieu-Clarke A (eds) The UN watercourses convention in force: strengthening international law for transboundary water management. Routledge, London, pp 231–242

Tanzi A (2015) The economic commission for Europe, water convention and the United Nations watercourses convention, an analysis of their harmonized contribution to international water law, Water Series no 6. Available at http://www.unwater.org/fileadmin/user_upload/unwater_new/docs/ece_mp.wat_42_eng_web.pdf. Accessed 15 June 2016

Tanzi A, Contartese C (2015) Dispute prevention, dispute settlement and implementation facilitation in international water law: the added value of the establishment of an implementation mechanism under the water convention. In: Tanzi A, McIntyre O, Kolliopoulos A, Rieu-Clarke A, Kinna R (eds) The UNECE convention on the protection and use of transboundary watercourses and international lakes, its contribution to international water cooperation. Koninklijke Brill NV, Leiden, pp 319–329

Tanzi A, Kolliopoulos A, Nikiforova N (2015) Normative features of the UNECE water convention. In: Tanzi A, McIntyre O, Kolliopoulos A, Rieu-Clarke A, Kinna R (eds) The UNECE convention on the protection and use of transboundary watercourses and international lakes, its contribution to international water cooperation. Koninklijke Brill NV, Leiden, pp 116–129

Trombitcaia I, Koeppel S (2015) From a regional towards a global instrument – the 2003 amendment to the UNECE water convention. In: Tanzi A, McIntyre O, Kolliopoulos A, Rieu-Clarke A, Kinna R (eds) The UNECE convention on the protection and use of transboundary watercourses and international lakes, its contribution to international water cooperation. Koninklijke Brill NV, Leiden, pp 15–31

United Nations (2016a) United Nations global issues, international law. Available at http://www.un.org/en/globalissues/internationallaw. Accessed 15 June 2016

United Nations (2016b) International decade for action "Water for Life" 2005–2015, Transboundary Waters. Available at http://www.un.org/waterforlifedecade/transboundary_waters.shtml. Accessed 15 June 2016

United Nations (2016c) United Nations treaty collection, convention on the protection and use of transboundary watercourses and international lakes, Helsinki, 17 March 1992. Available at https://treaties.un.org/Pages/ViewDetails.aspx?src=TREATY&mtdsg_no=XXVII-5&chapter=27&lang=en. Accessed 15 June 2016

United Nations (2016d) United Nations treaty collection, convention on the law of the non-navigational uses of international watercourses, New York, 21 May 1997. Available at https://treaties.un.org/Pages/ViewDetails.aspx?src=TREATY&mtdsg_no=XXVII-12&chapter=27&lang=en. Accessed 15 June 2016

United Nations (2016e) UN watercourses convention (UNWC) online user's guide. Available at http://www.unwatercoursesconvention.org/. Accessed 15 June 2016

United Nations Economic Commission for Europe (2011) Convention on the protection and use of transboundary watercourses and international lakes, the water convention: serving the planet. Available at https://www.unece.org/fileadmin/DAM/env/water/publications/brochure/Brochures_Leaflets/A4_trifold_en_web.pdf. Accessed 15 June 2016

United Nations Economic Commission for Europe (2012) The water convention: 20 years of successful water cooperation. Available at https://www.unece.org/fileadmin/DAM/env/water/mop_6_Rome/Background_docs/Timeline_A3_R3.pdf. Accessed 15 June 2016

United Nations Economic Commission for Europe (2016a) About the UNECE water convention. Available at http://www.unece.org/env/water/text/text.html. Accessed 15 June 2016

United Nations Economic Commission for Europe (2016b) Geographical scope. Available at http://www.unece.org/oes/nutshell/region.html. Accessed 15 June 2016

United Nations Environment Programme (2002) Atlas of international freshwater agreements. Available at http://transboundarywater.geo.orst.edu/publications/atlas/atlas_html/interagree.html. Accessed 15 June 2016

United Nations Food and Agriculture Organization (1978) Systematic index of international water resources: treaties, declarations, acts and cases, by basin, vol I. Legislative study #15

United Nations Food and Agriculture Organization (1984) Systematic index of international water resources: treaties, declarations, acts and cases, by basin. vol II. Legislative study #34

Vinogradov S, Wouters P, Jones P (2003) Transforming potential conflict into cooperation potential: the role of international water law, UNESCO, IHP, WWAP, IHP-VI, technical documents in hydrology, PCCP series, no 2, SC-2003/WS/67. Available at http://unesdoc.unesco.org/images/0013/001332/133258e.pdf. Accessed 15 June 2016

Wouters P, Vinogradov S (2003) Analysing the ECE water convention: what lessons for the regional management of transboundary water resources. In: Stokke OS, Thommessen ØB (eds) Yearbook of international co-operation on environment and development 2003/2004. Earthscan Publications, London, pp 55–63. Available at http://www.unwatercoursesconvention.org/images/2012/10/ECE-Convention-What-lessons-for-regional-management-of-transboundary-of-water-resources.pdf. Accessed 15 June 2016

Chapter 18
Ensuring Long-Term Cooperation Over Transboundary Water Resources Through Joint River Basin Management

Susanne Schmeier and Birgit Vogel

18.1 Introduction

Water resources are the basis for human well-being and development all over the world. At the same time, in their use and protection, they face numerous challenges stemming from different interests of different actors. Meeting these challenges requires integrated management that reconciles different interests in the use and protection of water resources and ensures the sustainable development of a river or lake basin as a whole. As soon as rivers and lakes[1] cross international borders and transboundary basins emerge, these challenges and the need to address them in a coordinated manner become even more complex. The links between interests in the use of their resources and the interests of nation-states add an international political dimension to the previously rather technical challenge of water resources management.

Such shared basins—more than 270 worldwide—cover more than 45% of the world's surface where more than 40% of the world's population resides (Delli Priscoli and Wolf 2009) (see Fig. 18.1). In total, 145 countries share basins with neighboring states (with 21 such states—including Hungary, Bangladesh, or Zambia—lying entirely in such international basins; Delli Priscoli and Wolf 2009).

---

[1]Transboundary watercourses refer to both river and lakes (as well as groundwater bodies) that transcend the boundaries of nation-states. However, for the sake of readability, we will refer—throughout this chapter—to rivers, nonetheless including lakes as well.

S. Schmeier (✉)
Transboundary Water Management, Gesellschaft für internationale Zusammenarbeit (GIZ), Eschborn, Germany
e-mail: susanne.schmeier@giz.de

B. Vogel
RBM Solutions – River Basin Management e.U., Vienna, Austria
e-mail: birgit.vogel@rbm-solutions.com

© The Author(s) 2018
S. Schmutz, J. Sendzimir (eds.), *Riverine Ecosystem Management*, Aquatic Ecology Series 8, https://doi.org/10.1007/978-3-319-73250-3_18

International River Basins

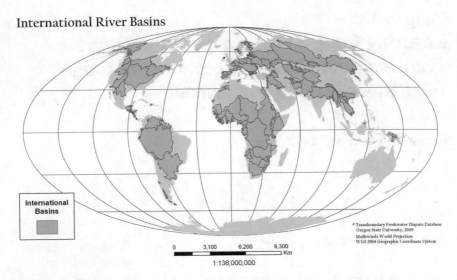

Fig. 18.1 International basins

Interests in the use and the development of these internationally shared basins naturally vary among riparian states. While one state might be interested in exploiting a river's hydropower potential, another state might fear a dam's negative impacts on the river's flow or its sediment load on its own agricultural development potential. Furthermore, while one state might use the river as an effluent for pollutants from its industries, another state might see its drinking water supply threatened by river pollution. As a consequence, disagreements and conflicts often emerge between riparian states, especially if states perceive the use of river's resources by their neighboring state (and the potential environmental and socioeconomic impacts that come with it) as a threat to its own water security. Examples include the recent disputes over hydropower projects in Central and Southeast Asia on the Syr Darya and the Mekong, respectively, as well as in Africa on the Nile River, but also less fierce conflicts such as the one over a navigation channel (the Bystroe Canal) in the Lower Danube Basin or on salmon migration in the Rhine.

Such disagreements—or even full-fledged conflicts—are increasingly on the agenda of politicians and the media. The Uzbek President, for instance, has threatened to go to war against upstream Tajikistan over the Rogun Hydropower Project, and the former Egyptian President Mohamed Morsi has emphasized that Egypt would be willing to confront any threats to Egypt's water security caused by the Grand Ethiopian Renaissance Dam (GERD). Likewise, international media have warned of water conflicts—either generally as a consequence of global climate change or with regard to a particular region—and expressed their fear that "water wars between countries could just be around the corner" (The Guardian 2012) and that "the world will soon be at war over water" (Newsweek 2015, similarly, Reuters 2012 or Spiegel 2012).

In reality, cooperation prevails in most basins shared between different states with their respective interests in the river's resources. Riparian states very often acknowledge that their interests in water resources management and development are better met through cooperation than through conflict (Wolf et al. 2003; De Stefano et al. 2010). This is because the benefits of cooperation in a shared basin tend to outweigh the gains of short-term unilateral action, which often come with considerable economic or political costs over the long term.

The mere acknowledgment of the benefits of cooperation does not, however, ensure cooperation over time. Nor does it allow for harvesting these benefits in the form of better water resource availability, strengthened socioeconomic development, or increased resilience to change. But it is these benefits that make states commit to cooperation over time. Therefore, riparian states in many international basins have committed to joint river basin management.

In this context, we understand "river basin management" in a broad sense, including not only river basin management plans themselves, but all actions and measures taken that aim to avoid or mitigate conflict over shared water resources, while increasing the benefits of cooperation and sharing them across the basin. In order to do so, they have established international water treaties and created river basin organizations as institutions that operationalize the principles of international water law and management and provide the platforms and instruments for joint river basin management.

The remainder of this chapter will focus on how riparian states to shared watercourses ensure their commitment to cooperation over time and implement joint activities that ensure that the benefits of cooperation outweigh potential short-term gains from unilateral action, thus also avoiding or mitigating conflicts among them. It does so by, firstly, taking a brief glance at the global level and at how specific principles for shared water resources are encoded in international water law and other instruments (Sect. 18.2). It then, secondly, moves to the basin level and provides an overview of basin-specific arrangements for cooperation over shared water resources (Sect. 18.3). It then digs deeper into basin management at the basin level by focusing on specific approaches and methods for basin management—namely, the development and implementation of river basin management plans (RBM plans), the management of data and information as a basis for informed decision-making, and the application of prior notification and consultation in the case of hydropower development (Sect. 18.4). The chapter then concludes with an outlook on how successful river basin management can be ensured under current conditions of environmental, climatic, and socioeconomic change (see Sect. 18.5).

18.2 The Global Legal Framework for Managing Shared Watercourses

International efforts to provide the global bases for managing water resources and river basins as constituted by international (water) law have only been established in the past century. Developed over time at different governance levels and in different regions and codified in international and regional conventions and agreements, the most important norms of international water law are the principle of equitable and reasonable utilization, the obligation to cooperation, and the obligation not to cause significant harm. They are based themselves on or are included in more general norms of international law, namely, prohibition of the use of force and the obligation to cooperate with other states, as well as more environment-specific norms, such as the principle of sustainable development.

Two important international instruments—the 1997 UN Convention on the Law of the Non-Navigational Uses of International Watercourses (UN Watercourses Convention) and the 1992 Convention on the Protection and Use of Transboundary Watercourses and International Lakes (Helsinki Convention/UNECE Water Convention)—codify the aforementioned principles in a legally binding manner (see Chap. 20 for more details on these conventions).

The principle of equitable and reasonable utilization is the basis for managing shared water resources. In order to establish equitable and reasonable utilization in a given context, a number of criteria can be applied (as established in Art. 6 of the UN Watercourses Convention). Equitable and reasonable utilization of shared water resources—as well as the general commitment toward sustainable development and the obligation not to cause transboundary environmental harm—implies the water-specific provision not to cause significant harm to co-riparian states. This is enshrined in Art. 7 of the UN Watercourses Convention. In order to comply with these substantive principles, riparian states of a shared watercourse have to cooperate (Art. 8 UN Watercourses Convention—namely, through the establishment of so-called joint bodies) (Art. 9 (2) UNECE Water Convention, United Nations 1992). Some guidance on how to ensure the implementation of these high-level principles in specific watercourses is provided in the form of procedural principles of international water law. They include, most importantly, the obligation to share data and information among riparian states (Art. 9 UN Watercourses Convention and Art. 6 and 13 UNECE Water Convention) and the obligation to notify co-riparian states of possible adverse effects of planned measures on a shared watercourse (Art. 11–19 UN Watercourses Convention, Art. 13 UNECE Water Convention).

The provisions in these global legal instruments only bind their respective parties. However, a number of norms for the management of shared water resources—including the ones mentioned above—have achieved the status of international customary law. Hence, they bind the entire international community. This has not only been confirmed by state practice in all regions of the world but also by international adjudication. Especially in recent years, international courts and tribunals—such as the International Court of Justice (ICJ) and the Permanent Court of Arbitration (PCA) in the 1997 Gabčíkovo-Nagymaros (Danube River), the 2010 Pulp Mills (Uruguay River), or the 2013 Kishenganga (Indus River) cases—have confirmed the importance of certain international water law norms, namely, the principle of equitable and reasonable use and the obligation not to cause transboundary harm, combined with the principle of prior notification (McIntyre 2011; Rieu-Clarke 2013). Moreover, they indicate how international water law and state practice itself develop toward more cooperation over shared rivers and lakes. This reflects an increasing global consensus that shared water resources can only be managed in a cooperative and coordinated manner to the mutual benefit of all riparians. Cooperative river basin management is part of this emerging consensus and is itself a means for ensuring it.

International water law norms thus provide a viable basis for the sustainable management of shared water resources. As with all international law norms, they do, however, face an enforcement challenge. That is, other than at the national level, the compliance of states with these norms cannot be enforced. Consequently, the key question currently is not whether and how a state will ensure that its use and development of water resources does not negatively affect another state in the basin. Rather it is whether the benefits of joint water resources management will be clear to all actors in the basin such that they emerge entirely from the interest of each state in such cooperation. River basin management is the means for ensuring such interest of riparian states by avoiding conflict (and the costs related to it) and increasing the benefits from water resources management through joint approaches, ultimately demonstrating the benefits of cooperation to riparian states.

In the next two subchapters, we will focus in more detail on the institutions that riparian states in different basins have established for ensuring long-term cooperation and the methods and approaches they apply for joint river basin management.

18.3 International Water Treaties and River Basin Organizations: Institutionalizing Cooperation Over Shared Watercourses at the Basin Level

In order to ensure the implementation of principles of water resources management over time in specific basins while taking into consideration their specific hydrological, environmental, socioeconomic, political, and cultural specificities, riparian states of many shared basins have signed international water treaties and established River Basin Organizations (RBOs).

The last 20 years have seen an increasing number of international water treaties being concluded in all regions of the world (TFDD Treaty Database)—with more than 100 agreements signed since the 1992 Rio Conference. Moreover, on average such treaties also seem to engage an increasing number of states (Fig. 18.2).

These treaties vary considerably in scope and content (TFDD Treaty Database). Some treaties have been concluded between two states only—either because they cover a bilateral river basin only or because compromise could only be achieved between a subset of riparians to a larger basin. The latter constellation is not uncommon. The Jordan River Basin, for instance, is largely covered by a set of bilateral agreements between Israel and Jordan, Israel and Palestine, Jordan and Palestine, and Jordan and Syria, respectively. Such situations, however, can lead to severe challenges in terms of integrated water resources management. Other treaties, on the other hand, have managed to commit a large number of riparian states to a basin, such as, for instance, the Danube River Protection Convention (ICPDR 1994), establishing the legal basis for joint river basin management between 14 states and the European Union.

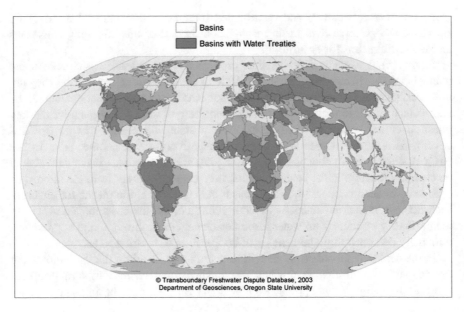

Fig. 18.2 International river basins with basin-specific international water treaties

International water treaties also vary in their content. Some treaties address water quantity challenges by clearly allocating specific amounts of water to the respective parties. Examples include the 1996 Ganges Treaty between India and Bangladesh or the 1959 Nile Waters Treaty between Egypt and Sudan (Agreement for the Full Utilization of the Nile Waters 1959), today the source of major conflict between these and the other Nile riparian states. Other treaties focus on similarly specific, yet rather narrow, issues at stake. The 1986 Lesotho Highlands Water Treaty, for instance, sets the legal basis for the development of a water infrastructure scheme between Lesotho and South Africa Treaty on the Lesotho Highlands (1986). Likewise, the 1992 Treaty on the Development and Utilization of the Water Resources of the Komati River Basin between South Africa and Swaziland defines the grounds for a basin development project consisting of one dam in each country. Similarly, but with a focus on water quality, two agreements between the USA and Mexico set specific standards for decreasing the salinity level of water entering from the USA into Mexico. Yet other treaties pursue an integrated approach and combine various interlinked water resources management issues in one legal document. Sometimes, these treaties even go beyond water resources management itself and focus on regional cooperation and integration more generally, with water being one of the driving forces. An example is the 1972 Convention concerning the Status of the Senegal River between Mali, Mauritania, and Senegal, aiming at promoting and intensifying economic cooperation and exchanges and joining efforts for economic development by developing the resources of the Senegal River (Senegal River Convention 1972).

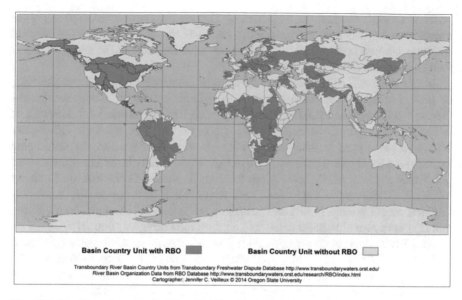

Basin Country Unit with RBO ▬   Basin Country Unit without RBO ☐

Transboundary River Basin Country Units from Transboundary Freshwater Dispute Database http://www.transboundarywaters.orst.edu/
River Basin Organization Data from RBO Database http://www.transboundarywaters.orst.edu/research/RBO/index.html
Cartographer: Jennifer C. Veilleux © 2014 Oregon State University

Fig. 18.3 Transboundary river basins with River Basin Organizations (RBOs)

Establishing and ensuring long-term cooperation that allows for implementing the legal and political commitments made in international water treaties often requires the establishment of institutions such as RBOs.[2] Worldwide, riparian states of more than 100 shared watercourses have opted to establish such RBOs (Schmeier 2013) (see Fig. 18.3). Interestingly, the distribution of RBOs varies considerably across continents: the coverage of basins by RBOs is highest in Africa and Europe (Schmeier 2013; Schmeier et al. 2015). In Asia and the Middle East/North Africa, however, a lot of transboundary basins remain without any institutionalized cooperation mechanism. This is due to a number of different reasons, including the nature of the collective action problem states face and to what extent it lends itself to institutionalized cooperation, the general level and intensity of regional cooperation and integration that can foster cooperation over water as well as, and, not least, the support of RBOs by the international donor community.

RBOs fulfill a number of important functions that provide the basis for cooperation and thus joint river basin management: RBOs provide negotiation and exchange platforms for states, allowing them to exchange on their interests in a predefined and constructive manner. This is done, for instance, through meetings at the ministerial or technical level—such as in the Mekong River Commission (MRC)'s Council and Joint Committee or the Orange-Senqu River Commission

---

[2]Most research on institutionalized cooperation and RBOs does not rely on a sound definition of what an RBO actually is. It therefore remains contested which institutions established for managing shared water resources can be considered as RBOs. In this chapter, we adopt a very broad approach, including a wide range of institutions most commonly figuring as RBOs, even if they do not meet all constitutive elements of RBOs as defined in Schmeier et al. (2015).

(ORASECOM)'s Council and Task Teams. This political and technical exchange in a regular and predefined manner ensures the basis for all cooperation—a constant dialogue between riparian states on their interests in the use and the development as well as the protection of the shared resources.

In this context, RBOs also ensure joint decision-making on how to use, develop, or protect shared water resources through specific decision-making mechanisms (Mostert 2003; Schmeier 2013). The ability of an RBO to ensure joint decisions of its member states on the shared resource is a key element of its success (Schmeier 2013). That is, if joint decision-making fails—in spite of required mechanisms being in place—an RBO easily turns into a paper tiger or eventually loses its legitimacy and potentially even ceases to exist. When the Joint Technical Committee on Regional Waters (JTCW) on the Euphrates-Tigris River failed to come to any consensus decision on the principles of water resources management between Turkey, Syria, and Iraq, it eventually ceased to exist (Islar and Ramasar 2009; Leb 2010).

And RBOs are crucial for mitigating disputes among riparian states (Giordano and Wolf 2003; Sohnle 2005; Blumstein and Schmeier 2016), which tend to (re-)emerge over time even if legal and institutional frameworks for cooperation exist. Recent severe conflicts over hydropower projects illustrate this, e.g., the Rogun Dam in Central Asia (Economist 2012) or the GERD on the Nile, but also less severe yet important conflicts on the Colorado River between the USA and Mexico or on the Indus between India and Pakistan. Examples of active RBO engagement in mitigating diverging interests and avoiding or resolving conflicts include the International Commission for the Protection of the Rhine (ICPR) in the Haringvliet sluice disagreement (ICPR 2011) and the International Commission for the Protection of the Elbe (ICPE) in a dispute between the Czech Republic and Germany over a barrage to be constructed just upstream of the German border (Sächsische Zeitung 2016; WSV 2016).

In addition to these basic functions of RBOs, providing the basis for any long-term, stable cooperation, cooperation needs to move to the next level—the level of joint management of the shared basin—in order to produce or increase joint benefits for all riparian states. The next and main part of this chapter will therefore focus on specific river basin management methods, approaches, and mechanisms.

18.4 Specific Mechanisms for River Basin Management: Implementing Cooperation Over Shared Rivers

International water treaties and RBOs do not only provide the legal and institutional basis for cooperation but also allow for the development and implementation of specific mechanisms, measures, and joint activities for river basin management. Through those mechanisms they reinforce cooperation over time by ensuring sustainable use and development of a basin's resources and creating benefits that transcend national boundaries and considerations. Below we focus on three specific mechanisms that are critical for the sustainable development and management of shared water resources: the sharing of data and information and their use for informed decision-making in a shared basin, the development and implementation of river basin management plans as the key tool for joint management, and the instrument of prior notification and consultation as a means for ensuring compliance with the obligation not to cause significant transboundary harm and its procedural element of prior information.

18.4.1 Data and Information Management and Sharing for River Basin Management

Data and information management, including the design and implementation of joint monitoring programs and networks and the sharing of data and information across riparian states, is a crucial prerequisite for river basin management and transboundary water cooperation more generally (Chenoweth and Feitelson 2001; Burton and Molden 2005; Pietersen and Beekman 2008). This prerequisite can keep cooperating countries well informed on general conditions in river basins they share as well as support objective impact assessment. Further, well-established systems of monitoring and data sharing can also support the effective operation of alarm systems for accidental pollution and flood events, supporting integrated responses across entire basins that benefit both humans and the environment in all riparian states involved. The UNECE Water Convention (United Nations 1992), for instance, foresees the establishment of monitoring programs to assess the conditions of transboundary waters and inform the public accordingly. Despite wide acknowledgment in international river basin management discourse, the actual implementation of joint monitoring and data sharing is still insufficient in many basins. This is largely due to three key challenges—the persistence of data gaps, incoherence across datasets, and a lack of willingness among riparian states to a basin to share data and information with the respective co-riparian states. As a consequence, insufficient transboundary monitoring and a lack of transparency in information sharing can significantly hamper river basin management and cooperation. Below we address these key challenges in data and information management.

Data gaps can significantly challenge river basin management and cooperation. Targeted assessments on the local, national, and basin-wide levels are often not or only partially undertaken, and the lack of evidence may cause critical tensions between riparian states. Sources of data gaps can vary and stem from personnel, technical, and/or financial capacity shortcomings but may also exist due to conflicts and political reasons. After the 2003 Rose Revolution separated the Republic of Georgia from Soviet leadership, the resulting political changes also affected monitoring and assessment of water resources, despite comprehensive reforms in the environmental sector. These effects included a substantial reduction in the number of monitoring sites and, hence, caused significant data gaps in the time series and assessment of water quality and hydrological conditions.

Today, Georgia sees energy generation through hydropower as a key political target, and related development is taking place rapidly (Vogel and Schmutz 2015). As a number of Georgia's rivers—e.g., the Kura River—are shared with neighboring countries, the aforementioned data gaps are particularly problematic. Existing data gaps, the lack of monitoring, and environmental baselines regarding water status hinder the comprehensive assessments of possible impacts from hydropower at both the national and transboundary levels. In addition, the development of Environmental Impact Assessments (EIAs) that are aligned to internationally recognized best practices also becomes a challenge. Data gaps and incomplete assessments result in hydropower development that is more based on assumptions than data and analysis. This increases possible risks of negative impacts on water resources and socioeconomics and of ineffective measures. Accumulation of negative impacts at transboundary scales that cause disputes with basin-sharing countries has so far not been sufficiently taken into account. Considering these challenges, Georgia is currently undertaking steps to close data gaps and to assess hydropower impacts more comprehensively (Vogel and Schmutz 2016). These activities aim to increase knowledge and minimize risks that might otherwise be irreversible. Another data gap example occurs currently in the Congo River Basin, where only nine functioning monitoring stations still exist. In this context, data gaps, ostensibly due to a lack of willingness and interest of riparian states, may ultimately result from lacking technical and financial capacity.

Incoherent datasets pose another challenge. Cooperating countries in transboundary basins often use different methodologies and independent monitoring approaches to collect and assess data (e.g., regarding water status, aquatic ecosystem health, socioeconomics). This usually leads to un-harmonized, incoherent assessment results that cannot be compared between the countries in a basin. This can cause misunderstandings between different countries concerning the interpretation of results—for instance, in interpreting potential transboundary impacts of projects in one country on another riparian country. As a consequence, this often affects cooperation between basin states as a whole.

A comparative study by the European Commission on coordination mechanisms in international river basins in Europe beyond EU territory (2012) showed that out of 75 European transboundary river basins, joint monitoring is fully or partially in place in 51 basins and joint databases are operated in 30 basins (European Commission 2012). In the remaining basins, monitoring and databases are either not jointly coordinated or not in place at all. In other parts of the world, data and information sharing also differs considerably across basins. While in some basins—for instance, in basins shared by Canada and the USA—data information sharing (e.g., through joint monitoring) is well advanced, many basins in the developing world do not have such mechanisms in place.

To tackle these challenges, many RBOs aim for and implement jointly agreed monitoring programs. Some also conduct joint river surveys, constituting a particularly advanced way of joint monitoring. For example, the Danube countries agreed, on the basis of the provisions of the Danube River Protection Convention (ICPDR 1994), to develop and implement transboundary monitoring through a Transnational Monitoring Network—TNMN (ICPDR 2007a, b). The TNMN also ensures regular quality assurance tests regarding monitoring results. This quality assurance aims to establish a solid basis toward trustworthy data that would also support objective discussions in case of transboundary impacts. As part of joint monitoring, the International Commission for the Protection of the Danube River (ICPDR) also organizes and implements Joint Danube Surveys since 2001 every 7 years (ICPDR 2002, 2008a, 2014). During these river surveys, a selected core team undertakes monitoring and sampling along the mainstream of the Danube, fully involving national teams of the 14 Danube countries. The implementation of coherent and commonly agreed sampling and assessment methods regarding chemical, hydro-morphological, and biological water status at all monitoring sites allow for full comparability of results and eases their interpretation. Besides the positive effect of capacity building in the field of monitoring, joint river surveys have the potential to increase trust between cooperating countries due to joint monitoring implementation, full transparency, and consolidation regarding assessment results (see Chaps. 15, 16, and 22 for how joint and participatory cooperation can build trust in river science and management).

A *lack of willingness of riparian states to share data* is a third challenge to joint river basin management. Data sharing becomes specifically challenging or even impossible when countries are not willing to exchange information—most often for political reasons situated outside of the water sector itself. Insufficient data and information sharing due to a lack of willingness of at least one riparian state often further intensifies differences between riparian states and easily leads to conflicts among them. An example can be found in the Mekong River Basin: China, as the most upstream country in the basin, is not a member to the Mekong River Commission (MRC) and, hence, does not participate in the MRC's river basin management activities. Moreover, China has only very limited data-sharing arrangements with its downstream neighbors, the MRC member states. When a severe drought hit the countries of the Lower Mekong Basin in 2010 (and again in 2012), China was quickly blamed for holding back water in its reservoirs behind large hydropower dams on the upper Mekong (Asia Times 2010; Vientiane Times 2010; Radio Free Asia 2012). This led to a significant conflict between China and some of its downstream neighbors. The conflict slowly de-escalated only when China started to share hydrological information on its share of the river and MRC, and Chinese hydrologists came together to exchange data and conduct analyses (MRC 2010). Up to the present, however, such cooperation has not been institutionalized, and regular formal data exchange remains extremely limited—not least due to a lack of interest on the Chinese side. Repeated accusations against China in the recent drought in mainland Southeast Asia have demonstrated once again that insufficient data sharing easily exacerbates existing disagreements or even conflicts (Voice of America 2016).

18.4.2   River Basin Management Plans

River basin management plans (RBMPs) provide the mechanisms and tools for countries to identify and jointly aim for strategic visions and to achieve defined management objectives regarding water resources and, hence, the sustainable management of shared water resources. RBMPs also form the basis of strategic and transparent planning, ideally involving the public, relevant stakeholders, and sectors. Most importantly, RBMPs address an entire river basin, leaving administrative borders aside in order to ensure integrated planning and management.

RBMPs define water resources management objectives to be achieved within a certain time frame. In some cases—especially in many basins in the developing world—RBMPs go beyond water resources management objectives and include coordinated efforts to develop water-related infrastructure (e.g., hydropower, irrigation schemes, water supply) as well as socioeconomic aspects. In addition to the objectives to be achieved, RBMPs have to include all steps of a river basin management cycle. This includes a basic characterization of river basins and the identification of pressures and impacts on water resources, followed by the validation with monitoring results and the definition of measures to achieve the objectives.

Globally, the development of RBMPs is increasingly becoming a best-practice approach, and RBMPs have been developed in a number of transboundary basins, especially in Europe and in Africa. At the same time, it has to be acknowledged that each basin is characterized by particular ecological, historic, socioeconomic, and political conditions, different water uses, and different impacts of such water uses. RBMPs, therefore, have to be adapted to the needs of each basin. Consequently, no single blueprint approach exists for RBMPs. In addition to the different characteristics of each basin, other dimensions also challenge the successful development and implementation of RBMPs. One challenge is related to different legal frameworks at both national and basin level (and potential incompatibilities between them). In the case of the EU, the development of national and international RBMPs is legally binding, according to the European Water Framework Directive (European Commission 2000). In other countries, however, the development of such plans—especially at the basin level—is often not foreseen in national and regional legal frameworks. The lack of a legal basis for joint basin planning often impedes the development of RBMPs and their implementation, including their translation into more specific regulatory acts concerning, for instance, permits and related monitoring and enforcement mechanisms, and their financing out of a national budget (GIZ 2017). And the existence of different legal bases in different basin states tends to further complicate matters. Another challenge is related to the effectiveness of river basin organizations and other joint bodies and, in particular, their respective technical bodies with regard to developing and implementing RBMPs at the different governance levels. And a third challenge is related to the timing of RBMP implementation, including the definition of clear implementation milestones and objectives along the river basin management cycle. The next paragraphs will address these challenges on the basis of one specific example—the Danube River Basin Management Plan (Danube RBMP), developed by the ICPDR.

18.4.2.1   Case Study: Danube River Basin Management Plan

So far, two Danube RBMPs have been developed under the coordination of the International Commission for the Protection of the Danube River (ICPDR) and have been published in 2009 and 2015, respectively. Danube RBM Plans guide and support the Danube countries that consist of EU and non-EU member states in their joint river basin management. The Danube RBMPs are management documents for achieving environmental objectives and protecting the aquatic environment as well as tools that provide orientation for the Danube River Basin's water sector and its actors.

The cooperation of the 14 countries under the Danube River Protection Convention (ICPDR 1994) is based on a high level of commitment, will, and trust. In addition, regulated cooperation and coordination mechanisms are essential tools for effective Danube river basin management. And the coordination between the national and international levels in the Danube River Basin is crucial to ensure the involvement of all contracting parties and stakeholders to the best possible extent. Specifically, for the coordination of the Danube RBMP, the basin-sharing countries follow an agreed, top-down/bottom-up approach. This approach ensures clear and functioning linkages between the national and international levels for the entire process of developing and implementing the plan. For example, most of the measures in international RBMPs are implemented at the national level and through national legal regulations. Therefore, the measures that are identified in the international Danube RBMP need to be reflected in the national RBMPs of the basin countries to ensure adequate implementation. If linkages between the national and international levels are lacking, the implementation of international RBMPs and actions may be at high risk.

The development of international RBMPs usually spans over longer periods of planning cycles and so requires well-coordinated planning. Therefore, for the compilation of the Danube RBMPs, aligned clear timelines were set, supporting the countries to jointly move toward the related aims in a strategic and coordinated way. Overall, according to the WFD objectives, the cooperating countries aim to achieve good status for all waters by 2027 through implementing four river basin planning cycles (ICPDR 2009, 2015). The first Danube RBMP has been developed within a period of 9 years until its adoption on the ministerial level. The second one has been compiled within 6 years.

The variety of technical aspects and their meaningful coordination and reflection in international RBMPs is demanding. The coordinated work of joint technical bodies can significantly facilitate the work of basin-sharing countries toward a final and consolidated RBMP. In this context, the ICPDR has seven permanent Expert Groups (EG) in place that consist of nominated country representatives and

ICPDR observers. Technical experts from the ICPDR Secretariat coordinate these EGs. The EGs have terms of reference and mandates adopted by each country cooperating within the ICPDR framework and meet several times a year. The Expert Groups discuss issues related to their terms of reference (ICPDR 2016) and prepare reports and recommendations for coordinated action toward the final Danube RBMP. On demand, time-limited Task Groups (TGs) may also be established for specific actions and in which not necessarily all countries are represented. In practice, the River Basin Management EG coordinates the overall development of Danube RBMPs, steering the other relevant EGs and compiling all information into one final plan. Draft versions are shared with all EGs as well as presented and consolidated at the ICPDR's plenary meetings that take place twice a year, until the final Danube RBM Plan is adopted for implementation. In the end, the coordinated technical work and joint management efforts of the EGs and TGs support countries in consolidating the content of the Danube RBMPs due to their full involvement during the entire development process.

Besides the coordination aspects that need to be in place to develop international RBMPs, the achievement of defined milestones that are aligned to a river basin management cycle is important. The Danube Basin Analysis (ICPDR 2005), which is equal to State of Basin Reports, was the first milestone within the Danube RBM Plan development. Four basin-wide Significant Water Management Issues (SWMI) have been identified, which can directly or indirectly affect the status of both surface waters and transboundary groundwaters. The SWMIs address pollution by organic substances and nutrients, pollution by hazardous substances, as well as hydromorphological alterations. The identification of these SWMIs has been an important step within Danube RBM planning. Over one and a half years, a joint process of the ICPDR and its River Basin Management Expert Group elaborated a strategic document on the four SWMIs. The SWMI document includes visions and management objectives for each SWMI (ICPDR 2008b, 2009, 2015) that have been adopted by the contracting parties in the ICPDR's Ordinary Meeting.

The SWMI visions and, specifically, the management objectives describe the basin-wide implementation steps toward the environmental objectives that need to be achieved by 2027 at the latest. The management objectives guide the Danube countries toward agreed joint aims. Basically, the management objectives for the SWMIs describe the measures that need to be taken to reduce or eliminate existing pressures for each SWMI on a basin-wide scale. In addition, the management objectives help bridge the gap that can exist between the national level and their agreed coordination on a basin-wide level to achieve the environmental objectives. Although SWMIs and their management objectives can be flexibly adjusted to the needs of the Danube River Basin from RBMP to RBMP, their adoption by the contacting parties of the ICPDR is crucial to ensure fixed joint aims and, hence, a solid implementation basis.

As a next milestone, a specific and coordinated analysis has been undertaken for each of the identified SWMIs within the Danube RBMP that enables targeted management on the basin-wide scale. For each SWMI, pressure types that may impact the water status are presented, which are then addressed with corresponding monitoring programs and measures as part of the Danube RBMP's Joint Programme of Measures. The Pressure and Impact Analysis for each SWMI is based on the Driver-Pressure-State-Impact-Response approach that is integral part of the Guidance Document No 3 (Analysis of Pressures and Impacts) of the Common Implementation Strategy for the EU WFD (European Commission 2003) and which includes the steps below:

1. Water uses that may have environmental effects are allocated to each SWMI.
2. Pressures that can stem from each driver are identified and clearly allocated to surface water bodies and groundwater bodies.
3. Identified pressures are assessed for their significance in possibly putting the respective water bodies at risk of failing the environmental objectives. The significance of pressures and the risk assessment is based on the application of clear criteria that have been developed and agreed upon by all 14 Danube countries. The agreement on criteria is an essential step to achieve results on a consolidated basis that is also aligned to the national river basin management in each country. This consolidation creates a common understanding from the beginning between all Danube River Basin (DRB) countries and applies one, single assessment approach for all Danube basin-wide purposes. Further, consolidation ensures a harmonized, transboundary approach that allows comparability of results across borders and, hence, facilitates the discussion of assessment results and counteracts potential areas of conflict from the beginning.
4. In a final step, the results of the risk assessment are responded to with the development of tailor-made, comprehensive monitoring programs in order to validate the findings with real water status assessment. The monitoring programs in the DRB and, therefore, the Danube RBMP are based upon (1) monitoring data of the ICPDR's Transnational Monitoring Network (TNMN—ICPDR 2007a, b), (2) on specifically collected national data provided through the Danube countries, as well as (3) on monitoring assessment results from the Joint Danube Surveys (ICPDR 2002, 2008a, 2014). All data is compiled in a joint database (DanubeGIS) and used for the analysis as well as GIS mapping within the Danube RBM Plan.

In case good water status is at risk or assessed to fail, measures as part of the DRB Joint Programme of Measures (JPM; ICPDR 2009, 2015) are developed to eliminate or at least limit the negative impacts. The JPM is a core piece of each Danube RBMP and is adopted by all contracting parties of the ICPDR for implementation. The JPM aims to ensure the basin-wide achievement of the joint management objectives, even in the case of future infrastructure development that could potentially deteriorate the state of the river again. The JPM outlines what steps need to be taken to achieve environmental objectives on a basin-wide scale, making fully coordinated use of national activities. Interim assessments on the effectiveness of the JPM are undertaken regularly for the basin-wide and the national scales. However, the implementation of the JPM is still a challenging task, and it is clear that high efforts by all Danube countries will be needed to achieve the aims at the latest by 2027.

18.4.3   Prior Notification and Consultation Mechanisms

As it has been mentioned already in the previous subchapter, the obligation not to cause significant harm to co-riparian states is a key element of international water law. It is not only embodied in global conventions and regional agreements (such as the SADC Protocol on Shared Watercourses 2000) but also in a considerable number of basin-specific treaties.[3] Its implementation is, however, challenging—as is its enforcement.

In order to ensure that measures taken in one riparian country do not have negative transboundary effects on other riparian countries, countries potentially affected by such measures have to, first and foremost, be informed about the respective measures and be enabled to evaluate its potential impacts. This is done through prior notification and consultation mechanisms. Such mechanisms are foreseen in some basin-specific treaties[4]—although the majority of treaties remain silent on this rather complex matter. This is largely due to the fact that the establishment of a prior notification mechanism is often perceived as a significant limitation to unilateral national development interests, since it implies the inclusion of other states' interests in the planning and development of water resources development projects.

Most often, treaty provisions follow general guidance provided by international conventions, the adjudication of international courts, and, more generally, customary international law. The process then consists of a first notification about a planned measure, the provision of relevant information (such as the environmental impact assessment (EIA) and related documents), a certain time frame during which a potentially affected state can respond to the information received (6 months according to Art. 13 of the 1997 UN Watercourses Convention), and a response by the respective state, potentially leading to consultations and negotiations or any other following action. The process does, however, not imply a veto right for the potentially affected state.

While international- and basin-specific agreements already provide detailed guidance on the process of prior notification, its implementation remains challenging

---

[3]Examples include the 2002 Framework Agreement on the Sava River Basin; the 2003 Convention on the Sustainable Management of Lake Tanganyika between Burundi, DR Congo, Tanzania, and Zambia; and the 2004 Agreement on the Establishment of the Zambezi Watercourse Commission between eight Southern African states.

[4]Provisions on prior notification and/or consultation (or similar means for informing co-riparians about planned measures and their potential transboundary effects) are included, for instance, in the 1975 Statute of the River Uruguay, an international water treaty governing the Uruguay River between Uruguay and Argentina; Annex II to the 1994 Treaty of Peace between the State of Israel and the Hashemite Kingdom of Jordan, which governs their relations concerning the Jordan River; the 2003 Protocol for Sustainable Development of the Lake Victoria Basin between Kenya, Tanzania, and Uganda; and the 1960 Indus Waters Treaty between India and Pakistan (for more details, refer to the TFDD RBO Database and Schmeier 2013).

(Farajota 2005; Rieu-Clarke et al. 2012, 135–163; Rieu-Clarke 2013; Blumstein and Schmeier 2016). This is largely due to a number of questions that remain open or tend to be interpreted differently by different riparian states in line with their respective interests. The first such challenge concerns questions under which conditions a potentially affected state has to be notified of a planned measure. It relates to the broader question of what actually constitutes significant, transboundary harm. A second challenge relates to the question as to when a notification is to be made—at the planning stage of a project or later during its implementation. And, third, challenges often relate to how to treat the results of a notification and consultation process—both with regard to when such process is finished and developments can go ahead and with regard to what happens if disagreements prevail.

18.4.3.1   Case Study: Xayaburi Hydropower Project in the Mekong River Basin

A particularly interesting case for prior notification and consultation mechanisms relates to how the Mekong River Commission (MRC) handled the Xayaburi Hydropower Project (XHP) in the Mekong River Basin: based on the 1995 Agreement on the Cooperation for the Sustainable Development of the Mekong River Basin (Mekong Agreement), signed by Cambodia, Laos, Thailand, and Vietnam, specific principles and provisions for water resource use and development have been defined (Mekong Agreement, Art. 5–7). They relate largely to intra- and inter-basin use of the Mekong's water, related harmful effects, and when and under which conditions these are to be discussed with co-riparian states. In order to implement these principles and provisions, Procedures for Notification, Prior Consultation and Agreement (PNPCA) have been developed. These PNPCA foresee notification of consultation with or agreement by other riparian states on water resources projects developed by one of the MRC member states, depending on its respective influences on the river.

The PNPCA were brought to a serious test in 2010, when the National Mekong Committee of the People's Democratic Republic of Lao submitted the XHP for notification and consultation to the MRC. The XHP was the first mainstream hydropower project submitted to the MRC for consideration. It was therefore the first instance at which the consultation dimension of the PNPCA—and hence a rather complex process—was triggered.

With the official notification by Laos, a period of 6 months was open for analyzing the documents submitted by Laos and gathering additional information that would allow to come to a conclusion concerning the potential impacts of the project on the Mekong River and other riparian states (and thus the question of potential transboundary harm) and determining the responsibility of Laos for such harmful transboundary effects (Mekong Agreement, Arts. 7 and 8).

The MRC Secretariat facilitated this process (for the roadmap see Fig. 18.4) leading to a Prior Consultation Review Report (MRC 2011a). The process was led by the PNPCA Task Group, bringing together the MRC Secretariat and MRC

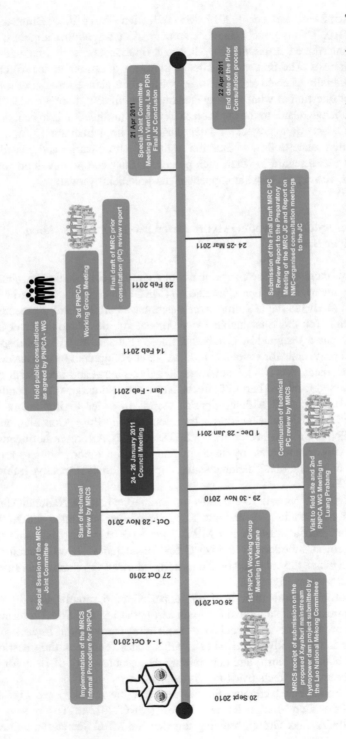

Fig. 18.4 Road map of procedures for notification, Prior Consultation and Agreement (PNPCA) regarding the Mekong mainstream hydropower scheme of Xayaburi—Lao PDR (MRC 2011c)

Programs in their work on the report. The Task Group was supported by external Expert Groups comprising regional and international experts, focusing on fisheries and sediments—the most critical issues—respectively. MRC and external experts based their assessment on the *Preliminary Design Guidance for Proposed Mainstream Dams in the Lower Mekong Basin* (PDG; MRC 2009). As endorsed by the MRC Joint Committee in 2009, the PDG allowed an objective cross-check regarding possible impacts of the XHP on the key elements of (1) hydrology; (2) fish passage and fisheries ecology; (3) sediment transport, morphology, and nutrient balance; (4) water quality, ecosystem health, and environmental flows; (5) navigation; and (6) safety of dams. In addition, social aspects as well as the dam design itself have been investigated, reviewed, and analyzed regarding eventual impacts. The PNPCA process also ensured the involvement of MRC member countries as well as a consultation of public stakeholders (in all but one countries).

The resulting Prior Consultation Review Report aimed to provide support and technical information on the project and its potential impacts on the river to the MRC's Joint Committee (JC), which is responsible for reaching a joint conclusion regarding the outcome of the notification process and, thus, the project. It presents a comprehensive analysis of all dimensions of potential transboundary impacts—including areas that could not be studied sufficiently and thus require further investigations. The report came to the conclusion that there were significant shortcomings of the XHP with regard to various dimensions. This included, for instance, bypasses for fish migration, mechanisms for ensuring sediment transport, dam safety, and other issues relating to the design of the dam. The report also identified significant data gaps and recommended the establishment of environmental baselines and continuous monitoring of the project's impacts before, during, and after its potential construction.

During its development, the report was discussed in three MRC JC Working Group Meetings before becoming part of the JC's formal meeting in April 2011. The JC could, however, not come to a joint decision in line with the procedural requirements of the MRC's PNPCA. Specifically, the JC did not jointly decide on whether to agree to the project, decline it, or extend the consultation process. It therefore transferred the decision to the MRC Council, the MRC's highest governing body. The Council, at its meeting in December 2011, could also not come to a conclusion. Since then, the process has neither been officially terminated nor did it continue to actively exist. This left both the process itself as well as much of MRC in a limbo situation.

Laos did, however, consider the process as completed and initiated the construction of the project in November 2011. This led to severe protest from other riparian states, namely, Cambodia and Vietnam. Vietnam had already stated in its official reply to the Prior Consultation Review Report that it "strongly suggests that the decision on the XHP as well as other planned hydropower projects on the Mekong mainstream [should] be deferred for at least 10 years" (MRC 2011b) and continued to raise this issue repeatedly since then.

Until today, the process has not been terminated, and the MRC member countries have not reached a formal decision yet. Instead, Laos drew its conclusion and continued the construction of the dam (with finalization expected for 2019), and Cambodia and, in particular, Vietnam continued their protests and engaged in other activities trying to assess and potentially mitigate the negative transboundary effects they expect. This clearly demonstrates the third challenge that prior notification and consultation processes often have to cope with. That is that the perceptions of the conclusions from a notification and consultation process (often not even officially terminated) tend to vary considerably across the countries involved. This is often linked to the fact that the possible outcomes of such processes are not clear to all parties beforehand—especially since downstream and potentially affected states tend to see such notification and consultation mechanisms as means for voicing their veto against a project, while upstream states often tend to be critical toward interference in their water resources development activities.

In addition to notification-specific challenges, this example from the Mekong also illustrates that joint river basin management and related activities are crucial for ensuring the long-term sustainable and cooperative development of shared river basins. Solid knowledge of the status of a river basin as well as joint basin planning activities can ensure that riparian states to a shared basin are informed about each country's development plans early on in the process and it can, moreover, help countries to choose projects that provide the most benefits to all riparian states and the basin while producing the least negative effects. And it demonstrates that insufficient processes of river basin management can significantly hamper the sustainable development of the basin and the relations between riparian countries more generally.

18.5 Conclusion

This chapter has shown that riparian states to transboundary rivers tend to choose cooperation over conflict. But it has also shown that they only do so—especially over longer periods of time—if they have functioning cooperation mechanisms in place that allow for generating and harvesting the benefits of cooperation for all. In addition to basic cooperation structures (international water treaties and RBOs), joint river basin management with its various dimensions and approaches is one of the key means for harvesting these cooperation benefits. Joint river basin management helps to establish joint visions and management objectives through which riparian states cooperate and streamline their activities toward common aims, thereby including both regional and national actors. Joint river basin management allows for identifying and defining key water management issues that need to be addressed by all riparian states, including issues relating to potential transboundary impacts of water resources development projects in one state on its neighbors and thus potential sources of future conflict. And joint river basin management supports monitoring and data acquisition and sharing, establishing a common and transparent ground for all riparian states (and potentially other actors) involved.

At the same time, transboundary river basin management is facing a number of challenges, not only relating to the mechanism of basin management itself (and thus, for instance, to the human, technical, and financial capacity required for effective basin management) but also to the broader context of transboundary water cooperation. Three challenges appear to be particularly problematic in this regard: firstly, the fact that national interests of basin-sharing countries often differ or even are in outright conflict makes the management of water resources across borders very problematic. Consequently and secondly, riparian states often pursue infrastructure development projects that exclusively meet their national (short-term) development needs and do not consider transboundary negative effects on other riparian states. And thirdly, even if states acknowledge that cooperation is more beneficial for their interests as well as the entire basin and, hence, establish legal and institutional mechanisms for cooperation, conflicts of interests may arise due to the different degrees to which agreements are legally binding at the national and the international levels. While national implementation in water resources management is usually legally binding, the enforcement of international RBMPs is vague. Hence, successful implementation and related compliance on the international level remains challenging.

In spite of these severe challenges, academic research as well as empirical evidence clearly demonstrates that long-term cooperation benefits prevail over potential short-term gains from noncooperative behavior—not only for the basin, its resources, and its ecosystems but also for riparian people and states as a whole. River basin management approaches and RBMPs that consider long-term planning and implementation can further improve cooperation. However, besides formally agreed mechanisms for cooperation and coordination, essential prerequisites are cooperation, will, and trust between countries sharing a basin. In case the latter aspects are lacking, transparent river basin management (see Chaps. 15 and 16) can be used as a tool to build these aspects as foundations for better formulation and implementation of policy, law, and, eventually, sustainable development of river basins and riparian states.

References

Agreement for the Full Utilization of the Nile Waters (1959) Agreement between the Republic of the Sudan and the United Arab Republic for the full utilization of the Nile waters, Cairo, Egypt, 8 Nov 1959

Asia Times (2010) Then the Mekong runs dry. Asia Times, 13 Mar 2010

Blumstein S, Schmeier S (2016) Disputes over international watercourses: can river basin organizations make a difference? In: Dinar A et al (eds) Management of transboundary water resources under scarcity: a multi-disciplinary approach. World Scientific, Hackensack, NJ

Burton M, Molden D (2005) Making sound decisions: information needs for basin water management. In: Svendsen M (ed) Irrigation and river basin management. Options for governance and institutions. CABI, Wallingford, pp 51–74

Chenoweth J, Feitelson E (2001) Analysis of factors influencing data and information exchange in international river basins: can such exchanges be used to build confidence in cooperative management? Water Int 26(4):499–512

De Stefano L, Edwards P, DeSilva L, Wolf A (2010) Tracking cooperation and conflict in international river basins: historic and recent trends. Water Policy 12(6):871–884

Delli Priscoli J, Wolf A (2009) Managing and transforming water conflicts. Cambridge University Press, Cambridge

Economist (2012) Water wars in Central Asia. Dammed if they do – spats over control of water oil an already unstable region, 29 Sept 2012

European Commission (2000) Directive 2000/60/EC of the European parliament and of the council of 23 October 2000 establishing a framework for community action in the field of water policy. http://eur-lex.europa.eu/legal-content/EN/TXT/?uri=CELEX:32000L0060. Accessed 15 June 2016

European Commission (2003) Guidance document no 3 – analysis of pressures and impacts of the common implementation strategy for the water framework directive. Produced by the working group 2.1 – impress. Office for Official Publications of the European Commission, Luxemburg, pp 13–14

European Commission (2012) Comparative study on pressures and measures in the major river basin management plans in the EU, Task 1 – governance (final report) including Task 1b on international coordination mechanisms 2012, European Commission Directorate General Environment and Climate Action

Farajota M (2005) Notification and consultation in the law applicable to international watercourses. In: Boisson de Chazournes L, Salman S (eds) Les ressources en eau et le droit international. Martinus Nijhoff, Leiden, pp 282–339

Giordano M, Wolf A (2003) Transboundary freshwater treaties. In: Nakayama M (ed) International waters in Southern Africa. UN University Press, Tokyo, pp 71–100

GIZ (2017) River basin management. The development and implementation of river basin management plans. GIZ, Eschborn

ICPDR (1994) Convention on cooperation for the protection and sustainable use of the Danube River (Danube River Protection Convention), 29 June 1994 in Sofia, Bulgaria. www.icpdr.org/icpdr-pages/legal.htm

ICPDR (2002) Technical report on the joint Danube survey, International Commission for the Protection of the Danube River, Vienna. www.icpdr.org

ICPDR (2005) The Danube river basin district – river basin characteristics, impact of human activities and economic analysis required under article 5, annex II and annex III, and inventory of protected areas required under article 6, annex IV of the EU water framework directive (2000/60/EC). Part A – basin wide overview. Short: Danube basin analysis (WFD Roof Report 2004), Document IC/084 of the International Commission for the Protection of the Danube River, Vienna. www.icpdr.org

ICPDR (2007a) Summary report to EU on monitoring programmes in the Danube river basin district designed under article 8 – part I; WFD roof report on monitoring – part I: surface waters. Development of WFD compliant monitoring programmes for the Danube river basin district document 122 of the International Commission for the Protection of the Danube River

ICPDR (2007b) Summary report to EU on monitoring programmes in the Danube river basin district designed under article 8 – part I; WFD roof report on monitoring – part II: groundwater. Development of WFD compliant monitoring programmes for the Danube river basin district document 122 of the International Commission for the Protection of the Danube River

ICPDR (2008a) Technical report on the joint Danube survey. International Commission for the Protection of the Danube River, Vienna. www.icpdr.org

ICPDR (2008b) Significant water management issues in the Danube river basin district: including visions and management objectives for each SWMI; document 132 of the International Commission for the Protection of the Danube River, Vienna

ICPDR (2009) Danube river basin management plan 2009. http://www.icpdr.org/main/activities-projects/danube-river-basin-management-plan-2009

ICPDR (2014) Technical report on the joint Danube survey. International Commission for the Protection of the Danube River, Vienna. www.icpdr.org

ICPDR (2015) Danube river basin management plan 2015. https://www.icpdr.org/main/activities-projects/river-basin-management-plan-update-2015

ICPDR (2016) Terms of references and work plans for the ICPDR expert groups. https://www.icpdr.org/main/publications/tor-workplans

ICPR (2011) Beschlussprotokoll der SG(1)-11-Sitzung, Koblenz, 2 Mar 2011 (restricted document)

Islar M, Ramasar V (2009) Security to all: allocating the waters of Euphrates and Tigris'. Paper presented at the 2009 Amsterdam conference on the human dimensions of global environmental change, Amsterdam, 2–4 Dec 2009

Leb C (2010) The Tigris-Euphrates joint technical committee – deadlocked, IUCN water programme: NEGOTIATE toolkit case studies, Gland

McIntyre O (2011) The world court's ongoing contribution to international water law: the pulp mills case between Argentina and Uruguay. Water Altern 4(2):124–144

Mostert E (2003) Conflict and cooperation in international freshwater management: a global review, Paris: UNESCO: IHP Technical Document in Hydrology No. 19

MRC (2009) Preliminary design guidance for proposed mainstream dams in the lower mekong basin (PDG). Vientiane, Lao PDR: MRC Secretariat. www.mrcmekong.org/assets/Publications/Consultations/SEA-Hydropower/Preliminary-DG-of-LMB-Mainstream-dams-FinalVersion-Sept09.pdf

MRC (2010) Op-Ed: low River levels caused by extreme low rainfall. Vientiane, Lao PDR, MRC Secretariat

MRC (2011a) Prior consultation review report regarding the Xayaburi hydropower dam project under www.mrcmekong.org/news-and-events/consultations

MRC (2011b) Reply to prior consultation. Submitted by the Vietnam National Mekong Committee, Hanoi, Vietnam, 15 Apr 2011

MRC (2011c) PNPCA road map regarding the Mekong mainstream hydropower scheme of Xayaburi. http://www.mrcmekong.org/news-and-events/consultations/xayaburi-hydropower-project-prior-consultation-process/

Newsweek (2015) The world will soon be at war over water. Newsweek, 24 Apr 2015. http://www.newsweek.com/2015/05/01/world-will-soon-be-war-over-water-324328.html

Pietersen K, Beekman H (2008) Strengthening cooperation among river basin organizations: a comparative study of the linkages between river/lake basin organizations and the respective cooperating national governments in seven major African basins. Deutsche Gesellschaft für Technische Zusammenarbeit (GTZ). Eschborn, Germany

Radio Free Asia (2012) Lao expert blames Chinese dams. Radio Free Asia/Voice of Asia, 30 Apr 2012

Reuters (2012) U.S. intelligence sees global water conflict risk rising, 22 Mar 2012

Rieu-Clarke A (2013) The obligation to notify and consult on planed measures concerning international watercourses – learning lessons from recent case law. Yearbook of International Environmental Law

Rieu-Clarke A, Moynihan R, Magsig G (2012) UN watercourses convention – user's guide (CWLPS 2012)

Sächsische Zeitung (2016) Neuer Anlauf für Elbe-Staustufe. Prags Behörden legen eine neue Dokumentation zum umstrittenen Projekt vor, 21 Mar 2016. http://www.sz-online.de/sachsen/neuer-anlauf-fuer-elbe-staustufe-3352694.html

SADC (2000) Revised protocol on shared watercourses in the Southern African Development Community, Windhoek, Namibia, 7 Aug 2000

Schmeier S (2013) Governing international watercourses. River basin organizations and the sustainable governance of internationally shared rivers and lakes. Routledge, London

Schmeier S, Gerlak A, Blumstein S (2015) Clearing the muddy waters of shared watercourses governance: conceptualizing international river basin organizations. Int Environ Agreem 16(4):597–619. https://doi.org/10.1007/s10784-015-9287-4

Senegal River Convention (1972) Convention Relative au Statut du Fleuve Sénégal/Convention concerning the Status of the Senegal River (Senegal River Convention)', Nouakchott, Mauritania, 11 Mar 1972

Sohnle J (2005) Nouvelles tendances en matière de règlement pacifique des différends relatifs aux ressources en eau douce internationales. In: Boisson de Chazournes L, Salman S (eds) Les ressources en eau et le droit international. Martinus Nijhoff, Leiden, pp 389–426

Spiegel (2012) US-Bericht warnt vor Ära der Kriege um Wasser, Spiegel Online, 22 Mar 2012

The Guardian (2012) Water wars between countries could be just around the corner. Davey warns, 22 Mar 2012

Treaty on the Lesotho Highlands (1986) Highlands treaty on the Lesotho Highlands Water Project between the Government of the Kingdom of Lesotho and the Government of the Republic of South Africa (LHWC Treaty)', Maseru, Lesotho, 24 Oct 1986

United Nations (1992) United Nations treaty collection, convention on the protection and use of transboundary watercourses and international lakes, Helsinki, 17 Mar 1992. https://treaties.un. org/Pages/ViewDetails.aspx?src=TREATY&mtdsg_no=XXVII-5&chapter=27&lang=en. Accessed 15 June 2016

United Nations (1997) United Nations treaty collection, convention on the law of the non-navigational uses of international watercourses, New York, 21 May 1997. https://treaties. un.org/Pages/ViewDetails.aspx?src=TREATY&mtdsg_no=XXVII-12&chapter=27&lang=en. Accessed 15 June 2016

Vientiane Times (2010) Shallow Mekong stops northern tourist, cargo boats. Vientiane Times, 2 Mar 2010

Vogel B, Schmutz S (2015) Environmental impact assessment for cumulative impacts of hydropower projects in Georgia, final report, project funded by the German Federal Ministry for the Environment, Nature Conservation, Building and Nuclear Safety and implemented by the Deutsche Gesellschaft für Internationale Zusammenarbeit (GIZ)

Vogel B, Schmutz S (2016) EU approximation in the field of environmental impact assessment for cumulative impacts of hydropower projects in Georgia. Project funded by the German Federal Ministry for the Environment, Nature Conservation, Building and Nuclear Safety and implemented by the Deutsche Gesellschaft für Internationale Zusammenarbeit (GIZ)

Voice of America (2016) Chinese dams blamed for exacerbating Southeast Asian drought. Voice of America, 1 Apr 2016

Wolf A, Yoffe S, Giordano M (2003) International waters: identifying basins at risk. Water Policy 5:29–60

WSV—Wasserstrassen- und Schifffahrtsverwaltung des Bundes (2016) Umweltverträglichkeitsprüfung für das Vorhaben Staustufe Děčín. Tschechische Republik, Magdeburg. http://www.ast-ost.gdws. wsv.de/aktuelles/Staustufe_Decin/index.html

Chapter 19
Biomonitoring and Bioassessment

Otto Moog, Stefan Schmutz, and Ilse Schwarzinger

19.1 Introduction

The water that we use, the air that we breathe, and the energy that we consume are limited resources. Among these, "water issues are one of the major problems that humanity must solve for its survival." This maxim was a key conclusion reached by top-level decision-makers at the first Asia Pacific Water Summit in December 2007, marking the first time in history that all Asian states met to discuss water issues. This statement does more than characterize the Asian situation. It applies to our entire globe. Water managers and scientists are aware that sustainably managing a fundamental resource like water requires rigorous scientific data and analysis to understand aquatic ecosystem functioning. Proper sustainable management requires that we know the "quantity" and "quality" of a water source. This chapter describes the basics of the "qualitative aspects" of water monitoring and management, namely, the biomonitoring and the assessment of "river quality."

Following centuries of human impacts on aquatic resources, monitoring of water chemistry has become a customary practice in many countries. The fundamental scientific aspects of sustainable water management in many areas of the world are still quite poorly understood, and biological monitoring tools are lacking for many developing and transitional countries (Bere and Nyamupingidza 2013). In recent decades, attempts to develop monitoring methodologies are increasing rapidly and have generated biological assessment and monitoring of aquatic resources that are reliable enough to be included in state monitoring programs in Europe, the United States, Australia, and South Africa. Following on this successful trend is a wave of

O. Moog (✉) · S. Schmutz · I. Schwarzinger
Institute of Hydrobiology and Aquatic Ecosystem Management, University of Natural Resources and Life Sciences, Vienna, Austria
e-mail: otto.moog@boku.ac.at; stefan.schmutz@boku.ac.at; ilse.schwarzinger@boku.ac.at

© The Author(s) 2018
S. Schmutz, J. Sendzimir (eds.), *Riverine Ecosystem Management*, Aquatic Ecology
Series 8, https://doi.org/10.1007/978-3-319-73250-3_19

increasing activity in developing, adapting, and testing of aquatic biomonitoring methods in Africa, Asia, and South America.

The need for aquatic biomonitoring is obvious, as rivers in most parts of the world are extremely overexploited and impacted in manifold ways. Assessing the status of rivers and identifying the threats is essential to develop adequate restoration and protection strategies. Each river or river reach is characterized by a unique signature of different biological and ecological characteristics as well as a large variety of pressures and impacts. Such a signature cannot be documented sufficiently by physical-chemical monitoring only. Biological monitoring covers a larger spectrum of pressures and multiple spatial scales over a longer time span. State-of-the-art environmental monitoring combines chemical and biological indicators in assessing the ecological conditions of aquatic systems.

19.2 History of Water Quality Assessment

The Ecological Assessment of River Quality Is Older than the Science of Ecology

There is a simple answer to the question "Since when do we need water quality assessment?": since humans destroyed their surface waters in a way that deteriorated drinking water quality. The first written record of water pollution was given about 350 years before Christ, when Aristotle reported on "black mud" and "red tubes"—as he called it—growing out of a "white slime" in brooks of the city Megara polluted by domestic sewage (Thienemann 1912). The famous Greek philosopher was the first who linked human pressures with observations of oxygen reduction (black decaying mud), a community of *Beggiatoa* sulfur bacteria (white slime), oligochaete sludge worms, and chironomids (red tubes). Aristotle's knowledge fell into oblivion, and the beginning of water quality assessment had to wait for about 1800 years. Anyhow, observations of a correlation between the composition/distribution of certain aquatic invertebrate species and different water pollution levels are not very recent findings. One could even say that this knowledge is older than "ecology" itself (as defined by Ernst Haeckel in 1866). As early as Kolenati (1848), it was already concluded by F. A. Kolenati that the absence of caddis larvae from a stream can be caused by the presence of factories upstream (*Stettiner Entomologische Zeitung 9*). Triggered by the severe cholera epidemics in Europe, two researchers, A. H. Hassal, London (1850), and F. Cohn, Breslau (1853), discovered and published the relationship between organic pollution, river fauna, and the quality of drinking water based on bioindicators. In the United States, the earliest biomonitoring research originates from Forbes (1887) who invented the biological community concept. Basically, using this concept plant and animal communities of a river were used to assess the degree of organic pollution. Around 1900 two German scientists (R. Kolkwitz & M. Marsson) studied polluted rivers around Berlin and described defined communities of organisms in different zones of organic enrichment. They developed the concept of "biological indicators of pollution" in their so-called saprobic system, which is still in use in several Central and Eastern European states.

19.3   The Saprobic System

Reliable and True and Frequently Used

Based on earlier research before the turn of the century, Kolkwitz and Marsson (1902) introduced the terms "Saprobien" for waste water organisms and "Katharobien" for organisms in clean rivers. From the stressor's point of view, saprobity is the state of the water quality resulting from organic enrichment as reflected by the species composition of the community. Kolkwitz and Marsson published indicator lists for benthic algae and invertebrates, which served as a valuable tool for water quality assessment for some decades.

The saprobic system was adapted after World War II by Liebmann (1951) who published a widespread manual on saprobiology and provided a substantial list of indicators. On this basis he introduced water quality mapping, which permits the visualization of a river's ecological status as "color-banded." The power of such mapping techniques to convey complex information in a convincing way is evident in how it has stimulated politicians, decision-makers, water managers, and other stakeholders and the interested public to combat pollution. The acceptance of the saprobic system was increased remarkably by the development of the saprobic index by Pantle and Buck (1955) enabling quantification of pollution intensity. This index ranging from 1 (very good quality) to 4 (extremely poor quality) could be easily interpreted by the end users. At about the same time in the United States, Beck (1954) created a biotic index to provide a simple measurement of stream pollution and its effects on the biology of the stream. This development very likely happened independently, because both authors did not cite each other's papers. Zelinka and Marvan (1961) modified the saprobic index by including the concept of saprobic valencies. They introduced a system reflecting the 100% (often bell-shaped) occurrence of a taxon among the water quality classes, i.e., ten points substituting the 100% are distributed among the four water quality classes. In Sládecek (1973) Sládecek summarized the knowledge in his book "System of Water Quality from the Biological Point of View" which served the following decades as a methodological bible for saprobiologists. In this time, the saprobic system was widely used in Central and Eastern Europe (e.g., Austria, Bulgaria, Czech Republic, Germany, Hungary, Romania, Slovakia, Slovenia, and former Yugoslavia).

The last update of the saprobic approach was precipitated around the millennium, when the European Water Framework Directive 2000/60/EC (WFD) substantially changed the biomonitoring approach of European aquatic ecosystems. Since then, the ecological status of water bodies needs to be defined based on type-specific approaches and reference conditions. Several countries decided to integrate the saprobic approach into the new integrative methodology for defining the ecological status of water bodies and thus to adjust the saprobic system. The revisions comprised alterations and additions to the list of indicator taxa, type-specific saprobic reference conditions (Rolauffs et al. 2004) and an adaptation to the ecological status classification of the WFD. Currently, the saprobic system is part of the multi-metric indices used in Austria (Ofenböck et al. 2004, 2010a), Czech Republic (Kokes et al. 2006), and Germany (Meier et al. 2006).

19.4   Biotic Indices and Scoring

Hundreds of Ways Toward One Goal

Since the early stages of water quality evaluation, some hundreds of methods for biological river status assessment have been developed (Birk et al. 2012). Unfortunately, the mathematical terms "index" and "score" in river status assessment are often used in a confusing way. A *biotic index* is a numerical expression of the sensitivity or tolerance of organism assemblages to anthropogenic stress. A *score* is a numeric expression of the ecological indicator status that can be used to calculate an index, which can be generated, e.g., as an average of scores of several indicators. The principle of biotic indices is to assign different types of taxa to different levels of disturbance. Sensitive taxa decrease or disappear, and tolerant taxa emerge or increase under stress.

The first indices were nearly simultaneously developed in the United States and Europe around 1950 (Beck 1954; Pantle and Buck 1955). The Trent Biotic Index (developed by Woodiwiss 1964, 1978) is seen as the origin of many biotic indices that are not following the saprobic approach, e.g., the "Indice biotique" (IB) in France (Verneaux and Tuffery 1967), the Belgian Biotic Index (BBI) (De Pauw and Vanhoren 1983), the "Indice biotico esteso" (IBE) in Italy (Ghetti 1986), and many others (Birk and Hering 2002). The Woodiwiss method combines quantitative measures of taxa richness with qualitative information on the sensitivity/tolerance of key indicator taxa.

From a mathematical point of view, the Woodiwiss approach (Trent Biotic Index) does not represent an index calculated with a formula. The biological quality value of a water body is accomplished through the use of the classification table (Table 19.1). The resultant "index" between 1 and 10 is the consequent number at the crossing point of two entrances in the fitting row and column: (1) the vertical row corresponding to the value of number of taxa in the sample and (2) the horizontal column with the fitting key indicator taxa. The biological quality value (1–10) can be transformed into quality class through conversion tables that may vary in different countries, river types, ecoregions, etc. Table 19.1 presents the original table from Woodiwiss (1964). For defining the number of taxa in the sample, i.e., the vertical entrance into the table, the determination level is given for each class or order, e.g., Ephemeroptera and Plecoptera are counted on genus level, Trichoptera and Diptera on family level. The sum of all genera and families in a sample reflects the number of taxa.

The basis of most currently used biotic systems is the Biological Monitoring Working Party system (BMWP) set up by the British Department of the Environment and recommended as biological classification system for national river pollution surveys (Armitage et al. 1983; Hawkes 1997). The BMWP sums up the tolerance scores of all macroinvertebrate families in the sample. Like the saprobic index, the BMWP is based on grouping benthic macroinvertebrates into categories depending on their response to organic pollution. Stoneflies or mayflies, for instance, indicate the cleanest waters and are given a tolerance score of 10. The lowest score (1) is allocated to Oligochaeta, which is regarded to be the most tolerant to pollution.

Table 19.1 Classification table for deriving the original Trent Biotic Index (Woodiwiss 1964)

Key indicator group	Diversity of fauna	Total number of groups present				
		0–1	2–5	6–10	11–15	16+
Plecoptera nymphs present	More than one species present	–	7	8	9	10
	Only one species present	–	6	7	8	9
Ephemeroptera nymphs present	More than one species present[a]	–	6	7	8	9
	Only one species present[a]	–	5	6	7	8
Trichoptera larvae present	More than one species present[b]	–	5	6	7	8
	Only one species present[b]	4	4	5	6	7
Gammarus present	All above species absent	3	4	5	6	7
Asellus present	All above species absent	2	3	4	5	6
Tubificid worms and/or red Chironomid larvae present	All above species absent	1	2	3	4	–
All above types absent	Some organisms such as *Eristalis tenax* not requiring oxygen may be present	0	1	2	–	–

[a]*Baetis rhodani* excluded
[b]*Baetis rhodani* is counted in this section for the purpose of classification

Table 19.2 Original BMWP score table

Families	Scores
Siphlonuridae, Heptageniidae, Leptophlebiidae, Ephemerellidae, Potamanthidae, Ephemeridae, Taeniopterygidae, Leuctridae, Capniidae, Perlodidae, Perlidae, Chloroperlidae, Aphelocheridae, Phryganeidae, Molannidae, Beraeidae, Odontoceridae, Leptoceridae, Goeridae, Lepidostomatidae, Brachycentridae, Sericostomatidae	10
Astacidae, Lestidae, Agriidae, Gomphidae, Cordulegastridae, Aeshnidae, Corduliidae, Libellulidae	8
Caenidae, Nemouridae, Rhyacophilidae, Polycentropodidae, Limnephilidae	7
Neritidae, Viviparidae, Ancylidae, Hydroptilidae, Unionidae, Corophiidae, Gammaridae, Platycnemididae, Coenagrionidae	6
Mesoveliidae, Hydrometridae, Gerridae, Nepidae, Naucoridae, Notonectidae, Pleidae, Corixidae, Haliplidae, Hygrobiidae, Dytiscidae, Gyrinidae, Hydrophilidae, Clambidae, Helodidae, Dryopidae, Elmidae, Chrysomelidae, Curculionidae, Hydropsychidae, Tipulidae, Simuliidae, Planariidae, Dendrocoelida	5
Baetidae, Sialidae, Piscicolidae	4
Valvatidae, Hydrobiidae, Lymnaeidae, Physidae, Planorbidae, Sphaeriidae, Glossiphoniidae, Hirudidae, Erpobdellidae, Asellidae	3
Chironomidae	2
Oligochaeta (whole class)	1

The ASPT (Average Score per Taxon) equals the average of the tolerance scores of all macroinvertebrate families found and thus ranges from 1 to 10 (Table 19.2). The main difference between both indices is that the BMWP system represents the indicative value of taxa diversity while the ASPT does not depend on the family richness. Formerly used as a single or double metric method, nowadays, multimetric approaches often include national adaptations of the BMWP as a core metric.

The original BMWP method "works" at the family level. Methods developed more recently outside Europe are resolved at higher taxonomic resolutions (Table 19.3), such as the genus and species levels (e.g., HKHbios (Asia), Ofenböck et al. 2010b; ETHbios (Ethiopia), Aschalew and Moog 2015)

Table 19.3 Taxa and sensitivity scores of benthic macroinvertebrates used in ETHbios calculations (Aschalew and Moog 2015)

Common name	Taxon	Score
Stone flies	Perlidae (*Neoperla* sp.)	10
Caddis flies	Lepidostomatidae, Philopotamidae	10
Beetles	Scirtidae	10
Mayflies	Baetidae > 2 spp., *Acanthiop* sp., Heptageniidae (*Afronurus* sp.), Leptophlebiidae	9
Caddis flies	Hydropsychidae > 2 spp.	9
Mayflies	Tricorythidae	8
Caddis flies	Leptoceridae, Ecnomidae	8
Beetles	Psephenidae, *Stenelmis* sp., *Microdinodes* sp.	8
Water mites	Hydracarina	8
Crabs	Potamidae	7
Dragonflies, damselflies	Aeshnidae, Lestidae	7
Beetles	Elmidae	7
Crane-flies	Tipulidae	7
Mollusca	*Pisidium* sp.	7
	Limpets	6
Mayflies	Baetidae with 2 spp., Caenidae	6
Caddis flies	Hydropsychidae with 2 spp.	6
Dragonflies	Gomphidae	6
Water bugs	Naucoridae	6
Horse-flies	Tabanidae	6
Caddis flies	Hydropsychidae with 1 sp.	5
Dragonflies	Coenagrionidae, Libellulidae	5
Water striders	Mesoveliidae, Veliidae, Gerridae	5
Beetles	Hydrophilidae, Dytiscidae, Gyrinidae, Haliplidae	5
Flies	Ceratopogonidae excl. Bezzia-Gr.	5
Mayflies	Baetidae with 1 sp.	4
Water bugs	Corixidae, Pleidae	4
	Belostomatidae, Notonectidae, Nepidae	3
Leeches	Hirudinea	3
Snails	Physidae, *Bulimus* sp.	3
Midges and Flies	Bezzia-group	3
	Musidae, Chironomidae with predominantly Tanytarsini and Tanypodinae	2
	Psychodidae, Ephydridae, Culicidae, Red Chironomidae, *Chironomus* sp., Syrphidae	1
Worms	Oligochaeta	1

Many methodological textbooks mention diversity indices with respect to biotic indices, e.g., Hawkes (1982), Washington (1984), Metcalfe (1989), Johnson et al. (1993), and Resh and Jackson (1993). The diversity approach uses species richness (mostly measured as the total number of taxa), abundance (measured as the number of individuals of each taxon), and evenness (the degree to which each taxon is equally represented) as components of community structure. Unstressed communities are said to be characterized by high diversity (taxa richness) and even distribution of individuals among species. Although a variety of diversity indices exists, there is no normative procedure that can be used for river quality evaluation solely (Kohmann and Schmedtje 1986). In any case, diversity indices are often used in combination with other metrics as a component of multi-metric indices.

19.5 The Multivariate Approach

Sophisticated, but Laborious in Development
Multivariate or model-based procedures are predictive systems that assess the deviation between the observed aquatic community and reference conditions predicted from environmental parameters, (e.g., reference condition approach). Models are developed to explain the composition and variability in the aquatic communities among reference sites. The models include a range of environmental parameters. Based on multivariate procedures, the model then predicts what biota should be present at an undisturbed "target" site or river type with a given set of environmental attributes. A study site can be considered in a "very good" or "reference condition" if the aquatic community found at the test site is similar to the predicted one. A study site is considered disturbed if the benthic community observed at the test site is different from the prediction.

Three prerequisites are necessary to successfully apply a multivariate prediction system:

1. A sound knowledge of the species inventory and composition, as well as the spatial and seasonal distribution of the target biota under reference conditions
2. A clear understanding of the criteria that define reference conditions
3. Models that reliably predict the biota for a particular site or river type given the natural variability of environmental conditions

The first remarkable predictive bioassessment tool was RIVPACS, the "River Invertebrate Prediction and Classification System," developed in the United Kingdom (Wright et al. 1989; Wright et al. 2000; RIVPACS 2005). In the mid-1990s, the BEnthic Assessment of SedimenT (BEAST) was developed in Canada (Reynoldson et al. 1995). Based on the mathematic principles of RIVPACS, comparable systems have been created in some other countries. To assess the biological health of Australian rivers, the AUSRIVAS (Australian River Assessment Scheme) was developed under the National River Health Program (NRHP) by the federal government in 1994 (AUSRIVAS 2005). The Australian scheme is distinguished by several differences:

the major habitats are sampled and modeled separately, and different models are used for different bioregions over Australia (Simpson and Norris 2000). Since 2001 the PERLA system is in operation in the Czech Republic (Kokes et al. 2006).

Verdonschot (1990) described macrofaunal site groups (cenotypes) in surface waters in the Netherlands, which are recognized on the basis of environmental variables and the abundance of taxa. These cenotypes were described as groups of taxa lumped together based on their limited internal variation. They are distinguished not by zones of overlap in their tolerances or occurrence, since no clear boundaries were provided, but only by a recognizable centroid. The cenotypes are mutually related in terms of key factors, which represent major ecological processes. The cenotypes and their mutual relationships form a web that offers an ecological basis for the daily practice of water and nature management. The web allows the development of water quality objectives, provides a tool to monitor and assess, indicates targets, and guides the management and restoration of water bodies (Verdonschot and Nijboer 2000).

The European Fish Index (EFI) was the first pan-continental model-based index developed for assessing the ecological status of European rivers (Pont et al. 2006). The EFI employs ten metrics describing conditions of fish assemblage regarding feeding, migration, habitat and spawning preferences and tolerance to anthropogenic stress. Site-specific reference conditions are predicted using multiple regression models. An updated version (EFI+) also considers fish length and river-type-specific responses in trout and cyprinid rivers (EFI+ Consortium 2009).

19.6   The Multi-metric Approach

Simple, but Virtuous

The multi-metric approach is currently the most common method among the sophisticated procedures of river status assessment. In the late 1970s limnologists came face to face with the fact that mechanisms of environmental degradation are usually complex, and their combined effects are not easy to measure. Estimates of the biotic integrity of a water body may be the best tool to assess the effect of multiple stressors in aquatic environments. Biotic integrity was defined as "the ability to support and maintain a balanced, integrated, adaptive community of organisms having a species composition, diversity, and functional organization comparable to that of the natural habitat of the region" (Frey 1977; Karr and Dudley 1981). To communicate biological information in a meaningful way, Jim Karr developed the Index of Biotic Integrity (IBI) in the early 1980s and published one of the most cited papers "Assessment of biotic integrity using fish communities" (Karr 1981). The IBI has proven to be very adaptable (Karr and Chu 1999; Simon and Lyons 1995), and quite soon, it has subsequently been adapted for use throughout many states of the United States, and later in many countries of all other continents. Initially developed to monitor fish, the multi-metric approach became extended for aquatic macroinvertebrates, terrestrial macroinvertebrates, macrophytes, algae, wetlands, riparian zones, large rivers, lakes, reservoirs, estuaries, and brackish water ecosystems.

The much-noted experiences of the EU-funded AQEM and STAR projects show that the multi-metric approach is a valuable procedure for bridging the gap between the methodologies and the needs for evaluating the ecological status of water bodies (Hering et al. 2004, 2006; Furse et al. 2006). The multi-metric approach fits quite well to the methodological demands of the European Water Framework Directive, since it attempts to provide an integrated analysis of the biological community of a site by deriving a variety of quantifiable biological characteristics (*metrics*) and knowledge of a site's fauna (Karr and Chu 1999).

Within a multi-metric index, each single metric is related predictably and reasonably to specific impacts caused by environmental alterations. For example, while the proportion of different feeding types is suited to assess the trophic integrity of an ecosystem, saprobic or acid indices provide a measure to directly assess the impact of certain pollutants and acidification. Thus, the multi-metric index considers multiple impacts and combines individual metrics (e.g., saprobic indices, diversity indices, feeding type composition, current preferences, etc.) into a nondimensional index, which can be used to assess a site's overall condition. By combining different categories of metrics (e.g., taxa richness, diversity measures, proportion of sensitive and tolerant species, trophic structure) reflecting different environmental conditions and aspects of the community, the multi-metric assessment is regarded as a more reliable tool than assessment methods based on single metrics.

The following metric types can be distinguished:

- Composition/abundance metrics. Metrics giving the relative proportion of a taxon or taxonomic group with respect to total abundance. Abundance (or biomass) of a taxon or taxonomic group and/or total abundance (or biomass).
- Richness/diversity metrics. Metrics giving the number of species, genera, or higher taxa within a certain taxonomical entity, including the total number of taxa or diversity indices.
- Sensitivity/tolerance metrics. Metrics related to taxa known to respond sensitively or tolerantly to a stressor or a single aspect of the stressor, either using presence/absence or abundance information.
- Functional metrics. Metrics addressing the ecological function of taxa (other than their sensitivity to stress), such as feeding types/guilds, habitat and flow velocity preferences, ecosystem type preferences, life cycle parameters, and biometric parameters. They can be based on taxa or abundance.

The procedure of data analysis during the development of a multi-metric index typically involves the following steps (Fig. 19.1):

- Selection of the most suitable form of a multi-metric index
- Metric calculation and selection
  - Selection of candidate metrics
  - Exclusion of numerically unsuitable metrics
  - Definition of a stressor gradient
  - Correlation of stressor gradients and metrics
  - Selection of core metrics

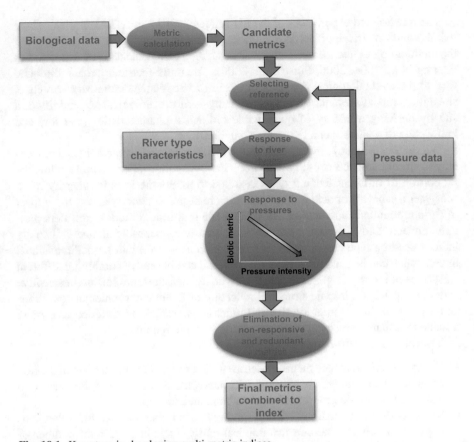

Fig. 19.1 Key steps in developing multi-metric indices

- Distribution of core metrics within the metric types and exclusion of redundant metrics
- Definition of upper and lower anchors and scaling

- Generation of a multi-metric index

  - Development of a general multi-metric index
  - Development of a stressor-specific multi-metric index

- Setting class boundaries
- Interpretation of results

Depending on purpose, ecosystem type, organism group, and available data multi-metric indices may be designed differently. In many cases, a general assessment reliably reflecting the integrity of an ecosystem is sufficient. In other cases, more specific data on the causes of deterioration is required. Thus, we distinguish two main forms of multi-metric index: (1) the general approach and (2) the stressor-specific approach (e.g., Ofenböck et al. 2004). Stressor-specific multi-metric indices can only be derived if data reflecting different specific stress types and environmental gradients are available and the autecology of the targeted organism group is well known.

The multi-metric index concept has proven to be very adaptable (Karr and Chu 1999), and many of the same metrics have been used successfully throughout different regions of the world in a variety of stream types (Simon and Lyons 1995). Metrics such as species richness (the total number of taxa) or the EPT approach (number or individuals or % share of Ephemeroptera, Plecoptera, and Trichoptera taxa) are common to most benthic invertebrate-based multi-metric indices.

19.7 Integrative Assessment Systems

The most sophisticated evaluation approaches are based on the use of a wide range of organisms that allows an integrated assessment of rivers. In the United States, the Rapid Bioassessment Protocols (RBPs) use biological indicators to infer data about running water quality. RBPs were introduced on a national level in the mid-to-late 1980s (Barbour et al. 1999). There are three main types of RBPs for streams⊠fish surveys, periphyton surveys, and macroinvertebrate surveys⊠each with detailed method descriptions. The macroinvertebrate survey is most commonly used, because it requires reasonable expertise or equipment. The EPA encourages the use of RBPs because they provide quick and valid results while being cost effective, time efficient, and minimally invasive (Barbour et al. 1999).

In the southern part of Africa, the SASS (Southern African Scoring System) is seen suitable for the assessment of the ecological integrity of river ecosystems (Dallas 1995, 2007; Dickens and Graham 2002). The South African Assessment Scheme (SAFRASS) protocols use three biotic indicator groups (diatoms, macroinvertebrates, and macrophytes) that respond to changes in river conditions (Lowe et al. 2013).

Since 2000 the Water Framework Directive (WFD) provides a common legal framework for water management in the European Union. The major aim of the WFD is to achieve good ecological status of all European waters (lakes, rivers, and groundwater bodies, transitional, and coastal waters) by 2027 at the latest. Based on annexes II and V of this directive, the EU member states use an integrated system to evaluate the "ecological status" of rivers based on various environmental and biotic features, the so-called quality elements (QE): water chemistry, hydro-morphology, algae, macrophytes, phytoplankton, benthic invertebrates, and fish. The classification scheme for the ecological status of water bodies includes five status classes: (1) very good; (2) good; (3) moderate; (4) poor; and (5) bad. Based on the assessment results of the single QEs, the worst assessment result for a BQE determines the overall assessment result (the "one-out-all-out" principle, Fig. 19.2).

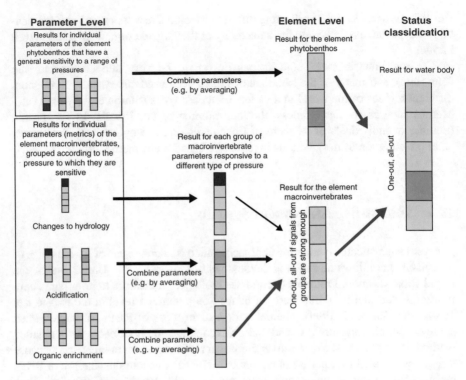

Fig. 19.2 The one-out-all-out principle of the EU Water Framework Directive (*WFD CIS Guidance Document No. 13: Overall Approach to the Classification of Ecological Status and Ecological Potential*)

19.8 Indicator Groups

Various organisms have been used in the assessment of the water quality and ecological integrity of aquatic ecosystems, including bacteria, protozoans, algae, macrophytes, benthic invertebrates, fish, and birds (Roux et al. 1993; Barbour et al. 1999; Bryce et al. 2002). The most frequently used groups are benthic invertebrates, algae, macrophytes, and fish. Current integrative methodologies, such as the US EPA bioassessment protocols, the European Water Framework Directive, or the South African SAFRASS approach, make use of more than one indicator group to evaluate the ecological quality of a water body. To avoid redundant information and thus unnecessary costs, those groups are used to indicate effects of specific stressors on the environment in the most effective way. Algae are perfect indicators to describe the effects of nutrients and eutrophication. The benefit of macrophyte bioindication is to document the effects of long-term nutrient aspects and hydro-morphological impairments. Benthic invertebrates are ideal indicators of (organic) pollution and hydro-morphological deficits at the micro-habitat scale. Fish are supreme indicators to study the effects of hydro-morphological deficits on the meso-habitat and reach scale, including lateral and longitudinal connectivity up to the basin scale.

19.8.1 Periphyton (Contributed by Peter Pfister, ARGE Limnologie, Innsbruck, Austria)

Tiny, Quick, and Beautiful

Algae and cyanobacteria are valuable indicators of environmental conditions in running and standing water bodies. As primary producers, periphyton acts as an important foundation of food webs in river ecosystems (Li et al. 2010). Because the assemblages usually attach to substrate, their growing and prospering can respond directly and sensitively to many kinds of physical, chemical, and biological variation occurring in the river reach, including temperature, nutrient levels, current regimes and grazing, etc. Algae are omnipresent in all types of water bodies, and species-rich communities can be found from pristine spring brooks to the effluents of wastewater treatment plants. In Central European running waters, Hürlimann and Niederhauser (2007) recorded densities between 10^3 and 10^6 individuals/cm^2 on stony surfaces. Their cosmopolitan character and worldwide distribution predestine them as ideal group for nationwide applicable assessment systems. There is a rich knowledge about their ecological requirements, tolerances, and preferences compared to other indicators (Arzet 1987; Hürlimann and Niederhauser 2007; Oemke and Burton 1986; Coring et al. 1999; Rott et al. 1997, 1999; Schmedtje et al. 1998; Tümpling and Friedrich 1999). Rapid reproduction rates and very short life cycles allow algae to react quickly to environmental change. Therefore, periphyton can be expected to reflect short-term impacts and sudden changes in the environment.

From a methodological point of view, the diatoms have a lot of practical advantages: they are comparatively easy to be identified in any stage of their life, one doesn't need a permit to take samples, and the storage of samples or preparations mounted on slides is cheap. On the other hand (which also might be a methodological benefit as it reduces the statistical "noise"), most algae show only little dependency from physical or hydro-morphological factors like flow velocity, substrate type, stream modification, residual flows, interruption of the continuum, and others.

19.8.2 Macrophytes (Contributed by Karin Pall, Systema GmbH, Vienna, Austria)

Habitat and Food in One

Macrophytes as autotrophic organisms, first of all, are highly sensitive to nutrient enrichment. This also applies to phytoplankton or phytobenthos. However, two important differences distinguish macrophytes from the latter: their reaction to changes in the trophic state as well as the conclusions which can be drawn from the assemblages found. In principle, all these groups of organisms respond to changes in the trophic level with changes in species spectrum and abundance, though phytoplankton and phytobenthos react much more quickly than macrophytes. Therefore, the former can serve as excellent short-term indicators for rapid detection of changes in the trophic condition of lakes or rivers. However, repeated investigations are necessary to derive reliable results.

In contrast to phytoplankton and phytobenthos assemblages, macrophyte communities do not show a sudden reaction to changes of the trophic level. They integrate the prevailing conditions over a longer time period. The analysis of the species composition and other features of the macrophyte vegetation thus is a particularly well-suited tool for monitoring long-term trends in trophic conditions. Even from a unique mapping, sound and temporally integrated information about the nutrient conditions in lakes or rivers can be derived. For this reason, the use of macrophytes as indicator organisms for nutrient enrichment already has a long tradition. Usually, as help for water protection institutes, the focus was on the exact localization of organic or nutrient pollution sources along lakeshores or river courses.

However, macrophytes do not solely reflect trophic conditions. They also respond very sensitively to other environmental impacts. In lakes they respond specifically to changes of the hydrological regime (alteration of the natural water level fluctuations) or hydrodynamic conditions (e.g., changes of the wave frequency and intensity by motorboats or navigation). In rivers they have proven to be excellent indicators for changes in the flow regime (e.g., potamalisation or rhithralisation). Furthermore, the specific composition of the macrophyte community is a pronounced reflection of the structural conditions found along the shores and in the water body of lakes and rivers, such as e.g., substrate diversity and dynamic or the degree of embankments. Therefore, last but not least, the macrophyte vegetation can serve as indicator for the structural alterations of the shoreline and the quality of the water-land-linkage.

In particular, two properties make macrophytes highly valuable indicators. On the one hand, it is their longevity. They remain at the same sites over several vegetation periods and thus can integrate the site conditions over a considerably longer time period than other quality elements as e.g. phytoplankton and phytobenthos. On the other hand, macrophytes always remain in the same place and are thus not able to avoid pressures and other environmental impacts, e.g., benthic invertebrates or fish. This enables an accurate localization of the sources and the spheres of impact of pressures.

19.8.3 Aquatic Macroinvertebrates

Tiny but Many: Helpful Creepy-Crawlers
Benthic macroinvertebrates are the most widely used indicator group for lotic systems. There are several advantages to using benthic macroinvertebrates in bioassessment, because they constitute a substantial proportion of freshwater biodiversity and are critical to ecosystem function. The following list summarizes briefly the advantages of benthic macroinvertebrate bioindication (Danecker 1986; Hellawell 1986; Moog 1988; Metcalfe 1989; Rosenberg and Resh 1992; Metcalfe-Smith 1994; Ollis et al. 2006).

- Benthic macroinvertebrates are widespread and can be found in most aquatic habitats.
- There are a large number (thousands) of species.
- From the point of systematics and phylogeny, they are a highly diverse group, which makes them excellent candidates for studies of changes in biodiversity.
- Different systematic groups of macroinvertebrates have different environmental needs and tolerances to pollution or other kinds of stressors.
- Benthic invertebrates cover a broad range of micro- and meso-habitats, ecotones, biocoenotic regions, trophic position (trophic interaction), etc.
- Macroinvertebrates feed on micro-/mesofauna as well as on algae and are the primary food source for fish. Therefore, an impact on macroinvertebrates impacts the food web and designated uses of the water resource.
- Small order streams often do not support fish but do support rich macro-invertebrate communities.
- Macroinvertebrates are to some extent mobile and can actively select habitats that fulfill their environmental needs.
- On the other hand, benthic invertebrates have limited mobility, and thus they are indicators of local environmental conditions.
- Since benthic invertebrates retain (bioaccumulate) toxic substances, chemical analysis will allow detection in them where levels are undetectable in the water resource.
- A biologist experienced in macroinvertebrate identification will be able to determine relatively quickly whether the environment has been degraded by identifying changes in the benthic community structure.
- Benthic macroinvertebrates have the ideal size to be easily collected and identified.
- Sampling of macroinvertebrates under a rapid assessment protocol is easy, requires few people and minimal equipment, and does not adversely affect other organisms.
- In the industrialized world, there is a good knowledge on identification, procession, and evaluation of benthic invertebrates.

19.8.4 Fish

Tasty and Valuable Indicators

Fish communities respond significantly and predictably to many kinds of anthropogenic disturbances, including eutrophication, acidification, chemical pollution, flow regulation, physical habitat alteration, fragmentation and introduced species (Li et al. 2010). Their sensitivities to the health of surrounding aquatic environments form the basis for using fishes to monitor environmental degradation. Over the last three decades, a variety of fish-based indices have been widely used to assess river quality, and the use of multi-metric indices, inspired by the index of biotic integrity (IBI), has grown rapidly.

- Fishes are present in most surface waters except in cases of stream size and migration restrictions.
- The identification of fishes is relatively easy, and their taxonomy, ecological requirements, and life histories are generally better known than in other species groups.
- Fish presence corresponds strongly to changes in hydrological and environmental flow patterns, while other biological quality elements (e.g., macroinvertebrates) hardly indicate these impacts.
- Fishes have evolved complex migration patterns, making them sensitive to continuum interruptions.
- The longevity of many fish species enables assessments to be sensitive to disturbance over relatively long time scales.
- The natural history and sensitivity to disturbances are well documented for many species, and their responses to environmental stressors are often known.
- Fishes generally occupy high trophic levels and thus integrate conditions of lower trophic levels. In addition, different fish species represent distinct trophic levels: omnivores, herbivores, insectivores, planktivores, and piscivores.
- Fishes occupy a variety of habitats in rivers: benthic, pelagic, rheophilic, limnophilic, etc. Species have specific habitat requirements and thus exhibit predictable responses to human-induced habitat alterations.
- Depressed growth and recruitment are easily assessed and reflect stress.
- Fishes are valuable economic resources and are of public concern. Using fishes as indicators confers an easy and intuitive understanding of cause effect relationships to stakeholders beyond the scientific community.

There is a common agreement that the performance of any biological assessment approach increases with the quality rating of its ecological background (Verdonschot and Moog 2006). Consequentially there was a remarkable increase of taxa lists that associated ecological information with indicator taxa in the last 10–15 years. These taxa lists include functional ecosystem characteristics, species traits, and others more in ecological assessment (see Chap. 20).

References

Armitage PD, Moss D, Wright JF, Furse M (1983) The performance of a new biological water quality score system based on macroinvertebrates over a wide range of unpolluted running water sites. Water Res 17:333–347

Arzet K (1987) Diatomeen als pH-Indikatoren in subrezenten Sedimenten von Weichwasserseen. Diss Abt Limnol Innsbruck 24:1–266

Aschalew L, Moog O (2015) Benthic macroinvertebrates based new biotic score "ETHbios" for assessing ecological conditions of highland streams and rivers in Ethiopia. Limnologica 52:11–19

AUSRIVAS – Australian River Assessment System (2005) AUSRIVAS bioassessment: macroinvertebrates. http://ausrivas.canberra.edu.au/Bioassessment/Macroinvertebrates/

Barbour MT, Gerritsen J, Snyder BD, Stribling JB (1999) Rapid bioassessment protocols for use in streams and wadeable rivers: periphyton, benthic macroinvertebrates and fish, 2nd edn. Washington, DC: US Environmental Protection Agency. www.epa.gov/owow/monitoring/rbp/. Viewed 8 Oct 2005. European Union, 2000. Directive 2000/60/EC of the European Parliament and of the Council. Official Journal of the European Communities, 72 pp

Beck WM (1954) Studies in stream pollution biology. I. A simplified ecological classification of organisms. Q J Florida Acad Sci 17(1954):211–227

Bere T, Nyamupingidza BB (2013) Use of biological monitoring tools beyond their country of origin: a case study of the South African Scoring System version 5 (SASS5). Hydrobiologia 722:223–232

Birk S, Hering D (2002) Waterview web-database – a comprehensive review of European assessment methods for rivers. FBA news 20(4)

Birk S, Bonne W, Borja A, Brucet S, Courrat A, Poikane S, Solimini A, van de Bund W, Zampoukas N, Hering D (2012) Three hundred ways to assess Europe's surface waters: an almost complete overview of biological methods to implement the Water Framework Directive. Ecol Indic 18:31–41

Bryce SA, Hughes RM, Kaufmann PR (2002) Development of a bird integrity index: using bird assemblages as indicators of riparian condition. Environ Manag 30(2):294–310. https://doi.org/10.1007/s00267-002-2702-y

Cohn F (1853) Über lebendige Organismen im Trinkwasser. Z klin Medizin 4:229–237

Coring E, Schneider S, Hamm A, Hofmann G (1999) Durchgehendes Trophiesystem auf der Grundlage der Trophieindikation mit Kieselalgen. Deutscher Verband für Wasserwirtschaft und Kulturbau e.V. (DVWK). 219 S. + Anhang

Dallas HF (1995) An evaluation of SASS (South African Scoring System) as a tool for the rapid bioassessment of water quality. MSc thesis, University of Cape Town, Cape Town

Dallas HF (2007) The influence of biotope availability on macroinvertebrate assemblages in South African rivers: implications for aquatic bioassessment. Freshw Biol 52(2):370–380

Danecker E (1986) Makrozoobenthos-Proben in der biologischen Gewässeranalyse. Wasser und Abwasser 30:325–406

De Pauw N, Vanhoren G (1983) Method for biological quality assessment of watercourses in Belgium. Hydrobiologia 100:153–168

Dickens CWS, Graham PM (2002) The South African Scoring System (SASS) version 5 rapid bioassessment method for rivers. Afr J Aquat Sci 27:1–10

EFI+ Consortium (2009) Manual for the application of the new European fish index – EFI+, EU project improvement and spatial extension of the European fish Index. Vienna

Forbes SA (1887) The lake as a microcosm. Bull Sci Assoc, Peoria, IL, 1887, S. 77–87, wiederveröffentlicht. Illinois Nat Hist Survey Bulletin, 15/9/Jahrgang, S. 537–550

Frey DG (1977) Biological integrity of water—a historical approach. In: Ballentine RK, Guarria LJ (eds) The integrity of water. Proceedings of a symposium. U.S. Environmental Protection Agency, Washington, pp 127–140

Furse M, Hering D, Moog O, Verdonschot P, Sandin L, Brabec K, Gritzalis K, Buffagni A, Pinto P, Friberg N, Murray-Bligh J, Kokes J, Alber R, Usseglio-Polatera P, Haase P, Sweeting R, Bis B, Szoszkiewicz K, Soszka H, Springe G, Sporka F, Krno I (2006) The STAR project: context, objectives and approaches. Hydrobiologia 566:3–29

Ghetti PF (1986) I macroinvertebrati nell'analisi di qualitá dei corsi d'acqua. Trento: Provincia Autonoma di Trento. 169 p. Manuale di applicazione. Volume allegato agli atti del convegno "esperienze e confronti nell'applicazione degli indici biotici in corsi d'acqua italiani"

Hassal A (1850) A microscopic examination of the water supplied to the inhabitants of London and suburban districts. Samuel Highley, London

Hawkes HA (1982) Biological surveillance of rivers. Water Pollut Control 81(3):329–342

Hawkes HA (1997) Origin and development of the biological monitoring working party score system. Water Res 32(3):964–968

Hellawell JM (1986) Biological indicators of freshwater pollution and environmental management. Elsevier Applied Science, London, 546 pp

Hering D, Moog O, Sandin L, Verdonschot PFM (2004) Overview and application of the AQEM assessment system. Hydrobiologia 516:1–20

Hering D, Feld CK, Moog O, Ofenboeck T (2006) Cook book for the development of a multimetric index for biological condition of aquatic ecosystems: experiences from the European AQEM and STAR projects and related initiatives. Hydrobiologia 566:311–324

Hürlimann J, Niederhauser P (2007) Methoden zur Untersuchung und Beurteilung der Fliessgewässer. Kieselalgen Stufe F (flächendeckend). Umwelt-Vollzug Nr. 0740. Bundesamt für Umwelt, Bern. 130 S

Johnson RK, Wiederholm T, Rosenberg DM (1993) Freshwater biomonitoring using individual organisms, populations and species assemblages of benthic macroinvertebrates. In: Rosenberg DM, Resh VH (eds) Freshwater biomonitoring and benthic macroinvertebrates. Chapman and Hall, New York, pp 40–125

Karr JR (1981) Assessment of biological integrity using fish communities. Fisheries 6(6):21–27

Karr JR, Chu EW (1999) Restoring life in running waters: better biological monitoring. Island Press, Washington, DC

Karr JR, Dudley DR (1981) Ecological perspectives on water quality goals. Environ Manag 5:55–68

Kohmann F, Schmedtje U (1986) Diversität und Diversitäts-Indices - eine brauchbare Methode zur Quantifizierung der Auswirkungen von Abwasserbelastungen auf aquatische Fließwasser-Zönosen? In: Bewertung der Gewässerqualität und Gewässergüteanforderungen. Bayer. Landesanstalt f. Wasserforschung, Oldenburg Verlag, München. Beiträge für Abwasser-, Fischerei-, und Flußbiologie 40:135–166

Kokes J, Zahradkova S, Nemejcova D, Hodovsky J, Jarkovsky J, Soldan T (2006) The PERLA system in the Czech Republic: a multivariate approach for assessing the ecological status of running waters. Hydrobiologia 566:343–354. ISSN 0018-8158

Kolenati FA (1848) Über den Nutzen und Schaden der Trichiopteren. Stettiner Entomologische Zeitung 9:51–52

Kolkwitz R, Marsson M (1902) Grundsätze für die biologische Beurteilung des Wassers nach seiner Flora und Fauna. Mitt. Aus d. Kgl. Prüfungsanstalt für Wasserversorgung u. Abwasserbeseitigung Berlin 1:33–72

Li L, Zheng B, Liu L (2010) Biomonitoring and bioindicators used for river ecosystems: definitions, approaches and trends. Procedia Environ Sci 2:1510–1524

Liebmann H (1951) Handbuch der Frischwasser- und Abwasserbiologie.- 1. Auflage, Verlag Oldenburg, München, 539 pp

Lowe S, Dallas H, Kennedy M, Taylor JC, Gibbins C, Lang P (2013) The SAFRASS biomonitoring scheme: general aspects, macrophytes (ZMTR) and benthic macroinvertebrates (ZISS) protocols. SAFRASS Deliverable Report to the ACP Group Science and Technology Programme, Contract No.AFS/2009/219013. University of Glasgow, Scotland

Meier C, Haase P, Rolauffs P, Schindehütte K, Schöll F, Sundermann A, Hering D (2006) Methodisches Handbuch Fließgewässerbewertung zur Untersuchung und Bewertung von Fließgewässern auf der Basis des Makrozoobenthos vor dem Hintergrund der EG Wasserrahmenrichtlinie. http://www.fliessgewaesserbewertung.de

Metcalfe JL (1989) Biological water quality assessment of running waters based on macro-invertebrate communities: history and present status in Europe. Environ Pollut 60(1–2):101–139

Metcalfe-Smith JL (1994) Biological water quality assessment of rivers: use of macroinvertebrate communities. In: Callow P, Petts GE (eds) The rivers handbook volume 2. Blackwell Science, London

Moog O (1988) Überlegungen zur Gütebeurteilung von Flussstauen.- Schriftenreihe der oberösterreichischen Kraftwerke AG.- Umweltforschung am Traunfluss 3, 110 pp

Oemke M, Burton TM (1986) Diatom colonization dynamics in a lotic system. Hydrobiologia 139:153–166

Ofenböck T, Moog O, Gerritsen J, Barbour M (2004) A stressor specific multimetric approach for monitoring running waters in Austria using benthic macro-invertebrates. Hydrobiologia 516:251–268

Ofenböck T, Moog O, Hartmann A, Stubauer I (2010a) Leitfaden zur Erhebung der Biologischen Qualitätselemente – Teil A2 Makrozoobenthos. Bundesministerium für Land- und Forstwirtschaft, Umwelt und Wasserwirtschaft, 214 pp

Ofenböck T, Moog O, Sharma S, Korte T (2010b) Development of the HKHbios: a new biotic score to assess the river quality in the Hindu Kush-Himalaya. Hydrobiologia 651(1):39–58

Ollis DJ, Dallas HF, Esler KJ, Boucher C (2006) Rapid bioassessment of the ecological integrity of river ecosystems using aquatic macroinvertebrates: review with a focus on South Africa. Afr J Aquat Sci 31:205–227

Pantle R, Buck H (1955) Die biologische Überwachung der Gewässer und die Darstellung der Ergebnisse. Bes Mitt dt Gewässerkundl Jb 12:135–143

Pont D, Hugueny B, Beier U, Goffaux D, Melcher A, Noble R, Rogers C, Roset N, Schmutz S (2006) Assessing river biotic condition at a continental scale: a European approach using functional metrics and fish assemblages. J Appl Ecol 43:70–80

Resh VH, Jackson JK (1993) Rapid assessment approaches to biomonitoring using benthic macroinvertebrates. In: Rosenberg DM, Resh VH (eds) Freshwater biomonitoring and benthic macroinvertebrates. Chapman and Hall, New York, pp 195–233

Reynoldson TB, Bailey RC, Day KE, Norris RH (1995) Biological guidelines for freshwater sediment based on BEnthic Assessment of SedimenT (the BEAST) using a multivariate approach for predicting biological state. Aust J Ecol 20:198–219

RIVPACS 2005. http://dorset.ceh.ac.uk/River_Ecology/River_Communities/Rivpacs_2003/rivpacs_introduction.htm

Rolauffs P, Stubauer I, Zahrádková S, Brabec K, Moog O (2004) Integration of the saprobic system into the European Union water framework directive – case studies in Austria, Germany and Czech Republic. Hydrobiologia 516:285–298

Rosenberg DM, Resh AP (eds) (1992) Freshwater biomonitoring and benthic macroinvertebrates. Chapman and Hall, New York

Rott E, Hofmann G, Pall K, Pfister P, Pipp E (1997) Indikationslisten für Aufwuchsalgen in Fließgewässern in Österreich. Teil 1: Saprobielle Indikation. Bundesministerium für Land- und Forstwirtschaft Wien, Wien, pp 1–73

Rott E, Pfister P, Van Dam H, Pall K, Pipp E, Binder N, Ortler K (1999) Indikationslisten für Aufwuchsalgen. Teil 2: Trophieindikation und autökologische Anmerkungen. Bundesministerium für Land- und Forstwirtschaft Wien, Wien, pp 1–248

Roux DJ, Van Vliet HR, Van Veelen M (1993) Towards integrated water quality monitoring: assessment of ecosystem health. Water SA 19(4):275–280

Schmedtje U, Bauer A, Gutowski A, Hofmann G, Leukart P, Melzer A, Mollenhauer D, Schneider S, Tremp H (1998) Trophiekartierung von aufwuchs- und makrophytendominierten Fliessgewässern. Bayerisches Landesamt für Wasserwirtschaft, München. Informationsberichte Heft 4/99, 516 Seiten

Simon TP, Lyons J (1995) Application of the index of biotic integrity to evaluate water resources integrity in freshwater ecosystems. In: Davis WS, Simon TP (eds) Biological assessment and criteria: tools for water resource planning and decision making. Lewis Press, Boca Raton, pp 245–262

Simpson J, Norris RH (2000) Biological assessment of water quality: development of AUSRIVAS models and outputs. In: Wright JF, Sutcliffe DW, Furse MT (eds) Assessing the biological quality of freshwaters. RIVPACS and other techniques. Freshwater Biological Association, Ambleside, pp 125–142

Sládecek V (1973) System of water quality from the biological point of view. Arch Hydrobiol Beih Ergebnisse Limnol 7:1–218

Thienemann A (1912) Aristotles und die Abwasserbiologie. Festschrift Medizinisch-Naturwissenschaftlichen Gesellschaft Münster. Commissionsverlag, Universitäts Buchhandlung Franz Coppenrath Münster

Tümpling W, Friedrich G (Hrsg.) (1999) Methoden der Biologischen Gewässeruntersuchung 2

Verdonschot PFM (1990) Ecological characterization of surface waters in the province of Overijssel (The Netherlands). Thesis, Agricultural University, Wageningen

Verdonschot PFM, Moog O (2006) Tools for assessing European streams with macroinvertebrates: major results and conclusions from the STAR project. Hydrobiologia 566:299–309

Verdonschot PFM, Nijboer RC (2000) Typology of macrofaunal assemblages applied to water and nature management: a Dutch approach. In: Wright JF, Sutcliffe DW, Furse MT (eds) Assessing the biological quality of fresh waters: RIVPACS and other techniques. Freshwater Biological Association, Ambleside. The RIVPACS International Workshop, 16–18 Sept 1997, Oxford, Chapter 17: 241–262

Verneaux J, Tuffery G (1967) Une methode zoologique pratique de determination de la qualite biologique des eaux courantes. Indices Biotiques. Ann Univ Besancon Biol Anim 3:79–90

Washington HG (1984) Diversity, biotic and similarity indices. A review with special relevance to aquatic ecosystems. Water Res 18(6):653–694

Woodiwiss FS (1964) The biological system of stream classification used by the Trent river board. Chem Ind:443–447

Woodiwiss FS (1978) Comparative study of biological ecological water quality assessment methods, Summary report. Commission of the European communities. Severn Trent Water Authority, UK, p 45

Wright JF, Armitage PD, Furse MT (1989) Prediction of invertebrate communities using stream measurements. Regul Rivers Res Manag 4:147–155

Wright JF, Sutcliffe DW, Furse MT (eds) (2000) Assessing the biological quality of fresh waters: RIVPACS and other techniques. Freshwater Biological Association, Ambleside, 373 pp

Zelinka M, Marvan P (1961) Zur Präzisierung der biologischen Klassifikation der Reinheit fließender Gewässer. Arch Hydrobiol 57:389–407

Chapter 20
Biodiversity and Freshwater Information Systems

Astrid Schmidt-Kloiber and Aaike De Wever

20.1 Data in Freshwater Research

Species observed in freshwaters are typically good indicators of the health/status of these ecosystems and are therefore frequently analyzed as part of ecological monitoring programs (see Chap. 19). The biodiversity data generated during such monitoring routines, in combination with data from other ecological studies in freshwaters, can form an invaluable source of information to support sustainable management and conservation of aquatic ecosystems (see Chap. 15). However, a large part of these data still remains scattered on individual researchers' computers and institute servers. Pressured by funding agencies such as the EU, the call for open access to data (e.g., Reichman et al. 2011), which enables the reuse of data for addressing large-scale and/or transdisciplinary research problems, is becoming increasingly important. Additionally, new data-intensive technologies in science, such as remote sensing and next-generation sequencing, demand effective management of the increasingly vast amount of data.

In addition to monitoring data, observational data generated in freshwater research typically also comprise experimental data (Fig. 20.1). If such data are adequately described through metadata (see Sect. 20.2), they can be integrated into processed data products and tools that can support management decisions, conservation priorities, or other policy-relevant issues.

A. Schmidt-Kloiber (✉)
Institute of Hydrobiology and Aquatic Ecosystem Management, University of Natural Resources and Life Sciences, Vienna, Austria
e-mail: astrid.schmidt-kloiber@boku.ac.at

A. De Wever
OD Nature, Royal Belgian Institute of Natural Sciences, Brussels, Belgium
e-mail: aaike.dewever@naturalsciences.be

© The Author(s) 2018
S. Schmutz, J. Sendzimir (eds.), *Riverine Ecosystem Management*, Aquatic Ecology
Series 8, https://doi.org/10.1007/978-3-319-73250-3_20

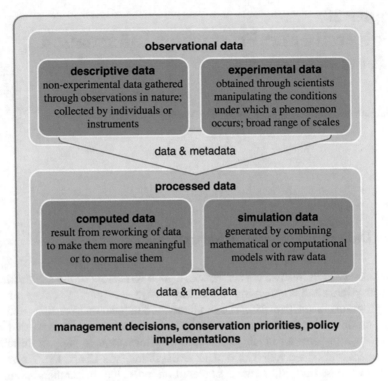

Fig. 20.1 Data types arising in freshwater research and management (inspired by Thessen and Patterson 2011), including the importance of metadata in the data flow

In this chapter we discuss the importance of documenting and describing data and making these metadata available to improve the understanding and discoverability of datasets and specifically examine different facets of biodiversity data. We provide an overview of existing freshwater (biodiversity) information systems that enable data holders to adequately publish their data and find appropriate data for their research. Finally, we offer recommendations on how to implement open data publishing practices as a way to support sustainable management and conservation.

20.2 Documenting Generated Data and Information: The Concept of Metadata

In order to appropriately reuse biological data, it is important to understand the context in which they were acquired or generated (Thessen and Patterson 2011). In this connection, the generation of metadata plays a significant and absolutely essential role.

Metadata are loosely defined as "data about data". More specifically, metadata should document and describe all aspects of a specific dataset (i.e., the who, why, what, when, and where) that would allow understanding of the physical format, content, and context of the data, as well as how to acquire, use, and cite the data (e.g., Michener 2006). For the data producer or provider, generating metadata presents an opportunity to document a dataset for possible own future use as well as for informing prospective users of its existence and characteristics. From the perspective of the data consumer or user, metadata enable both discovering data and assessing their appropriateness for particular use—their so-called fitness for purpose (Schmidt-Kloiber et al. 2012).

Basic and applied ecological research requires the availability of "high-quality" data, the definition of which varies and often depends on the specific purpose of a study. Scientists frequently reuse their own old data, but the use of data created by others and/or shared within large work groups (e.g., within EU-funded projects) remains limited. This is due to the fact that these research datasets are often not made publicly available nor deposited in permanent archives and therefore risk being lost over time (Shorish 2010; Vines et al. 2014). The scientific value of reusing a dataset for multiple (other) purposes than foreseen by the data originator(s) exceeds the perceived value by far. Documenting primary research datasets in metadata collections allows people to discover and understand these data and is therefore an important step forward to increase the usefulness and prolong the lifespan of a dataset.

In ecological sciences, the role of metadata in facilitating a scientist's work has been increasingly recognized since the 1980s (Michener 2006), and collecting meta-data datasets in dedicated databases is becoming more and more common. This is especially true for biodiversity-related (occurrence) datasets for which the importance of broad data compilations is already widely accepted. The Global Biodiversity Information Facility (GBIF; see Sect. 20.4.2.1), for example, collates and centralizes not only primary biodiversity data but also offers standards and tools for (meta)data collection. More specifically, for surface waters, the Freshwater Metadatabase was developed in the framework of a series of EU-funded freshwater research projects. In connection with this metadatabase, the Freshwater Metadata Journal was founded, in order to give the publication of metadata more scientific weight and bring about a change of perception in the freshwater community (Schmidt-Kloiber et al. 2014). The aforementioned database and journal offer an easy publishing process that—together with the possibility for citation—should make data generation and compilation efforts more visible to other researchers. Both resources are available through the Freshwater Information Platform (FIP; see Sect. 20.4.1).

20.3 Biodiversity Data

Biodiversity—beside its intrinsic value—supports essential ecosystem functions and, consequently, many ecosystem services that are key to human well-being (Cardinale et al. 2012). It is well known that freshwater ecosystems harbor a rich diversity of species and habitats, despite their comparatively small share of the world's surface (less than 1%). On the other hand, there is also evidence that the decline in freshwater biodiversity has been greater during the last few decades than that of marine or terrestrial counterparts (Garcia-Moreno et al. 2014; Darwall et al. 2009). The high level of connectivity of freshwater systems implies that fragmentation can have profoundly negative effects (Revenga et al. 2005). Multiple other interacting stresses, combining effects of intense agriculture, industry, or domestic activities, form a further, compounded risk for freshwater ecosystems. These impacts include water extraction, the introduction of exotic species, alteration of hydrological dynamics through the construction of dams and reservoirs, channelization, overexploitation, and increasing levels of organic and inorganic pollution (Dudgeon et al. 2006; Strayer and Dudgeon 2010; Vörösmarty et al. 2010). Climate change is anticipated to increase the intensity of these threats to freshwaters (Garcia-Moreno et al. 2014; Woodward et al. 2010).

A current estimate states that freshwater ecosystems provide suitable habitats for at least 126,000 plant and animal species (Balian et al. 2007). These species contribute to a wide range of critical goods and services for humans, including flood protection and food or water filtration (see Chap. 21) to name just a few. Securing these ecosystem services and understanding the underlying ecosystem processes require knowledge about the taxonomic, phylogenetic, genetic, and functional diversity of nature (e.g., Kissling et al. 2015). The urgency of this matter was recognized by the Parties to the United Nations Convention on Biological Diversity (CBD) who established the Aichi Targets for 2020, which aim to halt biodiversity loss, protect various levels of life forms, and implement sustainable use of natural resources (http://www.cbd.int/sp/elements/).

A recent review of these targets shows that many of them are unlikely to be met (Tittensor et al. 2014), leading to an increasing demand for comprehensive, sound, and up-to-date biodiversity data (Wetzel et al. 2015). Key gaps were identified in the knowledge about status and trends of biodiversity and associated ecosystem services. These gaps mostly arise because of barriers that prevent existing data from being discoverable, accessible, and digestible (Wetzel et al. 2015). The importance of the availability of large-scale datasets for analyzing and understanding the broad-scale patterns of spatial variation in richness and endemism is highlighted by Collen et al. (2013). This again is central to understanding the origin of diversity and the potential impacts of environmental change on current biodiversity patterns and allows for prioritization of conservation areas (Collen et al. 2013).

An earlier review of the Aichi Targets already found that considerable data on freshwater species and populations are available, but often are not accessible or harmonized in a way that they could be appropriately used to support management decisions (Revenga et al. 2005). This calls for an urgent paradigm shift with regard to how biodiversity data are collected, stored, published, and streamlined, so that many sustainable development challenges ahead can be successfully tackled (Wetzel et al. 2015). The authors therefore suggest that biodiversity data should be discoverable, accessible, and digestible in order to—together with a certain expertise—more effectively inform and implement environmental policies (see Fig. 20.2).

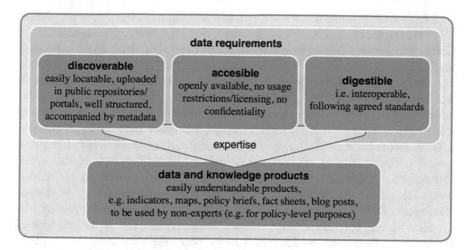

Fig. 20.2 Biodiversity data requirements (inspired by Wetzel et al. 2015)

20.3.1 Biodiversity Observation Networks and Essential Biodiversity Variables

So-called Biodiversity Observation Networks (BONs) can contribute to address these challenges by helping to coordinate data collections across large areas. They play a major role towards mobilizing biodiversity information for use by policy and decision-makers. In 2013 the Group on Earth Observation (GEO) BON introduced the concept of essential biodiversity variables (EBVs) with the aim to identify key measurements that are required to study, report, and manage biodiversity changes (Pereira et al. 2013; Turak et al. 2016b). The six broad EBV classes include genetic composition (1), species populations (2), species traits (3), community composition (4), ecosystem structure (5), and ecosystem function (6). Each of these classes needs a different approach to how data are collected and structured. In the following we give a few examples for selected EBV classes and their respective data availability and accessibility.

Knowledge of the genetic composition (1) plays an important role in freshwater research, as river catchments and lakes can be spatially separated and isolated from each other, which might limit gene flow such that populations of the same species may vary considerably in their genetic composition. This variability has, for example, particular applications for the management of freshwater fisheries, where loss of genetic variants may have major consequences for ecosystem service provision (Turak et al. 2016a). Most genetic data for freshwater species are accessible through the International Nucleotide Sequence Database Collaboration initiative (INSDC; see Sect. 20.4.2.4), of which GenBank is the best-known database. DNA barcoding and recent advances in environmental DNA technology greatly increase the potential to assess genetic biodiversity in freshwaters, especially in relation to conservation of rare and threatened species (Thomsen et al. 2011).

Information about the distribution and, as a consequence, the size of populations (2) of freshwater species has greatly improved in recent years. The activities of the Global Biodiversity Information Facility (GBIF; see Sect. 20.4.2.1) constitute an essential contribution to increase this knowledge. Specifically focused on freshwaters, the Freshwater Animal Diversity Assessment (FADA; see Sect. 20.4.2.2) for the first time assessed the diversity of fauna in inland waters. FADA provides a comprehensive overview on freshwater species richness of major taxa groups, also revealing many obvious taxonomic and geographic gaps, and hence the need to collect more data (Balian et al. 2007, 2008). Based on these insights, the EU-funded BioFresh project ("Biodiversity of Freshwater Ecosystems: Status, Trends, Pressures, and Conservation Priorities"; http://freshwaterbiodiversity.eu) created an online data portal focused on the mobilization of freshwater occurrence data, the Freshwater Biodiversity Data Portal (see Sect. 20.4.2.1). Another major initiative dealing with the EBV "species population" is the Freshwater Biodiversity Unit of the International Union for the Conservation of Nature (IUCN), which has been developing a global assessment of the distribution and conservation status of freshwater organisms (Darwall et al. 2009). Generally, the IUCN Red Lists (see Sect. 20.4.2.5) summarize the current knowledge on the state (including population size) and threat condition of species belonging to selected organism groups.

Species traits (3) seek to functionally classify taxa grouped by comparable biological profiles. In general, they are a powerful approach to understand community functioning through characterizing assemblages according to aspects of morphology, function, physiology, behavior, habitat use, reproduction, or life history. Databases collecting trait information were first established in the terrestrial realm, e.g., the TRY database for global plant traits (Kattge et al. 2011) or the PanTHERIA mammal database (Jones et al. 2009). While the concept of species traits is already widely recognized in freshwater ecological assessment too, comprehensive and publicly available databases only exist in some regions and for restricted taxa groups. These compilations frequently provide information at the genus or family levels, since the knowledge about traits on the species level often remains limited. In America the US EPA developed a trait database for about 3800 macroinvertebrate taxa (https://www.epa.gov/risk/freshwater-biological-traits-data base-traits). In Europe the freshwaterecology.info database offers trait information on about 20,000 taxa (with focus on species) belonging to five aquatic organism groups (fishes, macroinvertebrates, macrophytes, diatoms, phytoplankton). Freshwaterecology. info serves as base for bioassessment and monitoring and is a service for basic research, applied scientists, water managers, or other stakeholders. However, such trait-specific data are still lacking for many taxa and for most parts of the world, and even fundamental facts about the ecology of many common species are not known and require more basic research (Turak et al. 2016a; Schmidt-Kloiber and Hering 2015).

Regarding community composition (4), information on the structure of freshwater assemblages is already used with success for assessing freshwater ecosystems (Friberg et al. 2011). In Europe, environmental legislation aiming to protect and restore freshwater ecosystems and their biodiversity is based both on the Habitats Directive (HD; Council Directive 92/43/EEC) and the Water Framework Directive (WFD; Directive 2000/60/EC). The latter has placed aquatic organisms in a unique position, as the composition of freshwater biota defines the status of surface waterbodies, and thus determines the needs for restoration and associated investments. As part of implementing the WFD, monitoring of waterbodies generates much more data than the information on the "ecological status", since most of the national assessment systems also provide a variety of measures (so-called metrics) such as the community composition. WFD data, therefore, could contribute significantly to other objectives (e.g., monitoring the effects of emerging stressors, improving the knowledge of species distributions and species invasions, or understanding broad-scale drivers shaping species assemblages). However, the lack of detailed data in central storage systems makes accessibility difficult. Central availability is hampered by the impracticality of combining data stemming from different collection methods and formulated following different data structures and storage methods. It is also hampered by concerns about a consistent data quality regarding the underlying taxonomy, identification, and taxonomic resolution (Hering et al. 2010). Currently, the Water Information System for Europe (WISE; see Sect. 20.4.1) produces Europe-wide maps of water quality and ecological status of waterbodies, but original data (e.g., taxa lists) are not stored centrally so far.

For the two remaining EBVs—the ecosystem structure (5) and ecosystem function (6)—the availability of centrally stored data is rather limited. Observations of ecosystem structure for tracking change in freshwater ecosystems include measuring changes in the extent of inland-water habitats such as wetlands, lakes, rivers, and aquifers (Turak et al. 2016a). Though remote sensing technology for mapping the extent of wetlands and lakes is advancing rapidly, a central repository or entry point to consult the processed information of such analyses is not available yet. The use of indicators of ecosystem function other than those that may result from water quality or ecological status assessment data is rare. Analyzing the relationship between biodiversity and ecosystem functioning is still a growing research area but will need considerable further development before it will be possible to include measures of ecosystem function in freshwater biodiversity observations (Turak et al. 2016a, b).

Alongside these six EBV categories, there are also other widely used methods to assess components of freshwater biodiversity or indicate conditions of freshwater ecosystems that do not fit neatly into these classes (see review by Friberg et al. 2011). As these do not underlie any common Europe-wide legislation, data are even more scattered and inaccessible.

20.4 Main Freshwater-Related Information Systems

In many cases, data and information relevant for freshwater science, management, and policy can be found on platforms that sometimes also cover other realms. In the following we introduce three rough categories: "general data portals", "biodiversity-related data sources", and "spatial data sources". We further subdivide "biodiversity-related data sources" into those dealing with occurrence data, taxonomy, traits, genes, and others. An overview including examples is given in Table 20.1.

Table 20.1 Overview on different freshwater information systems including a rough classification, examples, and web links

Category/ subcategory	Name	URL	Type	Scale	Data included
General data portals					
WISE—Water Information System for Europe		http://water.europa.eu	Data portal	EU	
	EEA Water Data Centre (see separate entry)	http://www.eea.europa.eu/themes/water/dc	Data centre linked to from WISE	EU	Datasets, (interactive) maps, indicators, graphs and figures, related content
	Eurostat—Water Statistics	http://ec.europa.eu/eurostat/statistics-explained/index.php/Water_statistics	Data centre linked to from WISE	EU	Datasets, maps, graphs and figures, publications
	Eurostat—Water Database	http://ec.europa.eu/eurostat/web/environment/water/database	Dataset example	EU	Data, metadata
	FATE and impact of pollutants in terrestrial and aquatic ecosystems	http://fate.jrc.ec.europa.eu/	Data centre linked to from WISE	EU	Datsets, interactive maps
BISE—Biodiversity Information System for Europe		http://www.biodiversity.europa.eu	Data portal	EU	
	EEA Biodiversity Data Centre (BDC)	http://www.eea.europa.eu/themes/biodiversity/dc	Data centre linked to from BISE	EU	Datasets, (interactive) maps, indicators, graphs and figures, related content
	EUNIS—European Nature Information System	http://eunis.eea.europa.eu	Data centre linked to from BISE	EU	Species, maps, datasets, related content
EEA Data Centre			Data portal	EU	
	Waterbase—Lakes	http://www.eea.europa.eu/data-and-maps/data/waterbase-lakes-10	Dataset example	EU	Data, metadata, related content
	Waterbase—Rivers	http://www.eea.europa.eu/data-and-maps/data/waterbase-rivers-10	Dataset example	EU	Data, metadata, related content

	URL			
Waterbase—Groundwater	http://www.eea.europa.eu/data-and-maps/data/waterbase-groundwater-10	Dataset example	EU	Data, metadata, related content
Waterbase—Water Quantity	http://www.eea.europa.eu/data-and-maps/data/waterbase-water-quantity-8	Dataset example	EU	Data, metadata, related content
Waterbase—Emission to water	http://www.eea.europa.eu/data-and-maps/data/waterbase-emissions-4#tab-european-data	Dataset example	EU	Data, metadata
European past floods	http://www.eea.europa.eu/data-and-maps/data/european-past-floods	Dataset example	EU	Data, metadata, related content
Bathing Water Directive—Status of bathing water	http://www.eea.europa.eu/data-and-maps/data/bathing-water-directive-status-of-bathing-water-7	Dataset example	EU	Data, metadata, related content
WISE WFD Database	http://www.eea.europa.eu/data-and-maps/data/wise_wfd	Dataset example	EU	Data, metadata, related content
WISE River Basin Districts (RBDs)	http://www.eea.europa.eu/data-and-maps/data/wise-river-basin-districts-rbds-1	Dataset example	EU	Data, metadata, related content
WISE Large rivers and large lakes	http://www.eea.europa.eu/data-and-maps/data/wise-large-rivers-and-large-lakes	Dataset example	EU	Data, metadata, related content
WISE Groundwater	http://www.eea.europa.eu/data-and-maps/data/wise-groundwater	Dataset example	EU	Data, metadata, related content
Freshwater abstraction and hydropower	http://www.eea.europa.eu/data-and-maps/data/freshwater-abstraction-and-hydropower-2013	Dataset example	EU	Data, metadata
European catchments and Rivers network system (ECRINS)	http://www.eea.europa.eu/data-and-maps/data/european-catchments-and-rivers-network	Dataset example	EU	Data, metadata, related content
European river catchments	http://www.eea.europa.eu/data-and-maps/data/european-river-catchments-1	Dataset example	EU	Data, metadata, related content
Ecoregions for rivers and lakes	http://www.eea.europa.eu/data-and-maps/data/ecoregions-for-rivers-and-lakes	Dataset example	EU	Data, metadata, related content

(continued)

Table 20.1 (continued)

Category/ subcategory	Name	URL	Type	Scale	Data included
	Dams with reservoirs on rivers in Europe	http://www.eea.europa.eu/data-and-maps/figures/dams-with-reservoirs-on-rivers	Dataset example	EU	Map, metadata
	Water exploitation index plus for river basin districts	http://www.eea.europa.eu/data-and-maps/figures/water-exploitation-index-based-on-1	Dataset example	EU	Interactive map, metadata
	Natura 2000 data—the European network of protected sites	http://www.eea.europa.eu/data-and-maps/data/natura-7	Dataset example	EU	GIS data, metadata, related content
	Biogeographical regions	http://www.eea.europa.eu/data-and-maps/data/biogeographical-regions-europe-3	Dataset example	EU	Data, GIS data, metadata, related content
	European Red Lists	http://www.eea.europa.eu/data-and-maps/data/european-red-lists-4	Dataset example	EU	Species, data, metadata
	EU research projects on biodiversity and ecosystems	http://www.eea.europa.eu/data-and-maps/data/eu-research-projects-on-biodiversity	Dataset example	EU	Data, metadata
	Nationally designated areas	http://www.eea.europa.eu/data-and-maps/data/nationally-designated-areas-national-cdda-10	Dataset example	EU	Data, GIS data, metadata, related content
	CORINE Land Cover	http://www.eea.europa.eu/data-and-maps/data/corine-land-cover-2006-raster-3	Dataset example	EU	GIS data, metadata, related content
	Landscape fragmentation per 1 km² grid	http://www.eea.europa.eu/data-and-maps/figures/landscape-fragmentation-per-1-km2-3	Dataset example	EU	Map, metadata, related content
JRC Science Hub		https://ec.europa.eu/jrc/	Data portal	EU	
	Water Portal	http://water.jrc.ec.europa.eu/waterportal	Data centre	EU	Interactive maps
	WFD Ecological methods database	http://wfdmethods.jrc.ec.europa.eu	Dataset example	EU	
	Floods	https://ec.europa.eu/jrc/en/research-topic/floods	Data centre	Global, EU focus	Related content

FIP—Freshwater Information Platform	http://www.freshwaterplatform.eu	Data portal	Global, EU focus	
Freshwater Resources	http://www.freshwaterplatform.eu/index.php/resources.html	Dataset example	Global, EU focus	Related content
Freshwater Policies	http://www.freshwaterplatform.eu/index.php/overview.html	Dataset example	EU	Related content
Freshwater Blog	https://freshwaterblog.net	Dataset example	Global	Related content
EU BON (under construction)	http://beta.eubon.ebd.csic.es/	Data portal	EU	
Biodiversity related data sources				
Occurrence GBIF—Global Biodiversity Information Facility	http://www.gbif.org	Data portal	Global	Species data, occurrence data, datasets, metadata
Freshwater Metadatabase and Biodiversity Data Portal	http://data.freshwaterbiodiversity.eu/	Data centre linked to from FIP	Global, EU focus	Species data, occurrence data, datasets, metadata
Taxonomy FADA—Freshwater Animal Diversity Assessment	http://fada.biodiversity.be/	Data centre	Global	Species data, datasets
PESI—Pan-European Species directories Infrastructure	http://www.eu-nomen.eu	Data centre	EU	Species data
Traits freshwaterecology.info	http://www.freshwaterecology.info	Data centre linked to from FIP	EU	Species data, occurrence data, ecological data, related content
Genes BOLDSYSTEMS—Barcode of Life Data Systems	http://www.boldsystems.org	Data portal	Global	Species data, gene data
INSDC—International Nucleotide Sequence Database Collaboration	http://www.insdc.org	Data portal	Global	Species data, gene data
GenBank	http://www.ncbi.nlm.nih.gov/genbank/	Data centre linked to from INSDC	Global	Species data, gene data

(continued)

Table 20.1 (continued)

Category/subcategory	Name	URL	Type	Scale	Data included
Others	IUCN Red Lists	http://www.iucnredlist.org/	Data portal	Global	Species data, occurrence data, metadata
	EASIN—European Alien Species Information Network	http://easin.jrc.ec.europa.eu	Data portal	EU	Species data, occurrence data
Spatial data sources					
	FEOW—Freshwater Ecoregions of the World	http://www.feow.org	Dataset	Global	Maps, interactive map, related content
	Freshwater Key Biodiversity Areas	http://www.birdlife.org/datazone/freshwater	Dataset	EU	Interactive map
	Global Freshwater Biodiversity Atlas	http://atlas.freshwaterbiodiversity.eu/	Data portal linked to rom FIP	Global	Interactive maps, metadata, related content
	Protected Planet	http://www.protectedplanet.net	Dataset	Global	Interactive map, related content

20.4.1   General Data Portals

Water Information System for Europe
The Water Information System for Europe (WISE; http://water.europa.eu) is a partnership between the European Commission (formed by DG Environment, Joint Research Centre, and Eurostat) and the European Environment Agency (EEA). It is a gateway to information on European water issues, divided into four sections: EU water policies (e.g., directives, implementation reports, and supporting activities), data and themes (e.g., reported datasets, interactive maps, statistics, indicators), modeling (e.g., current and forecasting services across Europe), and projects and research (e.g., inventory of links to recently completed and ongoing water-related projects and research activities). WISE aims at reaching a wide audience covering EU, national, regional, and local administrations working in water policy development, as well as scientists, professionals, and the general public interested in water issues.

The WISE portal comprises a wide range of data and information collected by EU institutions and redirects visitors to three portals: the EEA Water Data Centre (see below), the Eurostat Water Statistics website, and the FATE website related to pollutants monitoring campaigns.

Biodiversity Information System for Europe
The Biodiversity Information System for Europe (BISE; http://www.biodiversity.europa.eu) offers information on biological diversity in general and covers all realms including freshwater. It is a partnership between the European Commission (DG Environment) and the European Environment Agency and is supported by the collaboration of the European Clearing House Mechanism network and the CBD Secretariat.

BISE is a gateway for data and information on biodiversity supporting the implementation of the EU 2020 Biodiversity Strategy and the Aichi Targets in Europe. It focuses on bringing together facts and figures about biodiversity and ecosystem services as well as linking to related policies, environmental data centers, assessments, and research findings from various sources. The portal offers five entry points: policy (e.g., policy, legislation, and supporting activities related to the Common Implementation Framework of the EU strategy), topics (e.g., state of species, habitats, ecosystems, genetic diversity, threats to biodiversity, impacts of biodiversity loss), data (e.g., data sources, statistics, and maps related to land, water, soil, air, marine), research (e.g., important EU-wide research projects related to biodiversity and ecosystem services), countries (e.g., links to information available from European countries), and networks (e.g., links to Europe-wide networks supporting information sharing across national borders).

Also BISE does not host actual data, but links to major sources of data and information including the EEA Biodiversity Data Centre (see below) and the European Nature Information System (EUNIS).

European Environment Agency Data Centres
The European Environment Agency hosts the main data sources linked to from WISE and BISE, namely, the Water Data Centre and the Biodiversity Data Centre.

Both data centers are major sources for a variety of datasets with relevance to managers and policy makers. In addition to raw data and metadata, several data products are made available in a more digestible way, such as interactive maps and summary graphs. For both domains, water and biodiversity, users can browse through and access a wide range of spatial data using the "EEA interactive maps and data viewers". Additionally, datasets are linked with "related content" if available.

Datasets in the Water Data Centre include the "ECRINS (European Catchments and Rivers Network System)", the "Waterbase", or the "WISE WFD Database". One of the main datasets hosted by the Biodiversity Data Centre is the "Natura 2000 data—the European network of protected sites". In addition and more generally, the EEA web portal also provides several reference datasets, such as "biogeographical regions", "CORINE land cover", "ecosystem types of Europe", or "nationally designated areas" (for more examples and links, see Table 20.1).

Most EEA spatial data are offered as map service on DiscoMap (http://discomap.eea.europa.eu/).

Joint Research Centre Science Hub
The Joint Research Centre provides a Water Portal (http://water.jrc.ec.europa.eu/) with visualization and download options for JRC products on freshwater and marine water resources and offers tools to calculate summary statistics for the available data. Additionally, JRC also maintains the "WFD Ecological Methods Database", which gives access to information about the national assessment methods used to classify the ecological status of rivers, lakes, and coastal and transitional waters as applied by the member states of the European Union in their monitoring programs according to the EU Water Framework Directive.

Freshwater Information Platform
The Freshwater Information Platform (FIP; http://freshwaterplatform.eu) represents an effort to regroup web products from several freshwater-related European research projects addressing freshwater biodiversity, ecology, and water management. It was initiated by four leading partners from the FP7 EU BioFresh project, which focused on raising awareness around freshwater biodiversity data, collating and mobilizing freshwater occurrence data, and using those data in large-scale analyses. The platform serves as one single gateway to different resources relevant for water managers, policy makers, scientists, and the interested public. It contains several complementary sections, either providing access to original data or summarizing research results in an easily explorable way. The most relevant sections are the Freshwater Biodiversity Data Portal and connected to it the Freshwater Metadatabase (see Sect. 20.4.2.1), the Global Freshwater Biodiversity Atlas (see Sect. 20.4.3), and the European freshwater species traits database freshwaterecology.info (see Sect. 20.4.2.3). Another relevant part of the FIP is the widely read Freshwater Blog, which publishes features, research highlights, interviews, and podcasts on freshwater science, policy, and conservation. The remaining sections of the FIP focus on freshwater resources (e.g., a glossary, fact sheets, "how-to" guides, etc.), freshwater-related policies (e.g., policy briefs), and freshwater networks. A specific section presents freshwater stressor-related tools. The FIP is designed as an open platform inviting contributions from a variety of aquatic ecology research fields and will be updated continuously.

Group on Earth Observations Biodiversity Observation Network
The Group on Earth Observations Biodiversity Observation Network (GEO BON; http://geobon.org) is a voluntary partnership of governments and organizations, which aims at improving the acquisition, coordination, and delivery of biodiversity observations and related services to users including decision-makers and the scientific community. The European contribution to GEO BON, the FP7 EU BON project (Building the European Biodiversity Observation Network; http://www.eubon.eu/), is developing a data platform (currently in beta stage) targeting at being a central access point for biodiversity data from different sources, e.g., processed data and remote sensing imaginary products. In addition to data from the GBIF network, it will link to the Long-Term Ecological Research (LTER) network, the Global Earth Observation System of Systems (GEOSS), and the Pan-European Species directories Infrastructure (PESI; see Sect. 20.4.2.2).

20.4.2 Biodiversity-Related Data Sources

20.4.2.1 Occurrence Data

Global Biodiversity Information Facility
The Global Biodiversity Information Facility (GBIF; http://www.gbif.org) is an international open data infrastructure, which is funded by governments and supported by member countries and other associated participants. GBIF started its efforts to collate global diversity data back in 2001 with the aim to provide free and open access to species occurrence data from one single online gateway. Currently GBIF offers more than 680 million files of occurrence data related to 1.6 million species provided by about 810 data publishers. The data portal covers all realms and represents a major source of occurrence data. Freshwater data can be extracted via species search.

Freshwater Metadatabase and Freshwater Biodiversity Data Portal
The Freshwater Metadatabase and Freshwater Biodiversity Data Portal provide access to information on freshwater datasets, species, and occurrence data. The metadatabase collects descriptions of datasets, thus making them discoverable regardless whether the data are publicly available or not. The database provides an overview on hundreds of major data sources related to freshwater research and management, and it offers the option to easily explore access rights of relevant datasets. Connected to it, the Freshwater Metadata Journal (www.freshwaterjournal. eu) allows straightforward metadata publishing. The data portal focuses on species and occurrence data. For species data, it links with the Freshwater Animal Diversity Assessment database (see below), whereas for occurrence data, it provides access to freshwater data on GBIF and acts as a data publishing platform for freshwater data.

Both the metadatabase and data portal are meant to help scientists in advertising and publishing their datasets, and they provide tools for the discovery, integration, and analysis of open and freely accessible freshwater biodiversity data. Both parts are integrated into the Freshwater Information Platform.

20.4.2.2 Taxonomy Data

Freshwater Animal Diversity Assessment
The Freshwater Animal Diversity Assessment (FADA; http://fada.biodiversity.be) is
an informal network of scientists specialized in freshwater biodiversity. The FADA
database is an information system dedicated to freshwater animal species diversity.
The system provides access to authoritative species lists and global distributions
compiled by world experts. The data are also integrated in the Freshwater Biodiversity
Data Portal, to which it acts as a taxonomic backbone.

Pan-European Species directories Infrastructure
The Pan-European Species directories Infrastructure (PESI; http://eu-nomen.eu/)
aims at delivering an integrated, annotated checklist of species occurring in Europe.
The PESI checklist (also called EU-nomen) serves as a taxonomic standard and
backbone for Europe. Databases from Euro+Med PlantBase, Fauna Europaea,
European Register of Marine Species, and Species Fungorum Europe are the base
of the PESI web portal. PESI includes interactions with the geographic focal point
networks, a network of taxonomic experts, and global species databases. Freshwater
information is available via dedicated species search. Results link to GBIF, the
Biodiversity Heritage Library (BHL, http://www.biodiversitylibrary.org), GenBank,
and BOLDSYSTEMS (see below).

20.4.2.3 Trait Data

freshwaterecology.info
The freshwaterecology.info database (http://www.freshwaterecology.info) has been
established during several EU-funded research projects and compiles information on
taxonomy, ecology, and distribution of species based on extensive literature surveys
performed by experts for the targeted organism groups. For five aquatic organism
groups (fishes, macroinvertebrates, macrophytes, diatoms, phytoplankton), ecological
preferences and biological traits (such as habitat preferences, pollution tolerance, life
cycle, etc.) are available online with various options and tools for extracting these data.
The freshwaterecology.info database is a part of the Freshwater Information Platform.

20.4.2.4 Genetic Data

Barcode of Life Data Systems
The Barcode of Life Data Systems (BOLDSYSTEMS; http://www.boldsystems.org)
aims at supporting the generation and application of DNA barcode data by aiding
the acquisition, storage, analysis, and publication of DNA barcode records. It
assembles molecular, morphological, and distributional data. The platform consists
of four main modules: a data portal, a database of barcode clusters, an educational
portal, and a data collection workbench. Freshwater data are available through
species names.

International Nucleotide Sequence Database Collaboration
The International Nucleotide Sequence Database Collaboration (INSDC; http://www. insdc.org) is a long-standing, foundational initiative that operates between three major genetic resources, namely, the DNA Data Bank of Japan (DDBJ), the European Molecular Biology Laboratory (EMBL), and the GenBank at the National Center for Biotechnology Information (NCBI) in the United States. These three organizations exchange data on a daily basis. Freshwater data can be extracted through species search.

20.4.2.5   Other Data

IUCN Red List
The IUCN has been working on its Red List of Threatened Species (http://www. iucnredlist.org) to assess the conservation status of species, subspecies, and varieties on a global scale for the past 50 years in order to highlight taxa threatened with extinction and thereby promote their conservation. It provides taxonomic, conservation status and distribution information on plants, fungi, and animals that have been globally evaluated using specifically defined categories and criteria. The Red List assessments bring together extensive knowledge of thousands of regional experts regarding the status of and threats to species. Regarding freshwaters, most comprehensive assessments are currently available for fishes, molluscs (mainly unionid bivalves), decapods (crabs, crayfish, and shrimps), Odonata (dragonflies and damselflies), and selected plant families.

European Alien Species Information Network
The European Alien Species Information Network (EASIN; http://easin.jrc.ec.europa.eu) is a platform developed by the European Commission's Joint Research Centre that enables easy access to data on alien species reported in Europe. It facilitates the exploration of existing alien species information from a variety of distributed sources through freely available tools and interoperable web services. It aims to assist policy makers and scientists in their efforts to tackle alien species invasions.

20.4.3   Spatial Data Sources

20.4.3.1   Freshwater Ecoregions of the World

The Freshwater Ecoregions of the World (FEOW; http://www.feow.org) represents a global biogeographic regionalization of the earth's freshwater biodiversity. The FEOW were developed by the WWF Conservation Science Program in partnership with the Nature Conservancy and 200 freshwater scientists from institutions around the world. The FEOW can be used for underpinning global and regional conservation planning efforts, particularly to identify outstanding and threatened freshwater systems, for serving as a logical framework of large-scale conservation strategies as well as for providing a global-scale knowledge base for increasing freshwater biogeographic knowledge.

20.4.3.2 Freshwater Key Biodiversity Areas

The Freshwater Key Biodiversity Areas website (KBAs; http://www.birdlife.org/datazone/freshwater), which is part of the BirdLife data zone, was supported through the BioFresh project and includes the results of assessments in Europe, the Mediterranean hotspot, and Kerala and Tamil Nadu (India). Key Biodiversity Areas are globally important areas for the persistence of biodiversity as identified using standard criteria. For freshwaters they are developed through spatial analysis of species information as assessed for the IUCN Red List of Threatened Species.

20.4.3.3 Global Freshwater Biodiversity Atlas

The Global Freshwater Biodiversity Atlas (http://atlas.freshwaterbiodiversity.eu) is another major component of the Freshwater Information Platform. The atlas features spatial information generated through freshwater research. It provides a series of interactive maps with different data layers on freshwater biodiversity richness, threats to freshwaters, and the effects of global change on freshwater ecosystems.

20.4.3.4 Protected Planet: World Database on Protected Areas

The Protected Planet webpage (http://www.protectedplanet.net) is the online interface for the World Database on Protected Areas (WDPA), which is a joint project of IUCN and UNEP. The WDPA features the most comprehensive global database on terrestrial and marine protected areas, whereby freshwaters are included in the terrestrial areas. The Protected Planet webpage provides maps and searching options with additional information from the WDPA, photos from Panoramio, and text descriptions from Wikipedia.

20.5 Challenges of Data Mobilization and the Way Forward

Freshwater data, especially biodiversity data, often remain difficult to access, despite the wide range of freshwater-specific information platforms and data portals, such as the ones mentioned above. This is due to the fact that a large number of smaller datasets or individual observations of occurrences are not integrated into public repositories, even though these data may already have been used in scientific papers.

While in molecular sciences, open access to primary data is already common practice, as sequences must be submitted to GenBank prior to publication, this is not the case in other areas of freshwater research. Reasons for the reluctance to publish data or even metadata include time and financial constraints, the concern that data could be used for impropriate purposes and the fear of abandoning intellectual property rights (Schmidt-Kloiber et al. 2012).

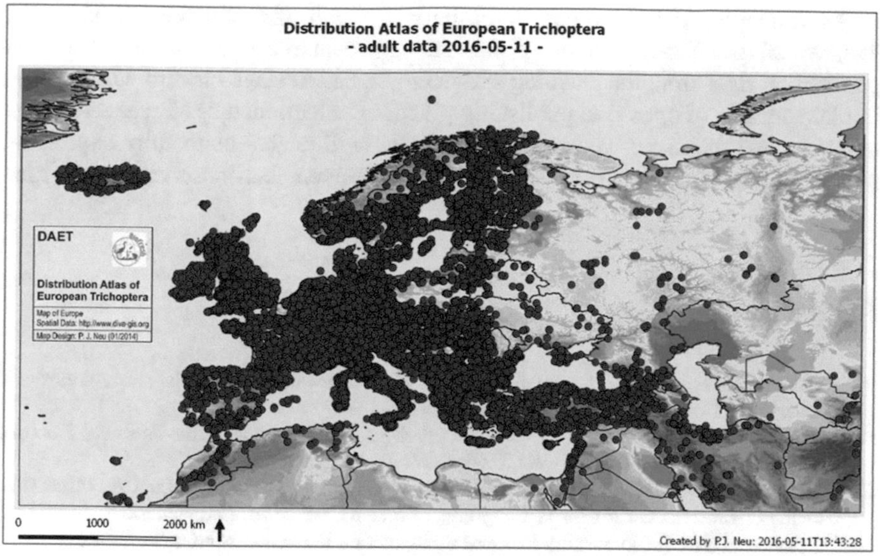

Fig. 20.3 Occurrence data of Trichoptera, initiated through BioFresh and collected by European caddisfly experts (Schmidt-Kloiber et al. 2017, Neu et al. 2018)

Recent efforts to make freshwater data easily available (see Penev et al. 2011 for an overview) include an initiative together with editors of leading freshwater journals encouraging the deposition of occurrence data in public repositories when publishing in one of the participating journals (De Wever et al. 2012). Similar efforts are undertaken by Dryad (http://datadryad.org), an international repository of data underlying peer-reviewed articles in basic and applied biosciences. Another approach to mobilizing biodiversity data was the creation of a dedicated journal to encourage data publication and to specifically address small datasets ("Biodiversity Data Journal"; Smith et al. 2013). Several authors extensively summarize the benefits of online data publication or the value of "data papers" as incentive for data publishing (e.g., Chavan and Penev 2011; Costello 2009), including the argument that papers connected to publicly available data get significantly more citations (Piwowar et al. 2007).

The importance of small datasets and their significance when compiled together can be illustrated by an initiative started within the EU-funded BioFresh project: more than 80 caddisfly experts from all over Europe were invited to contribute their personal species records to the "Distribution Atlas of European Trichoptera." This resulted in the collection of more than 450,000 point data of adult caddisflies (Schmidt-Kloiber et al. 2015, Neu et al. 2018). Only such a broad, common effort can develop such a holistic picture of the distribution patterns of different species. Biodiversity hotspots as well as sensitive areas of endemic species can be identified (Schmidt-Kloiber et al. 2017). Trying to include a broad time line in such a compilation offers views on the effects of historic events, such as glaciation, the origin of species, or the establishment of refugial areas. This may finally provide the baseline data against which to definitively measure changes in biodiversity due to different anthropogenic stressors or climate change and to establish effective management and/or conservation strategies such as the establishment of an IUCN Red List (Fig. 20.3).

As recovering data becomes increasingly difficult and resource-intensive with age, we advocate the adoption of data management practices that envisage data publication right from the planning and data generation stage onward. Only a wide implementation of open data publishing practices, which includes the generation of metadata and the use of community standards, will enable us to fully exploit the potential of the existing data for supporting management and conservation activities.

References

Balian EV, Segers H, Lévèque C, Martens K (2007) The Freshwater Animal Diversity Assessment: an overview of the results. Hydrobiologia 595:627–637

Balian EV, Segers H, Martens K, Lévèque C (2008) An introduction to the freshwater animal diversity assessment (FADA) project. Hydrobiologia 595:3–8

Cardinale BJ, Duffy JE, Gonzalez A, Hooper DU, Perrings C, Venail P, Narwani A, Mace GM, Tilman D, Wardle DA, Kinzig AP, Daily GC, Loreau M, Grace JB, Larigauderie A, Srivastava DS, Naeem S (2012) Biodiversity loss and its impact on humanity. Nature 486:59–67

Chavan V, Penev L (2011) The data paper: a mechanism to incentivize data publishing in biodiversity science. BMC Bioinf 12:S2

Collen B, Whitton F, Dyer EE, Baillie JEM, Cumberlidge N, Darwall WRT, Pollock C, Richman NI, Soulsby A-M, Böhm M (2013) Global patterns of freshwater species diversity, threat and endemism. Glob Ecol Biogeogr 23:40–51

Costello MJ (2009) Motivating online publication of data. Bioscience 59:418–427

Darwall WRT, Smith KG, Allen D, Seddon MB, McGregor RG, Clausnitzer V, Kalkman VJ (2009) Freshwater biodiversity: a hidden resource under threat. In: Vié J-C, Hilton-Taylor C, Stuart SN (eds) Wildlife in a changing world. IUCN, Gland, Switzerland, pp 43–54

De Wever A, Schmidt-Kloiber A, Gessner MO, Tockner K (2012) Freshwater journals unite to boost primary biodiversity data publication. Bioscience 62:529–530

Dudgeon D, Arthington AH, Gessner MO, Kawabata Z-I, Knowler DJ, Lévèque C, Naiman RJ, Prieur-Richard A-H, Soto D, Stiassny MLJ, Sullivan CA (2006) Freshwater biodiversity: importance, threats, status and conservation challenges. Biol Rev Camb Philos Soc 81:163–182

Friberg N, Bonada N, Bradley DC, Dunbar MJ, Edwards FK, Grey J, Hayes RB, Hildrew AG, Lamouroux N, Trimmer M, Woodward G (2011) Biomonitoring of human impacts in freshwater ecosystems: the good, the bad and the ugly (Chapter 1). In: Woodward G (ed) Advances in ecological research, vol 44. The Netherlands, Amsterdam, pp 1–68

Garcia-Moreno J, Harrison IJ, Dudgeon D (2014) Sustaining freshwater biodiversity in the anthropocene. The Global Water System in the Anthropocene – Challenges for Science and Governance, pp 247–270

Hering D, Borja A, Carstensen J, Carvalho L, Elliott M, Feld CK, Heiskanen A-S, Johnson RK, Moe J, Pont D, Solheim AL, van de Bund W (2010) The European Water Framework Directive at the age of 10: a critical review of the achievements with recommendations for the future. Sci Total Environ 408:1–13

Jones KE, Bielby J, Cardillo M, Fritz SA, O'Dell J, Orme CDL, Safi K, Sechrest W, Boakes EH, Carbone C, Connolly C, Cutts MJ, Foster JK, Grenyer R, Habib M, Plaster CA, Price SA, Rigby EA, Rist J, Teacher A, Bininda-Emonds ORP, Gittleman JL, Mace GM, Purvis A (2009) PanTHERIA: a species-level database of life history, ecology, and geography of extant and recently extinct mammals. Ecology 90:2648–2648

Kattge J, Díaz S, Lavorel S, Prentice IC, Leadley P, Bönisch G, Garnier E, Westoby M, Reich PB, Wright IJ, Cornelissen JHC, Violle C, Harrison SP, Van Bodegom PM, Reichstein M, Enquist BJ, Soudzilovskaia NA, Ackerly DD, Anand M, Atkin O, Bahn M, Baker TR, Baldocchi D, Bekker R, Blanco CC, Blonder B, Bond WJ, Bradstock R, Bunker DE, Casanoves F, Cavender-Bares J, Chambers JQ, Chapin FS III, Chave J, Coomes D, Cornwell WK, Craine JM, Dobrin BH, Duarte L, Durka W, Elser J, Esser G, Estiarte M, Fagan WF, Fang J, Fernández-Méndez F, Fidelis A, Finegan B, Flores O, Ford H, Frank D, Freschet GT, Fyllas NM, Gallagher RV, Green WA, Gutierrez AG, Hickler T, Higgins SI, Hodgson JG, Jalili A, Jansen S, Joly CA, Kerkhoff AJ, Kirkup D, Kitajima K, Kleyer M, Klotz S, Knops JMH, Kramer K, Kühn I, Kurokawa H, Laughlin D, Lee TD, Leishman M, Lens F, Lenz T, Lewis SL, Lloyd J, Llusià J, Louault F, Ma S, Mahecha MD, Manning P, Massad T, Medlyn BE, Messier J, Moles AT, Müller SC, Nadrowski K, Naeem S, Niinemets Ü, Nöllert S, Nüske A, Ogaya R, Oleksyn J, Onipchenko VG, Onoda Y, Ordoñez J, Overbeck G, Ozinga WA, Patiño S, Paula S, Pausas JG, Peñuelas J, Phillips OL, Pillar V, Poorter H, Poorter L, Poschlod P, Prinzing A, Proulx R, Rammig A, Reinsch S, Reu B, Sack L, Salgado-Negret B, Sardans J, Shiodera S, Shipley B, Siefert A, Sosinski E, Soussana JF, Swaine E, Swenson N, Thompson K, Thornton P, Waldram M, Weiher E, White M, White S, Wright SJ, Yguel B, Zaehle S, Zanne AE, Wirth C (2011) TRY – a global database of plant traits. Glob Chang Biol 17:2905–2935

Kissling WD, Hardisty A, García EA, Santamaria M, De Leo F, Pesole G, Freyhof J, Manset D, Wissel S, Konijn J, Los W (2015) Towards global interoperability for supporting biodiversity research on essential biodiversity variables (EBVs). Biodiversity 16:99–107

Michener W (2006) Meta-information concepts for ecological data management. Eco Inform 1:3–7

Neu PJ, Malicky H, Graf W, Schmidt-Kloiber A (2018) Distribution atlas of European trichoptera. Die Tierwelt Deutschlands, 84.Teil. ConchBooks, Hackenheim. 890 pp

Penev L, Mietchen D, Chavan V, Hagedorn G, Remsen D (2011) Pensoft Data Publishing Policies and Guidelines for Biodiversity Data

Pereira HM, Ferrier S, Walters M, Geller GN, Jongman RHG, Scholes RJ, Bruford MW, Brummitt N, Butchart SHM, Cardoso AC, Coops NC, Dulloo E, Faith DP, Freyhof J, Gregory RD, Heip C, Hoft R, Hurtt G, Jetz W, Karp DS, McGeoch MA, Obura D, Onoda Y, Pettorelli N, Reyers B, Sayre R, Scharlemann JPW, Stuart SN, Turak E, Walpole M, Wegmann M (2013) Essential biodiversity variables. Science 339:277–278

Piwowar HA, Day RS, Fridsma DB (2007) Sharing detailed research data is associated with increased citation rate. PLoS One 2:e308

Reichman OJ, Jones MB, Schildhauer MP (2011) Challenges and opportunities of open data in ecology. Science 331:703–705

Revenga C, Campbell I, Abell R, de Villiers P, Bryer M (2005) Prospects for monitoring freshwater ecosystems towards the 2010 targets. Philos Trans R Soc B Biol Sci 360:397–413

Schmidt-Kloiber A, Hering D (2015) www.freshwaterecology.info – An online tool that unifies, standardises and codifies more than 20,000 European freshwater organisms and their ecological preferences. Ecol Indic 53:271–282

Schmidt-Kloiber A, Moe SJ, Dudley B, Strackbein J, Vogl R (2012) The WISER metadatabase: the key to more than 100 ecological datasets from European rivers, lakes and coastal waters. Hydrobiologia 704:29–38

Schmidt-Kloiber A, Vogl R, De Wever A, Martens K (2014) Editorial – Launch of the Freshwater Metadata Journal (FMJ). Freshw Metadata J 1:1–4

Schmidt-Kloiber A, Neu PJ, Graf W (2015) Metadata to the distribution Atlas of European Trichoptera. Freshw Metadata J 9:1–6

Schmidt-Kloiber A, Neu PJ, Malicky M, Pletterbauer F, Malicky H, Graf W (2017) Aquatic biodiversity in Europe: a unique dataset on the distribution of Trichoptera species with important implications for conservation et al. Hydrobiologia 797:11–27. https://doi.org/10.1007/s10750-017-3116-4

Shorish Y (2010) The challenges to data sharing in the sciences: implications for data curation. Term paper, Graduate School of Library and Information Science, University of Illinois at Urbana-Champaign.   Available   on   http://people.lis.illinois.edu/*shorish1/courses.html. Accessed 5 Dec 2011

412 A. Schmidt-Kloiber and A. De Wever

Smith V, Georgiev T, Stoev P, Biserkov J, Miller J, Livermore L, Baker E, Mietchen D, Couvreur T, Mueller G, Dikow T, Helgen KM, Frank J, Agosti D, Roberts D, Penev L (2013) Beyond dead trees: integrating the scientific process in the Biodiversity Data Journal. Biodivers Data J 1:e995

Strayer DL, Dudgeon D (2010) Freshwater biodiversity conservation: recent progress and future challenges. J N Am Benthol Soc 29:344–358

Thessen A, Patterson D (2011) Data issues in the life sciences. ZooKeys 150:15

Thomsen PF, Kielgast J, Iversen LL, Wiuf C, Rasmussen M, Gilbert MTP, Orlando L, Willerslev E (2011) Monitoring endangered freshwater biodiversity using environmental DNA. Mol Ecol 21:2565–2573

Tittensor DP, Walpole M, Hill SLL, Boyce DG, Britten GL, Burgess ND, Butchart SHM, Leadley PW, Regan EC, Alkemade R, Baumung R, Bellard C, Bouwman L, Bowles-Newark NJ, Chenery AM, Cheung WWL, Christensen V, Cooper HD, Crowther AR, Dixon MJR, Galli A, Gaveau V, Gregory RD, Gutierrez NL, Hirsch TL, Höft R, Januchowski-Hartley SR, Karmann M, Krug CB, Leverington FJ, Loh J, Lojenga RK, Malsch K, Marques A, Morgan DHW, Mumby PJ, Newbold T, Noonan-Mooney K, Pagad SN, Parks BC, Pereira HM, Robertson T, Rondinini C, Santini L, Scharlemann JPW, Schindler S, Sumaila UR, Teh LSL, van Kolck J, Visconti P, Ye Y (2014) A mid-term analysis of progress toward international biodiversity targets. Science 346:241–244

Turak E, Dudgeon D, Harrison IJ, Freyhof J, De Wever A, Revenga C, Garcia-Moreno J, Abell R, Culp JM, Lento J, Mora B, Hilarides L, Flink S (2016) Observations of inland water biodiversity: progress, needs and priorities (Chapter 7). In: Walters M, Scholes R (eds) The GEO handbook on biodiversity observation networks. Springer, Cham, pp 165–186

Turak E, Harrison I, Dudgeon D, Abell R, Bush A, Darwall W, Finlayson CM, Ferrier S, Freyhof J, Hermoso V, Juffe-Bignoli D, Linke S, Nel J, Patricio HC, Pittock J, Raghavan R, Revenga C, Simaika JP, De Wever A (2016b) Essential biodiversity variables for measuring change in global freshwater biodiversity. Biol Conserv 213:272–279. https://doi.org/10.1016/j.biocon.2016.09.005

Vines TH, Albert AYK, Andrew RL, Débarre F, Bock DG, Franklin MT, Gilbert KJ, Moore J-S, Renaut S, Rennison DJ (2014) The availability of research data declines rapidly with article age. Curr Biol 24:94–97

Vörösmarty CJ, McIntyre PB, Gessner MO, Dudgeon D, Prusevich A, Green P, Glidden S, Bunn SE, Sullivan CA, Liermann CR, Davies PM (2010) Global threats to human water security and river biodiversity. Nature 467:555–561

Wetzel FT, Saarenmaa H, Regan E, Martin CS, Mergen P, Smirnova L, Tuama ÉÓ, García Camacho FA, Hoffmann A, Vohland K, Häuser CL (2015) The roles and contributions of Biodiversity Observation Networks (BONs) in better tracking progress to 2020 biodiversity targets: a European case study. Biodiversity 16:137–149

Woodward G, Perkins DM, Brown LE (2010) Climate change and freshwater ecosystems: impacts across multiple levels of organization. Philos Trans R Soc B Biol Sci 365:2093–2106

Chapter 21
Ecosystem Services in River Landscapes

Kerstin Böck, Renate Polt, and Lisa Schülting

21.1 What Are Ecosystem Services?

River landscapes have served as areas for settlements, infrastructure, and production for several thousand years. They provide water for drinking, cooling, and irrigation, fish as food supply or for recreational fishing, areas for flood protection, and they can have cultural and esthetic value. The increasing intensification of land use and the associated channelization, damming, and other radical changes (e.g., through operation of hydropower plants) led to a shift of the functions and related services available in river landscapes. To counteract this trend, one first step is to enhance public awareness of their importance.

One possible way to raise awareness about the importance of unimpaired river landscapes for the provision of services and to consider them more easily within decision-making processes is the ecosystem services (ES) concept. This concept highlights the relationship between different influences on ecosystems and the availability of their functions as they relate to provision of services for humans.

Ecosystem services refer to the interface between ecosystems and human well-being and are described as the many different benefits ecosystems provide to people (MEA 2003). The cascade model originally published by Haines-Young and Potschin (2010) is used as a basis in many studies dealing with the ES concept. It distinguishes between ecological structures, processes, and benefits that humans derive from ecosystems (Fig. 21.1). While ecosystem functions describe the capacity to provide goods or services for human society, the extent of ES and consequently the benefit for humans are determined by the actual demand.

K. Böck (✉) · R. Polt · L. Schülting
Institute of Hydrobiology and Aquatic Ecosystem Management, University of Natural Resources and Life Sciences, Vienna, Austria
e-mail: kerstin.boeck@boku.ac.at; renate.polt@boku.ac.at; lisa.schuelting@boku.ac.at

© The Author(s) 2018
S. Schmutz, J. Sendzimir (eds.), *Riverine Ecosystem Management*, Aquatic Ecology
Series 8, https://doi.org/10.1007/978-3-319-73250-3_21

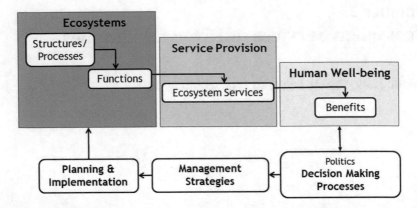

Fig. 21.1 Cascade model showing the link between ES and human well-being (Böck et al. 2015 based on Haines-Young and Potschin 2010; De Groot et al. 2010; Van Oudenhoven et al. 2012)

Originating in the 1970s, the ES concept became an issue in the international environmental discussions in the 1990s. Since then its influence has continued to rise. An important milestone was Robert Costanza's publication on the value of world's natural capital and ES that calculated the total value of world's ES with $16–$54 trillion per year (Costanza et al. 1997). A more recent publication by Costanza et al. (2014) gives an updated estimate for the total global ES in 2011 of $125 trillion per year. Other important steps were the Millennium Ecosystem Assessment (MEA 2003) that assessed the consequences of ecosystem change for human well-being, the TEEB initiative (TEEB 2010) that highlighted the global economic benefits of biodiversity, and the establishment of the Intergovernmental Science-Policy Platform on Biodiversity and Ecosystem Services (IPBES 2015) that acts as an interface between scientists and policy-makers. Furthermore, the Convention on Biological Diversity (CBD 2010) has a strong focus on ES and includes this issue as one of its strategic goals. A comprehensive review on the history of the ES concept is given by Gómez-Baggethun et al. (2010).

In the field of landscape planning, the ES approach is comparable with the idea of "landscape functionality" that has already been an issue for decades (Kienast 2010; Von Haaren and Albert 2011; Grunewald and Bastian 2013). Although this concept overlaps to a certain extent with the understanding of ES, many open questions remain before the ES concept is fully integrated in landscape research and decision-making (Hermann et al. 2014).

Depending on the research question and the context, there are different ways to categorize ES. An overview of these different classifications is given by, e.g., Häyhä and Franzese (2014). The most commonly used classification was developed in the frame of the Millennium Ecosystem Assessment (MEA 2003) that divided ES into four categories: supporting, provisioning, regulating, and cultural services. Below, some examples are listed for the freshwater context (based on Aylward et al. 2005):

- *Supporting services*—needed as a basis for almost all other services. Ecosystems provide living spaces for plants and animals and support their maintenance. Examples for the freshwater context are:

  • Role in nutrient cycling—maintenance of floodplain fertility
  • Primary production

- *Provisioning services*—material/"tangible" outputs from ecosystems including food, water, and other resources:

  • Water for consumptive use—drinking, domestic use, agriculture, and industrial use
  • Water for nonconsumptive use—generating power, transport, and navigation
  • Aquatic organisms—food and medicines

- *Regulating services*—services that ecosystems provide based on their regulating capacity:

  • Maintenance of water quality—natural filtration and water treatment
  • Buffering of flood flows, erosion control through water/land interactions, and flood control infrastructure

- *Cultural services*—nonmaterial benefits people obtain from ecosystems including esthetic, spiritual, and psychological benefits:

  • Recreation—river rafting, kayaking, hiking, and fishing
  • Tourism—river viewing
  • Existence values—personal satisfaction from free-flowing rivers

Within the scientific debate, also another term—the so-called ecosystem disservices—is discussed. This describes the negative values of ecosystems such as diseases, parasites, predators, or certain insects that are often overlooked in valuation attempts of ES (Dunn 2010). In the context of river landscape management, the ecosystem disservice "flooding" is of specific importance. While flooding also provides valuable services, such as supporting fish nurseries in the floodplain or storing water in the floodplain and the aquifer, especially in areas where people settled or built too close to water bodies or in previous floodplains, flood events are considered as "bads" (Nedkov and Burkhard 2012). However, there is often little awareness of the fact that many of these disservices are actually caused by human activities in the first place.

Although the ES concept is highly popular in the scientific realm and is discussed in countless research articles and policy papers, its application in practice lags behind (Portman 2013; Hauck et al. 2013; Albert et al. 2014). One reason for this research-practice gap is seen in the lack of any guidance for policy-makers to define, measure, and value ES and to integrate them in policy and governance (Bouma and van Beukering 2015). This points to the need for more direct science-policy interaction between researchers and stakeholders and a better communication to stakeholders and the public (Neßhöver et al. 2013; de Groot et al. 2010). Also, Böck et al. (2015)

found that many stakeholders detected a certain redundancy between the ES concept and already existing guidelines and legal frameworks. Adding ES rather seemed like an additional instead of a reduced workload.

Despite these limitations, the ES concept can be very useful and offers many potential areas of application. They include (1) raising awareness about the importance of conserving ecosystems and their biodiversity; (2) understanding their significance in relation to human activities and well-being; (3) providing a new communication framework between policy-makers, scientists, and the public on nature and society interlinkages; and (4) promoting the idea that maintaining natural capital through conservation and restoration will help sustain the provision of ES that we depend on (Wallis et al. 2011). It can also be helpful in the course of the evaluation of restoration projects to examine the effect of restoration on biodiversity and consequently the provision of ES (Rey Benayas et al. 2009).

21.2 Evaluation and Assessment Approaches for Water-Related ES

The increasing consumption of natural resources and the related loss of biodiversity are often used as an argument for the valuation of ES. Thereby, their visibility and their consideration in political and economic decision-making processes can be improved (Schwaiger et al. 2015). Several quantitative and qualitative, monetary and nonmonetary, assessment methods and techniques have been developed to systematically assess the multitude of ES and the importance of biodiversity (Schröter-Schlaack et al. 2014). In addition to these newly developed approaches, already existing methods for data collection and assessment are also used for ES studies. This includes mapping and monitoring activities, (expert) interviews, or statistical analyses (Grunewald and Bastian 2013).

Deciding on the most suitable assessment method depends on the research question and the assessment goal. Such a decision may hinge on the fact that the range of ES that can be assessed in monetary terms is rather small in comparison to the larger number of ES that can only be assessed qualitatively.

From the full range of ES, a large part can be assessed qualitatively, a smaller part can be assessed quantitatively, and even smaller parts can be assessed in monetary terms. In particular, cultural ES are often given less consideration in quantitative and monetary assessments as their classification and measuring are difficult (Satz et al. 2013). To ensure a representative picture, we need to combine monetary with other quantitative and qualitative assessments (ten Brink and Bräuer 2008). Therefore, Häyhä and Franzese (2014) suggest an interdisciplinary and system perspective for ES assessments to avoid partially informed decisions and a consequent mismanagement of natural resources. Also Kumar and Kumar (2008) and Gómez-Baggethun et al. (2014) point to the need for integrated assessment approaches that also consider the social and ecological aspects of ecosystem service values. This need has already been considered by several recent initiatives such as the MEA, the TEEB initiative, or the IPBES framework (Kelemen et al. 2014).

Several researchers point to the need of not only considering the utilitarian value of ecosystems for humans but also bearing in mind that ecosystems can also have intrinsic values, irrespective of their utility for human well-being (MEA 2005). Already in the past, several researchers questioned whether a human-centered utilitarian perspective is sufficient to protect the environment or whether it is necessary to consider the needs of the environment apart from its usefulness to humans (Seligman 1989). Giddings et al. (2002) raised the question of how money can compensate a tree for acid rain or an animal for its loss of habitat.

It also needs to be considered that people are not only utility maximizers or satisficers but have several other conflicting objectives (reciprocity, relational and ecological identity and similar processes) that influence decision-making processes (Kumar and Kumar 2008). Gómez-Baggethun et al. (2010) discuss the uncertainties of the side effects of mainstreaming market-based conservation approaches in terms of possible changes in people's motivation for conservation as well as in their human-nature relationship.

It should be noted here that evaluation and assessment approaches are not limited to applications within the specific context of the ES concept. Even though many recent studies focus on the valuation of the benefits humans derive from ecosystems, the valuation approach itself can be also applied to other aspects of nature, such as whole ecosystems, individual species, or habitats. Ecological valuation methods, for instance, usually value ecosystems, rather than ES (Kronenberg and Andersson 2016). The approach of assessing nature is also important in the field of biomonitoring and bioassessment (see Chap. 19).

21.2.1 Monetary ES Assessment Approaches

The monetary evaluation of ES is seen particularly critically by various authors (e.g., de Groot et al. 2010; Kosoy and Corbera 2010; Spash 2008). They highlight technical difficulties and ethical implications. Norgaard (2010), for instance, argues that "the metaphor of nature as a stock that provides a flow of services" is not sufficient to face today's challenges but can only work as part of a larger solution. Also a large share of interviewed stakeholders in a study of Böck et al. (2015) had a negative view toward a monetary ES assessment and voiced the fear of a commodification of nature.

Despite these limitations, economic arguments are more and more frequently used in nature protection practice. A prominent result of this emerging trend is the international TEEB initiative (TEEB 2010), which has initiated several follow-up projects at the national level (Schröter-Schlaack et al. 2014). An economic view can improve the visibility of nature's functions and services and stress the related values to critical decision-making bodies, e.g., the World Bank, which currently follows the tenets of neoliberal economics. It tries to support decision-making processes by revealing, in monetary terms, the benefit of protecting and the consequences of using nature (Schröter-Schlaack et al. 2014). Van Beukering et al. (2015) suggest four main reasons for an economic valuation of ES:

1. Advocacy—using the economic valuation of ES to promote the economic importance of the environment
2. Assisting decision- and policy-makers to make better informed decisions
3. Assessing the compensation required after the damage of an ecosystem
4. Setting taxes, fees, or charges for the use of ES

As we can see from these arguments, monetary valuation is not always undesirable. For this reason, Kallis et al. (2013) propose a normative framework as a decision support when to choose monetary valuation that considers four questions/criteria:

1. Additionality—Will it improve the environmental conditions at stake?
2. Equality—Will it reduce inequalities and redistribute power?
3. Complexity blinding—Will it suppress other valuations?
4. Neoliberalism—Will it serve processes of enclosure of the commons?

According to the authors, a monetary valuation is reasonable if the answers to the first two questions are "yes" and the answers to questions (3) and (4) are "no."

A large number of various valuation methods are available to estimate the value of different ES. This is because no single economic valuation technique is applicable to all ES, but the methods vary depending on the ES' characteristics and data availability (DEFRA 2007).

A basic distinction is made between market-based and nonmarket-based valuation methods (Fig. 21.2). The first method derives economic values from market prices, while in the second case, ES are valued indirectly via revealed preference methods. The most commonly applied (indirect) market valuation methods are the hedonic pricing and the travel cost method. The hedonic pricing method compares sales prices of two commodities (usually houses). The commodities need to be identical in most respects, except in regard to a certain environmental characteristic (e.g., traffic noise). The difference of the commodities' sale prices can then be interpreted as a revealed "willingness to pay" for the ES, resulting in a price for the ES. The travel cost method observes the travel expenses (e.g., travel costs, time, admittance fees) of people visiting, for instance, a recreation site, which implicitly represent the economic value of the site (Koetse et al. 2015). If there is no market price available and the application of revealed preference methods is not possible, nonmarket valuation methods (i.e., stated preference methods) are used. The most important approaches in this context are the contingent valuation and the choice experiment method. In that case, surveys are used to ask people for their preferences for hypothetical changes in the provision of ES. Thereby, the values that people attach to them are estimated (DEFRA 2007).

Two alternative methods are the meta-analysis and the value transfer. Although they are not valuation methods in themselves, they are relevant to mention as they are often used to derive ES values.

Building on the work of Turner et al. (2004) and Young (2005), Brouwer et al. (2009, p.35f) summarized commonly used monetary valuation methodologies for water resources and differentiated them regarding the assessed water use. All

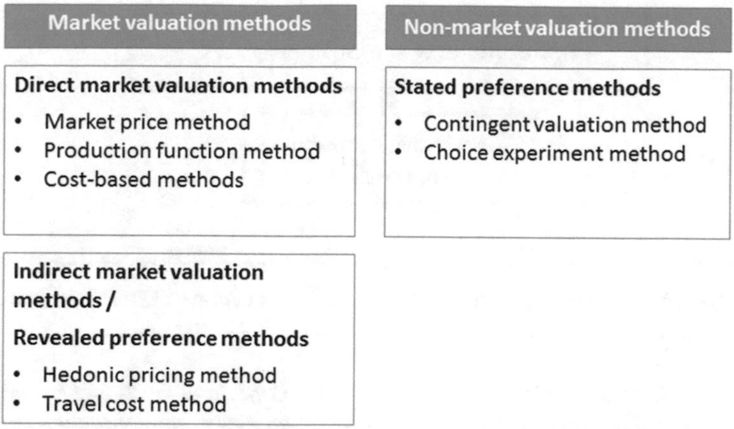

Market valuation methods	Non-market valuation methods
Direct market valuation methods • Market price method • Production function method • Cost-based methods	**Stated preference methods** • Contingent valuation method • Choice experiment method
Indirect market valuation methods / **Revealed preference methods** • Hedonic pricing method • Travel cost method	

Fig. 21.2 Methods for the valuation of ES based on Koetse et al. (2015)

techniques have certain weaknesses and strengths. The decision on which one to use depends on several factors (based on Brouwer et al. 2009):

- Type of ecosystem good/service to be valued.
- Type of values—use values can be estimated by all valuation techniques, while nonuse values can only be estimated by the stated preference method.
- Valuation purpose.
- Data availability.
- Required accuracy of results.
- Available resources and time.

21.2.2 Nonmonetary Assessment Approaches

Because only a small part of nature's services can actually be assessed in monetary terms (Fig. 21.2), there is a need for nonmonetary assessment approaches that also consider those services that are difficult to quantify. This is particularly the case for cultural services that are difficult to integrate in decision-making due to the challenge of assigning a monetary value to them (Chan et al. 2012). This can lead to a limited awareness of the variety of services that are provided by ecosystems and can be a challenge for mainstreaming ES across different societal actors (Martín-López et al. 2012). Although they are highly valued by different stakeholder groups, they are not reflected by economic indicators and therefore often sacrificed for economic or ecological reasons (Milcu et al. 2013; Chan et al. 2011).

To address these services, sociocultural valuation approaches are increasingly gaining attention (Chan et al. 2006, 2012). These approaches consider services that are related to nonuse values, such as local identity or the intrinsic value of ecosystems, and cannot be addressed using economic techniques (Castro Martínez et al. 2013).

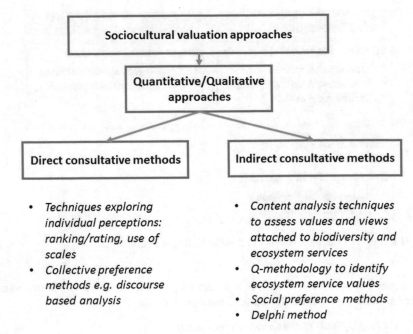

Fig. 21.3 Overview on sociocultural valuation approaches (based on Kelemen et al. 2014; Castro Martínez et al. 2013)

They can be divided into qualitative and quantitative methods, both involving direct and indirect consultative methods (Fig. 21.3).

Oteros-Rozas et al. (2014), for instance, describe a sociocultural valuation approach of ES that investigates people's perceptions of the importance of ES in a cultural landscape.

There are several other creative ideas from different researchers regarding nonmonetary assessment approaches. For instance, García-Llorente et al. (2016) suggest that to analyze the social support for biodiversity conservation activities and the related delivery of ES, one calculates the willingness to give up time. Kumar and Kumar (2008) point to the need of considering psychological and sociological aspects in ecosystem service valuation. They argue that there is a need for approaches that combine natural and social science research.

21.3 Mapping and Assessment of ES

An alternative to valuation that has become increasingly popular is the spatial representation of ES. This approach is often associated with participatory mapping- or photo-based methods (Milcu et al. 2013; Raymond et al. 2009; Eder and Arnberger 2016) and is also frequently combined with monetary and nonmonetary valuation methods.

The approach of mapping ES has rapidly increased in the last years. It can be particularly helpful to identify highly valuable areas for conservation, ES' supply and demand in a specific area, and trade-offs between different services (Chan et al. 2006; Häyhä and Franzese 2014). Mapping ES can visualize the effect of different management strategies on their supply and can therefore be a starting point for developing strategies for management and conservation (Naidoo et al. 2008). It is also very useful with regard to communication and visualization purposes and can thereby support decision-making processes (Alkemade et al. 2014), especially participatory science processes that involve scientists, policy-makers, and local practitioners.

Verhagen et al. (2015) identified three main applications of ES maps, namely (1), the identification of "hotspots of change," i.e., areas that need to be protected from changes that affect the service supply, (2) the visualization of trade-offs and synergies between ES, and (3) the active management of landscapes to "optimize ES to locations," i.e., spatial configurations to optimize ES supply.

Different authors have proposed a variety of approaches for mapping ES, including the use of biophysical metrics or monetary valuations (Häyhä and Franzese 2014). Martínez-Harms and Balvanera (2012) classify ES mapping approaches into three main approaches: (1) valuation of ES through benefit transfer (transferring the monetary value of a similar previous study to the current land cover map), (2) community value methods that integrate survey-based perceptions of place with biophysical data, and (3) different social-ecological assessment approaches that model the relationship between ecological and social variables to map ES supply. Which of the different proposed methods is applied depends on data availability, the scope of the study, and time constraints (Verhagen et al. 2015).

Regulating services are the most commonly mapped group of ES. When looking at the individual services, carbon storage, carbon sequestration, food production, and recreation are most frequently studied (Martínez-Harms and Balvanera 2012).

Several tools and frameworks are already available to map ES (Castro Martínez et al. 2013):

- The *InVEST-tool* was developed within the frame of the Natural Capital Project (http://www.naturalcapitalproject.org/invest/). Analyses can be conducted at local, regional, or global scale. The results are either returned in biophysical or economic terms.
- The web-based *ARIES* (Artificial Intelligence for Ecosystem Services) technology (http://ariesonline.org/) combines a series of applications that assist in mapping ecosystem service provision, use, and benefit through utilizing GIS data from global through local scale.
- Remote sensing—Land use/land cover can be used as a proxy for quantifying and mapping ES. A review on relevant remote sensing approaches can be found in Ayanu et al. (2012).
- Polyscape—A GIS framework that is designed to explore spatially explicit synergies and trade-offs among ES to support landscape management (Jackson et al. 2013).

21.4 ES Indicators

The assessment of ES generally requires the identification and application of a suitable and comprehensive set of indicators. Indicators are variables that provide aggregated information on certain phenomena (Wiggering and Müller 2004), provide a means of measuring service provision (Norton et al. 2015), and serve as communication tools to simplify the complexity of human-environmental systems (Müller and Burkhard 2012). Heink and Kowarik (2010, p. 590) provide the following general definition: "An indicator in ecology and environmental planning is a component or a measure of environmentally relevant phenomena used to depict or evaluate environmental conditions or changes or to set environmental goals." Some examples for indicators describing freshwater ES are the area occupied by riparian forests, the amount of fish produced (catch in tonnes by commercial and recreational fisheries), the ecological status of the water body in question, or the number of visitors to specific sites (Maes et al. 2016).

When selecting indicators for an ES valuation, a careful and critical approach is essential. The framing and selection process not only defines what is being assessed (Hauck et al. 2015) but it also has significant influence on how the subsequent assessment can help policy- and decision-makers to take appropriate steps to counter undesirable ecosystem changes (Feld et al. 2010; Niemeijer and de Groot 2008).

An important factor to consider when choosing ES indicators is practicability and, in this regard, especially data availability. Some authors argue to use indicators that can be described by data that was already collected for other purposes. Others say this approach focuses too much on the currently most visible and easily accessible services (Hauck et al. 2015) and at the same time causes other ES to recede from view and to be neglected in policy decisions (Heink et al. 2015; Maes et al. 2012). Another essential aspect of the indicator selection process is to make a distinction between potential and actual use of ES, because it results in different requirements for indicators. This circumstance is best explained with an example: When referring to the actual use of fish, one would choose a flow indicator (measured per unit of time) like tonnes of fish caught per year. When referring to the potential use of fish however, e.g., to find out about the reproductive potential of the fish population, a stock indicator (measured at a particular point in time) like the fish population size would be more suitable (Boyd and Banzhaf 2007).

Wiggering and Müller (2004) provide a list of scientific- and management-related factors, which are relevant for successful indicator development, including, among others, a clear representation of the phenomenon of interest by the indicator, a clear proof of relevant cause-effect relations, a high degree of comparability in and with indicator sets, a good fulfillment of statistical requirements, a high political relevance concerning the decision process, a high comprehensibility and public transparency, a satisfying measurability, and a high degree of data availability. Keeping all these factors in mind, the indicator selection depends heavily on the specific context of application and the characteristics of the investigated ecosystem (Reyers et al. 2010; Fisher et al. 2009).

21.5   The ES Concept: Applications in Research and Practice

The ES concept offers a number of potential applications in research as well as in the practice realm. In the following, the application of the ES concept is shown at the example of different case studies.

21.5.1   Case Study: Monetary Quantification of ES Within the EU Project "REFORM"

An example of a project dealing with the quantification of ES is the EU-funded project REFORM (REstoring rivers FOR effective catchment Management) where, among many other issues, 20 pairs of restored and unrestored river reaches were investigated throughout Europe. The main project objective was to provide a framework for successful and cost-effective river restoration and to monitor the biological responses to hydromorphological changes, in order to reach the good ecological status or potential of rivers (REFORM 2015).

To investigate and quantitatively assess the success of the applied river restoration measures, REFORM used different response variables, such as hydromorphological attributes, habitat composition of the river and its floodplain, aquatic- and floodplain-related organism groups (fish, invertebrates, floodplain vegetation, etc.), and stable isotopes. For eight pairs of restored and unrestored reaches, however, the project team additionally applied the ES approach to estimate restoration success (Muhar et al. 2016). Provisioning (agricultural products, wood, infiltrated drinking water), regulating (flooding, nutrient retention, carbon sequestration), and cultural (recreational hunting and fishing, kayaking, biodiversity conservation, appreciation of scenic landscapes) services were quantified and monetized by means of locally available data and literature, by conducting surveys among inhabitants and visitors and by using a selection of economic methods (e.g., market value, willingness-to-pay survey). Afterward, the resulting numbers were summed up to provide an estimate of annual economic ecosystem service value, normalized per area (Vermaat et al. 2015).

The authors of this study acknowledge that a monetary quantification of ES may not depict the fullness and diversity of societal appreciation (Westman 1977). Nevertheless they used this approach because it enables the comparison and evaluation of trade-offs and provides tangible information that is understandable for the general public and policy-makers (Vermaat et al. 2015).

The results of the analysis show that river restoration indeed enhances overall societal benefits. The restored reaches and their floodplains provided significantly higher service delivery and higher total value than the paired, unrestored reaches did (Vermaat et al. 2015).

21.5.2 Case Study: Application of the ES Concept in the Context of the Renaturalization of the River Emscher

One of the largest river restoration projects in Europe is currently being carried out in North Rhine-Westphalia, Germany with a total budget of 4.5 billion Euros (RWI 2013). The Emscher River is a tributary of the River Rhine. Its river system drains the Ruhr region, and its catchment of around 865 km^2 is home to some 2.5 million inhabitants. Thus, it plays an important role for economy and recreation in the area (Busch et al. 2001). As with most other rivers in Europe, it has experienced major alterations during industrialization since the beginning of the nineteenth century. The Emscher system was transformed into a system of concrete sewage channels due to straightening, embankment, and sewage discharge (Gerner et al. 2015; EG/LV 2015; Sommerhäuser and Gerner 2015; Winking et al. 2014).

A 30-year project to restore the river system was started in the 1990s. The restoration measures include the construction of four decentralized sewage plants and 400 km of new, separated sewers, as well as restructuring of 350 river km in order to obtain a near-natural state (Sommerhäuser and Gerner 2015).

Even though the project is still ongoing, the benefits from several ecosystem services, such as biodiversity, climate regulation, water quality, flood retention, recreation, and regional attractiveness, have already been recognized. Since the 1990s, the flood retention area has doubled, and a substantial increase of plant and aquatic macroinvertebrate species was measured. The increase of green areas (around 1 km^2) improves climate regulation in the urban area. Additionally, the creation of 120 km of bike and walk ways strongly contributes to the enhancement of the region's attractiveness and educational value as well as to the increase of its monetary value (Sommerhäuser and Gerner 2015). Besides the direct effects of the restoration results, the construction measures themselves have to be considered. A study from 2013 revealed vast socioeconomic effects from the project's implementation. On average, the project directly creates or saves 1400 jobs per year, resulting in 41,554 person-years of useful work. According to the study, the numbers are even higher (109,787 person-years and 3700 jobs/year) when including indirectly connected production and employment effects. Additionally, the project generates tax incomes for the municipalities (around 50 m. Euros), federal states (around 91 m. Euros), and the German state (around 1.1 bn. Euros) over the whole project period (RWI 2013). The effect of the restoration activities on ES in this case study is specifically investigated within the frame of the "DESSIN" project (IWW Water Centre 2014) that aims to demonstrate a methodology for the valuation of ES.

21.5.3   Case Study: The ES Approach as a Way to Address Different Stakeholders' Perspectives in River Landscapes

Besides its potential applicability for assessment and communication purposes, the ES approach can be used to address different perspectives toward river landscapes. Based on this consideration, the ES concept was applied in three study cases in Austria as a basis to investigate people's perceptions of the availability and importance of various services provided by near-natural, restored, and degraded river landscapes. The focus lay on stakeholders with a certain decision-making competency and recreational river users in the case studies of the Enns and Drau rivers (Böck et al. 2013). In the third case study, the river Traisen, the focus lay on children and young adults (Poppe et al. 2015).

The results of the semiquantitative questionnaire-based surveys revealed that in all three case study surveys, participants perceived cultural and supporting services the highest. In the Enns and Drau river case studies, survey participants regarded nature experience, recreation, and tourism as well as recreational fishery and water sports as specifically relevant (Böck et al. 2013). The surveyed young adults in the Traisen river case study most often associated structural elements, such as water or stones, with cultural functions, e.g., recreation possibilities within the river landscape. Similar to the first two case studies, they regarded cultural, regulating, and supporting services—specifically room for free movement, restfulness, and reduction of pollutants—as highly important (Poppe et al. 2015).

All three case studies applied a strong focus on nonmonetary river landscape uses. This contrasts with the limited consideration of these services in practice in the formulation and application of policy. There, provisioning services tend to be given priority as they can be quantified and evaluated more easily and are therefore better comparable with economic values like jobs and property values.

The results of the investigations shall contribute to gaining a comprehensive view of river landscapes and thereby improving future restoration planning and management. The insights into people's perceptions have the potential to foster awareness for the importance of conserving a wide range of different river ES and assist in estimating future educational needs.

21.6   Policy Context: ES Concept as Decision-Making Tool

The ES concept has the potential to contribute to already available management approaches. Its integration can help to support the evaluation of policy impacts, e.g., through the application of combined quantitative and qualitative valuation approaches. Its implementation into existing policy frameworks is discussed by several authors (e.g., Wallis et al. 2011; Vlachopoulou et al. 2014) and has the potential to be an added value in future decision-making processes.

21.6.1   Integration of the ES Approach into the WFD

Although the ES concept is not yet explicitly mentioned in the EU Water Framework Directive, the aspect of ensuring the provision of ES is implicitly linked with the WFD objective of "reaching a good ecological status" (Wallis et al. 2011). The following WFD articles specifically refer to the valuation of ES (based on Wallis et al. 2011):

- Article 5: assessment of the economic significance of water use, current level of cost recovery
- Article 4: decisions on derogations
- Article 9: assessment of the level of cost recovery and incentive pricing
- Article 11: selection of the most cost-effective sets of measures for achieving good ecological status/potential for the programs of measures

The clear linkage between the WFD and its principles and the ecosystem approach is also shown by Vlachopoulou et al. (2014). They argue that it has the potential to act in a complementary way. It could, for instance, be applied for the evaluation of different management scenarios and thereby support decision-making processes. In particular, the more holistic management approaches that are supported by the ES approach, such as the spatial mapping of ES or the comprehensive evaluation of multiple benefits, can potentially assist in achieving the goals of the WFD (Vlachopoulou et al. 2014).

In this context, Koundouri et al. (2015) propose an integrated methodology that regards the ES approach as the core aspect to achieve a more sustainable and efficient water management. It combines the ES framework with traditional economic frameworks and consists of a socioeconomic characterization of the river basin area, an assessment of water use costs that are recovered, and a suggestion for appropriate measures for sustainable water management.

In order to enhance the WFD implementation, Reyjol et al. (2014) established a list of research needs that also includes the reinforcement of the knowledge on relationships between good ecological status, biodiversity, and ES. They recommend enhancing understanding of ecological processes through developing further research on the links between good ecological status, biodiversity, and ecosystem functioning. As the ES concept is still not very well known among water policy-makers and managers, they point to the need for easily comprehensible guidelines for these actor groups.

21.6.2   Integration of ES into Biodiversity Policy: 2020

In 2011 the European Commission adopted the "Biodiversity Strategy to 2020" aiming to "halt the loss of biodiversity and the degradation of ES in the EU by 2020 and restore them in so far as feasible, while stepping up the EU contribution to averting global biodiversity loss" (European Commission 2015). Besides the first target of creating a habitat network by ensuring the implementation of the Birds and Habitat Directive, the second target of the strategy specifically addresses the protection and restoration of ecosystems and their services, since it is assumed that the restoration of ecosystems goes hand in hand with the protection and provision of ES. More specifically, Action 5 of the strategy recommends to "assess the state of ecosystems and their services in their national territory by 2014, assess the economic value of such services, and promote the integration of these values into accounting and reporting systems at EU and national level by 2020" (European Commission 2011). For fulfilling these aims, the working group "Mapping and Assessment of Ecosystems and their Services" (MAES) was established and delivered an approach for mapping and assessment of ES. In a technical report from 2014, the working group stresses that several conceptual issues regarding ES remain unexplained and that the links between biodiversity, ecosystem functioning, and the provision of ES are still not well understood (Maes et al. 2014).

A midterm review of the Biodiversity Strategy to 2020 states that since the strategy was adopted, some local improvements have been made in terms of ecosystem restoration but at a rate so low that it leads to further ecosystem degradation and loss of ES. Human pressures on freshwater systems and other ecosystems remain unfavorably high, such that high impacts on biodiversity in freshwater systems persist. However, some important enhancements in the knowledge base have been made, and the collected data of assessed and mapped ecosystem and related services will be available for the support of decision-makers and private stakeholders in planning processes (European Commission 2015).

21.7 Opportunities of the ES Concept in River Landscape Management

Despite the aforementioned limitations and the fact that the ES concept has not yet "taken off" in river landscape management practice, it has the potential to raise awareness in society in general and for administrative actors and political representatives in particular (Böck et al. 2015). It could help to improve societal and political acceptance of river restoration projects (Vermaat et al. 2015) and serve as a tool to improve environmental communication and education (Böck et al. 2015; Rewitzer et al. 2014). Due to its integrative character, actors in river landscape management also regarded the ES concept as a valuable support for planning and decision-making processes (Böck et al. 2015). For landscape planning processes, Kienast (2010) stresses conceptually strong points of the ES concept, such as the systematic approach of determining services and interdisciplinary, holistic approaches to supporting decisions based on integrating values generated from different perspectives (see Chap. 16).

A major benefit from the application of the ES concept can be gained through its integration into ongoing programs, tools, processes, and policies. Through the recognition and quantification of ecosystems' benefits for society, they are no longer deemed as worthless in decision-making processes (Everard 2009). On the contrary, the worth of ES will be increasingly recognized as society begins to decarbonize our economies to mitigate the impacts of climate change. This will involve lowering or eliminating the use of fossil fuels, whose services were used to substitute for the loss of ES over the past two centuries. Increasing acknowledgement of the vital role that ES play in the functioning of river social-ecological systems will be formalized not only in policy but in practice. This transition will be challenging, if only for its novelty, but earlier incorporation of ES into our economic and political practice will make it easier.

References

Albert C, Hauck J, Buhr N, von Haaren C (2014) What ecosystem services information do users want? Investigating interests and requirements among landscape and regional planners in Germany. Landsc Ecol 29:1–13. http://www.scopus.com/inward/record.url?eid=2-s2.0-84893006728&partnerID=40&md5=baf78f54f0ae84393c4c69302e13cb0d

Alkemade R, Burkhard B, Crossman ND, Nedkov S, Petz K (2014) Quantifying ecosystem services and indicators for science, policy and practice. Ecol Indic 37(PART A):161–162. https://doi.org/10.1016/j.ecolind.2013.11.014

Ayanu YZ, Conrad C, Nauss T, Wegmann M, Koellner T (2012) Quantifying and mapping ecosystem services supplies and demands: a review of remote sensing applications. Environ Sci Technol 46(16):8529–8541. https://doi.org/10.1021/es300157u

Aylward B, Bandyopadhyay J, Belausteguigotia J-C (2005) Freshwater Ecosystem Services. Ecosyst Hum Well Being Policy Responses 3:213–254

Böck K, Muhar A, Oberdiek J, Muhar S (2013) Die Wahrnehmung von Fließgewässerbezogenen "Ökosystemleistungen" Und Konfliktpotenzialen Am Fallbeispiel "Flusslandschaft Enns". Österr Wasser- Und Abfallw 65:418–428

Böck K, Muhar S, Muhar A, Polt R (2015) The ecosystem services concept: gaps between science and practice in river landscape management. GAIA Ecol Perspect Sci Soc 24(1):32–40. https://doi.org/10.14512/gaia.24.1.8

Bouma JA, van Beukering PJH (2015) Ecosystem services: from concept to practice. In: Bouma JA, van Beukering PJH (eds) Ecosystem services. From concept to practice. Cambridge University Press, Cambridge, pp 3–21

Boyd J, Banzhaf S (2007) What are ecosystem services? The need for standardized environmental accounting units. Ecol Econ 63(2-3):616–626. https://doi.org/10.1016/j.ecolecon.2007.01.002

Brouwer R, Barton D, Bateman I, Brander L, Georgiou S, Martin-Ortega J, Navrud S, Pulido-Velazquez M, Schaafsma M, Wagtendonk A (2009) Economic valuation of environmental and resource costs and benefits in the water framework directive: technical guidelines for practitioners. Institute for Environmental Studies, VU University Amsterdam, Amsterdam

Busch D, Büther H, Rahm H, Ostermann K, Thiel A (2001) Emscher-PLUS, Projekt zur Langzeit-Untersuchung des Sanierungserfolges. Staatliches Umweltamt Herten, Eigenverlag

Castro Martínez AJ, García-Llorente M, Martín-López B, Palomo I, Iniesta-Arandia I (2013) Multidimensional approaches in ecosystem services assessment. In: Alcaraz-Segura D, Di Bella CD, Straschnoy YV (eds) Earth observation of ecosystem services. CRC Press, Boca Raton, pp 441–468

CBD (2010) The strategic plan for biodiversity 2011–2020 and the aichi biodiversity targets. Nagoya, Japan

Chan KMA, Rebecca Shaw M, Cameron DR, Underwood EC, Daily GC (2006) Conservation planning for ecosystem services. PLoS Biol 4(11):e379. https://doi.org/10.1371/journal.pbio.0040379

Chan KMA, Goldstein J, Satterfield T, Hannahs N, Kikiloi K, Naidoo R, Vadeboncoeur N, Woodside U (2011) Cultural servives and non-use values. In: Kareiva P, Tallis H, Ricketts TH, Daily GC, Polasky S (eds) Natural capital: theory and practice of mapping ecosystem services. Oxford University Press, Oxford, pp 206–228

Chan KMA, Guerry AD, Balvanera P, Klain S, Satterfield T, Basurto X, Bostrom A et al (2012) Where are cultural and social in ecosystem services? A framework for constructive engagement. Bioscience 62(8):744–756. https://doi.org/10.1525/bio.2012.62.8.7

Costanza R, D'Arge R, de Groot R, Farber S, Grasso M, Hannon B, Limburg K et al (1997) The value of the world's ecosystem services and natural capital. Nature 387(6630):253–260. https://doi.org/10.1038/387253a0

Costanza R, de Groot R, Sutton P, van der Ploeg S, Anderson SJ, Kubiszewski I, Farber S, Kerry Turner R (2014) Changes in the global value of ecosystem services. Glob Environ Chang 26 (May):152–158. https://doi.org/10.1016/j.gloenvcha.2014.04.002

de Groot RS, Alkemade R, Braat L, Hein L, Willemen L (2010) Challenges in integrating the concept of ecosystem services and values in landscape planning, management and decision making. Ecol Complex 7(3):260–272. https://doi.org/10.1016/j.ecocom.2009.10.006

DEFRA (2007) An introductory guide to valuing ecosystem services. https://www.gov.uk/government/uploads/system/uploads/attachment_data/file/69192/pb12852-eco-valuing-071205.pdf

Dunn RR (2010) Global mapping of ecosystem disservices: the unspoken reality that nature sometimes kills us. Biotropica 42(5):555–557. https://doi.org/10.1111/j.1744-7429.2010.00698.x

Eder R, Arnberger A (2016) How heterogeneous are adolescents' preferences for natural and semi-natural riverscapes as recreational settings? Landsc Res 41(5):555–568. https://doi.org/10.1080/01426397.2015.1117063

EG/LV (2015) Emscher Umbau. http://www.eglv.de/

European Commission (2015) Report from the commission to the European Parliament and the council: the mid term review of the EU biodiversity strategy to 2020, Brussels

European Commission (2011) EU biodiversity strategy to 2020 – fact sheet. doi:10.277924101

Everard M (2009) Using science to create a better place: ecosystem services case studies. Bristol: Environment Agency. http://catalog.ipbes.net/assessments/194

Feld CK, Paulo Sousa J, da Silva PM, Dawson TP (2010) Indicators for biodiversity and ecosystem services: towards an improved framework for ecosystems assessment. Biodivers Conserv 19 (10):2895–2919. https://doi.org/10.1007/s10531-010-9875-0

Fisher B, Kerry Turner R, Morling P (2009) Defining and classifying ecosystem services for decision making. Ecol Econ 68(3):643–653. https://doi.org/10.1016/j.ecolecon.2008.09.014

García-Llorente M, Castro AJ, Quintas-Soriano C, López I, Castro H, Montes C, Martín-López B (2016) The value of time in biological conservation and supplied ecosystem services: a willingness to give up time exercise. J Arid Environ 124:13–21

Gerner N, Birk S, Winking C, Nafo I (2015) Welche Ökosystemleistungen bringen Renaturierungen in urbanen Räumen mit sich? In DGL 2015 – Essen – Jahrestagung Der Deutschen Gesellschaft für Limnologie und der deutschsprachigen Sektionen der SIL

Giddings B, Hopwood B, O'Brien G (2002) Environment, economy and society: fitting them together into sustainable development. Sustain Dev 10:187–196

Gómez-Baggethun E, de Groot R, Lomas PL, Montes C (2010) The history of ecosystem services in economic theory and practice: from early notions to markets and payment schemes. Ecol Econ 69(6):1209–1218. https://doi.org/10.1016/j.ecolecon.2009.11.007

Gómez-Baggethun E, Martín-López B, Barton D, Braat L, Kelemen E, García-Llorente M, Saarikoski H, et al (2014) State-of-the-Art Report on Integrated Valuation of Ecosystem Services. EU FP7 OpenNESS Project Deliverable D.4.1/WP4

Grunewald K, Bastian O (2013) Ökosystemleistungen. Konzept, Methoden und Fallbeispiele. Springer Spektrum, Heidelberg

Haines-Young R, Potschin M (2010) The links between biodiversity, ecosystem services and human well-being. In: Raffaelli DG, Frid CLJ (eds) Ecosystem ecology: a new synthesis. Cambridge University Press, Cambridge

Hauck J, Schweppe-Kraft B, Albert C, Görg C, Jax K, Jensen R, Fürst C et al (2013) The promise of the ecosystem services concept for planning and decision-making. Gaia 22(4):232–236

Hauck J, Albert C, Fürst C, Geneletti D, La Rosa D, Lorz C, Spyra M (2015) Developing and applying ecosystem service indicators in decision-support at various scales. Ecol Indic 61:1–5. https://doi.org/10.1016/j.ecolind.2015.09.037

Häyhä T, Franzese PP (2014) Ecosystem services assessment: a review under an ecological-economic and systems perspective. Ecol Model 289(October):124–132. https://doi.org/10.1016/j.ecolmodel.2014.07.002

Heink U, Kowarik I (2010) What are indicators? On the definition of indicators in ecology and environmental planning. Ecol Indic 10(3):584–593. https://doi.org/10.1016/j.ecolind.2009.09.009

Heink U, Hauck J, Jax K, Sukopp U (2015) Requirements for the selection of ecosystem service indicators – the case of MAES indicators. Ecol Indic 61:18–26. https://doi.org/10.1016/j.ecolind.2015.09.031

Hermann A, Kuttner M, Hainz-Renetzeder C, Konkoly-Gyuró É, Tirászi Á, Brandenburg C, Allex B, Ziener K, Wrbka T (2014) Assessment framework for landscape services in European cultural landscapes: an Austrian Hungarian case study. Ecol Indic 37(February):229–240. https://doi.org/10.1016/j.ecolind.2013.01.019

IPBES (2015) IPBES. http://www.ipbes.net/

IWW Water Centre (2014) DESSIN-Demonstrate Ecosystem Services Enabling Innovation in the Water Sector. https://dessin-project.eu/

Jackson B, Pagella T, Sinclair F, Orellana B, Henshaw A, Reynolds B, Mcintyre N, Wheater H, Eycott A (2013) Polyscape: a GIS mapping framework providing efficient and spatially explicit landscape-scale valuation of multiple ecosystem services. Landsc Urban Plan 112(April):74–88. https://doi.org/10.1016/j.landurbplan.2012.12.014

Kallis G, Gómez-Baggethun E, Zografos C (2013) To value or not to value? That is not the question. Ecol Econ 94:97–105. https://doi.org/10.1016/j.ecolecon.2013.07.002

Kelemen E, García-Llorente M, Pataki G, Martín-López B, Gómez-Baggethun E (2014) Non-monetary techniques for the valuation of ecosystem service. In OpenNESS Reference Book. EC FP7 Grant Agreement No. 308428, 4. www.openness-project.eu/library/reference-book

Kienast F (2010) Landschaftsdienstleistungen: Ein taugliches Konzept für Forschung und Praxis? Forum für Wissen 2010:7–12

Koetse MJ, Brouwer R, van Beukering PJH (2015) Economic valuation methods for ecosystem services. In: Bouma JA, van Beukering PJH (eds) Ecosystem services: from concept to practice. Cambridge University Press, Cambridge, pp 108–131

Kosoy N, Corbera E (2010) Payments for ecosystem services as commodity fetishism. Ecol Econ 69(6):1228–1236. https://doi.org/10.1016/j.ecolecon.2009.11.002

Koundouri P, Ker Rault P, Pergamalis V, Skianis V, Souliotis I (2015) Development of an integrated methodology for the sustainable environmental and socio-economic management of river eco-systems. Sci Total Environ 540:–90, 100. https://doi.org/10.1016/j.scitotenv.2015.07.082

Kronenberg J, Andersson E (2016) Integrated valuation: integrating value dimensions and valuation methods. FP 7 project "Green Surge", WP 4, Milestone 32. ULOD, Poland; SRC, Sweden

Kumar M, Kumar P (2008) Valuation of the ecosystem services: a psycho-cultural perspective. Ecol Econ 64(4):808–819. https://doi.org/10.1016/j.ecolecon.2007.05.008

Maes J, Egoh B, Willemen L, Liquete C, Vihervaara P, Schägner JP, Grizzetti B et al (2012) Mapping ecosystem services for policy support and decision making in the European Union. Ecosyst Serv 1(1):31–39. https://doi.org/10.1016/j.ecoser.2012.06.004

Maes J, Teller A, Erhard M, Murphy P, Paracchini ML, Barredo JI, Grizzetti B et al (2014) Mapping and assessment of ecosystems and their services in the EU. The Swedish forest pilot. https://doi.org/10.2779/75203

Maes J, Liquete C, Teller A, Erhard M, Paracchini ML, Barredo JI, Grizzetti B et al (2016) An indicator framework for assessing ecosystem services in support of the EU biodiversity strategy to 2020. Ecosyst Serv 17:14–23. https://doi.org/10.1016/j.ecoser.2015.10.023

Martínez-Harms MJ, Balvanera P (2012) Methods for mapping ecosystem service supply: a review. Int J Biodiv Sci Ecosyst Serv Manag 8(1–2):17–25. https://doi.org/10.1080/21513732.2012.663792

Martín-López B, Iniesta-Arandia I, García-Llorente M, Palomo I, Casado-Arzuaga I, Del Amo DG, Gómez-Baggethun E et al (2012) Uncovering ecosystem service bundles through social preferences. PLoS One 7(6):e38970. https://doi.org/10.1371/journal.pone.0038970

MEA (2003) Ecosystems and human well-being: a framework for assessment. Washington, DC. http://www.millenniumassessment.org/en/Framework.html

MEA (2005) Ecosystems and human well-being: a framework for assessment. Millenium ecosystem assessment. Washington, DC

Milcu AI, Hanspach J, Abson D, Fischer J (2013) Cultural ecosystem services: a literature review and prospects for future research. Ecol Soc 18(3). https://doi.org/10.5751/ES-05790-180344

Muhar S, Januschke K, Kail J, Poppe M, Schmutz S, Hering D, Buijse AD (2016) Evaluating good-practice cases for river restoration across Europe: context, methodological framework, selected results and recommendations. Hydrobiologia 769(1):3–19. https://doi.org/10.1007/s10750-016-2652-7

Müller F, Burkhard B (2012) The indicator side of ecosystem services. Ecosyst Serv 1(1):26–30. https://doi.org/10.1016/j.ecoser.2012.06.001

Naidoo R, Balmford A, Costanza R, Fisher B, Green RE, Lehner B, Malcolm TR, Ricketts TH (2008) Global mapping of ecosystem services and conservation priorities. Proc Natl Acad Sci USA 105(28):9495–9500. https://doi.org/10.1073/pnas.0707823105

Nedkov S, Burkhard B (2012) Flood regulating ecosystem services – mapping supply and demand, in the Etropole Municipality, Bulgaria. Ecol Indic 21:67–79. https://doi.org/10.1016/j.ecolind.2011.06.022

Neßhöver C, Timaeus J, Wittmer H, Krieg A, Geamana N, Van Den Hove S, Young J, Watt A (2013) Improving the science-policy interface of biodiversity research projects. Gaia 22(2):99–103

Niemeijer D, de Groot RS (2008) A conceptual framework for selecting environmental indicator sets. Ecol Indic 8(1):14–25. https://doi.org/10.1016/j.ecolind.2006.11.012

Norgaard RB (2010) Ecosystem services: from eye-opening metaphor to complexity blinder. Ecol Econ 69(6):1219–1227. https://doi.org/10.1016/j.ecolecon.2009.11.009

Norton L, Greene S, Scholefield P, Dunbar M (2015) The importance of scale in the development of ecosystem service indicators? Ecol Indic 61:130–140. https://doi.org/10.1016/j.ecolind.2015.08.051

Oteros-Rozas E, Martín-López B, González JA, Plieninger T, López CA, Montes C (2014) Socio-cultural valuation of ecosystem services in a transhumance social-ecological network. Reg Environ Chang 14(4):1269–1289. https://doi.org/10.1007/s10113-013-0571-y

Poppe M, Muhar S, Scheikl S, Böck K, Loach A, Zitek A, Heidenreich A, Schrittwieser M, Kurz-Aigner R (2015) Traisen.w3. Traisen. WasWieWarum? Identifizierung und Wahrnehmung von Funktionen in Flusslandschaften und Verstehen einzugsgebietsbezogener Prozesse am Beispiel der Traisen. Zwischenverwendungsnachweis, Wien

Portman ME (2013) Ecosystem services in practice: challenges to real world implementation of ecosystem services across multiple landscapes – a critical review. Appl Geogr 45:185–192

Raymond CM, Bryan BA, MacDonald DH, Cast A, Strathearn S, Grandgirard A, Kalivas T (2009) Mapping community values for natural capital and ecosystem services. Ecol Econ 68(5):1301–1315. https://doi.org/10.1016/j.ecolecon.2008.12.006

REFORM (2015) REstoring Rivers FOR Effective Catchment Management. http://www.reformrivers.eu/

Rewitzer S, Matzdorf B, Trampnau S (2014) Das Konzept der Ökosystemleistungen aus Sicht der deutschen Umweltverbände. Natur und Landschaft 89(2):61–65

Rey Benayas JM, Newton AC, Diaz A, Bullock JM (2009) Enhancement of biodiversity and ecosystem services by ecological restoration: a meta-analysis. Science 325(5944):1121–1124. http://www.sciencemag.org/content/325/5944/1121

Reyers B, Bidoglio G, Dhar U, Gundimeda H, O'Farrell P, Paracchini ML, Prieto OG, Schutyser F (2010) Measuring biophysical quantities and the use of indicators. Econ Ecosyst Biodiv Ecol Econ Found, no. June: 47. doi:10.4324/9781849775489

Rheinisch-Westfälisches Institut für Wirtschaftsforschung (RWI) (2013) Regionalökonomische Effekte des Emscherumbaus. Projekt im Auftrag der Emschergenossenschaft. Endbericht, Essen

RWI (2013) Regionalökonomische Effekte Des Emscherumbaus. Germany, Essen.

Satz D, Gould RK, Chan KMA, Guerry A, Norton B, Satterfield T, Halpern BS et al (2013) The challenges of incorporating cultural ecosystem services into environmental assessment. Ambio 42(6):675–684. https://doi.org/10.1007/s13280-013-0386-6

Schröter-Schlaack C, Wittmer H, Mewes M, Schniewind I (2014) Der Nutzen von Ökonomie und Ökosystemleistungen für die Naturschutzpraxis. Workshop IV: Landwirtschaft. BfN-Skripten 359. Bonn-Bad Godesberg

Schwaiger E, Berthold A, Gaugitsch H, Götzl M, Milota E, Mirtl M, Peterseil G, Sonderegger J, Stix S (2015) Wirtschaftliche Bedeutung von Ökosystemleistungen. Monetäre Bewertung: Risiken und Potenziale, Wien. http://www.umweltbundesamt.at/umweltsituation/landnutzung/landnutzungumweltressourcen/oekonomischebewertung/

Seligman C (1989) Environmental ethics. J Soc Issues 45(1):169–184

Sommerhäuser M, Gerner N (2015) Ökosystemleistungen als Instrument der Wasserwirtschaft dargestellt am Beispiel des Emscherumbaus. In: 5. Ökologisches Kolloquium der BfG: Ökosystemleistungen – Herausforderungen und Chancen im Management von Fließgewässern and PIANCSeminar: Ecosystem Services: Identification, Assessment and Benefits for Navigation Infrastructure Projects

Spash CL (2008) How much is that ecosystem in the window? The one with the bio-diverse trail. Environ Values 17(2):259–284. https://doi.org/10.3197/096327108X303882

TEEB (2010) The economics of ecosystems and biodiversity: mainstreaming the economics of nature: a synthesis of the approach, conclusions and recommendations of TEEB

ten Brink P, Bräuer I (2008) Proceedings of the workshop on the economics of the global loss of biological diversity, with inputs from Kuik O, Markandya A, Nunes P, and Rayment M, Kettunen M, Neuville A, Vakrou A and Schröter-Schlaack. In: Brussels, Belgium. http://ec.europa.eu/environment/nature/biodiversity/economics/teeb_en.htm

Turner K, Georgiou S, Clark R, Brouwer R, Burke J (2004) Economic valuation of water resources in agriculture. From the sectoral to a functional perspective of natural resource management. FAO, Rome

van Beukering PJH, Brouwer R, Koetse MJ (2015) Economic values of ecosystem services. In: Bouma JA, van Beukering PJH (eds) Ecosystem services. From concept to practice. Cambridge University Press, Cambridge, pp 89–107

van Oudenhoven APE, Petz K, Alkemade R, Hein L, de Groot RS (2012) Framework for systematic indicator selection to assess effects of land management on ecosystem services. Ecol Indic 21:110–122

Verhagen W, Verburg PH, Schulp N, Stürck J (2015) Mapping ecosystem services. In: Bouma JA, van Beukering PJH (eds) Ecosystem services. From concept to practice. Cambridge University Press, Cambridge, pp 65–86

Vermaat JE, Wagtendonk AJ, Brouwer R, Sheremet O, Ansink E, Brockhoff T, Plug M et al (2015) Assessing the societal benefits of river restoration using the ecosystem services approach. Hydrobiologia. https://doi.org/10.1007/s10750-015-2482-z

Vlachopoulou M, Coughlin D, Forrow D, Kirk S, Logan P, Voulvoulis N (2014) The potential of using the ecosystem approach in the implementation of the EU water framework directive. Sci Total Environ 470–471:684–694. https://doi.org/10.1016/j.scitotenv.2013.09.072

Von Haaren C, Albert C (2011) Integrating ecosystem services and environmental planning: limitations and synergies. Int J Biodiv Sci Ecosyst Serv Manag 7(3):150–167. https://doi.org/10.1080/21513732.2011.616534

Wallis C, Séon-Massin N, Martini F, Schouppe M (2011) Implementation of the Water Framework Directive. When ecosystem services come into play. In 2nd "Water Science Meets Policy" Event. Brussels, 29 & 30 September 2011, 212. http://www.onema.fr/IMG/EV/meetings/ecosystem-services.pdf

Westman WE (1977) How much are nature's services worth? Science 197(4307):960–964. http://science.sciencemag.org/content/197/4307/960.abstract

Wiggering H, Müller F (2004) Umweltziele und Indikatoren - Wissenschaftliche Anforderungen an ihre Festlegung und Fallbeispiele. Edited by Hubert Wiggering and Felix Müller. Geowissenschaften + Umwelt. Springer, Heidelberg. https://doi.org/10.1007/978-3-642-18940-1

Winking C, Lorenz AW, Sures B, Hering D (2014) Recolonisation patterns of benthic invertebrates: a field investigation of restored former sewage channels. Freshw Biol 59(9):1932–1944. https://doi.org/10.1111/fwb.12397

Young RA (2005) Determining the economic value of water: concepts and methods. Resources for the Future, Washington. ISBN 1891853988

Chapter 22
Public Participation and Environmental Education

Michaela Poppe, Gabriele Weigelhofer, and Gerold Winkler

Learning together to manage together (Ridder et al. 2005)

Public participation can generally be defined as allowing people (stakeholders, interested parties, public) to influence the outcome of plans and working processes that constitute the operations of governance (CIS 2003). It can be practiced in different phases of integrated river basin management, but the public's environmental understanding forms one basis for participation. Environmental education is the process of recognizing values and clarifying concepts in order to develop skills and attitudes necessary to understand and appreciate the interrelation among people, their culture, and their biophysical surroundings (Palmer 2003). In this chapter, we discuss how environmental education and public participation interact with and are influenced by each other and need to be embedded in all areas and levels of societal processes.

Since participation is a principle of sustainable development (Costanza et al. 2000; Wagner et al. 2002), participatory decision-making is seen as key element for sustainable river basin management (Hedelin 2008). PP can be practiced in different phases of integrated river basin management (IRBM), from the involvement in decision-making process over the actual implementation of measures to the participation in environmental monitoring and research. An additional objective of PP is to increase public awareness of environmental issues and water management and to strengthen the commitment and support of decisions. Important for PP is the public's environmental understanding.

M. Poppe (✉) · G. Winkler
Institute of Hydrobiology and Aquatic Ecosystem Management, University of Natural Resources and Life Sciences, Vienna, Austria
e-mail: michaela.poppe@boku.ac.at; gerold.winkler@boku.ac.at

G. Weigelhofer
Institute of Hydrobiology and Aquatic Ecosystem Management, University of Natural Resources and Life Sciences, Vienna, Austria

WasserCluster Lunz, Biological Station GmbH, Lunz am See, Austria
e-mail: gabriele.weigelhofer@boku.ac.at

© The Author(s) 2018
S. Schmutz, J. Sendzimir (eds.), *Riverine Ecosystem Management*, Aquatic Ecology
Series 8, https://doi.org/10.1007/978-3-319-73250-3_22

Environmental education (EE) is the process of recognizing values and clarifying concepts in order to develop skills and attitudes necessary to understand and appreciate the interrelation among people, their culture, and their biophysical surroundings (Palmer 2003). EE also entails practice in decision-making and formulation at the individual and group levels of a code of behavior about issues concerning environmental quality (IUCN 1970; Howe 2009). The goal of EE is to develop a citizenry that is aware of, and concerned about, the environment and its associated problems and which has the knowledge, skills, attitudes, motivations, and commitment to work individually and collectively toward solutions of current problems and the prevention of new ones (UNESCO 1975).

For these reasons, EE forms one basis for PP, since participatory activities enhance environmental knowledge.

The structure of the chapter follows the discussion of the most addressed issues regarding PP and EE in IRBM. Based on the legal background for participatory processes, aims and potential benefits of PP documented by many authors are summarized. A theoretical paragraph deals with different levels of participation, their associated techniques and a basic scheme highlights the most important processes. Case studies show the potential of participatory processes in IRBM at different participation levels which reveal possible challenges and lead to potential solutions. Sustainable and participatory decision-making in IRBM results in the need for enhanced EE. Therefore, the integration of EE in different educational concepts is demonstrated in some examples. Finally, some promising and novel activities and methods to foster EE and PP, such as citizen science, are presented.

22.1 Legal Background for PP in Integrated River Basin Management

Much international legislation and policy has been developed to encourage PP (Bell et al. 2012). Indeed, the importance of community participation in sustainable development was enshrined within Principle 10 of the 1992 Rio Declaration of Environment and Development (UN 1992: Preamble of Chap. 23 "Strengthening the role of major groups"). Over 170 governments assembled at the Earth Summit in Rio de Janeiro and affirmed the importance of public access to information, participation, and justice in decision-making and produced the global action plan for sustainable development laid down in Agenda 21. The implementation of Agenda 21 was intended to involve action at international, national, regional, and local levels with regard to sustainable development (UN 1992). Some national and state, i.e., provincial, governments have legislated or advised that local authorities take steps to implement the plan locally, as recommended in Chap. 28 of the document. These programs are often known as "Local Agenda 21" or "LA21" aiming, among others, to protect freshwater resources (CIS 2003). Additionally, the International Conference on Water and the Environment in Dublin in 1992 set out the four Dublin

Principles that are still relevant today (Principle 2: "Water development and management should be based on a participatory approach, involving users, planners").

In Europe, many agreements and EU directives have included provisions for participatory input to environmental decisions. The Water Framework Directive (WFD, Directive 2000/60/EC; EC 2000) is one of the first European regulations that explicitly demands a high degree of involvement of non-state actors in the implementation (Newig et al. 2005). PP is stated in the preambles 14 and 46 and in Article 14. In the EU Floods Directive (EFD, Directive 2007/60/EC; EC 2007), public information and consultation are stipulated in Article 9 and Article 10. Member States are required take appropriate steps to coordinate the application of both directives, focusing on opportunities for information exchange and for consulting the public at key stages. The active involvement of all interested parties under Article 14 of the WFD and Article 10 of the EFD shall be coordinated, and the Member States shall encourage all interested parties in the production, review, and updating of the river basin management plans (Annex VII) and the flood risk management plans. According to the philosophy of these two articles, decisions must be taken with maximum transparency. The text of the WFD is supplemented by guidance provided as part of the Common Implementation Strategy of the European Commission and the Member States (CIS 2003). This document describes the concept of PP and indicates also how to organize PP in IRBM, which actors to involve, when and how to organize PP.

In addition, the participation approach is also reflected in a range of further international agreements such as the European Landscape Convention (Council of Europe 2000), the UNECE Aarhus Convention (UNECE 1998) on Access to Information, Public Participation in Decision-making and Access to Justice in Environmental Matters, the SEA directive (2001/42/EC; EC 2001a) on the assessment of the effects of certain plans and programs on the environment, the Public Access to Information Directive (2003/04/EC; EC 2003a), and the Public Participation Directive (2003/35/EC; EC 2003b).

22.2   The Benefits of Public Participation

Many publications highlight the benefits of PP in IRBM (Mostert 2003; CIS 2003; Mostert et al. 2007; Reed 2008; Ker-Rault and Jeffrey 2008; Pahl-Wostl et al. 2008; de Stefano 2010; Demetropoulou et al. 2010; von Korff et al. 2010, 2012; Luyet et al. 2012; Hassenforder et al. 2015). These benefits are seen as:

1. Raising public awareness on environmental issues and specifically on each river basin's environmental situation and local catchment by information and consultation processes
2. Making use of the different stakeholders' knowledge, experience, and initiatives and thus improving the quality of plans, measures, and river basin management
3. Public acceptance, commitment, and support of decision-making processes
4. More transparent and creative decision-making

438 M. Poppe et al.

5. Fewer misunderstandings and delays and more effective implementation
6. Avoiding potential conflicts, problems of management, and costs in the long term
7. Increased social learning and experience are leading to enhanced democratic legitimacy of competent authorities and increased accountability
8. Capacity building among stakeholders and competent authorities
9. Strengthening of decision-making procedures
10. Improvement in the quality of river resources
11. Promotion of goals associated with sustainable development

Focusing on educational aspects, additional benefits are reported in terms of social learning, systems thinking, and systems understanding in participatory processes (Pahl-Wostl et al. 2008; Hassenforder et al. 2016).

1. Through the potentially intensive interaction in a participation process, participants can build new networks and work to resolve conflicts, thus having an opportunity to increase their social capital, which in turn may enable them to more easily solve problems in a wider range of contexts and new conflicts in the future.
2. Decision-makers, planners, or community members can directly experience a systems understanding that is understood through praxis and can therefore be readily translated into improved actions and decisions.
3. Participants are more likely to apply the new systems' understanding over the long term, beyond the temporal and planning targets of the initial participatory processes.
4. Participation can facilitate systems learning and thereby "implant" a foundational environmental understanding, tailored to solve similar long-term contested decision areas.

22.3 Participation Levels, Techniques, and Basic Framework of Public Participation

The "degree of participation" is one of the most addressed categories for process description in the literature (Hassenforder et al. 2015). Arnstein (1969), who wrote the first main contribution related to this topic, set out a ladder for citizen participation based on eight steps. Based on adaptations and revisions of Arnstein's typology (Vroom 2003; Ridder et al. 2005; IAP2 2015), we use the five following degrees of participation from lower to higher levels of participation and impact:

Information: provide the public with objective information.
Consultation: obtain public feedback on analyses and/or decisions and provide feedback to the public how the decision was influenced.
Involvement: cooperation with the public to ensure that public concerns are considered and reflected in the alternatives developed.

Collaboration: partner with the public in identification of the preferred solution and incorporate the public recommendations into the decisions.

Empowerment: place decision-making (and sometimes implementation) in the hands of the public.

The degree of involvement is a critical point in PP, because it influences all processes, in particular, the choice of participative technique. The lower levels of PP consist of one-way flows of information (either from the parties involved to decision-makers or vice versa), whereas the upper steps are characterized by the existence of two-way flows (Videira et al. 2006; Demetropoulou et al. 2010). At least 30 different participatory techniques have been applied in IRBM (Mostert 2003).

The first step of PP involves informing the public to create a foundation for participation. This can be achieved by leaflets, brochures, websites, and maps (CIS 2003; Ridder et al. 2005). Through consultation and collaboration processes, the government makes documents available for written comments, organizes a public hearing, or actively seeks the comments and opinions of the public through, for instance, surveys and interviews, (online) questionnaires, or public hearings or role playing games (CIS 2003; Ridder et al. 2005). Interactive Web GIS applications are promising tools to record public reactions that reflect local knowledge, thereby linking public comments with geographic positions or spatial coordinates that connect discussion with the specific reality of local culture and environment (CIS 2003).

The higher levels of PP imply that the public is invited to contribute actively to the planning process by discussing issues and contributing to the solution. Shared decision-making implies that interested parties not only participate actively in the planning process but also become (partly) responsible for the outcome. Appropriate tools for these levels of PP are, among others, workshops, review sessions, scenario building, and round table conferences (CIS 2003; Ridder et al. 2005; Maurel et al. 2007). Group sessions to define the problem can be mediated through model building to identify key relationships related to the problem. Modeling can be done at the conceptual level, or with mathematical simulation to look at problem dynamics, or with social simulation or role-playing of social relations (see Chap. 16). As an example of the latter approach, "citizen jury," is an interesting participatory technique where a group of randomly selected people, who represent a microcosm of their community, are paid to attend a series of meetings to learn about and discuss a specific issue and make public their conclusions. This method aims to strengthen the democratic process and to enhance social learning and environmental knowledge by including within it the considered views of a cross section of members of the public (CIS 2003).

The choice of PP level, or which sequence of levels to pursue, depends on balancing a range of factors that include goals (supporting decision process and/or education, the timing of PP, the stage of the planning process, the (political and historical) context for PP, available resources, objectives or benefits of PP, and the stakeholders identified to be involved (CIS 2003). A basic participation framework in IRBM is pictured in Fig. 22.1.

Fig. 22.1 Scheme of a
basic participation process
in IRBM (based after CIS
2003; Luyet et al. 2012;
Ridder et al. 2005)

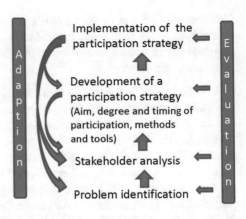

Following the problem identification, stakeholder analysis is an appropriate technique for getting to know the people, groups, and organizations that may influence the success of a project or may be affected by it. Defining the participation strategy involves determining the adequate methods and tools as well as the aim and the time of participation. To promote participation, the participation strategy should be discussed with the public, taking into account their concerns and interests (Ridder et al. 2005). During the implementation process, it is necessary to observe the participatory process and adapt tools and strategies if necessary. The monitoring and the evaluation of a participation process continue throughout the whole project and can be organized as a participation process in itself (Ridder et al. 2005). The evaluation is important in providing information to improve future, similar applications, enhancing the understanding of its impacts on stakeholders, and documenting experiences and outcomes (Luyet et al. 2012; Hassenforder et al. 2015).

Designing a participatory process and choosing the appropriate methods in IRBM depend on so many variables that it is difficult to undertake them in a standardized and linear manner. Rather, it requires an open and adaptive process (Dionnet et al. 2013). In the following section, we present current examples of PP on different levels of impact and spatial scales.

22.4 Applications of PP in Integrated River Basin Management

22.4.1 Public Information and Consultation

At the international scale, the ICPDR (International Commission for the Protection of the Danube River) is an example of an international RBM commission that values the importance of stakeholder involvement. The PP expert group deals with ICPDR activities concerning public information and consultation, outreach and awareness raising, and environmental education (ICPDR 2003). It plans the collection and consideration of comments related to WFD requirements. In 2004, based on a first hearing, the ICPDR developed formal consultation mechanisms, including a call for applications and selection of stakeholders. After a second hearing, the consultation of stakeholders in workshops and a dialogue on the RBM planning began. Online surveys and public calls for the submission of comments on draft documents on the river basin management plan (RBMP) and the flood risk management plan (FRMP) were conducted (ICPDR 2012; Frank 2015).

At the national scale, the "Round Table Water—*Runder Tisch Wasser*" was introduced in Austria in 2005, to interlink and consult national organizations of different sectors, e.g., economy, agriculture, municipalities, fisheries, NGOs, water supply, and water protection, focusing on ongoing developments in water management issues. Referring to the WFD, the Austrian public was consulted through online questionnaires for 6 months in 2015 to comment on the draft of the first FRMP and the second RBMP as part of a public hearing process (BMLFUW 2015a, b). All results are accessible online for the public at the "Water Information System Austria—*Wasserinformationssystem Austria*" (WISA—https://www.bmnt.gv.at/wasser/wisa).

At regional scale, online consultation projects such as "*Mitreden-U*," conducted by the German Federal Ministry of the Environment (BMU) in Germany in 2010 (Schulz and Newig 2015) and the Austrian *Flussdialog* introduced in 2008, proved that online participation is a useful participatory technique for including large numbers of residents at regional scale. Questions regarding local flood risk management or the potential development of tourism in the river landscape yielded a high respondent rate (Tragner 2009, 2010; Plansinn 2011). Selected results were integrated into the local planning process or in RBMPs (Plansinn 2011).

The concept of the *Flussraumbetreuung* was applied in pilot project on the Austrian river Traun in 2007 (Nikowitz and Ernst 2011). Partners of the government and several stakeholders established a 4-year river management position to foster integrative and sustainable decisions and plans for the river Traun at catchment scale. The aims were to develop concepts for RBMP and FRMP, the information and consultation of stakeholders within the catchment, accompanied by activities to promote environmental education. The *Flussraumbetreuung* could be stated as cost-effective and cost-efficient support of the WFD and EFD at regional scale (Nikowitz and Ernst 2011).

22.4.2 Involvement and Collaboration

At the international scale, the ICPDR supports the active involvement of stakeholders and the public in the governance of the Danube river basin through observer organizations on the levels of both expert group meetings and plenary meetings. Stakeholders were actively involved in defining environmental objectives and developing the "program of measures" (PoM) from 2005 onward (Frank 2015). Furthermore, the ICPDR develops the regional framework for water councils at the sub-basin and national levels, guarantees information dissemination, national and sub-basin consultation, and supports active involvement at the national scale.

Within the "River Basin Agenda Alpine Space," the "river basin dialogue" concept was developed to enhance the public's knowledge and collaboration and tested in 11 model river basins in the Alpine region (Revital & Freiland umweltcosulting 2007). The "river basin forums" and "river basin platforms" were tested in several model river basins as ways to join planners and responsible persons in public discussions and proved to be thoroughly suitable ways to develop RBMP through public participation (Revital & Freiland umweltcosulting 2007).

Within the regional project "Sustainable development of the Kamptal riverine landscape—*Nachhaltige Entwicklung der Kamptal Flusslandschaft*" (Preis et al. 2006; Stickler 2008), the residents of the river valley were actively involved in elaborating a citizens' guiding view to define the region's development goals (Muhar et al. 2006). The project team integrated this citizens' guiding view with other sectoral goals into a common guiding view of the Kamp river landscape (Renner et al. 2013).

At local scales, Local Agenda 21 activities are actively promoted by the national government in most European countries (FOSP 2005), among others, in Germany, Austria, Denmark, France, the UK, Italy, the Netherlands, Switzerland, and Spain. Within these countries, informal activities and local projects, such as on water protection, have been implemented and are considered as a multi-sectoral integration and networking tool (Prado Lorenzo and Garcia Sanchez 2007).

Several flood protection projects in Austria, such as at the river Golling or Großache, strike out in a new direction to actively involve the residents at early stages in the elaboration of the FRMP by co-planning activities (Stickler 2008). This collaboration is formally required as the landowners have to approve the flood risk plans.

22.4.3   Empowerment

The River Contract model was officially recognized by the French government in 1981 and was included in the 1992 Water Act (EEA 2014). It has increasingly been used as a tool to restore, improve, or conserve a river through a series of actions that are agreed in a broad participatory process involving all basin residents, and private and public entities involved in water management. In 2010 the process of creating a River Contract for the Matarraña in Spain began as a stakeholder dialogue that matured into their developing their own initiative. This horizontal engagement later informed stakeholder input to the RBMPs. This model of decentralized participative management was tested for more than a decade in the Walloon Region in Belgium (Rosillon et al. 2005). Some 16 ongoing projects have joined 43% of the Walloon Region and 48% of the region's 262 municipalities in the mutual responsibilities of a river contract (Rosillon et al. 2005). In Italy, Lombardy and Piedmont were pioneering regions, implementing a number of river contracts for the protection of spring systems, the environmental rehabilitation of flood detention basins, the enhancement of secondary hydrographic networks (e.g., channels, creeks), and the improvement of agricultural systems (WWAP 2015). Other Italian regions followed these examples and adopted their own river contract model. The projects vary from structural restoration measures such as construction of retention basins for flood control, while others include "social" measures such as environmental education and training (WWAP 2015).

22.5 Challenges of Public Participation and Possible Solutions

Below we present (Table 22.1) an overview of challenges and risks of implementing PP in IRBM as well as potential solutions (based on Luyet et al. 2012; Ridder et al. 2005; CIS 2003).

22.6 Environmental Education as Foundation for Sustainable Development

Sustainability for IRBM relies on environmental education (Pahl-Wostl 2002; Ison et al. 2007; Ilbury 2010) based on a dialogue between policy-makers, scientists, stakeholders, and the public at large. As early as September 1965, a meeting of the International Union for Conservation of Nature (IUCN) Education Commission's North West Europe Committee called for "environmental education in schools, in higher education, and in training for the land-linked professions" (Palmer 2003). In 1968, the United Nations Educational, Scientific and Cultural Organization (UNESCO) organized a Biosphere Conference in Paris, and in a later report on the event, the IUCN declared that "perhaps for the first time, world awareness of environmental education was fully evidenced" (IUCN 1970).

Environmental education (EE) is a transgenerational process with the long-term objective of imparting environmental awareness, ecological knowledge, attitudes, values, commitments for actions, and ethical responsibilities for the rational use of resources and for sound and sustainable development (UN 1992). Agenda 21 also looks beyond basic educational needs, outlining the necessity of using formal and informal education as tools for achieving environment and development awareness and building the skills necessary for sustainable lifestyles (UN 1992; Le Blanc et al. 2012). The United Nations Environmental Program states that young people will face major challenges in providing sufficient water and food, controlling diseases, generating sufficient energy, and adapting to climate change in near future. Therefore the UN-Decade "education for sustainable development (ESD)" was established between 2005 and 2014, and the teaching of skills related to global environmental

Table 22.1 Challenges and risks of PP in IRBM and their potential solutions

	Challenges	Potential solutions
Societal background	Centralized and hierarchical administrative structures may hamper the development of participatory processes (Demetropoulou et al. 2010; De Stefano 2010)	Support countries with centralized and hierarchical administrative structures in developing participatory mechanism (Demetropoulou et al. 2010)
Reservations about PP	Expensive process (Vroom 2000; Mostert 2003)	Plan PP and involve the public as early as possible; allow financial resources for participatory processes, particularly in early stages; balance the costs in terms of time and money and potential benefits
	Time-consuming process (Vroom 2000; Luyet et al. 2012)	Define a participatory plan and develop a realistic timetable; plan PP from the beginning of the project and involve the public as early as possible (CIS 2003) to minimize delays from repetition or misunderstandings
Stakeholder identification and degree of involvement	Involvement of stakeholders who are not representative (Junker et al. 2007; Reed 2008)	Implement a soundly based stakeholder analysis (CIS 2003)
	Potential stakeholder frustration (Germain 2001; Reed 2008; Pahl-Wostl et al. 2008)	Define the aim of PP and specify the extent that participation influences the final decisions (Videira et al. 2006; CIS 2003); give feedback to the stakeholders and specify clearly their role (Ridder et al. 2005)
	Empowerment of an already important stakeholder and the misuse of a stakeholder group to get advantage by using their influence (elite capture; Buttoud and Yunusova 2002; Platteau 2004)	Define an appropriate level of participation and balance out power pressures during the project (Ridder et al. 2005); involve experienced moderators at meetings (Reed 2008)
	In both the WFD and the EFD, the term "encourage active involvement" leaves substantial leeway for member states to implement a wide variety of forms of public involvement, including non-participatory forms of drafting plans (Newig et al. 2014)	Consult the public in draft FRMPs (EFD) and PoMs (WFD) as part of a strategic environmental assessment according to the SEA Directive (Carter and Howe 2006)
	Impacts of participation stand only for the minimum required level of public information. PP is perceived as a charade with no true involvement and collaboration of the interested parties (Videira et al. 2006)	Create of a new cadre of "river basin district managers" as key actors at the core of water management regimes. To succeed, this cadre will need particular skills and competences (Irvine et al. 2015), conducive organizational

(continued)

Table 22.1 (continued)

	Challenges	Potential solutions
		arrangements, and scope to build relationships and networks within and out with their organizations (Collins et al. 2007)
	There is no comprehensive, standardized scheme of PP project design available (Luyet et al. 2012). Every participatory process is essentially unique, so that there are no classes of solutions that can be readily and applied a priori to a specific case (Ker-Rault and Jeffrey 2008)	Adapt strategies based on open, flexible discussion during the implementation process. Explore the links between active public involvement in the preparation of RBMPs and FRP, and the subsequent role of these stakeholders in implementation, or more specifically, the influence of PP on certain categories of measures and elements in the management cycle (EEA 2014)
Adequate scale	Identifying the right scale for PP. Referring to the WFD, PP is required at many different scales: river basin district, river basin, sub-basin, water body, national level, national part of an international river basin district, regional and local government level, etc. (CIS 2003)	Adapt the participatory methods and techniques on the defined scale (s) and according to the number of people (CIS 2003)
Evaluation of PP	No standardized approach related to participation evaluation (Luyet et al. 2012)	Plan the participation evaluation at an early stage, well organized, and choose its criteria depending not only on the goals of the project but also on the focus, purpose, and timing of the evaluation (Luyet et el. 2012)

challenges and sustainability as well as lifelong learning are seen as increasingly relevant (Irvine et al. 2015).

22.7   Implementation of Environmental Education in Educational Concepts

At national scales various agencies are responsible for overseeing national strategies for integrating environmental education into formal and informal education (Le Blanc et al. 2012). For example, the incorporation of sustainable development into the curricula of primary and secondary education systems has generally been managed in one of three ways: (1) EE has been integrated directly into existing curricula (e.g., as in France, Germany, and the UK) and requires an interdisciplinary approach; (2) EE has been introduced as an additional educational activity, e.g., as part of an after school club (this approach has been taken in Israel, Bulgaria and Monaco); or (3) decision-making about the incorporation of EE has been devolved to school leadership. In Europe, most northern and mid-European countries have successfully incorporated EE in existing education systems (e.g., Austria, Denmark, France, Germany), whereas eastern and southern European countries have had little success in promoting EE (Leal-Filho 2010; Le Blanc et al. 2012).

22.7.1 Implementation of Environmental Education and Education for Sustainable Development in Higher Education

For more than 800 years, higher education institutions (HEIs) in Europe play an important role in educating professionals who lead, manage, and teach in society's institutions. Research and services to the public have become additional key missions in the early twentieth century (Scott 2006). Currently, HEIs are in the process of transformation, triggered by several global trends and challenges such as internationalization, global financial crises, commodification of education, and the "2030 Agenda for Sustainable Development."

The educational domain of HEIs started to put emphasis on the agenda of ESD from 2005 to 2014, wherein HEI's have been recognized as a key factor for sustainable societal and economic development. The sustainable development goals (SDGs) came into effect in January 2016, setting 17 goals and 169 targets to achieve a sustainable global development that meets the needs of the present generations, without compromising the ability of future generations to meet their own needs (UN 2015). Science is perceived to play an important role for the achievement of the SDGs, in terms of reflecting the interconnection of goals and targets, formulating evidence-based targets and indicators, assessing progress, testing solutions, and identifying emerging risks and opportunities (ICSU – ISSC 2015; Costanza et al. 2016; McKinnon et al. 2016; Lu et al. 2015). These global trends are paving the way to a new era for HEIs, suggesting fundamental changes of their structures, functions, societal roles, and collaborative activities with a wide array of stakeholders.

The integration of EE and ESD in higher education is presented via the example of the University of Natural Resources and Life Sciences Vienna (BOKU). BOKU is positioning itself as a teaching and research center for renewable resources, which are necessary for human life. BOKU's main objective is to make a considerable contribution to the conservation and protection of resources for future generations. As one of the first universities in Austria, BOKU finalized its sustainability strategy in 2014. Since then, sustainability objectives, measures, and timelines have been defined and are being implemented for the areas research, education, identity and society (organizational culture, strategy, and interactions with society), and operations (environmental management and social responsibility).

Building and preserving the integrity of riverine ecosystems requires sustainability-driven approaches, due to their ecological sensitivity and crucial value for human beings via the provision of resources and ecosystems services. Sustainable management mandates balancing ecological needs with the socioeconomic needs of growing human populations, based upon a sound understanding of the multifaceted functioning of riverine ecosystems. While several educational programs (both BSc and MSc levels) at the BOKU include riverine ecosystem management components, BOKU took a further step toward placing sustainable riverine ecosystem management as core learning outcome within the curricula of two master's programs, the "Applied Limnology (AL)" and the "international joint degree program Limnology and Wetland Management (LWM)". Both programs emphasize EE and ESD elements, such as:

- Core learning outcomes of the curricula and teaching/learning methods are emphasizing holistic, integrative, and critical approaches to tackle environmental sustainability issues.
- Interdisciplinary and transdisciplinary courses and teaching staff.
- Curriculum includes several courses wherein the ESD approach is integrated, plus specific courses on ESD.
- Integration of the global dimension and various temporal scales, different sociocultural environments and fostering intercultural communication.
- Participatory quality assurance mechanism (inclusion of stakeholders and students into quality assurance and curricula development).

Both, the AL and the LWM program are taught in the English language and are designed for scientists, technologists, engineers, conservationists, and environmental managers from all over the world. Both master's programs provide interdisciplinary expertise and professional skills for determining the ecological status of aquatic ecosystems, the assessment of human impacts, the conservation of endangered species and habitats, the development of appropriate restoration measures, and transdisciplinary expertise to interact with stakeholders, managers, and policymakers to achieve sustainable ecosystem management solutions. The LWM program is highlighting the global dimension in particular, by providing first-hand experience on the sustainable management of aquatic ecosystems in different climatic and sociocultural environments.

> **Info box**
> **Master's programs "Applied Limnology (AL)" and "international joint degree program Limnology and Wetland Management (LWM)"**
> The 2-year AL program can be studied either at BOKU throughout or spliced-up with modules/courses offered by a large number of partner

(continued)

universities within and outside of Europe. The international 18-month Master's program LWM comprises one study semester at each of the LWM partner universities: BOKU (Vienna, Austria), Egerton University (Egerton, Kenya), and UNESCO-IHE (Delft, The Netherlands). The three taught program semesters are followed by a 6-month MSc research period at any of the partner universities, and the MSc degree is awarded as a joint degree (Weblink http://www.wau.boku.ac.at/ihg/master-program).

22.8 Activities and Methods to Promote Environmental Education and Public Participation

22.8.1 Lifelong Learning Initiative

The European Lifelong Learning Initiative defines lifelong learning as "a continuously supportive process which stimulates and empowers individuals to acquire all the knowledge, values, skills and understanding they will require throughout their lifetimes and to apply them with confidence, creativity and enjoyment, in all roles circumstances, and environments" (Watson 2003). The term recognizes that learning is not confined to childhood or the classroom but takes place throughout life and in a range of situations. The objectives of lifelong learning include active citizenship, personal fulfillment, and social inclusion, as well as employment-related aspects (EC 2001b). A key principle of lifelong learning is the improved access to learning activities including higher education institutions. This aspect opens educational opportunities to students and returning professionals, including short-term or part-time study programs, moreover supported by EU educational programs, such as Erasmus+.[1] The lifelong learning approach can well be applied to develop the capacity for sustainable water management (Irvine et al. 2015).

Governmental organizations and NGOs carry out a huge variety of individual activities to promote EE and participatory processes at different scales to raise the awareness on water issues. Focusing on IRBM the "World Water Day" and the "Danube Day" are examples for activities at international scale repeated at yearly intervals, where the public is informed about freshwater topics and invited to "water" events. As local authorities are closest to the citizens in their jurisdiction, they play an important role in the education and mobilization of the public in favor of sustainable development (Prado Lorenz and Garcia Sanchez 2007). Linked to Agenda 21, local and regional *Lernfeste* or "Learning Regions—*Lernende Regionen*" have been run in Germany and in Austria since the late 1990s (Adolf Grimme Institut 2001; BMLFUW 2012; Erler et al. 2014).

---

[1]http://ec.europa.eu/programmes/erasmus-plus/.

22.8.2 Educating Young Students in Systems Thinking and Systems Modeling

Equipping young people with the skills to participate successfully in increasingly complex environments and societies is a central issue for policy makers around the world. The OECD Program for International Student Assessment[2] thoroughly investigated the science competencies of 15-year-old students in 2006. The report documents that teenagers in OECD countries are mostly well aware of environmental issues but often know little about their causes and options to tackle those challenges in the future. Only the understanding of complex socioenvironmental systems establishes a basis for making decisions leading to sustainable development. Systems thinking paradigms provide useful tools for rethinking the relationship between humans and their environment and developing practical solutions to embody these new relationships (Tippett 2005; Oderquist and Verakker 2010; Zitek et al. 2013). Focusing on IRBM, there are several case studies where systems modeling enhanced systems understanding (Halbe et al. 2013; Hare 2011; Zitek et al. 2007). Additionally, it has shown that mental models underlying decision-making can be influenced by participatory modeling processes (Tippett 2005; Videira et al. 2005; see Chap. 16).

Within two Sparkling Science projects "FlussAuWOW and Traisen.w3," scientists worked together with 15- to 18-year-old students of Austrian Secondary Schools over 4 years on river basin management issues (Poppe et al. 2013, 2016). One of the aims of these projects was to apply multimodal school activities to foster systems understanding. To support the development of causal systems thinking, students developed qualitative causal models on processes in the catchment of the river Traisen within an interactive, hierarchically structured, learning environment that was developed within the EU-FP7 project "DynaLearn" (http://www.dynalearn. eu), which was based on qualitative reasoning (Bredeweg et al. 2013). Evaluations of students' pretests highlighted that the students did not know about the environmental problems on their doorsteps (Poppe et al. 2013). The comparison of students pre- and posttests proved that students' systems thinking and motivation for learning could be increased (Poppe et al. 2013, 2016). Ensuring that young people are proficient in system knowledge and understanding also in relation to their own surrounding environment makes it more likely that sustainable considerations are soundly addressed in the future.

---

[2]PISA – http://www.pisa.oecd.org.

22.8.3   Citizen Science

Citizen science provides a combination of environmental education and public participation in scientific research. It includes all activities that involve nonscientists (so-called citizen scientists (CS)) in authentic scientific research (Bonney et al. 2009; Dickinson et al. 2010; Wiggins and Crowston 2011). Currently, several hundred projects, mostly in the USA, Australia, India, Canada, and the Russian Federation, are conducted in different fields of science with over 1000 volunteers (Conrad and Hilchey 2011). The high number of CS projects can be attributed to an increasing public interest in nature conservation during the last century combined with decreasing financial support for ecological monitoring and the development of online resources and communication techniques, such as GIS-based information systems and graphical user interfaces, which facilitate the management of citizen science projects (Bonney et al. 2009; Dickinson et al. 2012). CS may provide a crucial part of IRBM through participatory processes, starting from detecting environmental problems via environmental monitoring data collection and assessment to the building a science basis for planning as well as supervising and evaluating projects. Citizen science projects can be classified according to public involvement in (1) *contributory* (projects led by scientists, CS only collect data), (2) *collaborative* (CS collect and analyze data, involvement in study design and dissemination possible), and (3) *co-created* projects (participant-driven projects; CS are involved in all phases of the research) (Wiggins and Crowston 2011; Dickinson et al. 2012; Miller-Rushing et al. 2012).

The strengths of citizen science lie in the provision of large data sets with long-time series (including also historic data) and better spatial and temporal resolution. The financial and personal restrictions of traditional research projects cannot support data collection of such quantity (Dickinson et al. 2012; Miller-Rushing et al. 2012). Citizen science projects can make distinctive contributions to ecological research in the fields of global climate change, landscape ecology, evolution, and macroecology, providing information about large-scale patterns and changes in, e.g., phenology, the distribution of rare and invasive species, or diseases (Dickinson et al. 2012). In addition, CS projects can function as early warning systems, e.g., extinction or invasion of species and failure in water quality. Through the inclusion of the public, citizen science also provides a framework for studying human biomes such as urban, agricultural, and residential areas (Dickinson et al. 2012).

Besides scientific research, citizen science projects usually aim to increase the public's environmental awareness by addressing nature conservation issues at multiple scales (local or regional), thereby providing key information for both managers and decision-makers (Dickinson et al. 2012). Both environmental education and scientific literacy are generated through the (mostly electronic) supply of educational background materials, easily understandable explanations of the underlying research questions, and clearly described working protocols. Some projects even aim at including formal science education with learning goals defined *a priori* by providing inquiry-based curricula for schools (Zoellick et al. 2012). Although the educational

Table 22.2 Challenges and possible solutions in citizen science projects (Conrad and Hilchey 2011; Dickinson et al. 2012)

Challenges	Solutions
Lack of volunteer interest/dropout of volunteers	Motivate participants via incentives such as contests, games, and certificates of recognition; ensure regular press release, newsletters, and blogs, which highlight the achievements of the participants; provide opportunities for social interaction and outdoor activities
Lack of funding	Emphasize need of large data sets; emphasize benefits, including environmental learning; guarantee data quality and usability
Inability of participants to access appropriate information	Use different electronic and analogous media (e.g., local newspapers) to advertise; enable easy access to information and data delivery also for participants not familiar or equipped with modern technologies (e.g., schools)
Data fragmentation, data incompleteness, sampling bias (e.g., different sampling time)	Aim at large sample size; use special statistical analyses; provide standardized sampling procedures
Insufficient expertise, lack of objectivity, observer variability (different skills, effort, etc.)	Train volunteers; provide professional assistance during data collection; provide well-designed protocols; simplify tasks; validate data via comparisons with professional data
Substandard experimental design	Include experimental design with better trained subgroups; use CS data as starting point for experimental studies
Utility of data for the management	Focus on outcomes that serve society; ensure monitoring data will be relevant to the policies; ensure data quality
Publication of data in peer-reviewed scientific journals	Ensure data quality; provide data validation; include scientific hypotheses

impact of citizen science projects has yet to be assessed, studies show that the collaboration with scientists is highly motivating for the participants, enhances their scientific literacy, and often results in robust learning outcomes that can influence career choices (Zoellick et al. 2012).

Despite the many advantages of involving the public in scientific projects, scientists are confronted with various challenges regarding participant recruitment, funding, data quality, and data usability (Table 22.2; Conrad and Hilchey 2011; Dickinson et al. 2012). Today, a huge number of websites (e.g., www.cits.ci.org, www.dataone.org, and www.citizen-science.at) provide cyberinfrastructure, tools, guidelines (e.g., Citizen Science Central Toolkit), and resources for the initiation and administration of citizen science projects (see, e.g., Silvertown 2009; Dickinson et al. 2012). Numerous free web services support scientists in the installation of project websites, the development of smartphone apps (e.g., Cyber tracker, EpiCollect), and the creation of online games and databases for GIS data (e.g., Open Street Map, Experimental Tribe; see, e.g., https://www.zentrumfuercitizenscience.at).

Nevertheless, long-term involvement of the public in scientific research may still be a challenge, especially in Europe. Incentives such as contests, games, certificates, and official partnerships can help to recruit and keep participants (Conrad and Hilchey 2011). Another major challenge in citizen science projects is the quality of the collected data. Data bias can occur due to variations in the skills and knowledge of the volunteers, different sampling efforts (e.g., oversampling of "interesting" species), and the unequal distribution of observations (e.g., increased data collections around centers of human activities) (Cohn 2008). However, studies have found that the quality of data collected by volunteers was comparable to those of professionals, if volunteers were properly trained and guided throughout the project, and research tasks were kept simple (Fore et al. 2001; Canfield et al. 2002). Data validation is a crucial step in scientific research and, thus, needs to be addressed thoroughly in citizen science projects to increase the acceptance of volunteer-collected data by both governmental authorities and editors of scientific journals (Cohn 2008; Dickinson et al. 2012).

22.9 Conclusions

In an era of rising environmental uncertainty, the need for scientific understanding of riverine systems and their processes represents an important rationale for including local residents in effective and sustainable river management. Despite all the international agreements to promote EE, the public is often not aware of the environmental problems on their doorstep. EE should serve as the basis for and be implemented specifically as part of the organization of any participatory science events. EE and PP interact with and are influenced by each other and need to be embedded in all areas and levels of societal processes. This will ultimately facilitate better overall understanding of environmental issues and should aid proper realization and implementation of policies supporting sustainable decision-making.

A key concept of social innovation is the involvement and empowerment of citizens. Both PP and EE are working to develop a number of such innovations, including social learning, system modeling, citizen science, as well as new approaches to communication. These innovations can be seen as experiments to build more democratic and adaptive processes for developing science and policy. As governance becomes more inclusive of citizens, these emerging processes that integrate science, policy, and practice are like developmental stages of a society in transition toward sustainable development. To sustain this transition, a great deal of attention is needed to study the role and importance of education and learning implications of participatory processes and environmental governance. These activities have the potential to transform behavior and may help change current patterns in water resource management toward a more sustainable social-ecological system.

References

Adolf Grimme Institut (2001) Lernfeste: Brücken in neue Lernwelten. Bonn

Arnstein SR (1969) A ladder of citizen participation. AIP J 35:216–224

Bell S, Morse S, Shah RA (2012) Understanding stakeholder participation in research as part of sustainable development. J Environ Manag 101:13–22. https://doi.org/10.1016/j.jenvman.2012. 02.004

BMLFUW – Bundesministerium für Land- und Forstwirtschaft, Wasser und Umwelt (2012) Ein praktischer Leitfaden für Lernfeste. Lernende Regionen. 58 p. Wien

BMLFUW – Bundesministerium für Land- und Forstwirtschaft, Wasser und Umwelt (2015a) 1. Nationaler Hochwasserrisikomanagementplan. Sicher leben mit der Natur. 24p. Wien

BMLFUW – Bundesministerium für Land- und Forstwirtschaft, Wasser und Umwelt (2015b) Nationaler Gewässerbewirtschaftungsplan 2015. Entwurf. 344p. Wien

Bonney R, Cooper CB, Dickinson J, Kelling S, Phillips T, Rosenberg KV (2009) Citizen science: a developing tool for expanding science knowledge and scientific literacy. Bioscience 59:977–984

Bredeweg B, Liem J, Beek W, Linnebank F, del Río JG, Lozano E, Wißner M, Bühling R, Salles P, Noble R, Zitek A, Borisova P, Mioduser D (2013) DynaLearn-an intelligent learning environment for learning conceptual knowledge. AI Mag 34(4):9–35

Buttoud G, Yunusova I (2002) A "mixed model" for the formulation of a multipurpose mountain forest policy. Theory vs. practice on the example of Kyrgyzstan. Forest Policy Econ 4 (2):149–160. https://doi.org/10.1016/S1389-9341(02)00014-X

Canfield DE, Brown CD, Bachmann RW, Hoyer MV (2002) Volunteer lake monitoring: testing the reliability of data collected by the Florida LAKEWATCH Program. Lake Reserv Manage 18:1–9

Carter J, Howe J (2006) Stakeholder participation and the water framework directive: the case of the Ribble. Local Environ 11(2):217–231. https://doi.org/10.1080/13549830600558564

CIS – Common Implementation Strategy (2003) Common implementation strategy for the water framework directive (2000/60/EC). Public Participation in relation to the Water Framework Directive

Cohn JP (2008) Citizen science: can volunteers do real research? Bioscience 58:192–197

Collins K, Blackmore C, Morris D, Watson D (2007) A systemic approach to managing multiple perspectives and stakeholding in water catchments: some findings from three UK case studies. Environ Sci Policy:564–574. https://doi.org/10.1016/j.envsci.2006.12.005

Conrad CC, Hilchey KG (2011) A review of citizen science and community-based environmental monitoring: issues and opportunities. Environ Monit Assess 176:273–291

Costanza R, Daly H, Folke C, Hawken P, Holling CS, McMichael AJ, Pimentel D, Rapport D (2000) Managing our environmental portfolio. BioScience Roundtable 50(2):149–155. https:// doi.org/10.1641/0006-3568(2000)050[0149:MOEP]2.3.CO;2

Costanza R, Fioramonti L, Kubiszewski I (2016) The UN Sustainable Development Goals and the dynamics of well-being. Front Ecol Environ 14(2):59–59

Council of Europe (2000) European landscape convention. European Treaty Series, 176. Florence, 20.X.2000

De Stefano L (2010) Facing the water framework directive challenges: a baseline of stakeholder participation in the European Union. J Environ Manag 91:1332–1340. https://doi.org/10.1016/j. jenvman.2010.02.014

Demetropoulou L, Nikolaidis N, Papadoulakis V, Tsakiris K, Koussouris T, Kalogerakis N, Koukaras K, Chatzinikolaou A, Theodoropoulos K (2010) Water framework directive implementation in Greece: introducing participation in water governance – the case of the evrotas river basin management plan. Environ Policy Gov 20:336–349. https://doi.org/10.1002/eet.553

Dickinson JL, Zuckerberg B, Bonter DN (2010) Citizen science as an ecological research tool: challenges and benefits. Annu Rev Ecol Evol Syst 41:149–172

Dickinson JL, Shirk J, Bonter D, Bonney R, Crain RL, Martin J (2012) The current state of citizen science as a tool for ecological research and public engagement. Front Ecol Environ 10:291–297

Dionnet M, Daniell KA, Imache A, Von Korff Y, Bouarfa S, Garin P, Jamin J-Y, Rollin D, Rougier J-E (2013) Improving participatory processes through collective simulation: use of a community of practice. Ecol Soc 18(1). https://doi.org/10.5751/ES-05244-180136

EC – European Commission (2000) Directive 2000/60/EC of the European Parliament and the Council of 23 October 2000 establishing a framework for Community action in the field of water policy. Off J Eur Union L327:1–72

EC – European Commission (2001a) Directive 2001/42/EC on the assessment of the effects of certain plans and programmes on the environment. Off J Eur Communities 197:30–37

EC – European Commission (2001b) Making a European area of lifelong learning a reality. COM (2001),678 final, 40p

EC – European Commission (2003a). Directive 2003/4/EC of the European Parliament and of the Council of 28 January 2003 on public access to environmental information and repealing Council Directive 90/313/EEC. Off J Eur Union, L41/26, 26–32

EC – European Commission (2003b) Directive 2003/35/EC providing for public participation in respect of the drawing up of certain plans and programmes relating to the environment and amending with regard to public participation. Off J Eur Union L156:17–24

EC – European Commission (2007) Directive 2007/60/EC of the European Parliament and the Council of 23 October 2007 on the assessment and management of flood risks. Off J Eur Union L288:27–34

EEA – European Environmental Agency (2014) Public participation: contributing to better water management. Experiences from eight case studies across Europe. 58p. Copenhagen

Erler I, Fidlschuster L, Fischer M, Thien K (2014) Lebenslanges Lernen als Thema für LEADER-Regionen 2014–2020. Österreichisches Institut für ErwachsenenbildungWien. 64p

Fore LS, Paulsen K, O'Laughlin K (2001) Assessing the performance of volunteers in monitoring streams. Freshw Biol 46:109–123

FOSP – Federal Office for Spatial Development (2005) National Promotion of Local Agenda 21 in Europe. Zürich, 33p

Frank G (2015) Danube river basin management plan – update 2015. Contribution by Danubeparks within the Public Consultation Process. 9p. Orth an der Donau

Germain R (2001) Public perceptions of the USDA Forest Service public participation process. Forest Policy Econ 3(3-4):113–124. https://doi.org/10.1016/S1389-9341(01)00065-X

Halbe J, Pahl-Wostl C, Sendzimir J, Adamowski J (2013) Towards adaptive and integrated management paradigms to meet the challenges of water governance. Water Sci Technol 67 (11):2251–2260. https://doi.org/10.2166/wst.2013.146

Hare M (2011) Forms of participatory modelling and its potential for widespread adoption in the water sector. Environ Policy Gov 21:386–402. https://doi.org/10.1002/eet.590

Hassenforder E, Smajgl A, Ward J (2015) Towards understanding participatory processes: framework, application and results. J Environ Manag 157:84–95. https://doi.org/10.1016/j.jenvman.2015.04.012

Hassenforder E, Pittock J, Barreteau O, Daniell KA, Ferrand N (2016) The MEPPP framework: a framework for monitoring and evaluating participatory planning processes. Environ Manag 57:79–96. https://doi.org/10.1007/s00267-015-0599-5

Hedelin B (2008) Criteria for the assessment of processes for sustainable river basin management and their congruence with the EU water framework directive. Eur Environ 18:228–242. https://doi.org/10.1002/eet.481

Howe C (2009). The role of education as a tool for environmental conservation and sustainable development. PhD dissertation. 187p. Retrieved from http://www.iccs.org.uk/thesis/phd-howe.caroline09.pdf

IAP2 – International Association for Public Participation (2015) Core values awards. Louisville, US. 44p

ICPDR – International Commission for the Protection of the Danube River (2003) Danube river basin strategy for public participation in river basin management. Elaborated synthesis report of the public participation workshop April 4–5, 2003 – Bratislava. 26p

ICPDR – International Commission for the Protection of the Danube River (2012) WFD & EFD: Public participation plan. IC WD 517, 6p

ICSU – ISSC (2015) Review of the sustainable development goals: the science perspective. Paris, International Council for Science (ICSU). ISBN: 978-0-930357-97-9

Ilbury DT (2010) Are we learning to change? Mapping global progress in education for sustainable development in the lead up to "Rio Plus 20". Global Environ Res 14:101–107

Irvine K, Weigelhofer G, Popescu I, Pfeiffer E, Păun A, Drobot R, Gettel G, Staska B, Stanca A, Hein T, Habersack H (2015) Educating for action: aligning skills with policies for sustainable development in the Danube river basin. Sci Total Environ 543:765–777. https://doi.org/10.1016/j.scitotenv.2015.09.072

Ison R, Blackmore C, Jigins J (2007) Social learning: an alternative policy instrument for managing in the context of Europe's water. Environ Sci Policy:493–498. https://doi.org/10.1016/j.envsci.2007.04.003

IUCN – International Union for Conservation of Nature (1970) International working meeting on environmental education in the school curriculum. Final Report. Gland. 42p

Junker B, Buchecker M, Müller-Böker U (2007) Objectives of public participation: which actors should be involved in the decision making for river restorations? Water Resour Res 43 (10):1–11. https://doi.org/10.1029/2006WR005584

Ker-Rault PA, Jeffrey PJ (2008) Deconstructing public participation in the Water Framework Directive: implementation and compliance with the letter or with the spirit of the law? Water Environ J 22:241–249. https://doi.org/10.1111/j.1747-6593.2008.00125.x

Le Blanc D, Allen N, Cornforth J, Stoddart H, Ullah F (eds) (2012) Review of implementation of Agenda 21. Division for Sustainable Development of the United Nations Department of Economic and Social Affairs. Report 249p

Leal-Filho W (2010) An overview of ESD in European countries: what is the role of national governments? Global Environ Res 14:119–124

Lu Y, Nakicenovic N, Visbeck M, Stevance AS (2015) Policy: five priorities for the UN sustainable development goals-comment. Nature 520(7548):432–433

Luyet V, Schlaepfer R, Parlange MB, Buttler A (2012) A framework to implement stakeholder participation in environmental projects. J Environ Manag 111:213–219. https://doi.org/10.1016/j.jenvman.2012.06.026

Maurel P, Craps M, Cernesson F, Raymond R, Valkering P, Ferrand N (2007) Concepts and methods for analysing the role of Information and Communication tools (IC-tools) in social learning processes for river basin management. Environ Model Softw 22:630–639. https://doi.org/10.1016/j.envsoft.2005.12.016

McKinnon MC, Cheng SH, Dupre S, Edmond J, Garside R, Glew L, Holland MB, Levine E, Masuda MJ, Miller DC, Oliveira I, Revenaz J, Roe D, Shamer S, Wilkie D, Wongbusarakum S, Woodhouse E (2016) What are the effects of nature conservation on human well-being? A systematic map of empirical evidence from developing countries. Environ Evid 5(1):1

Miller-Rushing A, Primack R, Bonney R (2012) The history of public participation in ecological research. Front Ecol Environ 10:285–290

Mostert E (ed) (2003) Public participation and the European Water Framework Directive — A framework for analysis — Inception report of the HarmoniCOP project — Harmonising COllaborative Planning. RBA Centre, Delft University of Technology. 47p. Retrieved from http://www.harmonicop.uni-osnabrueck.de/_files/_down/HarmoniCOPinception.pdf

Mostert E, Pahl-Wostl C, Rees Y, Searle B, Tàbara D, Tippett J (2007) Management social learning in european river-basin management: barriers and fostering mechanisms from 10 river basins. Ecol Soc 12, 19. Retrieved from http://www.ecologyandsociety.org/vol12/iss1/art19/

Muhar S, Preis S, Hinterhofer M, Jungwirth M, Habersack H, Hauer C, Hofbauer S, Hittinger H (2006) Partizipationsprozesse im Rahmen des Projektes "Nachhaltige Entwicklung der Kamptal-Flusslandschaft". Österreichische Wasser- und Abfallwirtschaft 58(11–12):169–173

Newig J, Pahl-Wostl C, Sigel K (2005) The role of public participation in managing uncertainty in the implementation of the water framework directive. Eur Environ 15:333–343. https://doi.org/10.1002/eet.398

Newig J, Challies E, Jager N, Kochskämper E (2014) What role for public participation in implementing the EU Floods Directive? A comparison with the water framework directive, early evidence from germany and a research agenda. Environ Policy Gov 24:275–288. https://doi.org/10.1002/eet.1650

Nikowitz T, Ernst V (2011) Leitfaden Flussraumbetreuung in Österreich. Bundsministerium für Land- und Forstwirtschaft,Wasser und Umwelt und WWF Osterreich, Wien, 120p

Oderquist CS, Verakker SO (2010) Education for sustainable development : a systems thinking approach. Global Environ Res 14:193–202

Pahl-Wostl C (2002) Participative and stakeholder-based policy design, evaluation and modeling processes. Integr Assess 3(1):3–14. https://doi.org/10.1076/iaij.3.1.3.7409

Pahl-Wostl C, Tàbara D, Bouwen R, Craps M, Dewulf A, Mostert E, Ridder D, Taillieu T (2008) The importance of social learning and culture for sustainable water management. Ecol Econ 64:484–495. https://doi.org/10.1016/j.ecolecon.2007.08.007

Palmer JA (2003) Environmental education in the 21st century: theory, practice, progress and promise. Routledge, London and New York, p 280

PlanSinn (2011) Flussdialog Obere Mur. Protokoll. 14p, Judenburg

Platteau JP (2004) Monitoring elite capture in community driven development. Dev Chang 35 (2):223–246

Poppe M, Zitek A, Scheikl S, Preis S, Mansberger R, Grillmayer R, Muhar S (2013) Erfassen von Ursache-Wirkungs-Beziehungen in Flusslandschaften: Vermittlung von Systemwissen in Schulen als Beitrag für ein nachhaltiges Flussgebietsmanagement. Österreichische Wasser- Und Abfallwirtschaft 65:429–438. https://doi.org/10.1007/s00506-013-0119-x

Poppe M, Böck K, Zitek A, Scheikl S, Loach A, Muhar S (2016) Was? Wie? Warum? Jugendliche erforschen Flusslandschaften – Förderung des Systemverständnisses als Basis für gelebte Partizipation im Flussgebietsmanagement. Österreichische Wasser- und Abfallwirtschaft 68:342–353. https://doi.org/10.1007/s00506-016-0325-4

Prado Lorenzo JM, Garcia Sanchez IM (2007) The effect of participation in the development of local agenda 21 in the European Union. Proceedings of the international conference of territorial intelligence, pp 523–546. Retrieved from https://halshs.archives-ouvertes.fr/halshs-00519902

Preis S, Muhar S, Habersack H, Huaer C, Hofbauer S, Jungwirth M (2006) Nachhaltige Entwicklung der Flussslandschaft Kamp: Darstellung eines Managementprozesses in Hinblick auf die Vorgaben der EU-Wasserrahmenrichtlinie. Österreichische Wasser- und Abfallwirtschaft 58(11–12):159–167

Reed MS (2008) Stakeholder participation for environmental management: a literature review. Biol Conserv 141:2417–2431. https://doi.org/10.1016/j.biocon.2008.07.014

Renner R, Schneider F, Hohenwallner D, Kopeinig C, Kruse S, Lienert J, Link S, Muhar S (2013) Meeting the challenges of transdisciplinary knowledge production for sustainable water governance. Mt Res Dev 33(3):234–247. Retrieved from http://www.bioone.org/doi/abs/10.1659/MRD-JOURNAL-D-13-00002.1

Revital & Freiland Umweltconsulting (2007) Flussraum Agenda Alpenraum. Modell und Beispiele für eine nachhaltige Entwicklung alpiner Flussräume. Kurzbericht. 52p. München

Ridder D, Mostert E, Wolters HA (eds) (2005) HarmoniCOPHandbook. Learning together to manage together. 115p. Osnabrueck

Rosillon F, Vander Borght P, Bado Sama H (2005) River contract in Wallonia (Belgium) and its application for water management in the Sourou valley (Burkina Faso). Water Sci Technol 52 (9):85–93. Retrieved from http://www.ncbi.nlm.nih.gov/pubmed/16445177

Schulz D, Newig J (2015) Assessing online consultation in participatory governance: conceptual framework and a case study of a national sustainability-related consultation platform in Germany. Environ Policy Gov 25:55–69. https://doi.org/10.1002/eet.1655

Scott JC (2006) The Mission of the University: medieval to postmodern transformations. J High Educ 77(1):1–39

Silvertown J (2009) A new dawn for citizen science. Trends Ecol Evol 24:467–471

Stickler T (2008) FloodRisk II. Leitfaden: Öffentlichkeitsbeteiligung im Hochwasserschutz. Bundesministerium für Verkehr, Innovation und Technologie & Umweltbundesamt Wien, 141p. Wien

Tippett J (2005) Participatory planning in river catchments, an innovative toolkit tested in Southern Africa and North West England. Water Sci Technol 52(9):95–105. Retrieved from http://www.scopus.com/scopus/inward/record.url?eid=2-s2.0-31344438464&partnerID=40&rel=R5.6.0

Tragner F (2009) Flussdialog Obere Traun. Presseinformation

Tragner F (2010) Flussdialog Die Krems im Gespräch. Presseinformation

UN – United Nations (1992) Agenda 21: Konferenz der Vereinten Nationen für Umwelt und Entwicklung. Rio de Janeiro. 361p. Retrieved from http://www.un.org/depts/german/conf/agenda21/agenda_21.pdf

UN – United Nations (2015) Transforming our World: the 2030 agenda for sustainable development: A/RES/70/1

UNECE – United Nations Economic Commission for Europe (1998) Convention on access to information, public participation in decision-making and access to justice in environmental matters. Aarhus Convention. doi: https://doi.org/10.1017/CBO9780511494345.010

UNESCO – United Nations Educational, Scientific and Cultural Organization (1975) The belgrade charter: a framework for environmental education. Adopted by the UNESCO-UNEP international environmental workshop in October 1975. Belgrad

Videira N, Antunes MP, Santos R (2005) Building up the science in the art of participatory modeling for sustainability. Proceedings of the 23rd international conference of the system dynamics society. 32p

Videira N, Antunes P, Santos R, Lobo G (2006) Public and stakeholder participation in European water policy: a critical review of project evaluation processes. Eur Environ 16:19–31. https://doi.org/10.1002/eet.401

Von Korff Y, D'Aquino P, Daniell KA, Bijlsma R (2010) Designing participation processes for water management and beyond. Synthesis. Ecol Soc 15(3). http://www.ecologyandsociety.org/vol15/iss3/art1

Von Korff Y, Daniell KA, Moellenkamp S, Bots P, Bijlsma RM (2012) Implementing participatory water management: recent advances in theory, practice, and evaluation. Ecol Soc 17(1):30–44. https://doi.org/10.5751/ES-04733-170130

Vroom V (2000) Leadership and the decision-making process. Organ Dyn 28(4):82–94

Vroom V (2003) Educating managers for decision making and leadership. Manag Decis 41(10):968–978

Wagner W, Gawel J, Furuma H, De Souza MP, Teixeira D, Rios L, Ohgaki S, Zehnder A, Hemond HF (2002) Sustainable watershed management: an international multi-watershed case study. Ambio 31(1):2–13. https://doi.org/10.1639/0044-7447(2002)031[0002:swmaim]2.0.co;2

Watson L (2003) Lifelong learning in Australia. Canberra: Department of Education, Science and Training. Retrieved from http://www.forschungsnetzwerk.at/downloadpub/australia_lll_03_13.pdf

Wiggins A, Crowston K (2011) From conservation to crowdsourcing: a typology of citizen science. System Sciences (HICSS), 2011 44th Hawaii international conference on, pp 1–10

WWAP – World Water Assessment Programme (2015) Facing the challenges: casestudies and indicators. UNESCO's contribution to the united nations world water development report 2015. United Nations Educational, Scientific and Cultural Organisation. Paris. 75p

Zitek A, Muhar S, Preis S, Schmutz S (2007) The riverine landscape Kamp (Austria): an integrative case study for qualitative modelling of sustainable development. In: Price C (ed), Proceedings of the 21st international workshop on qualitative reasoning, pp. 212–217. Aberystwyth

Zitek A, Poppe M, Stelzhammer M, Muhar S, Bredeweg B (2013) Learning by conceptual modelling—changes in knowledge structure and content. IEEE Trans Learn Technol 6 (3):217–227. https://doi.org/10.1109/TLT.2013.7

Zoellick B, Nelson SJ, Schauffler M (2012) Participatory science and education: bringing both views into focus. Front Ecol Environ 10:310–313

Chapter 23
NGOs in Freshwater Resource Management

Christoph Litschauer, Christoph Walder, Irene Lucius, Sigrid Scheikl, S. V. Suresh Babu, and Archana Nirmal Kumar

23.1 Why Do NGOs Work in Freshwater Conservation?

Freshwater ecosystems, such as rivers, lakes, active floodplains, and marshes, are crucial to our existence. They provide the water needed to support human lives and livelihoods and are vital to key economic sectors such as agriculture, fisheries, and tourism. Freshwater ecosystems are also home to an astonishing diversity of plants and animals. Indeed, many of the socioeconomic functions of freshwater ecosystems—food production and water purification, for instance—are dependent on this biodiversity.

Despite these enormous values, freshwater ecosystems are under threat throughout the world. Outright destruction and more insidious degradation mean that fewer and fewer areas are able to function naturally and provide the goods and services upon which so many people depend, particularly the rural poor.

C. Litschauer (✉)
National Park Donau-Auen, Orth/Donau, Austria
e-mail: c.litschauer@donauauen.at

C. Walder
WWF, Vienna, Austria
e-mail: christoph.walder@wwf.at

I. Lucius
WWF Danube-Carpathian Programme, Vienna, Austria
e-mail: ilucius@wwf.panda.org

S. Scheikl
Institute of Hydrobiology and Aquatic Ecosystem Management, University of Natural Resources and Life Sciences, Vienna, Austria
e-mail: sigrid.scheikl@boku.ac.at

S. V. Suresh Babu · A. N. Kumar
WWF India, New Delhi, India
e-mail: suresh@wwfindia.net; ankumar@wwfindia.net

© The Author(s) 2018
S. Schmutz, J. Sendzimir (eds.), *Riverine Ecosystem Management*, Aquatic Ecology Series 8, https://doi.org/10.1007/978-3-319-73250-3_23

This becomes even more alarming when it is realized just how scarce freshwater ecosystems are in the first place, covering for less than 1% of the Earth's surface.

Freshwater conservation has therefore been a priority among environmental NGOs for decades. In fact, many local or national NGOs and initiatives were formed as a direct consequence of too little to no governmental action toward sustainable management of freshwater resources. Many international NGOs, such as WWF, also started to act because of the crucial role water resources play in the context of human development globally. They function as advocates for integration of local, regional, and international water policies, for a proper science–policy interface, and for adequate stakeholder involvement in decision-making in order to ensure that the needs of both nature and local communities are being met.

In the following text, we will present three case studies that illustrate how environmental NGOs can make a difference and what challenges they face.

23.2 Showcase India

Suresh Babu SV and Archana Nirmal Kumar, Rivers, Wetlands and Water Policy Team, WWF-India

India's rivers, wetlands, and lakes account for only 4% of the world's freshwater resources but sustain more than 16% of the world's human population. WWF-India works in the Ganga and Brahmaputra basins (Behera et al. 2011), where 50% of the country's population depend on for water, food, and livelihood security. Community-led conservation of urban, peri-urban, and high conservation wetlands in Rajasthan, Punjab, and Karnataka is another area of focus to fulfill the goal of protecting, managing, and restoring rivers and wetlands to retain biodiversity values, sustain ecosystem services, and provide long-term water security for people and nature.

23.2.1 Rivers for Life, Life for Rivers Initiative

Covering roughly 30% of India's land area, the basin of the Ganges river is home to 500 million people and thousands of aquatic species (Behera et al. 2011). However, the river is also highly polluted and overexploited. Huge volumes of water are used for irrigation, and untreated sewage and toxic effluents are dumped into it daily. At the same time, issues of unsustainable hydropower continue to be a major threat, affecting the flows, the connectivity, and the health of the river. The 600-km-long Ramganga River is one of the most polluted tributaries of the Ganga, flowing through Moradabad (the *Pital Nagri*). The issues of the Ramganga are characteristic of the whole basin: over abstraction; a growing footprint of industries, cities, and agriculture; degradation of habitats; and decline of key aquatic species (Fig. 23.1).

River conservation is complex and requires a multidisciplinary, multi-stakeholder approach. Hence, WWF-India's Rivers for Life, Life for Rivers initiative (RfLLfR, 2012–2017), supported by the HSBC Water Program, is structured around four

Fig. 23.1 Location of the
Ganga River (Ganges)

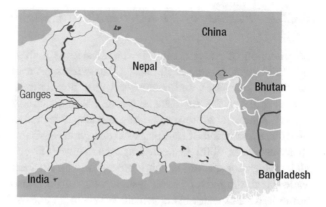

pillars: sustainable water management, urban and industrial water stewardship, climate change adaptation, and habitat and biodiversity conservation. Implemented across seven districts of Uttar Pradesh with a geographical focus on the 300 km stretch of the Ramganga (Kalagarh to Hardoi) and the 900 km stretch of the Ganga (Bijnor to Varanasi), this program envisions Ganga and Ramganga as healthy river systems rich in biodiversity and aims to provide long-term water security to communities, businesses, and nature. This builds on WWF-India's work in the Ganga basin since 1997 with conservation of aquatic biodiversity and the Living Ganga Program (2007–2012), which developed strategies for sustainable energy and water resource management in a critical stretch of 800 km from Gangotri to Kanpur.

To restore the ecological health of the rivers, the program works with a diverse range of stakeholders across academia, government institutions, policy makers, civil society members, local communities, and businesses to implement its multidisciplinary conservation strategy.

The involvement of various partners, including religious leaders, government departments, and NGOs, in the project, the improvement in the livelihood of the local people, and the riverine habitat have provided additional acceptability for WWF-India's work. Awareness of the local community toward the environment has helped greatly in motivating the community to work for conservation and climate change adaptation. Studies on water availability and water quality in the Ganga and Ramganga Rivers are being carried out in order to understand the factors determining the flexibility and resilience of ecosystems, including the habitat preference of species, river connectivity and quality, flow regimes, disturbances, and mortality of the aquatic biodiversity, especially of Ganges river dolphins, gharials, and freshwater turtles.

Under the RfLLfR program, the development of a data base and research through GIS and remote sensing is an extremely valuable asset that will support further research and will help to formulate adaptation strategies where required, with the long-term goal to save the ecosystem.

The lessons learned from WWF-India's intervention, particularly its strategy with regard to garnering support from the local communities, has been widely disseminated and accepted by the government and local people as a replicable model.

23.2.2   Friends of the River Multi-stakeholder Groups

Degradation of water quality has manifested as a shared risk to people (polluted, unreliable water, huge health costs), businesses (cost of clean water and increased regulatory pressure), and nature (declining populations and distribution of species and overall health). As a shared resource and a shared risk, water demands collaborative action to address some of the basin-level issues. It is this philosophy that drove the formation of multi-stakeholder groups—Ganga/Ramganga Mitras. Today there are over 6000 Ganga/Ramganga Mitras (Friends of the River)—a multi-stakeholder consortium which has been trained on various aspects of river conservation including river health assessments, turtle conservation, better practices in agriculture, and water management. WWF-India and WWF UK jointly conducted a mapping of buyers from leather cluster of Kanpur and created a UK Leather Buyers platform to work with tanneries in promotion of water stewardship.

Advocacy work to encourage scale-up and buy-in by the government has been an important component of the project. WWF-India is now in the process of advocating and scaling up its work to the national level with the active involvement of the government and the concerned local authorities—a recent example being the mainstreaming of environmental flows in the Ganga River Basin Environment Management Plan (GRBEMP) submitted by the consortia of seven Indian Institute of Technologies to the Ministry of Water Resources, River Development and Ganga Rejuvenation in 2015. In 2014, jointly with the National Mission for Clean Ganga, WWF-India prepared an operational strategy for the implementation of Dolphin Action Plan 2010–2020. The My Ganga, My Dolphin campaign (October 5–7, 2012), in partnership with the Uttar Pradesh Government, 18 NGOs, and HSBC, aimed at the assessment of distribution and population status of the Ganges river dolphin (India's National Aquatic Animal) using a unified methodology—for the first time in India. This will now be scaled up to the national level by the Government. WWF-India is engaging with the UP Government to develop an aquatic biodiversity management plan for 200 km stretch of the Ganga. Similarly, at the district level, jointly with the local administration, WWF is working with six districts to demonstrate alternative water and agriculture management to enhance adaptive capacities of people and reduce threats to the river.

23.3   Showcase Austria

Eco Masterplan III—Strategic considerations for sufficient water protection and ecologically sustainable expansion of hydropower in Austria (WWF Austria 2014; Seliger et al. 2016; Scheikl et al. 2016)

The three WWF Austria Eco Masterplans are prime examples of successful synergies between science and NGOs, since each of them is based on scientific studies carried out by the Institute of Hydrobiology and Aquatic Ecosystem Management (IHG) at the University of Natural Resources and Life Sciences (BOKU) in Vienna.

Both Eco Masterplans I and II were designed to designate the ecological sensitivity of Austrian rivers and subsequently identify protection priorities with respect to the future use of water resources. The four underlying criteria were the ecological status, the situation in a protected area, the hydromorphological status, and the length of the free-flowing section (WWF Austria 2009, 2010). Springing from these initiatives, the Eco Masterplan III (WWF Austria 2014) provides the fundamentals for an effective "Hydropower Masterplan," in which the options and limitations for the expansion of hydropower in Austria are investigated from an ecological and energy-economic perspective. Due to subsequent scenario development, different degrees of expansion with the corresponding ecological consequences and energy-economic implications can be illustrated and discussed on a national scale for the first time.

One key element of the Eco Masterplan III is the decision support tool "HY: CON" (HYdropower and CONservation) developed by the IHG (Seliger et al. 2016; Scheikl et al. 2016): HY:CON represents a strategic and transparent methodological approach to balance the energy-economic attractiveness of planned hydropower projects and the conservation needs of the river stretches that are affected by these projects.

In total, 39 ecological criteria were used to assess the ecological sensitivity of sites earmarked for power plant construction. Due to the high number of single criteria, almost all of the 102 investigated hydropower projects overlapped with at least one criterion. In order to avoid an overestimation of the ecological sensitivity, related criteria was assigned to one out of eight thematic groups: (1) ecological status; (2) hydromorphological condition; (3) river continuity; (4) floodplains; (5) situation in a legally binding protection area, where new hydropower projects are prohibited by law; (6) situation in another designated protected area; (7) actual habitats of key species (endangered and/or indicator species); and (8) key habitats (see Table 23.1).

In order to cover possible future trends regarding the prioritization of conservation and river ecological aspects, different scenarios were developed. To evaluate the conflict potential of hydropower projects with regard to the conservation need of a river stretch, each criterion was classified according to its protection priority. These scorings were a function of the priority given to conservation objectives in the scenarios (minimal conservation up to high conservation). Depending on the scenario, the scoring ranged from "low conservation value" to "exclusion of further hydropower development." If no "exclusion" criterion was affected, the final evaluation was conducted by calculating the mean of the highest-rated criteria per thematic group.

The energy-economic assessment of recorded hydropower project is geared to the "classic" energy-economic, optimization principles of economic, safe, and environmentally friendly electricity supply and is based on the criteria for assessing hydropower projects in energy-economic terms.

Finally, the results of the energy-economic assessment were interrelated with the results of each conservation scenario (Fig. 23.2). The results show that, not only in the "high conservation" scenario but even in the "moderate conservation" scenario, a large share of planned hydropower projects was in conflict with exclusion criteria

464   C. Litschauer et al.

Table 23.1 Thematic groups (bold) and related ecological criteria used for the identification and evaluation of possible conservation conflicts of planned hydropower projects

1. Ecological status	7. Actual habitat of key species
Ecological status	*Austropotamobius pallipes*
2. Hydromorphological (hy-mo) condition	*Margaritifera margaritifera*
Hy-mo status	*Myricaria germanica*
3. River continuity	*Hucho hucho*
Connected habitat	*Carassius carassius*
Free-flowing section (small river)	*Leucaspius delineatus*
Free-flowing section (medium and large rivers)	*Leuciscus idus*
Migration corridor of medium-distance migrating fish species	*Coregonus* sp.
4. Floodplains	*Sander volgensis*
Conservation value of remaining connected floodplains	*Thymallus thymallus*
5. Legally binding protection sites	*Chondrostoma nasus*
National park	**8. Key habitats**
Special protection area	Glacial river
Wilderness area	Large river
6. Other protection sites	Lake outflow
Protected landscape	Rare river types
Natural monument	Type-specific river sections
Nature reserve	First km (Epirhithral) as far as passable from river mouth
Resting area	First 5 km (metarhithral, hyporhithral small, and epipotamal small) as far as passable from river mouth
Ramsar area	Tributary to rivers >500 km^2 catchment size until first impassable barrier
River sanctuary	Potential reproduction area of *Salmo trutta lacustris*
WFD-relevant Natura 2000 area	
Other Natura 2000 area	
Other protection areas	

and therefore was highly in conflict with conservation needs. Even in the "minimal conservation" scenario, several projects conflict with "high" conservation needs. Additionally, from the energy-economic point of view, the attractiveness of small hydropower projects (<10 MW) were rated as "low" or "moderate," yet at the same time, they affect sensitive river stretches despite their small size.

The HY:CON study formed the scientific basis for the WWF Eco Masterplan III where the main outcomes were finally summarized as follows:

1. The expansion of hydropower in Austria happens uncoordinated. Projects do not originate from an elaborate national plan—in fact, most of the projects derive

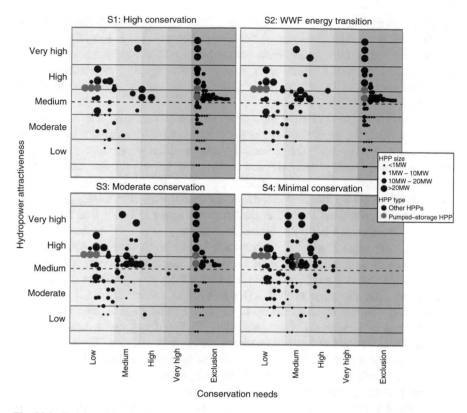

Fig. 23.2 Results of the four conservation scenarios. The size of the dots indicates the size of the assessed hydropower project (depending on the installed capacity in megawatts). The further right a dot is placed, the higher the conservation conflict. Projects above the dashed line were considered as attractive, whereas projects below the line were evaluated as less attractive from the energy-economic point of view (Modified from Seliger et al. 2016)

from private interests of (private) investors or entrepreneurial targets of single energy suppliers and other companies.

2. Expansion targets for hydropower in Austria are far too ambitious (7 TWh/a till 2020). Moreover, many of the projects identified within this study show massive conflicts of objectives with other European and national targets and/or legal regulations. A high share of the investigated projects is in conflict with conservation needs in almost all scenarios—even in the "minimal conservation scenario," numerous projects are in "high" conflict with conservation needs. Therefore, an implementation of the present projects appears to be questionable even from a legal perspective.

In the course of data collection, major data gaps became evident. There is no comprehensive, publicly accessible database regarding power plant projects in Austria that includes, in particular, energy-economic figures that follow a consistent system. Furthermore, a significant lack of ecological data exists; specific information on relevant water-related objects of protection (species and habitats) does not exist or is not publicly accessible. However, the study shows that it is indeed possible to create, in a timely manner and with reasonable efforts, an information basis that contributes to the development of an expansion strategy and to secure the most important ecological values. Strategic planning, also in the frame of IRBM, is feasible.

It is equally obvious that the limit of an expansion potential that is acceptable from an ecological perspective has already been reached. This is especially evident in the intensity of hydropower expansion along Austrian waterbodies (70% of the Austrian river network already affected by hydropower), on the one hand, and, on the other hand, when considering realistic expansion scenarios, half of all projects are in conflict with applicable European and national laws. Thus, they can only be approved by exemption. Therefore, this data confirms for WWF that the peak of hydropower expansion has been reached in Austria.

The results regarding small hydropower plants are sobering. Although this industry enjoys a public reputation of being a seemingly clean, acceptable form of energy (often touted as an alternative to large projects), this view needs to be put into perspective after thorough analysis. The number of micro and small projects in Austria must be examined critically from both an energy-economic perspective and in terms of nature and water protection. Half of the projects analyzed were evaluated as not attractive from the energy-economic point of view. The contribution of these projects to the overall energy production is negligible. It can be argued that small hydropower plants do not support climate protection as these only make an insignificant energy contribution and do little, from a national strategic perspective, to decarbonize the Austrian economy. Yet it is expected that the vast number of projects has a massive impact on nature, since so many small hydropower plants are located in ecologically sensitive areas.

From these outcomes, WWF Austria formulated three conclusions on where the national policy has to change:

Redefinition of Strategic Expansion Targets of Hydropower

When applying realistic expansion scenarios, the expansion potential aimed at 7 TWh/a by 2020 cannot be accomplished. Such a large expansion potential remains largely theoretical and cannot be implemented under the current framework conditions (duration of planning and approval procedures, electricity prices, etc.). Hence, WWF recommends to decrease the pressure on national energy suppliers, policy, and economics as soon as possible and to initiate an open, transparent discussion, also involving the public with regard to future options of expansion. Thereby, realistic targets and reasonable implementation strategies can be developed which could lead, with reasonable involvement of the public, to viable results.

Determining Exclusion Zones

Many years of experience in the discussion around the expansion of hydropower in Austria, also on international level, have shown that a clear and comprehensible designation of exclusion zones has been a smart and effective political initiative. It supports the conservation and protection of ecologically valuable waterbodies, while at the same time creating legal and planning security for e-economics.

Interlinking Subsidies with a Strategic Management Plan

In particular, small and micro hydropower plants are to be rejected, not only from an ecological perspective, but because these are also significantly less attractive in comparison with medium and large hydropower facilities from an energy-economic view. The support of hydropower plants should be bound to ecological standards. Hence, facilities located at ecologically sensitive waterbodies and which do not represent energy-economically attractive sites must not be subsidized.

This case study shows that an NGO can act as an effective intermediate between science and policy. Based on scientific studies, conclusions were drawn and then translated into policy action plans.

23.4 Showcase Danube–Carpathian Region

Partnership For a Living Danube (WWF DCP)

Floodplains of the Danube, the EU's longest river, and its tributaries have long been hotspots of biodiversity, providing a myriad of ecosystem services, including flood protection, drinking water provision, nutrient removal, biomass and food production, and landscapes for tourism and recreation. Despite this, the Danube alone has seen 80% of its floodplains and wetlands disappear over the past 150 years. It has been diked, dredged, and dammed for hydroelectric power production, shipping, and flood mitigation (see Chap. 24).

The effects of such industrial development of riverine landscapes have been wide-ranging and include plummeting fish and wildlife populations and decreases in water quality. Floodplains cut off hydrologically from the river channel can no longer act as natural water retention areas with consequences for flood risk.

This is why WWF promoted the Lower Danube Green Corridor Initiative, a framework for cooperation and coordination between the countries of the Lower Danube, Bulgaria, Romania, Moldova, and Ukraine, aiming to protect and restore floodplain ecosystems. As an NGO with observer status, WWF also managed to keep floodplain restoration high up on the agenda of the International Commission for the Protection of the Danube River (ICPDR) and its Basin Management Plans of 2009 and 2015.

Furthermore, WWF and the Coca-Cola Company (TCCC) entered a project partnership to restore vital wetlands and floodplains along the Danube River and its tributaries. The ambitious project aims to increase the river's flood water storage capacity by 12 million m^3 and to restore over 5300 ha of wetland habitat, which is a substantial contribution to what governments of the Danube River basin pledged to restore by 2021.

This example shows that NGOs can be influential cooperation partners of governments and businesses, able to initiate and pilot action toward ecosystem restoration and good water management.

23.5   The Role of NGOs in Freshwater Conservation

The three case studies above are typical for thousands of places where NGOs work with scientists, governments, companies, and the local communities to improve freshwater conservation and, as a last consequence, protection of biodiversity. How NGOs can make a difference and overcome the challenges they face in their freshwater conservation work can be summarized by six principles:

1. Vision
Management of river basins should be governed by a long-term vision that is agreed to by consensus between all major stakeholders. The vision must give equal weight to the three pillars of sustainable development—economic, social, and environmental concerns. NGOs stress the need to maintain and restore ecosystem services and biodiversity in order to enhance local livelihoods.

2. Integration
Global policies and decisions must be integrated into national and regional decision-making processes. In many cases, this integration will be required across administrative boundaries. To achieve effective integration at the scale of a whole river basin, NGOs are a valuable partner, as they can provide a direct link between the public and the authorities. NGOs can also support applied science by linking researchers with local practitioners, and this helps integrate science and policy.

3. Scale
The primary scale for strategic decision-making must be the whole river basin. Operational decisions must then be taken in accordance with the basin-wide strategy but can be made at subbasin or local levels. This principle applies in all cases, including transboundary river basins.

The enormous diversity in the size and characteristics of river basins means that approaches suited to one location are not automatically transferable to another. NGOs can help to guarantee as much coherence as possible between "top-down" and "bottom-up" approaches in the pursuit of common environmental and socioeconomic objectives.

4. Timing

Coordination is critical for ensuring that the different elements of Integrated River Basin Management (IRBM) (see Chaps. 15 and 16) are implemented in the right sequence. On the one hand, it is important to base management decisions on sound information, strong institutional mechanisms, and broad stakeholder participation. On the other hand, urgent action should not be postponed while tools, data, and processes are perfected. It may be better to begin implementing river basin management sooner rather than later with emphasis on low/no regret measures, using existing information and experience and applying the lessons learned to achieve continuous improvement. NGOs can provide knowledge as well as experience and help define the urgency of implementing first steps in IRBM (WWF International 2002).

5. Participation

High priority must be given to establishing effective mechanisms for active public participation in planning and decision-making, right from the start of the process. NGOs play a crucial role as observer in such processes, ensuring transparency, and stakeholder involvement with broad access to information.

6. Capacity and Knowledge

Investment of adequate financial and human resources into capacity building and participation processes is one of the keys to successful river basin management, especially in those parts of the world where existing capacity is likely to be most limited. IRBM must be based on sound scientific data and an understanding of freshwater ecosystems and their component key hydrological and ecological processes. Similarly, socioeconomic analyses are key to understanding the drivers behind water use and abuse. NGOs can provide capacity of knowledge by facilitating the transfer of scientific data analysis to the broader public in order to inform and raise awareness on critical issues. At the same time, the illustrations are used to improve and advance policy processes on national, regional, and global levels.

References

Behera S, Areendran G, Gautam P, Sagar V (2011) For a living Ganga-working with people and aquatic species. WWF-India, New Delhi

Scheikl S, Seliger C, Loach A, Preis S, Schinegger R, Walder C, Schmutz S, Muhar S (2016) Schutz ökologisch sensibler Fließgewässer: Konzepte und Fallbeispiele. Österreichische Wasser- und Abfallwirtschaft 68:288–300

Seliger C, Scheikl S, Schmutz S, Schinegger R, Fleck S, Neubarth J, Walder C, Muhar S (2016) Hy: Con: a strategic tool for balancing hydropower development and conservation needs. River Res Appl 32:1438–1449

WWF Austria (2009) Ökomasterplan: Schutz für Österreichs Flussjuwele! Österreichweite Untersuchung zu Zustand und Schutzwürdigkeit von Fließgewässern Darstellung der Ergebnisse anhand 53 ausgewählter Flüsse

WWF Austria (2010) Ökomasterplan Stufe II—Schutz für Österreichs Flussjuwele -Zustand und Schutzwürdigkeit der Österreichischen Fließgewässer mit einem Einzugsgebiet größer 10 km^2—Ergebnisse und Handlungsempfehlungen

WWF Austria (2014) Ökomasterplan Stufe III—Schutz für Österreichs Flussjuwele. Strategische Betrachtungen für einen ausreichenden Gewässerschutz sowie einen ökologisch verträglichen Ausbau der Wasserkraft in Österreich

WWF International (2002) Managing water wisely: Promoting sustainable development through integrated river basin management, WWF Living Waters Programme

Part III
Case Studies

Chapter 24
Danube Under Pressure: Hydropower Rules the Fish

Herwig Waidbacher, Silke-Silvia Drexler, and Paul Meulenbroek

24.1 Introduction

Major studies, conducted recently at some Danube hydropower impoundments and along the river itself, have pinpointed certain challenging ecological situations for certain faunal associations (Schiemer 2000; Jungwirth 1984; Waidbacher 1989; Herzig 1987; Bretschko1992). One of the important groups affected are riverine fish assemblages. Fish communities are good indicators of habitat structure as well as of the ecological integrity of river systems due to their complex habitat requirements at different stages of their life cycles (Schmutz et al. 2014; Schiemer 2000; Schmutz and Jungwirth 1999). The construction of impoundments changes river systems ecologically by disrupting the connection between the river and the lateral backwaters, by changing the shoreline, and by stabilizing previously dynamic water levels as well as other impacts (Schiemer and Waidbacher 1992).

Impoundments confront fish with new situations that present a challenging difference with the sets of parameters they have adapted to in unmodified river habitats. Due to reduced flow, increased depth, low water temperatures, short retention times, silty to muddy sediments resulting from increased sedimentation, and higher benthic biomass in the sediment depositions, these impoundments conform more to the habitat needs of lacustrine fish species. However, the relatively low average annual temperature of the river, the lack of shoreline structures, and low plankton density inhibit better development of such "backwater" fish associations. The original dominant riverine fish species can mainly be found only in free-flowing sections, except for a few individuals in the uppermost part of the impoundments (Waidbacher 1989).

H. Waidbacher (✉) · S.-S. Drexler · P. Meulenbroek
Institute of Hydrobiology and Aquatic Ecosystem Management, University of Natural Resources and Life Sciences, Vienna, Austria
e-mail: herwig.waidbacher@boku.ac.at; silke.drexler@boku.ac.at; paul.meulenbroek@boku.ac.at

In light of these results, strategies have been developed to counteract and minimize negative impacts caused by the construction of new hydropower dams. A more ecologically sustainable solution has been implemented during the construction of a low head dam (8.6 m height) for the impoundment at "Freudenau" in Vienna in 1998. A special attempt has been made here to maintain the ecological integrity of the river system by introducing a large number of mitigating measures. These include creating large gravel areas, improving the lateral integration between the river and the backwaters, and increasing the diversity of the inshore riverbed structures to improve the quality of spawning substrates and nurseries for fish (Waidbacher et al. 1996).

The results of the latest monitoring 2013–2015 are presented here and can be seen as a first indication of the response of the fish association to the innovative large-scale measures of "Freudenau" impoundment.

24.2 Historic Development of the Austrian Danube and Its Faunal Elements

The upper part of river Danube extends from the river's source in Germany to the Austrian/Slovakian border and is topographically well defined by its high slope (0.43‰ in Austria) and high bedload transport. Large tributaries from the Alps considerably increase river discharge, which reached a mean value of approx. 2000 m^3/s eastward from Vienna prior to river engineering (Liepolt 1967). The pristine morphological condition of the river alternated between canyons with narrow riparian zones to braided reaches with large alluvial areas, especially in the plains of Eastern Austria. A variety of river arms offered a rich diversity of ecological structures with gradients of flow velocity, substrate, and riparian vegetation. This provided ideal conditions for a typical Austrian Danube fish community (Hohensinner et al. 2005).

During the last 100 years, these ecological conditions have been considerably changed by river regulation and damming (Hohensinner et al. 2004). The main regulation started in the second half of the nineteenth century and resulted in substantial changes due to straightening and enforcement of most of the river's flow into one channel and an abandonment of side arms. This had major effects on:

(a) The ecological conditions of the river habitats (e.g., increase of flow velocity, bedload erosion, and deepening of the riverbed)
(b) The interactive dynamics between river and riparian zone
(c) The relative proportion of alluvial habitat types

The construction of large run-of-the-river hydropower plants started in the 1920s with the ultimate goal of forming a continuous chain of impoundments along the German/Austrian Danube section (Rathkolb et al. 2012). These developments resulted in severe ecological degradation due to an almost complete disconnection between river and lateral backwaters, mostly monotonous shoreline constructions and a stabilized water level over long distances. The characteristic limnological features of these impoundments are:

(a) Short retention times
(b) Low water temperatures
(c) Sedimentation of fine particles in the central impoundment
(d) Reduction of littoral gravel banks to the uppermost sections of the impoundments
(e) Low plankton density
(f) Higher densities of benthic invertebrates in the fine sediment deposits

24.3 Basic Scheme of Impacts of Danube Hydropower Impoundments on Native Fish Associations

Fish communities are good indicators of ecological integrity of river systems because of their complex habitat requirements that shift in the course of their life cycles. The changes in population structures and abundances induced by damming can be elucidated by comparing the fish fauna in free-flowing sections with that of impounded areas. The first such investigations were done in river Danube as part of an interdisciplinary study of the impoundment of "Altenwörth" (50 km upstream of Vienna) in the mid-1980s (Hary and Nachtnebel 1989; Waidbacher 1989; Schiemer and Waidbacher 1992). The fauna in the free-flowing river is characterized by a dominance of rheophilic species (i.e., their life cycle is bound to rapid-flowing water conditions). Species such as barbel (*Barbus barbus*) and nase (*Chondrostoma nasus*) occur in high abundances in the free-flowing section of the "Wachau," followed by a distinct predominance of eurytopic species [e.g., roach (*Rutilus rutilus*) and bleak (*Alburnus alburnus*)] in the impounded section. Data are based on electro-boat fishing along the shoreline (system Coffelt, attracting efficiency approximately 6 m width and 2.5 m depth) and additionally long-line fishing at the river bottom. The difference in the species composition between the uppermost part of the impounded river, with high flow velocity and coarse-grained sediments, and the central part of the impoundment, with reduced flow, fine substrates, and monotonous shoreline structures, is relatively low (Table 24.1). However, the population density of the characteristic riverine species, nase and barbel, declines noticeably in the main impoundment (Fig. 24.1).

Table 24.1 Number of adult and juvenile species in the different sections of the impoundment in the main channel of the Danube at "Altenwörth" and "Freudenau"

	Altenwörth		Freudenau	
	Adult	Juvenile	Adult	Juvenile
Free-flowing	32	21	24	21
Head of impoundment	35	18	23	17
Central impoundment	36	18	21	12

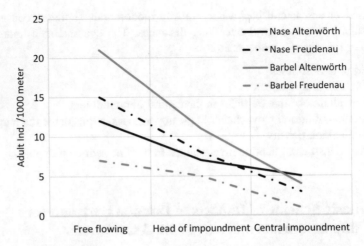

Fig. 24.1 Example for the distribution of two originally dominant fish species in the monotonous constructed Danube impoundment of "Altenwörth," 50 km westward of Vienna, and the latest constructed impoundment of "Freudenau"; adult nase and barbel individuals per 1000 m electrofishing in the riparian zones (own data, late spring/summer situation)

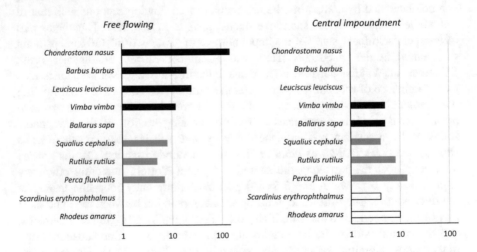

Fig. 24.2 Mean numerical composition of juvenile fish in three shore seine catches; free-flowing area is located in "Wachau"; central impoundment in the impoundment of "Altenwörth"; black, rheophilic; gray, eurytopic; white, limnophilic species (own data)

An analysis of the size structure of the characteristic riverine species shows that in the vicinity of the dam, only old age classes are represented, supported by abundant food supplies in the rich benthic deposits (Waidbacher 1989). Surveys of fish juveniles, as seen in Fig. 24.2, show that the overall density is low, and riverine species are rarely represented or are completely missing in the main impoundment zone. Flow velocity and the nature of littoral substrates (mainly riprap) are not adequate to function as spawning sites and rearing areas for riverine species.

24.4 Implementation of Mitigation Measures in the Latest Constructed Hydropower Dam and Impoundment of Vienna/Freudenau

Based on the results of research at the "Altenwörth" impoundment in the mid-1980s, strategies have been developed to improve the ecological conditions of affected areas. Ecological improvements were designed to counteract and reduce negative impacts caused by the hydropower dams over periods long enough to make such improvements sustainable.

As an example, the objectives for habitat improvements for characteristic Danube fish populations contain the creation of:

(a) Dynamic gravel banks
(b) Dynamic sand habitats
(c) Shelters in times of flood events
(d) Possibility for upstream migration
(e) Lateral connections of water bodies
(f) Riparian bays and channel systems

Various "ecologically sustainable" solutions have been implemented during the construction of the low head dam for the impoundment at "Freudenau" in Vienna. In this case, for the first time, a whole suite of mitigating measures has been introduced to maintain the ecological integrity of the Danube and especially to support the development of self-reproducing fish communities. Figure 24.3 gives a rough overview of the location of implemented measures in four sections, which are described in more detail below.

Section 1

Along riparian floodplains, the connection of lateral water bodies to the main river channel favors the migration of fish, especially lacustrine backwater fish species, and offers rearing and feeding areas (Fig. 24.4). Migration into riparian side arms is extremely important for different life stages of some endangered fish species, such as white-eye bream (*Ballerus sapa*) and zope (*Ballerus ballerus*).

Section 2

The original, dominant, rheophilic fish fauna is represented in Danube impoundments by adult individuals only. To mitigate these effects, gravel bank spawning grounds have been constructed in extended areas in the uppermost part of the "Freudenau" impoundment to support the reproduction of original faunal elements of the river (Fig. 24.5). This was done by the construction of an underwater riprap, which prevents the gravel bar from major erosion into the main channel.

Section 3

An extensive riparian channel and bay system in the central impoundment serves mainly as a spawning ground and rearing area for eurytopic species and as a feeding area for all fish associations. During flood events these zones act as refuges (Fig. 24.6).

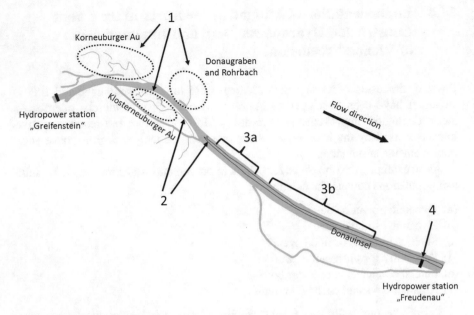

Fig. 24.3 Location of different ecologically sustainable solutions in four sections of the impoundment "Freudenau" (1. lateral connections; 2. gravel banks with riprap stabilization; 3. riparian channel and bay systems; 4. fish migration bypass)

Fig. 24.4 Lateral connection of Korneuburger Au with the main channel of the Danube via a fish bypass system (courtesy of Verbund AG)

Fig. 24.5 Extended underwater gravel bank inshore structure under construction; red arrows indicate "double riprap"; blue line indicates the water level nowadays after construction (Section 2)

Fig. 24.6 Constructed riparian channel and bay system in the central impoundment (Section 3)

Section 4

Fish migration in a bypass channel system supports genetic exchange (Fig. 24.7).

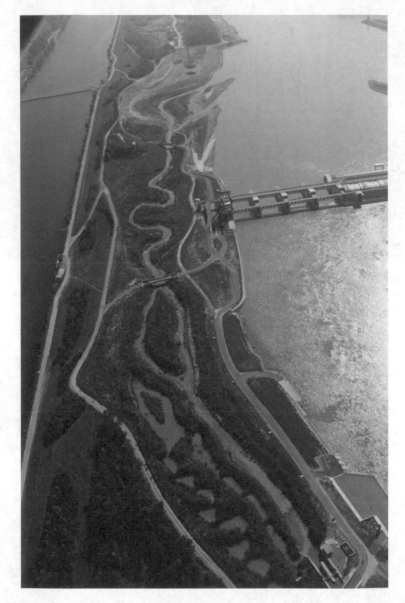

Fig. 24.7 Bypass channel system for fish migration at the power station "Freudenau" (Section 4) (courtesy of Verbund AG)

24.5 Ecological Response and Sustainability of the Constructed Habitat Improvements at "Freudenau"

A second round of research was conducted to monitor constructed habitat improvements some 18 years after construction. Without the influence of constructed measures, the fish assemblages in the main channel of Freudenau responded in the same pattern as already seen in "Altenwörth," namely:

- Decrease or lack of juvenile fish in the central impoundment
- Low number of species in the riparian part of the impounded area
- Low abundance of riverine assemblages in the central part of the impoundment

Considering the ecological improvements indicated by research some 18 years after construction, a clear positive sign for fish assemblages becomes visible:

In Section 1, a better connection has been constructed between the channel and the riparian floodplain waters of "Klosterneuburger Au" (right bank of the Danube). A pool pass allows fish migration at two different water levels of the backwater (summer and winter) and has been accepted by 29 fish species in the direction to the backwater and by 38 species in the direction to the main river channel. In addition to the movement pattern expected in times of spawning activities, the results (fish trap in the pool pass) from 2006 show a remarkably fast response of riverine fish, which were washed into the backwater system during a flood event, in finding again the migration pass for leaving the backwaters in the direction of the main Danube channel. Eighty-five percent of the composition of the sampled migrating rheophilic fish, which showed locomotion after the flood event, belongs to the species assemblage of nase, ide (*Leuciscus idus*), vimba (*Vimba vimba*), asp (*Leuciscus aspius*), and schrätzer (*Gymnocephalus schraetser*)—a classic river fish assemblage (Fig. 24.8).

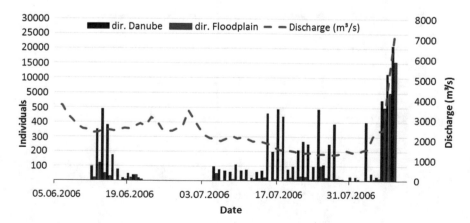

Fig. 24.8 Left axis: Number of migrating individuals into the floodplain system (dir. Floodplain) and vice versa in the direction of the main channel of the Danube (dir. Danube). Right axis: Discharge of the Danube (adapted after Schinninger 2008)

Expected peaks in fish migration are visible in the period of late spring, pinpointing migration activities for spawning and after reproduction (Schinninger 2008).

Investigation via a pool pass system of migration activities to the "Korneuburger Au"—situated on the left bank of the Danube—has shown similar results (Jungwirth and Schmutz 1988). In total, 32 species migrated in both directions. Bleak, roach, white bream, bream (*Abramis brama*), barbel, nase, and zope were the most frequently observed species in this study.

Where connection to the main channel is limited, i.e., lack of a pool pass or other migration facilities, fish communities in backwaters can show a high specialization and often are inhabited by rare species. Some species, such as the weatherfish (*Misgurnus fossilis*) recorded in this study, occur exclusively in disconnected flood-plain waters. The design concept implemented for ecological improvement at "Freudenau" supports such species by leaving small floodplain habitats discon-nected in years without natural flood events (e.g., "Rohrbach" habitat).

In Section 2, a large amount of gravel material was excavated from the riverbed and newly located in the riparian area to construct gravel bank spawning grounds. Extended shallow areas of several hectares have been artificially established and secured against abrasion by a massive underwater riprap structure (Fig. 24.5). Despite several flood events (up to a 200-year-flood event) over 18 years, no massive changes in the constructions are visible, and the gravel banks are still functioning as spawning grounds. Figure 24.9 shows results from 2013 where larval stages of fish

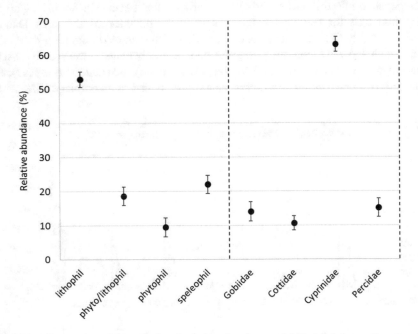

Fig. 24.9 Mean and confidence limits of relative abundance of drifted fish larvae for spawning guilds and families at an artificially built gravel bank, $n = 171$ (adapted after Meulenbroek et al. 2017a)

have been sampled downstream of a large artificial gravel bank situated at the upper most part of the "Viennese Donauinsel." The fish larvae have been identified via barcoding and displayed high shares of lithophilic riverine cyprinids in their abundances (Meulenbroek et al. 2017a). Although spawning activities of adult individuals could not be observed in turbid waters, the drifting of fish larvae in the presumed time period provides indications of successful reproduction activities at the artificially constructed gravel bank.

Section 3 is divided in two parts where one part (3a) is self-cleaning from fine sediments along the shoreline after flood events, while the other one (3b) is not.

In Section 3a, flow velocity is high enough at mean discharge (1 m/s) to wash out fine sediments, which are deposited during flood events in the riparian structures. Inside of the constructed riparian arms, a riverine fish assemblage has developed and persists even after 18 years (Table 24.2).

In Section 3b, the fine sediment deposition along the shoreline is not cleared at mean discharge. The constructed riparian bays are hotspots of biodiversity in the depauperated, i.e., species impoverished, central impoundment of "Freudenau." Their quick colonization by fish and benthic invertebrates just after their construction was documented by Chovanec et al. (2002) and Straif et al. (2003). The importance of such measures was highlighted in 2013–2015, by the high diversity, e.g., total 38 fish species, and high abundances of juvenile riverine species found in these areas (Fig. 24.10).

Recent findings of early life stage abundances suggest several colonization patterns for such riparian habitats. The most unlikely pattern is colonization only from the main channel via unidirectional drift. But there are three different drift patterns visible as described by Meulenbroek et al. (2017a):

(1) Larvae drift into the side arm over longer time periods with different densities and the use of the habitats as nursery grounds.
(2) There are spawning activities at different densities within these side systems.
(3) There is additional drifting of larvae in the direction of the main channel.

These factors identify the multiple functions of these habitats in providing suitable nursery and spawning grounds for an essential variety of Danube fish species.

Furthermore, the high abundance of juveniles in these riparian flat habitats with high sedimentation is additionally sustained by low predating pressure from fish-eating birds. Such water bodies are too shallow for cormorants or goosanders to hunt for prey, and it is most likely that a few herons, stepping, picking, and taking fish,

Table 24.2 Comparison of the number of species within the most upstream riparian side arm for each habitat guild between years 1999/2000 and 2014/2015

	1999/2000	2014/2015
Eurytopic	8	8
Limnophilic	1	1
Rheophilic A	5	4
Rheophilic B	2	3
Total	16	16

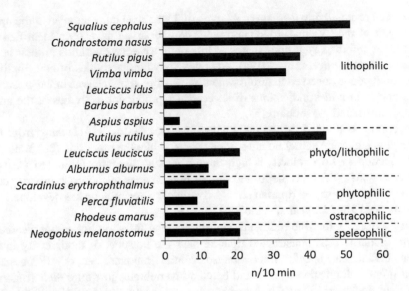

Fig. 24.10 Mean density (per 10-min electrofishing) of the 14 most abundant fish species for different reproductive guilds of one of the riparian side arms in 2014 (Section 3)

can be surely sustained by the system. The repeated validation over 18 years indicates that these mitigation measures are sustainable, even though some vegetation and sediment management needs to be established in the near future.

Results of the latest surveys, 18 years after construction, show that the fish assemblage in the impoundment of "Freudenau" follows the same pattern as in other Austrian Danube impoundments if the riparian mitigation measures are not taken under consideration (compare Figs. 24.2 and 24.11).

However, the mitigation measures show satisfactory improvements in the habitat conditions and support the functions of lost habitats essential for riverine fish. Fish association of juveniles found in a riparian side arm in the central impoundment in 2015 shows that nase and barbel as well as rudd (*Scardinius erythrophthalmus*) and bitterling (*Rhodeus amarus*) are part of the young-of-the-year assemblage and ready for building up new adult stocks.

Additionally, a new development became visible. It's the first time in major scientific Danube investigations that alien species are visible in extraordinarily high densities. Beside the racer goby (*Babka gymnotrachelus*) and the bighead goby (*Ponticola kessleri*), the round goby (*Neogobius melanostomus*) dominates the bottom fish fauna at least in the impounded area. The bottom of main channel and the bottom of side arms, especially close to ripraps, are completely "infected" with enormous ecological effects on food webs and the native fauna caused by competition (Ahnelt et al. 1998; Wiesner 2005; Ebm 2016).

In Section 4, a fish migration bypass system has been constructed with three major components that robustly complement each other in a sustainable way. It starts with a bay system in the tail water (Fig. 24.7) with calm, shallow waters over some 200 m.

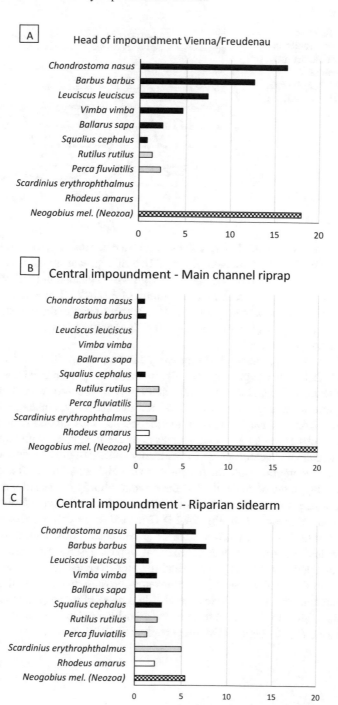

Fig. 24.11 Species composition of juvenile fish in the impoundment of "Freudenau"; black are rheophilic, gray are eurytopic, and white are limnophilic species; $n =$ individuals/10-min electrofishing

Fig. 24.12 Relative
occurrence of juvenile and
adult nase within the fish
bypass system of
"Freudenau" (adapted after
Meulenbroek et al. 2017b)

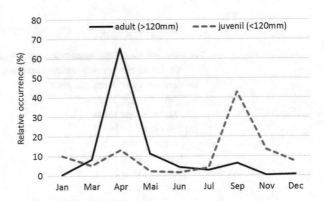

The subsequent, seminatural bypass channel with a mean discharge of 1.6 m³/s and an average slope of 0.7% is situated in a riverbed of 7 m width and a corresponding average current speed of 0.6 m/s. The discharge is not constant but follows the mean discharge of the Danube, reaching a maximum of 3.6 m³/s. The length of this free-flowing section is approximately 900 m. The uppermost part of the system is built as a pool pass of 19 pools with a minimum of 70 m² per pool and a water level difference of 11 cm from pool to pool.

Beside the systems function as migratory facility, shown in 2000 by Eberstaller and Pinka (2001), the bypass system also provides a spawning ground for all the guilds of Danube fish and therefore makes an important contribution to the maintenance of several endangered species. In a monitoring survey, conducted throughout 2013 and 2014, seasonal changes in abundances, species diversity, and spawning events were observed. A total of 41 species colonize the bypass with temporary and spatial fluctuations. In early spring, the indicator species of the free-flowing Danube, nase and barbel, migrated into the fish pass in very high quantities. After spawning in April and May, most of the adults left the system. Shortly afterward a massive drift of early life stages of riverine fish species was observed, followed a few months later by thousands of juvenile fish (Fig. 24.12) (Meulenbroek et al. 2017b).

Present studies at the bypass system of "Freudenau" show that, in contrast to a pure technical construction, the seminatural bypass system provides a migration function and performs like a Danube tributary. However, geodetic research showed a deepening of the riverbed caused by continuous erosion due to lack of gravel input from upstream. This demonstrates the absolute need of management actions from time to time (after 15–20 years) to secure the positive ecological values for fish and other riverine faunal elements (Meulenbroek et al. 2017b).

24.6   Conclusion

In the Austrian stretch of river Danube (approx. 350 km), ten hydropower stations/ impoundments have been implemented within the last 70 years. All of them massively affect the fish fauna. The most threatened fish are those of the rheophilic guild, which was dominant during pristine conditions. Straightening the river channel at larger scales started in the 1850s. Their further development favored lacustrine as well as eurytopic species at the same time that it decreased abundances and occurrences of riverine species by shortening free-flowing habitats and cutting off side arms.

Impoundments deny rheophilic fish a number of structures found in free-flowing river stretches: suitable gravel spawning grounds, small- and large-scale inshore structures for nursery and juvenile development, and shelters in times of flood events and winter situation as well as proper food security. As a consequence, fish ecological research shows an extreme decrease of riverine adults in the central impoundments, and successful reproduction is only possible in small, restricted areas of running waters with gravel habitats in the tail water of the dams.

However, in impoundments stronger development of eurytopic and lacustrine fish species is hampered by comparatively low water temperatures, low plankton density needed as starter feed for their larvae, a lack of macrophytes as spawning habitats, and a lack of structured refuge and nursery habitats.

Based on these abiotic and biotic conditions, a Danube impoundment does not serve the development of a proper life cycle for riverine fish or for lacustrine communities. Eurytopic species are most likely to accept suboptimal conditions, and therefore in most impoundments a very limited number of eurytopic species dominate the fish fauna.

Planning and constructing of the latest Danube hydropower plant at "Freudenau" (operation started 1998) considered a variety of ecological measures to improve the biotic integrity of the affected river section. Large-scale habitat constructions—based on the lessons learned at other impoundments—include double-riprap secured gravel banks, creation of massive inshore riverbed structures, a bypass system for fish migration, and creation/connection/integration to riparian backwaters and side arms. Results from the fish assemblages as seen in Fig. 24.11 pinpoint the positive ecological development of the central impounded area only when riparian side arms and structures are situated.

Because of "aging" of the constructed riparian elements, succession happens in the riparian vegetation as well as in the habitat morphology, and hence continuous human management and maintenance are vital to sustain the habitat's functioning. Given the scale that humans use the river's flow to satisfy such needs as electricity, in response habitat management has to secure the functioning of ecological improvements to guarantee future fish stocks for next generations. Hydropower rules the fish!

References

Ahnelt H, Banarescu P, Spolwind R, Harka A, Waidbacher H (1998) Occurrence and distribution of three gobiid species (Pisces, Gobiidae) in the middle and upper Danube region-examples of different dispersal patterns? BIOLOGIA-BRATISLAVA 53:665–678

Bretschko G (1992) The sedimentfauna in the uppermost parts of the impoundment. Veröffentlichungen der Arbeitsgemeinschaft Donauforschung 8(2-4):131–168

Chovanec A, Schiemer F, Waidbacher H, Spolwind R (2002) Rehabilitation of a heavily modified river section of the Danube in Vienna (Austria): biological assessment of landscape linkages on different scales. Int Rev Hydrobiol 87(2–3):183–195

Eberstaller J, Pinka P (2001) Überprüfung der Funktionsfähigkeit der FAH am KW Freudenau. Zusammenfassender Bericht, Verbund Austrian Hydropower

Ebm N (2016) The diet of Neogobius Melanostomus (Pallas, 1814) in the area of hydropower plant "Freudenau" (Danube River). Diplomarbeit/Masterarbeit—Institut für Hydrobiologie Gewässermanagement (IHG), BOKU-Universität für Bodenkultur, p 207

Hary N, Nachtnebel P (1989) Ökosystemstudie Donaustau Altenwörth. Veränderungen durch das Donaukraftwerk Altenwörth. ÖAW. Veröffentlichung des österreichischen MaB-Programmes, Bd. 14. Universitätsverlag Wagner. Innsbruck, p 445

Herzig A (1987) Donaustau Altenwörth—Zur Limnologie eines staureguierten Flusses. Wasser und Abwasser 31:215–237

Hohensinner S, Habersack H, Jungwirth M, Zauner G (2004) Reconstruction of the characteristics of a natural alluvial river–floodplain system and hydromorphological changes following human modifications: the Danube River (1812–1991). River Res Appl 20(1):25–41

Hohensinner S, Haidvogl G, Jungwirth M, Muhar S, Preis S, Schmutz S (2005) Historical analysis of habitat turnover and age distributions as a reference for restoration of Austrian Danube floodplains. WIT Trans Ecol Environ 83:489–502

Jungwirth M (1984) Die fischereilichen Veränderungen in Laufstauen alpiner Flüsse, aufgezeigt am Beispiel der österreichischen Donau. Wasser und Abwasser 26:103–110

Jungwirth M, Schmutz S (1988) Untersuchung der Fischaufstiegshilfe bei der Stauhaltung I im Gießgang Greifenstein. Wiener Mitteilungen 80:1–94

Liepolt R (1967) Limnologie der Donau. Schweizerbart'sche Verlag, Stuttgart

Meulenbroek P, Drexler S, Gstöttenmayer D, Gruber S, Krumböck S, Rauch P, Stauffer C, Waidbacher V, Zirgoi S, Zwettler M, Waidbacher H (2017a) Species specific fish larvae drift in constructed riparian zones at the Vienna impoundment of River Danube, Austria—species occurrence, frequencies and seasonal patterns based on DNA Barcoding (in prep)

Meulenbroek P, Drexler S, Geistler M, Rauch P, Waidbacher H (2017b) The Danube Fish bypass system of "Freudenau" and its importance as a lifecycle habitat (in prep)

Rathkolb O, Hufschmid R, Kuchler A, Leidinger H (2012) Wasserkaft. Elektrizität. Gesellschaft. Kraftwerksprojekte ab 1880 im Spannungsfeld. Band 104 der Schriftenreihe Forschung in der Verbund AG

Schiemer F (2000) Fish as indicators for the assessment of the ecological integrity of large rivers. In: Jungwirth M, Muhar S, Schmutz S (eds) Assessing the ecological integrity of running waters. Springer, Dordrecht, pp 271–278

Schiemer F, Waidbacher H (1992) Strategies for conservation of a Danubian fish fauna. River Conserv Manage 26:363–382

Schinninger I (2008) Fischökologische Untersuchung im Einflussbereich des Kraftwerkes Wien/ "Freudenau" unter besonderer Berücksichtigung der Konnektivität zwischen der Klosterneuburger Au und dem Donaustrom. Diplomarbeit/Masterarbeit, BOKU-Universität für Bodenkultur, p 126

Schmutz S, Jungwirth M (1999) Fish as indicators of large river connectivity: the Danube and its tributaries. Arch Hydrobiol Suppl Large Rivers Stuttgart 115(3):329–348

Schmutz S, Kremser H, Melcher A, Jungwirth M, Muhar S, Waidbacher H, Zauner G (2014) Ecological effects of rehabilitation measures at the Austrian Danube: a meta-analysis of fish assemblages. Hydrobiologia 729(1):49–60

Straif M, Waidbacher H, Spolwind R, Schönbauer B, Bretschko G (2003) Die Besiedelung neu geschaffener Uferstrukturen im Stauraum Wien-Freudenau (Donauinsel) durch Fisch-Benthosbiozönosen. In: Land Oberösterreich, Biologiezentrum der Oberösterreichischen Landesmuseen, DenisiaNeue Ufer Strukturierungsmaßnahmen im Stauraum Wien, 10, 34, Land Oberösterreich, Linz, ISSN 1608-8700

Waidbacher H (1989) Veränderungen der Fischfauna durch Errichtung des Donaukraftwerkes Altenwörth. In: Ökosystemstudie Donaustau Altenwörth. Austrian Academy of Science, Wien, pp S.123–S.161

Waidbacher H, Haidvogl G, Wimmer R (1996) Fischökologische Verhältnisse im Donaubereich Wien/Freudenau. In: Bretschko G, Waidbacher H (eds) Beschreibung der räumlichen und zeitlichen Verteilung der benthischen Lebensgemeinschaften und der Fischbiozönosen im Projektsbereich des KW Freudenau. Limnologische Beweissicherung. DOKW im Auftrag der obersten Wasserrechtsbehörde. Univ. für Bodenkultur, Wien, p 184

Wiesner C (2005) New records of non-indigenous gobies (Neogobius spp.) in the Austrian Danube. J Appl Ichthyol 21(4):324–324

Chapter 25
Danube Floodplain Lobau

Stefan Preiner, Gabriele Weigelhofer, Andrea Funk, Severin Hohensinner,
Walter Reckendorfer, Friedrich Schiemer, and Thomas Hein

Along the Upper Danube, almost all former floodplain areas have been lost due to river regulation, large-scale land-use changes, and terrestrialization processes. In the Lobau floodplain near the City of Vienna, ongoing terrestrialization leads to a dramatic loss of aquatic and semiaquatic habitats. Although the ecological values of the remaining floodplain area, such as high productivity and high biodiversity, are widely acknowledged, the implementation of restoration measures is difficult. In urban environments such as the Lobau, planning and decision-making for floodplain restoration inevitably involves trade-offs, uncertainties, and conflicting objectives and value judgments. Beyond ecological values, the main socioeconomic aspects are flood control, drinking water supply for Vienna, and recreation.

The aim of this chapter is to present the current ecological situation and the major development tendencies of the Lobau floodplain and to show the effects of potential management measures on the ecological situation.

S. Preiner (✉) · G. Weigelhofer · A. Funk · T. Hein
Institute of Hydrobiology and Aquatic Ecosystem Management, University of Natural Resources and Life Sciences, Vienna, Austria

WasserCluster Lunz Biological Station GmbH, Lunz am See, Austria
e-mail: stefan.preiner@boku.ac.at; gabriele.weigelhofer@boku.ac.at; andrea.funk@boku.ac.at; thomas.hein@boku.ac.at

S. Hohensinner
Institute of Hydrobiology and Aquatic Ecosystem Management, University of Natural Resources and Life Sciences, Vienna, Austria
e-mail: severin.hohensinner@boku.ac.at

W. Reckendorfer
VERBUND Hydro Power GmbH, Vienna, Austria
e-mail: walter.reckendorfer@verbund.com

F. Schiemer
Department of Limnology and Oceanography, University of Vienna, Vienna, Austria
e-mail: friedrich.schiemer@univie.ac.at

S. Schmutz, J. Sendzimir (eds.), *Riverine Ecosystem Management*, Aquatic Ecology Series 8, https://doi.org/10.1007/978-3-319-73250-3_25

25.1 Introduction

Floodplains are among the most productive habitats in the world and play a major role in the dynamics and the ecological integrity of riverine landscapes (Tockner et al. 2002). Due to the variety of different habitats, floodplains are key landscape elements for biogeochemical processing and hotspots for biodiversity (McClain et al. 2003; Tockner et al. 2010; Weigelhofer et al. 2015). Furthermore, floodplains provide a broad range of ecological and socioeconomic goods and services, including flood retention, groundwater recharge, and aesthetic and recreational values (Hein et al. 2006; Sanon et al. 2012; Rebelo et al. 2013).

However, river regulation, flow control, and large-scale land-use changes have altered nearly all central European riverine landscapes and have reduced the ecological and economic integrity of floodplains drastically. More than 68% of the floodplains of the Danube River have already been lost, with the highest reductions, e.g., up to 90%, of the former floodplain areas in the Upper Danube region (Tockner et al. 2009; Hein et al. 2016). The consequences of these changes are an increase in catastrophic flooding of urban areas, a reduction of in-river retention of nutrients, the loss of physical habitat diversity, and a correspondingly high percentage of endangered riverine species (ICPDR 2009; Hein et al. 2016). As a result, the Danube River Basin is among the most pressured large rivers in the world (Tockner et al. 2009).

Although strongly impacted by regulation measures, the free-flowing section of the Danube between Vienna and Bratislava is one of the last remnants of fluvial landscapes in the Upper Danube and in central Europe. It provides habitat for a diverse fauna and flora and was designated as a national park in 1996 (Reckendorfer et al. 2005). Within this stretch, the urban Lobau floodplain is located, which is a 2020 ha large floodplain area in the most western part of the national park right at the eastern border of the City of Vienna (Fig. 25.1).

During the major river engineering phase for the Danube between 1870 and 1885, this former dynamic floodplain has been disconnected from the main channel by the construction of a flood protection levee. Lateral embankments along the main river channel have severely altered the geomorphic and hydrological dynamics and have impeded the natural sequence of erosion and sedimentation (Hein et al. 2006; Hohensinner et al. 2008). Nowadays, the floodplain can be separated into two subareas that differ considerably in their ecological integrity due to differences in the degree of lateral hydrological connectivity with the main channel and the intensity of human use. The upstream section of the floodplain, the "Upper Lobau," has been completely disconnected from flood events and is integrated into the settlement area of the City of Vienna, thus severely impacted by human activities. The downstream section, called "Lower Lobau," is still included in the flood regime of the River Danube by a downstream opening and is kept as floodplain forest with water bodies of varying degrees of ecological integrity.

During recent decades, the vertical erosion in the main riverbed (incision) and the ongoing aggradation in the floodplain have further decoupled the wetland area from the main river channel, both hydrologically and ecologically (Reckendorfer et al. 2013b).

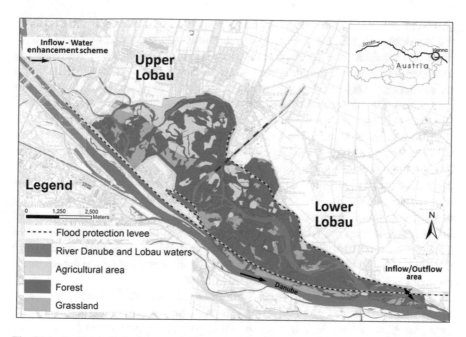

Fig. 25.1 Study area Lobau. Arrows indicate current inflow and outflow areas in the Lower Lobau and the inflow of the water enhancement scheme in the Upper Lobau

The consequences of this development are an ongoing decrease of the lateral hydrological connectivity, the loss of semiaquatic areas in the floodplain, a decreasing biodiversity, and the loss of typical river-floodplain species. Without any management activities, most aquatic and semiaquatic habitats are expected to disappear within the next decades (Schiemer et al. 1999; Hohensinner et al. 2008), with far-reaching ecological and socioeconomic consequences. The improvement of the current status of residual Danube floodplains through a basin-wide application of restoration measures is also one of the key water management issues identified in the Danube River Basin Management Plan (ICPDR 2009).

This chapter presents the current ecological situation of the urban Lobau floodplain, major development tendencies, the effects of management measures to improve the ecological situation, and constraints due to the intense human utilization.

25.2 Study Area

The Upper Lobau comprises an area of about 540 ha. It has been completely and permanently disconnected from the main channel even during floods (Fig. 25.1), thus oscillating hydrologically with the river only in response to groundwater fluctuations (Funk et al. 2009; Weigelhofer et al. 2011). The 12-km-long main side arm is separated into large, shallow, lake-like basins by check dams with culverts that prevent water losses to the river and secure the current water levels in the floodplain. Due to the prevailing lentic conditions, extended reed and macrophyte communities have developed in the backwaters. Before the implementation of management measures, the individual water bodies showed distinct differences in both water chemistry and trophic state, featuring highly eutrophic conditions in some backwaters. The Upper Lobau is situated within the settlement areas of the City of Vienna and is used intensively as a recreation area for hiking, cycling, and bathing.

In contrast to the Upper Lobau, the Lower Lobau (1480 ha, Fig. 25.1) is connected with the River Danube at least temporarily via back-flooding (Reckendorfer et al. 2013b). In the current situation, water from the Danube River enters the Lower Lobau during high water levels through a small opening in the main levee located at the downstream end (Schönauer Schlitz, Fig. 25.1) and drains during receding riverine water levels. With increasing Danube water levels, larger areas are inundated, including also terrestrial habitats. The connectivity of the various floodplain water bodies ranges from 140 days per year in the downstream parts of the main side arm adjacent to the inflow to less than 2 days per year in isolated ponds in the upstream parts of the Lower Lobau (Reckendorfer et al. 2006).

Sedimentation and terrestrialization processes prevail in the Lower Lobau due to decreasing surface and groundwater levels, siltation of sediments during floods, eutrophication, and the lack of erosion and export of fine sediments and nutrients to the river. Currently, terrestrialization accounts for 0.2–3.5% loss of aquatic habitats per year in the Lower Lobau. Most affected are shallow, semiaquatic habitats and isolated backwaters (Reckendorfer et al. 2013a).

Despite the hydrological deficits, the Lower Lobau still harbors a diverse and complex mosaic of aquatic, semiaquatic, and terrestrial habitats, resulting in a high biodiversity. Therefore, the Lower Lobau is an integral part of the "Nationalpark Donau-Auen" since 1996 and was designated a Natura 2000 area by the EU. Beyond the ecological value, the Lower Lobau plays a central role in the landscape water balance. It retains floodwater, recharges groundwater, and provides further socio-economic values, such as recreation (Rebelo et al. 2013). In addition, the Lower Lobau serves as an important drinking water reservoir for the City of Vienna. In extraordinary situations, such as drought or maintenance activities, the floodplain can provide drinking water for up to about 25% of Vienna's inhabitants.

25.3   Management and Restoration Approaches

A systematic evaluation of existing floodplain areas and an assessment of their restoration potential form the basis for specific management activities (Schwarz 2010). Among various possible measures, the lateral hydrological connection with the main channel plays a central role in floodplain restoration (Henry et al. 2002). It leads to the exchange of water, material, and species among the various floodplain water bodies and initiates the rejuvenation of floodplain habitats, thus maintaining a high biodiversity. The rehabilitation of the lateral hydrological connectivity is also a key issue for the further development of the Lobau to mitigate the continuous decoupling of floodplain habitats from the hydrological regime of the Danube.

However, the development of adequate management measures, which can compensate for historic and ongoing human impacts, is a subtle matter in urban floodplains, such as the Lobau. Sustainable management of urban floodplains needs to integrate the partly competing, socioeconomic aspects and requirements (Knoflacher and Gigler 2004) and to counteract further large-scale changes in the hydrological regime of the main river due to river regulation measures, impoundments, and climate changes. Consequently, different approaches for the Upper and Lower Lobau were developed in a step-by-step procedure, corresponding to the divergent development of these two areas (Weigelhofer et al. 2013). Due to the strong socioeconomic restrictions resulting from use as settlement and recreation area, a water enhancement scheme was implemented in the Upper Lobau as a first step that provided a subtle and strictly regulated water enrichment in this part of the floodplain. In the Lower Lobau, the absence of nearby settlements allows for an increased hydrological connection with the Danube. However, the high nature conservation value of this area requires careful planning. Thus, as a second step, several options for reconnection were developed for the Lower Lobau under consideration of existing ecological and socioeconomic demands and based on the experiences from the Upper Lobau. Hydro-ecological models were used to investigate and evaluate possible effects of these management options on the various floodplain habitats and functions.

25.3.1 Upper Lobau

In the Upper Lobau, a water enhancement scheme was initiated in the late 1990s to maintain minimum water levels in the backwaters during the vegetation season (Imhof et al. 1992), restore the surface water connection along the chain of backwater pools, ensure the enrichment of groundwater within the area, and discharge nutrients from the area's highly eutrophic water bodies (Imhof et al. 1992; Schiemer 1995; Weigelhofer et al. 2011).

The measure is a strictly controlled surface water supply to the Upper Lobau using predominantly nutrient-poor, bank filtrate water of the Danube discharged via the New Danube flood relief channel, an artificial side arm of the Danube (Funk et al. 2009). Water enhancement is limited to the vegetation season from March to October in order to allow the floodplain to dry up during winter (Funk et al. 2009). Additionally, the discharge is restricted to a maximum of 1.5 m^3 s^{-1} and is shut down during flooding periods to conform to water quality criteria (Weigelhofer et al. 2011). As a result, the mean discharge amounted to only 0.25 m^3 s^{-1} supplied to the Upper Lobau between 2001 and 2008 (Weigelhofer et al. 2011).

Despite the low discharges, the implementation of the water enhancement scheme resulted in stabilized high surface and subsurface water levels, sustaining the presence of the backwaters. The increased water exchange also stabilized the trophic conditions of the backwaters and reduced extreme trophic conditions by improving the supply of dissolved oxygen to the sediment surface. It also increased water exchange through dense macrophyte stocks as well as the resulting filtration of nutrients and algae (Bondar-Kunze et al. 2009; Weigelhofer et al. 2011).

In addition, the water enhancement scheme significantly increased local habitat diversity by creating temporary flowing conditions in narrow parts of the channel (Funk et al. 2009). This enabled a shift in the mollusk and dragonfly communities in these areas toward higher abundances of rheophilic species. Due to the restricted extent of the water exchange, however, the enhancement scheme has shown no effects on the fish community so far, especially regarding the recolonization of the main side arm by rare Danube species or rheophilic specialists.

25.3.2 Lower Lobau

Management options proposed for the Lower Lobau range from the conservation of the present status to rehabilitation toward pristine conditions of parts of the floodplain. The measures aim to rehabilitate rare aquatic and semiaquatic habitats, restore the ecological functioning of floodplain areas, and conserve the existing fauna, flora, and designated habitats based on European regulations (Council of the European Communities 1992; NP Donau-Auen 2009; Hein et al. 2016). Thus, ecological aims comprise both the rehabilitation of rheophilic floodplain habitats and the protection of newly established lentic and semiaquatic species. Among the most important socioeconomic demands in the Lower Lobau are an integrated flood protection scheme for the City of Vienna, based on the EU Floods Directive (Council of the European Communities 2007), and the guarantee of the current drinking water supply.

The identification of adequate management measures was based on several approaches: (1) the historical analyses of the floodplain development from the unregulated to the current state (Hohensinner et al. 2008; Baart et al. 2013), (2) long-term monitoring of hydrological and ecological properties of floodplain water bodies from 2005 to 2014 (Reckendorfer et al. 2013b; Weigelhofer et al. 2015), (3) the development of hydrological and ecological models predicting the further development of the floodplain under different hydrologic scenarios (Baart et al. 2010, 2013; Funk et al. 2013), and (4) a multi-criteria decision study about the effects of a wide range of management scenarios on the various ecological and socioeconomic demands (Hein et al. 2006; Sanon et al. 2012). The following sections show the main procedures and the results of the different approaches.

Historic Development of the Lower Lobau

In 1817, before the onset of the Danube regulation, the total water area in the Danube section of the Lower Lobau covered approximately 11.9 km^2, about 37% of which were formed by side arms and backwaters (Graf et al. 2013). The majority of the secondary water bodies were hydrologically dynamic side arms, with more than 66% belonging to the Eupotamon B habitat (Graf et al. 2013; typology of aquatic habitats according to Hohensinner et al. 2011, based on Amoros 2001). These lotic side arms were branched by vegetated islands and gravel bars. About 20% of the backwaters were classified as Parapotamon A (highly dynamic, connected above mean water level), while periodically connected (Parapotamon B) and isolated backwaters (Plesio-/Palaeopotamon) comprised about 14% of the total water area (excluding the area of the Danube main channel). As a consequence of river regulation, the total water area was reduced to about 4.2 km^2 at mean water level. Dynamic floodplain water bodies were displaced by plesio-/paleopotamal water bodies, accounting for up to 77% of the water area in 2011, while 9% were assigned to Parapotamon B. Only recently, due to the restoration of the most downstream part of the Lobau by lowering of the levee, Parapotamon A backwaters have reoccurred and currently add up to 14% (Graf et al. 2013). As a consequence of channel incision and deposition of sediments in the floodplain, the groundwater tables have declined from 1.9 m in 1849 to 3 m in 2003 at mean water level (Hohensinner et al. 2008). Furthermore, the habitat turnover rates and, accordingly, habitat rejuvenation have decreased dramatically between the pre-regulation and the post-regulation phase (Baart et al. 2013).

The changes in the availability of aquatic habitats significantly affected the macrophyte communities in the Lower Lobau (Baart et al. 2013). The total number of species increased slightly from 108 species before the regulation to 116 species around 2000. Furthermore, the composition of the macrophyte communities changed. Hydrophytes, which are typical for shallow eutrophic, lentic water bodies such as *Nymphoides peltata* and *Nymphaea alba*, were able to spread in the main side arm after the regulation. Additionally, oligotraphent species, such as Charophyceae, were able to colonize paleopotamal water bodies supplied by nutrient-poor groundwater. By contrast, many eu-/parapotamal species, such as pioneer plants, as well as marsh plants have disappeared.

The loss of aquatic and semiaquatic water bodies also decreased the available habitat and abundance of aquatic and amphibian species in the Lower Lobau.

Hydro-chemical Status Quo of the Lower Lobau

In general, the water bodies of the Lower Lobau show significantly lower concentrations of suspended solids and nutrients than the River Danube (Table 25.1, Hein et al. 1999). Thus, hydrological connectivity with the Danube determines the hydro-chemical character of the respective water bodies. Along the main side arm, a distinct hydro-chemical gradient has been established in both the water column and the sediments (Reckendorfer et al. 2013b). With increasing distance to the inflow, the concentrations of geochemical parameters (e.g., conductivity, alkalinity) increase, while the amounts of total suspended solids and dissolved nutrients decrease (Table 25.1). The fine sediment layer is significantly higher in water bodies near the inflow than in more distant parts of the main side arm, yielding a maximum thickness of up to 270 cm close to the first check dam compared to a maximum of 50 cm in distant parts. The organic content of the sediments shows the reverse pattern, with increased concentrations in the upstream parts of the Lower Lobau. During floods, the inflowing river water imports

Table 25.1 Mean concentrations of nutrients (N–NH$_4$, N–NO$_3$, P–PO$_4$, P-tot), chlorophyll-a, and suspended matter (total suspended solids, organic content) of different parts of the Lower Lobau, the Upper Lobau, as well as the Danube and the New Danube (an artificial flood relief channel of the Danube) as potential water sources for reconnection

Parameter		Lower Lobau downstream parts	Upstream parts	Isolated
N–NH$_4$	µg L^{-1}	36.3 ± 163.5 ($n = 234$)	26 ± 101.1 ($n =406$)	92.3 ± 454.9 ($n = 744$)
N–NO3	µg L^{-1}	543.3 ± 617.8 ($n = 234$)	202.1 ± 333.4 ($n = 406$)	287 ± 384.6 ($n = 744$)
P–PO$_4$	µg L^{-1}	6 ± 9.3 ($n = 234$)	1.2 ± 3.2 ($n = 406$)	15 ± 42.1 ($n = 744$)
P-tot	µg L^{-1}	44.5 ± 34.5 ($n = 234$)	16.1 ± 9.5 ($n = 406$)	49.7 ± 81.6 ($n = 744$)
Chlorophyll a	µg L^{-1}	17 ± 17 ($n = 506$)	5.7 ± 6.8 ($n = 745$)	14.5 ± 51.4 ($n = 1706$)
Total suspended solids	mg L^{-1}	27.1 ± 69.3 ($n = 233$)	3.4 ± 4.3 ($n = 404$)	3.8 ± 7.8 ($n = 766$)
Organic suspended solids	%	28.8 ± 16.3 ($n = 233$)	63.6 ± 20.7 ($n = 403$)	80.7 ± 17.5 ($n = 764$)
		Upper Lobau	New Danube	Danube
N–NH$_4$	µg L^{-1}	30.8 ± 25.6 ($n = 63$)	21.2 ± 13.4 ($n = 112$)	51.2 ± 229.9 ($n = 73$)
N–NO$_3$	µg L^{-1}	193 ± 368.4 ($n = 63$)	981.1 ± 548 ($n = 98$)	1537.2 ± 472.5 ($n = 73$)
P–PO$_4$	µg L^{-1}	0.5 ± 0.6 ($n = 63$)	2 ± 4.3 ($n = 98$)	18.2 ± 11.2 ($n = 73$)
P-tot	µg L^{-1}	12 ± 4.7 ($n = 63$)	14.6 ± 7.2 ($n = 98$)	60.5 ± 54 ($n = 73$)
Chlorophyll a	µg L^{-1}	4.2 ± 2.4 ($n = 102$)	7.8 ± 10.3 ($n = 28$)	10.2 ± 10.8 ($n = 161$)
Total suspended solids	mg L^{-1}	1.8 ± 1.2 ($n = 79$)	3 ± 2.5 ($n = 112$)	58.9 ± 119.4 ($n = 83$)
Organic suspended solids	%	87.1 ± 14.8 ($n = 79$)	58.2 ± 20.7 ($n = 89$)	15.7 ± 11 ($n = 83$)

Data obtained from a monthly long-term monitoring of hydrological and ecological properties of the Upper and the Lower Lobau from 2005 to 2014

nutrients from the Danube, thus leading to a hydro-chemical homogenization of the water bodies in the main side arm (Weigelhofer et al. 2015). The nutrient inputs also stimulate primary production, resulting in chlorophyll-a peaks in the connected parts of the main side arm after floods (Schiemer et al. 2006).

Isolated water bodies (Plesio-/Palaeopotamon) in the upstream part of the Lower Lobau have developed their own independent hydro-chemical characteristics (Table 25.1). In general, fine sediment layers are highest there, yielding also the highest organic matter and phosphorus concentrations of all water bodies in the Lobau (Reckendorfer et al. 2013b). Due to their low connectivity, isolated backwaters may partly dry out during periods of low precipitation in summer and winter. The repeated drying and rewetting of the sediments may result in the release of huge amounts of phosphorus from the sediments to the water column (Schoenbrunner et al. 2012), momentarily boosting phytoplankton and zooplankton production in these backwaters (Weigelhofer et al. 2013). Thus, isolated backwaters are usually characterized by high nutrient concentrations, occasional high peaks of phytoplankton biomass, and changes in the nutrient cycles (Welti et al. 2012).

In its present state, 50% of the water area is characterized by shallow mean depths of less than 1 m, and more than 75% of the water area shows maximum water velocities less than 0.1 $m s^{-1}$ (Baart et al. 2010). These conditions favor the establishment of species-rich, macrophyte communities. The highest macrophyte biomass and species richness can be currently found in deeper water bodies of the main side arm (Baart et al. 2010; Reckendorfer et al. 2013b).

Potential Development of the Lower Lobau Under Different Management Scenarios

In order to find an optimal ecologic solution for the Lower Lobau, various hydrological management options have been evaluated regarding their effects on the probable development of the floodplain compared to development without management measures. Here, we present the two most relevant management options to highlight the effects of different degrees of connection of the Lower Lobau with the Danube on the development of the floodplain's ecological state and functions, as well as on habitats. Option 1 ("controlled water supply") includes a limited water supply from the New Danube at the upstream part of the main side arm, comparable to the already existing water enhancement scheme in the Upper Lobau (Table 25.2, Fig. 25.2). Due to restricted water exchange, this option represents a conservative approach. Option 2 ("partial reconnection") includes an uncontrolled reconnection at the upstream part of the main side arm. Though limited in maximum discharge, it represents an approach to restore basic floodplain functions.

Table 25.2 Description of selected potential management options for the Lower Lobau

	Measures	Objectives
Controlled water supply	Water supply with nutrient- and sediment-poor water from the "new Danube" with maximum discharge of 5 m^3/s	Controlled surface water supply, control of trophic state of the system Conserve selected aquatic habitats
Partial reconnection	Uncontrolled upstream connection with the Danube with maximum discharge of 120 m^3/s Morphological adaptions of the affected main side arm necessary	Increase in hydrological dynamics, increase of nutrient inputs from the Danube, increase of ecological integrity of the floodplain

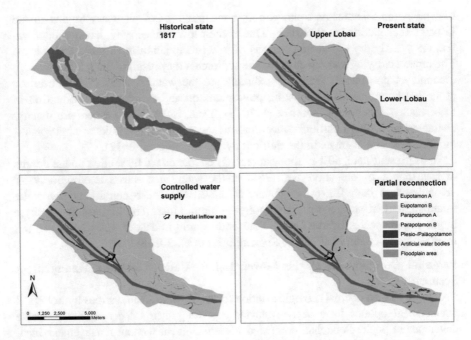

Fig. 25.2 Distribution of aquatic habitats in the Lobau for the historical situation 1817, the present state, and two management options "controlled water supply" and "partial reconnection." Potential inflow areas are marked by arrows

25.4   Methods and Results

The effects of the different management options on the development of the floodplain were assessed via detailed state-of-the-art hydrological and ecological predictive models, which were based on the results of the hydrological and ecological long-term monitoring of the Lower Lobau. In the following section, the most important results are shortly described.

Detailed descriptions of the model structure and setup can be found in Hein et al. (2006), Baart et al. (2010), and Funk et al. (2013).

For the management option "controlled water supply," the hydrological changes are predicted to be small, and lotic water bodies are only created in narrow passages (Fig. 25.2). Nevertheless, the water enhancement scheme will result in a steady transport of water through the system and in a slight increase of the water table and the water areas in ways similar to the development in the Upper Lobau.

By reconnecting the Lobau to the main river channel (management option "partial reconnection"), paleopotamal water bodies will decrease significantly in size, and, at the same time, lotic water bodies will reestablish. The area of habitat types Eupotamon B and Parapotamon A will be increased, and the rejuvenation ability of the floodplain will be enhanced, at least in some areas.

Table 25.3 Modeled values of indicator parameters for the management scenarios "present state," "controlled water supply," and "partial reconnection"

Indicator	Present state	Controlled water supply	Partial reconnection
Total area of connected water bodies (ha)	54.4	99.7	166.6
Share of highly productive water bodies (15 < Chl-a < 30 μg L^{-1}) (ha)	13.2	17.2	33.6
Potential areas for submerged macrophytes (ha)	96.3	100.9	114.4
Potential areas for emerged macrophytes (ha)	81.7	86.7	122.7
Total macrophyte species number	12.3	12.5	13

From the study "OptimaLobau—Optimised management of riverine landscapes based on a multi-criteria decision support system," Austrian Ministry of Science, ProVision 133–113

The management option "controlled water supply" will result in changes of nutrient concentrations and primary production patterns (Weigelhofer et al. 2011) due to the input of nutrient poor water from the New Danube channel and its steady transport through the system. Concurrently, changes in nutrients and primary production will be small in the upper part of the Lower Lobau (Table 25.1). In the lower parts of the Lower Lobau, the enhancement of connectivity will decrease the concentrations, especially after floods; the subsequent chlorophyll-a peak will clearly be reduced.

For the "partial reconnection" option, model results are different, because the main channel of the Danube will be the water source, and much higher amounts of water will be imported, leading to increased nutrient concentrations in the water columns of the whole side arm system. As a result, primary production will be much higher than at present and also higher than in the main channel of the Danube, because of better light availability. Side ditches and currently isolated ponds will be connected to the side arm, increasing the overall area of connected water bodies (Table 25.3) and also the share of highly productive water bodies.

Macrophyte species composition in floodplain areas is indicative of the degree of hydrological dynamics, and restoration measures are, thus, expected to result in alterations in the macrophyte composition (Reckendorfer et al. 2005). Both a controlled water supply and a reconnection will increase the potential habitats for emerged and submerged macrophytes by increasing the total surface water area (Table 25.3). Macrophytes will especially profit from the increase in shallow water areas (Baart et al. 2010, 2013).

In the "partial reconnection" option, the reversal of paleopotamal to parapotamal water bodies will lead to the reemergence of pioneer species and species tolerant to higher flow velocities (Baart et al. 2013). Otherwise, vegetation communities of oligo- and mesotrophic habitats, such as Characeae, which are currently quite abundant in paleopotamal areas of the Lobau, will decline dramatically. While the total number of macrophyte species will barely change, there will be a huge decrease of species-rich areas with paleopotamal character compared to the present state.

Fig. 25.3 Total sum of the normalized weighted usable area (WUA), a measure of the available habitat area, per management option over all selected, protected indicator species (European pond turtle, five species of amphibians, six species of fish, and one invertebrate species). Conservation/ restoration potential compared to the present state (modified after Funk et al. 2013)

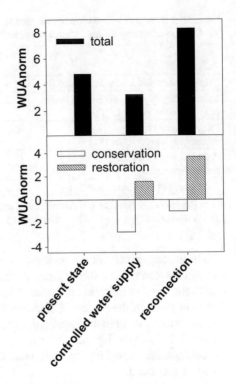

Regarding faunal indicator species, Funk et al. (2013) showed a reduction of available habitat area for selected species (WUA, weighted usable area after Bovee, 1986, Fig. 25.3) for the management option "controlled water supply" due to a decrease of permanent and temporary stagnant habitats. These are important for stagnophilic species, such as amphibians, while at the same time the potential for the creation of lotic habitats for rheophilic species remains low.

For the management option "partial reconnection," the developed model calculated an increase of the useable area. Habitats of stagnophilic species, such as *Emys orbicularis* or amphibian species, are conserved or even fostered due to increased water tables in the area. Due to partial reconnection, potential habitats for rheophilic species (e.g., the fish species *Vimba vimba*, *Romanogobio albipinnatus*, or *Aspius aspius*) are created. Funk et al. (2013) concluded that additionally to the preservation of most of the habitats of the present community, sufficient habitats for rheophilic species will be created by that management option.

Basic Assessment of Trade-Offs Among Ecosystem Functions and Services in the Lower Lobau

In a feasibility study, several potential management measures were developed and assessed for their effects on different ecosystem properties and relevant ecosystem services (Sanon et al. 2012). Multi-criteria decision analysis (MCDA) was applied to support floodplain managers to identify best-compromise solutions for areas of conflicting interests by highlighting possible trade-offs among the various ecological and socioeconomic demands. In order to guarantee a clear demarcation of possible

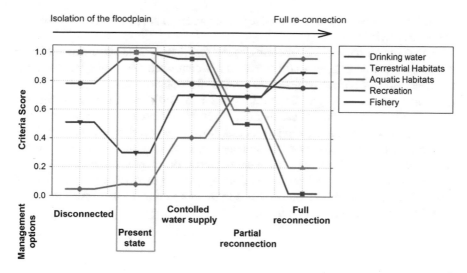

Fig. 25.4 Impacts of different management options on the potential drinking water production, the ecological conditions of terrestrial and aquatic habitats, recreation, and fishery. The x-axis represents hydrological options with increasing connectivity. Present state is circumscribed by the orange rectangle (Sanon et al. 2012 modified)

trade-offs, different hydrologic management scenarios were combined with future-use scenarios with either predominantly ecological or socioeconomic focus. The effects of these hypothetical management options were evaluated through state-of-the-art multi-criteria decision analysis (Sanon et al. 2012).

The results showed that the largest trade-offs occurred between the ecological aims and the aspects related to drinking water supply. An increased connection with the Danube will improve the hydrological and ecological status of the floodplain water bodies considerably through increases in the available aquatic area and the establishment of lotic conditions and will also enable other socioeconomic demands, such as fishery and recreation, within the restrictions given by the national park regulations. In addition, terrestrial habitats will benefit from the more frequent flood inundation that enables the establishment of typical floodplain vegetation. However, an increased hydrological connectivity will probably reduce the groundwater residence time in the area that may impair the potential for drinking water production (Fig. 25.4).

25.5   Conclusions

In the Upper Lobau restoration efforts are limited to a strongly controlled water enhancement scheme. However, it significantly changes nutrient dynamics and primary production patterns of the floodplain water bodies and increases habitat diversity and improves conditions for rheotolerant species at least locally. In summary, the water enhancement scheme in the Upper Lobau demonstrates that controlled management schemes can be successful in urban areas, where it is not possible to reconnect directly to the river main channel, and, therefore, to rehabilitate the natural hydrological regime.

In the Lower Lobau, basic processes, such as nutrient cycling and aquatic primary production, can be managed effectively with a controlled water supply. However, the restoration of typical habitats of dynamic floodplains and the resettlement of rheophilic species need greater efforts to steer toward pre-regulation conditions. Such significant investments are required to regain dynamic water level changes, including floods, and, thus, support natural floodplain rejuvenation. Although the Lower Lobau is not located within a settlement area, restoration is limited by current ecosystem services (including secondary habitat development for rare species, drinking water supply, and recreational uses) that are in conflict with ecological restoration efforts aiming to initiate basic hydromorphological processes that operated during pre-regulation conditions.

References

Amoros C (2001) The concept of habitat diversity between and within ecosystems applied to river side-arm restoration (abstract). Environ Manag 28:805–817

Baart I, Gschöpf C, Blaschke AP, Preiner S, Hein T (2010) Prediction of potential macrophyte development in response to restoration measures in an urban riverine wetland. Aquat Bot 93:153–162

Baart I, Hohensinner S, Zsuffa I, Hein T (2013) Supporting analysis of floodplain restoration options by historical analysis. Environ Sci Policy 34:92–102

Bondar-Kunze E, Preiner S, Schiemer F, Weigelhofer G, Hein T (2009) Effect of enhanced water exchange on ecosystem functions in backwaters of an urban floodplain. Aquat Sci Res Across Bound 71:437–447

Bovee KD (1986) Development and evaluation of habitat suitability criteria for use in the instream flow incremental methodology. Biol Rep 86:235. US Fish and Wildlife Service

Council of the European Communities (1992) Council Directive 92/43/EEC of 21 May 1992 on the conservation of natural habitats and of wild fauna and flora. Off J Eur Communities L206:7–50

Council of the European Communities (2007) Council directive 2007/60/EC of 23 October 2007 on the assessment and management of flood risks. Off J Eur Communities L288:27–34

Funk A, Reckendorfer W, Kucera-Hirzinger V, Raab R, Schiemer F (2009) Aquatic diversity in a former floodplain: remediation in an urban context. Ecol Eng 35:1476–1484

Funk A, Gschoepf C, Blaschke AP, Weigelhofer G, Reckendorfer W (2013) Ecological niche models for the evaluation of management options in an urban floodplain—conservation vs. restoration purposes. Environ Sci Pol 34:1–13

Graf W, Chovanec A, Hohensinner S, Leitner P, Schmidt-Kloiber A, Stubauer I, Waringer J, Ofenboeck G (2013) Macrozoobenthos as an indicator group for the assessment of major rivers incorporating riparian ecological aspects. Österreichische Wasser- und Abfallwirtschaft 65:386–399

Hein T, Heiler G, Pennetzdorfer D, Riedler P, Schagerl M, Schiemer F (1999) The Danube restoration project: functional aspects and planktonic productivity in the floodplain system. Regul Rivers Res Manage 15:259–270

Hein T, Blaschke AP, Haidvogl G, Hohensinner S, Kucera-Hirzinger V, Preiner S, Reiter K, Schuh B, Weigelhofer G, Zsuffa I (2006) Optimised management strategies for the biosphere reserve Lobau, Austria- based on a multi criteria decision support system. Int J Ecohydrol Hydrobiol 6:25–36

Hein T, Schwarz U, Habersack H, Nichersu I, Preiner S, Willby N, Weigelhofer G (2016) Current status and restoration options for floodplains along the Danube River. Sci Total Environ 543:778–790. https://doi.org/10.1016/j.scitotenv.2015.09.073

Henry CP, Amoros C, Roset N (2002) Restoration ecology of riverine wetlands: a 5-year post-operation survey on the Rhone River, France. Ecol Eng 18:543–554

Hohensinner S, Herrnegger M, Blaschke AP, Habereder C, Haidvogl G, Hein T, Jungwirth M, Wei M (2008) Type-specific reference conditions of fluvial landscapes: a search in the past by 3D-reconstruction. Catena 75:200–215

Hohensinner S, Jungwirth M, Muhar S, Schmutz S (2011) Spatio-temporal habitat dynamics in a changing Danube River landscape 1812–2006. River Res Appl 27:939–955

ICPDR—International Commission for the Protection of the Danube River (2009) Danube river basin management plan (DRBMP) according to EU Water Framwork Directive (WFD)

Imhof G, Schiemer F, Janauer A (1992) Dotation Lobau—Begleitendes ökologisches Versuchsprogramm. Artificial Water Enrichment for Lobau—Ecological Test Programme. Österr WasserWirtschaft 44(11–12):289–299

Knoflacher M, Gigler U (2004) A conceptual model about the application of adaptive management for sustainable development. In: Pahl-Wostl C, Schmidt S, Jakeman T (eds) iEMSs 2004 International Congress: "Complexity and integrated resources management". International Environmental Modelling and Software Society, Osnabrueck

McClain ME, Boyer EW, Dent CL, Gergel SE, Grimm NB, Groffman PM, Hart SC, Harvey JW, Johnston CA, Mayorga E, McDowell WH, Pinay G (2003) Biogeochemical hot spots and hot moments at the interface of terrestrial and aquatic ecosystems. Ecosystems 6:301–312

NP Donau-Auen (2009) NP Donau-Auen management plan

Rebelo LM, Johnston R, Hein T, Weigelhofer G, D'Haeyer T, Kone B, Cools J (2013) Challenges to the integration of wetlands into IWRM: the case of the inner Niger Delta (Mali) and the Lobau floodplain (Austria). Environ Sci Policy 34:58–68

Reckendorfer W., R.Schmalfuss, C.Baumgartner, Habersack H., Hohensinner S., Jungwirth M., & Schiemer F. (2005) The integrated river engineering project for the free-flowing Danube in the Austrian alluvial zone national park: framework conditions, decisions process and solutions. Arch Hydrobiol,

Reckendorfer W, Baranyi C, Funk A, Schiemer F (2006) Floodplain restoration by reinforcing hydrological connectivity: expected effects on aquatic mollusc communities. J Appl Ecol 43:474–484

Reckendorfer W, Boettinger M, Funk A, Hein T (2013a) Die Entwicklung der Donau-Auen bei Wien—Ursachen, Auswirkungen und naturschutzfachliche Folgen. Geogr Augustana 13:45–53

Reckendorfer W, Funk A, Gschöpf C, Hein T, Schiemer F (2013b) Aquatic ecosystem functions of an isolated floodplain and their implications for flood retention and management. J Appl Ecol 50:119–128

Sanon S, Hein T, Douven W, Winkler P (2012) Quantifying ecosystem service trade-offs: the case of an urban floodplain in Vienna, Austria. J Environ Manag 111:159–172

Schiemer F (1995) Revitalisierungsmaßnahmen für Augewässer - Möglichkeiten und Grenzen. ArchHydrobiolSuppl 101:383–398

Schiemer F, Baumgartner C, Tockner K (1999) The Danube restoration project: conceptual framework, monitoring program and predictions on hydrologically controlled changes. Reg Rivers Research & Management 15:231–244

Schiemer F, Hein T, Peduzzi P (2006) Hydrological control of system characteristics of floodplain lakes. Ecohydrol Hydrobiol 6:7–18

Schoenbrunner IM, Preiner S, Hein T (2012) Impact of drying and re-flooding of sediment on phosphorus dynamics of river-floodplain systems. Sci Total Environ 432:329–337

Schwarz U (2010) Assessment of the restoration potential along the Danube and main tributaries. For WWF International Danube-Carpathian Programme, Vienna, 58 p

Tockner K, Malard F, Uehlinger U, Ward JV (2002) Nutrients and organic matter in a glacial river-floodplain system (Val Roseg, Switzerland). Limnol Oceanogr 47:266–277

Tockner K, Uehlinger U, Robinson CT (2009) Rivers of Europe. Elsevier

Tockner K, Pusch M, Borchardt D, Lorang MS (2010) Multiple stressors in coupled river-floodplain ecosystems. Freshw Biol 55:135–151

Weigelhofer G, Hein T, Kucera-Hirzinger V, Zornig H, Schiemer F (2011) Hydrological improvement of a former floodplain in an urban area: potential and limits. Ecol Eng 37:1507–1514

Weigelhofer G, Reckendorfer W, Funk A, Hein T (2013) Auenrevitalisierung - Potential und Grenzen am Beispiel der Lobau, Nationalpark Donauauen. Österreichische Wasser- und Abfallwirtschaft2 65:400–407

Weigelhofer G, Preiner S, Funk A, Bondar-Kunze E, Hein T (2015) The hydrochemical response of small and shallow floodplain water bodies to temporary surface water connections with the main river. Freshw Biol 60:781–793

Welti N, Bondar-Kunze E, Tritthart M, Pinay G, Hein T (2012) Nitrogen dynamics in complex Danube River floodplain systems: effects of restoration. River Syst 20:71–85

Chapter 26
Danube Sturgeons: Past and Future

Thomas Friedrich

26.1 Introduction

Sturgeons are an ancient order of fish (Acipenseriformes), dating back in their occurrence to over 200 million years ago. The order comprises two families (Acipenseridae and Polyodontidae) and 27 species. Their natural range is restricted to the northern hemisphere. Sturgeons exhibit a very long life cycle (maximum lifespan up to over 150 years, depending on species). They are late-maturing species, and many grow to very large sizes (up to 6–7 m long). Most of the sturgeon species are anadromous. There are also potamodromous (landlocked) species and forms, spending their entire life cycle in freshwater (Fig. 26.1).

Within their natural range, sturgeons can be considered one of the best indicators for riverine ecosystem health, and their significant decline over the past century poses one of the ultimate challenges for river basin management. Worldwide, many sturgeon species are already considered extinct, highly endangered, or vulnerable, as they are extremely sensitive to a broad selection of anthropogenic impacts. Due to their highly valued caviar and meat, they were heavily overfished in the past, a pressure still continuing up to the present day. Because of their long generation intervals of up to 20 years and their irregular spawning patterns of 2–7 years, this family is extremely sensitive to overexploitation, and the recovery of stocks needs long time periods.

The life cycle of sturgeons includes long spawning migration, ranging between several hundred and several thousand kilometers. Obstacles within the river systems they use, therefore, pose a serious additional threat to sturgeon stocks by preventing them from reaching their spawning grounds. Furthermore, juveniles and spawned

T. Friedrich (✉)
Institute of Hydrobiology and Aquatic Ecosystem Management, University of Natural Resources and Life Sciences, Vienna, Austria
e-mail: thomas.friedrich@boku.ac.at

S. Schmutz, J. Sendzimir (eds.), *Riverine Ecosystem Management*, Aquatic Ecology
Series 8, https://doi.org/10.1007/978-3-319-73250-3_26

Fig. 26.1 A sterlet (Acipenser ruthenus), a pure freshwater sturgeon species, native in the Pontocaspian region. Sturgeons are unique with regard to external appearance and their traits

adults need a wide selection of habitats within the river and have to have the ability to migrate downstream after spawning.

While overexploitation and migration barriers are considered responsible for the diminishing stocks worldwide, additional threats cannot be overlooked. As sturgeons can produce fertile offspring through interspecific hybridization, the introduction of nonnative sturgeon species or genotypes can lead to an outbreeding depression of native stocks, as already to some extent described in Ludwig et al. (2009). Due to their longevity and their benthic habitat, sturgeons are also sensitive to pollution and the effects of accumulated heavy metals in the sediments, which may lead to organ dysfunctions, especially affecting the gonads and reducing fertility (Jarić et al. 2011; Poleksic et al. 2010).

Considering all of these different impacts, it becomes obvious why sturgeons pose such a difficult challenge for management. In the case of the Danube, this challenge is even further complicated, since the Danube is the most international river system worldwide, extending into territories of 19 countries (ICPDR 2015), with sturgeon stocks also using the coastal areas of three additional countries in the Black Sea Region.

26.2 Sturgeon Stocks in the Danube River Basin: A Historic View of Their Development up to the Present

Six sturgeon species are native in the Danube River Basin (Holčík 1989). In the past their stocks played an important economical role as a food resource for the human population in the Danube River Basin (Schmall and Friedrich 2014), and they might even be one driver for early human settlements in the Danube River Basin (Balon 1968). Fisheries for sturgeons are well documented since 3500 BCE (Hochleithner and Gessner 2012; Kirschbaum 2010). Intensive exploitation of sturgeon stocks continued well into the Middle Ages, with the construction of traps and so-called "sturgeon" fences, covering the whole river width as a highly effective method for catching whole migration runs (Schmall and Friedrich 2014). In the early modern times at the beginning of the eighteenth century, the stocks were already so damaged that catching large sturgeons in the Upper Danube and the upstream part of the Middle Danube was considered extraordinary by the inhabitants along the shores of the Danube. Such rare, large sturgeons were reserved for aristocrats and clergy (Balon 1968; Fitzinger and Heckel 1836). Fish market data from the year 1548 suggest a prodigious abundance of sturgeon, e.g., 50,000 kg of fresh sturgeons sold on some days in the Viennese fish market (Krisch 1900). Therefore, large, highly fecund sturgeons became exceptional catches in these Danube reaches within only two centuries (Heckel and Kner 1858; Schmall and Friedrich 2014, Fig. 26.2).

In recent times along the Upper Danube, only very small numbers of one freshwater, sturgeon species, the sterlet (*Acipenser ruthenus*), can still be found. Most sterlet stocks nowadays must be actively sustained by stocking, and the self-sustaining, reproducing stock is considered to be lower than 1000 specimens (Wolfram and Mikschi 2007; Friedrich 2013). This stock is still threatened by habitat degradation and the introduction of allochthonous sturgeon genotypes and species (Friedrich et al. 2014). In the Middle Danube, this species is more numerous but also had a sharp decline after the destruction of an important spawning habitat during the construction of Gabčíkovo Dam (Guti 2008).

The potamodromous ship sturgeon (*Acipenser nudiventris*) is considered functionally extinct (Reinartz and Slavcheva 2016) as only very few single specimens were caught within the last decade in the Middle Danube. There is no program for controlled propagation of the Danube stock, as no brood stock is available. There is an ongoing discussion about a resident form of the Russian sturgeon in the Middle Danube (Heckel and Kner 1858; Holčík et al. 1981; Hensel and Holčík 1997). If such a form existed, it would probably be already extinct, as catches within the last 20 years (Guti 2008) are more likely to be fish that have escaped from hatcheries, rather than indicators of a relict population.

The Russian sturgeon (*Acipenser gueldenstaedtii*), the stellate sturgeon (*Acipenser stellatus*), and the beluga sturgeon (*Huso huso*) are now restricted to the lower part of the Danube, as the construction of the Djerdap I and II dams at the Iron Gate in 1972 and 1984 blocks upstream migration. Within the Danube two different migration types are known to exist for, at least, the beluga sturgeon and the Russian sturgeon. The vernal form migrates and spawns in spring, and the hiemal form migrates during the fall, overwinters in the river, and finally migrates further upstream to spawn during the following spring (Khodorevskaya et al. 2009). All three species are also highly endangered, in particular the long-distance migratory hiemal forms.

Fisheries were regulated and enforced in communist times in the Lower Danube. Post-communism severe poaching and unregulated fisheries increased dramatically in several Lower Danube countries. Although a total ban was implemented by 2005 through the Romanian government for a time period of 10 years and Bulgaria following shortly thereafter, illegal fishery and poaching still pose a very high pressure on such low stocks. An extension of the ban for another 5 years was agreed

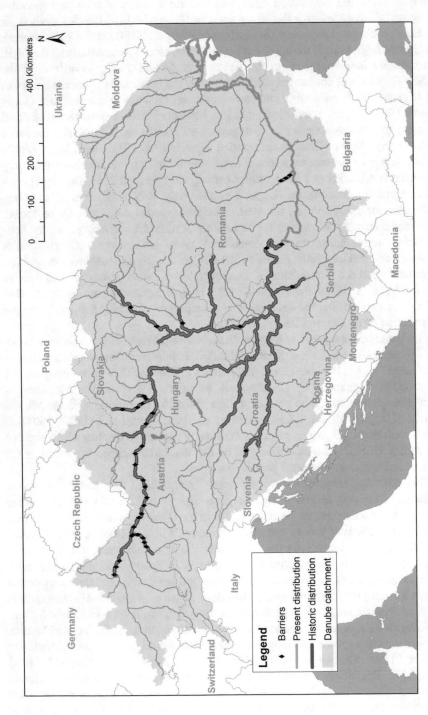

Fig. 26.2 Present and past distribution of anadromous sturgeons in the Danube River Basin

upon in 2015 after long discussions; however, due to their long generation intervals, it is obvious that a recovery of the stocks cannot be accomplished within this short time frame. For all three species, controlled propagation and stocking programs are carried out in the Lower Danube. These species can also be found in hatcheries and aquaculture facilities all over Europe. However, in most cases it is not clear whether the fish belong to Danubian or Caspian genotypes. Consequently, introduction of these fish into natural water bodies has to be considered as a potential risk to the autochthonous species.

Another sturgeon species, the common European sturgeon (*Acipenser sturio*), occurred historically only in the Lower Danube but is considered extinct in the Black Sea Basin (Bacalbasa-Dobrivici and Holčík 2000), its only remaining stock living in the Garonne-Dordogne-Gironde River Basin and, following reintroduction efforts, in the Rhine and Elbe Rivers (Gessner et al. 2010).

26.3 Integrated Approach for Sturgeon Restoration in the Danube River Basin: Sturgeon 2020

In 2005 the *Action plan for conservation of sturgeons (Acipenseridae) in the Danube River Basin* (SAP) was developed under the umbrella of the "Bern Convention" to foster sturgeon conservation (Bloesch et al. 2005). The plan was designed after existing action plans for *A. sturio* in France and Germany, which have been extended and updated over the years (Gessner et al. 2010; Rosenthal et al. 2009).

Although ratified by all Danube countries, the SAP was never truly implemented on a transnational level, with a few exceptions for stand-alone projects, e.g., for the sterlet in Bavaria (Reinartz 2008), despite being supported by a wide range of protective international legislation:

- CITES—Convention on International Trade with Endangered Species
- Bern Convention
- CMS—Convention on the Conservation of Migratory Species of Wild Animals
- CBD—Convention on Biological Diversity
- Ramsar Convention
- BSC—Black Sea Convention on the Protection of the Black Sea
- EU environmental directives
- WFD—Water Framework Directive
- Governmental and nongovernmental organizations (ICPDR, International Commission for Protection of the Danube River; IAD, International Association for Danube Research; WWF, World Wide Fund For Nature; WSCS, World Sturgeon Conservation Society)

To address dramatic declines in Danube sturgeon populations (IUCN 2010), in 2012 the Danube Sturgeon Task Force (DSTF) was established as a network of dedicated volunteers from scientific, governmental, and nongovernmental organizations. The aim was to develop a fully transboundary, overall strategy that would foster synergies between existing organizations to support the conservation of the native sturgeon species in the Danube River Basin and Black Sea. Therefore, the SAP was updated and streamlined into the program "Sturgeon 2020" (Sandu et al. 2013), with support of the EUSDR (European Strategy for the Danube Region). "Sturgeon 2020" aims to foster sturgeon conservation according to the EUSDR target "to ensure viable populations of sturgeon and other indigenous fish species by 2020." It is considered to be a living document since sturgeon conservation depends on long-term commitments by all countries in the Danube River Basin and along the Black Sea, and it requires transnational cooperation between stakeholders, governments, scientists, local communities, and NGOs (Sandu et al. 2013).

A range of different priorities with adjacent, necessary measures have been developed to address the high heterogeneity of different conditions and human impacts within the Danube River Basin and the Black Sea Region (Sandu et al. 2013). Actions envisaged within "Sturgeon 2020" are grouped into six interconnected key topics:

(1) Acquiring political support for sturgeon conservation
(2) Capacity building and law enforcement
(3) In situ conservation
(4) Ex situ conservation
(5) Socioeconomic measures in support of sturgeon conservation
(6) Raising public awareness

The six main actions of "Sturgeon 2020" can be sorted into different groups (e.g., No. 1, acquiring political support, and No. 6, raising public awareness). These are supporting actions, which focus on raising funds, long-term governmental commitment, and political support necessary for the implementation of concrete conservation actions and projects, also to be backed up by matching legislation.

One of the crucial steps for the conservation of sturgeons is to implement mechanisms that help to control and stop poaching in the Lower Danube countries and the Black Sea. As the stocks are still declining and in a critical condition, especially for the Russian sturgeon, as well as the long-distance migrating forms, illegal harvest is still one of the major threats for sturgeons in the Danube River Basin. Actions No. 2 (capacity building and law enforcement) and No. 5 (socioeconomic measures in support of sturgeon conservation) are intended to tackle this problem at both ends. It is necessary to build up the capacities for law enforcement and to design structures to control and combat illegal, unregulated, and undocumented fisheries, as well as illegal and mislabeled trade of sturgeons and their products. A recent study by WWF showed many irregularities in the caviar trade within the Danube River Basin (Jahrl 2013). Therefore, international and national legislation have to be harmonized to give a solid and consistent basis for transboundary enforcement. This process needs to be sustained continuously to overcome the inertia from different interests of various stakeholders as was seen with the tedious process of persuasion for the prolongation of the fishery ban in Romania.

However, due to the difficult economic situation in the Danube Delta, legislation and enforcement alone may only reduce poaching. To totally stop illegal fishery, local communities must be helped to find or develop alternative ways of income. Sturgeon fisheries represented a vital income source for many communities in this area. Even with intensified enforcement, the current fishing bans cannot prevent poaching if they are not accompanied by appropriate compensation measures for local fishermen. Therefore, a two-step effort is necessary. Before increasing any enforcement, first, we must increase awareness of local fishermen to the actual situation of sturgeon stocks in the Danube River and the consequences of ongoing poaching. Poaching will perpetuate the current decline, denying coming generations' access to this resource and preventing reestablishment of a sustainable fishery in the future.

A project by the WWF called "Joint actions to raise awareness on over-exploitation of Danube sturgeons in Romania and Bulgaria," which was funded by the European Commission under the LIFE program (L'Instrument Financier pour l'Environnement), acted as a pioneer for the development of communication and dissemination channels between fishing communities in Romania and Bulgaria and stakeholders working on the conservation of the Danube sturgeons. It also included measures that targeted education of enforcement agencies, as well as actions No. 1 and No. 6 so as to raise public awareness and gain political support for sturgeon conservation. The project ran from 2012 to 2015, and a successor was submitted and granted for the 2016 LIFE period. The second step, to generate alternate ways of income for local communities, has to be politically addressed but can be developed with strong support from other stakeholders of sturgeon conservation. Ideas range from ecotourism to the establishment of sturgeon aquacultures to provide sturgeon products without compromising wild stocks. They even embrace the use of local community knowledge on sturgeon ecology and include them actively in monitoring actions for conservation purposes. However, such innovations would need tight guidance by conservational institutions in order to minimize fraudulent actions that actually sustain poaching while masquerading as conservation.

Actions for in situ conservation target the preservation of the complete sturgeon life cycle as well as protection of its genetic diversity in its natural habitat. To meet such conservation targets, it is necessary to identify, protect, and restore the life cycle and habitats of sturgeons on the Danube River in order to prevent additional habitat alterations and to mitigate existing deficiencies. The major problem in this regard is the present lack of knowledge on habitat use, migration patterns, spawning site selection, and other autoecological traits of this family. These deficiencies make further research with standardized methodologies indispensable. Several local-scale projects currently address this field (e.g., Friedrich et al. 2016; Hontz et al. 2012; Ratschan et al. 2014; Suciu et al. 2015), but the research horizon must be extended to include more Danube stretches (Table 26.1). "Sturgeon 2020" recommends the establishment of a standardized monitoring network, working with the same predefined protocols, which would then be able to document changes in population dynamics as well as in habitat conditions. This could be the basis of an adaptive monitoring plan, with regard to ex situ (see Chap. 27) fishery regulations and actual habitat restoration and reconstruction.

Aside from fundamental research on population dynamics and in situ habitat actions, further observations are necessary to focus on migration and overcoming barriers. Facilitation of fish migration at the two Djerdap dams at the Iron Gates

Table 26.1 Prioritized measures for in situ conservation of sturgeons in the DRB

Species	Species-specific and region-specific requirements			
	Upper Danube	Middle Danube	Lower Danube	Black Sea
Acipenser gueldenstaedtii		**Iron Gate dam passage upstream/ downstream,** reintroduction	**Population analysis, autoecology research, applied in situ measures**	**Population analysis, autoecology research, applied in situ measures**
Acipenser nudiventris		**Population analysis, applied in situ measures**		
Acipenser ruthenus	**Population analysis, autoecology research, applied in situ measures**	**Population analysis, autoecology research, applied in situ measures**	**Population analysis, autoecology research, applied in situ measures**	
Acipenser stellatus		**Iron Gate dam passage upstream/ downstream,** reintroduction	**Population analysis, autoecology research, applied in situ measures**	**Population analysis, autoecology research, applied in situ measures**
Acipenser sturio		*Reintroduction*	*Reintroduction*	*Reintroduction*
Huso huso		**Iron Gate dam passage upstream/ downstream,** reintroduction	**Population analysis, autoecology research, applied in situ measures**	**Population analysis, autoecology research, applied in situ measures**

High priority in bold, medium priority in regular, least priority in italic (after Sandu et al. 2013)

would reopen over 1000 river kilometers of habitat and is one of the key priorities for sturgeon conservation in the Danube River Basin. A scoping mission by the FAO and the ICPDR (Comoglio 2011) led to a pre-feasibility study on flow velocities and sturgeon behavior downstream of the Iron Gate II Dam (Suciu et al. 2015). This project should be accompanied by further studies in the near future on turbine passage and sturgeon behavior on approach to the dam, leading to follow-up studies for both upstream and downstream migration facilities for juveniles and adults.

The construction of fish passages at the Iron Gates is slowly promoted. However, there is ongoing, intense debate about the construction of a submerged sill 550 km downstream. In its original execution, this sill sets up extremely high flow velocities that strongly hamper sturgeon migration or disrupt it entirely (Bloesch 2016). Acoustic telemetry showed that out of 315 tagged individuals, only 10 could pass the unfinished sill (Déak and Matei 2015). Such a tiny fraction cannot maintain a sturgeon population, especially given the other two passages at the Iron Gate dams in the future. As a result of this controversial situation, the construction of the sill has been stopped, and alternatives including decommission of overvalued, built constructions have to be discussed (Bloesch 2016). While in situ conservation of sturgeon populations is the overall goal, in the short- and midterm, accompanying ex situ actions (No. 4) are necessary.

The main idea of ex situ is to establish captive, life cycle units, which should serve as living gene banks. By rearing juvenile fish in a hatchery outside of the river ("off-site") and releasing them into the wild at a later stage, it is possible to stabilize and strengthen populations. However, ex situ measures are not a stand-alone activity, as they cannot be sustainable without in situ measures and therefore can only act as short- to midterm solutions until in situ habitats and populations are reestablished (Reinartz 2015). Conservation hatcheries always have to be linked with the natural population so as to maintain a natural gene pool and guarantee a broad genetic diversity, which is essential for long-term survival of species. Furthermore, seminatural conditions should be provided by the rearing facility to enable the sturgeons to adapt in the wild and to show homing behavior based on water chemistry, nutrition, flow velocities, temperatures, exposure to predators, etc. (Reinartz 2015; Friedrich et al. 2016).

A feasibility study for ex situ measures for Danube sturgeons (Reinartz 2015) and two hatchery manuals of the FAO (Chebanov et al. 2011; Chebanov and Galich 2011) provided the basis for two pilot projects to address the need for ex situ actions in two Danube sections. The LIFE project "LIFE- Sterlet: Restoration of sterlet populations in the Austrian Danube" targets ex situ actions for sterlet in the Upper Danube and applies a technique wherein eggs and juveniles will be reared under seminatural conditions in Danube water with natural diets. The juveniles will be released into suitable areas in several size classes and are partly tagged with external tags and hydroacoustic transmitters to supplement in situ research (Friedrich et al. 2016). In the Lower Danube, the project STURGENE evaluated different facilities for their suitability for two tasks: to keep wild brood stock and/or to raise juveniles for ex situ actions (Reinartz et al. 2016). This survey showed that there are few facilities available for either of the two tasks, and although some hatcheries can be adapted to fulfill at least the role to keep and spawn brood stock, it will be necessary to build new facilities to accomplish for both tasks. To minimize the risk to loose entire conservation units and genetic strains due to catastrophic events, it is necessary to spread them over different facilities (Friedrich et al. 2015). Both LIFE Sterlet and STURGENE are considered pilot projects, and the overall aim within "Sturgeon 2020" is to extend the measures taken within both projects to all species and areas within the Danube River Basin (Table 26.2). The next step is the development of an ex situ hatchery in the Lower Danube for the native species and the securing of funding for the construction and long-term operation of the facility.

26.4 Conclusions

Restoration of sturgeon stocks and habitats within the Danube River Basin can be considered a long-term multitask on a multinational level. The DSTF and its program "Sturgeon 2020," with the support of the ICPDR, the EUSDR, and several ministries, could establish several projects for the conservation of sturgeons in a rather short time frame and within one defined framework, finally implementing the

Table 26.2 Prioritized measures for ex situ conservation of sturgeons in the DRB

Species	Species-specific and region-specific requirements			
	Upper Danube	Middle Danube	Lower Danube	Black Sea
Acipenser gueldenstaedtii		Ex situ measures (after restoration of river continuity at Iron Gate dams)	**Rescue program, extensive ex situ measures**	**Rescue program, extensive ex situ measures**
Acipenser nudiventris		**Rescue program**		
Acipenser ruthenus	**Linking to in situ measures and population assessment**	**Linking to in situ measures and population assessment**	Linking to in situ measures and population assessment	
Acipenser stellatus		Ex situ measures (after restoration of river continuity at Iron Gate dams)	**Linking to in situ measures and population assessment**	**Linking to in situ measures and population assessment**
Acipenser sturio		*Linking to recovery programs in Europe*	*Linking to recovery programs in Europe*	*Linking to recovery programs in Europe*
Huso huso		Ex situ measures (after restoration of river continuity at Iron Gate dams)	**Extensive ex situ measures**	**Extensive ex situ measures**

High priority in bold, medium priority in regular, least priority in italic (after Sandu et al. 2013)

first coordinated actions nearly 10 years after the formulation of the first sturgeon action plan in 2005. However, more political and financial support and coordination of efforts in the Danube River Basin are still necessary. Therefore, an international project under the name MEASURES was envisioned in 2016 to be submitted to the Danube Transnational Programme in order to strengthen the sturgeon network and to implement the next steps within Sturgeon 2020.

References

Bacalbasa-Dobrivici N, Holčík J (2000) Distribution of *Acipenser sturio* L., 1758 in the Black Sea and its watershed. Bol Inst Esp Oceanogr 16:37–41

Balon EK (1968) Einfluß des Fischfanges auf die Fischgemeinschaften der Donau. Arch Hydrobiol 3:228–249

Bloesch J (2016) Major obstacles for Danube sturgeon spawning migration: the Iron Gate dams and the navigation project in the lower Danube. Danube News 33(18):11–13

Bloesch J, Jones T, Reinartz R, Striebel B (eds) (2005) Action plan for the conservation of the sturgeons (Acipenseridae) in the Danube River basin. Convention on the Conservation of European Wildlife and Natural Habitats, Strasbourg

Chebanov M, Galich E (2011) Sturgeon hatchery manual. FAO fisheries and aquaculture technical paper no. 558, Ankara

Chebanov M, Rosenthal H, Gessner J, Van Anrooy R, Doukakis P, Pourkazemi M, Williot P (2011) Sturgeon hatchery practices and management for release—guidelines FAO fisheries and aquaculture technical paper no. 570, Ankara

Comoglio C (2011) FAO scoping mission at iron gates I and II dams (Romania and Serbia). Preliminary assessment of the feasibility for providing free passage to migratory fish species, Mission report May 2011

Déak G, Matei M (2015) Methods, techniques and monitoring results regarding the sturgeon migration on Lower Danube. http://www.afdj.ro/sites/default/files/prezentari/presentation_incdpm_deak_bern_convention_0.pdf

Fitzinger LJ, Heckel J (1836) Monographische Darstellung der Gattung Acipenser. Ann Wien Mus 1:261–326

Friedrich T (2013) Sturgeons in Austrian rivers: historic distribution, current status and potential for their restoration. World sturgeon conservation society, special publication n°5, books on demand, Norderstedt

Friedrich T, Schmall B, Ratschan C, Zauner G (2014) Die Störarten der Donau Teil 3: Sterlet (Acipenser ruthenus) und aktuelle Schutzprojekte im Donauraum. Österreichs Fischerei 67:167–183

Friedrich T, Reinartz R, Peterí A (2015) First screening of facilities and broodstock in captivity with regard to ex- situ conservation of Danube sturgeons. Project report

Friedrich T, Pekarík L, Reinartz R, Ratschan C (2016) Restoration programs for the Sterlet (Acipenser ruthenus) in the upper and middle Danube. Danube News 33(18):4–5

Gessner J, Tautenhaun M, von Nordheim H, Borchers T (2010) German action plan for the conservation and restoration of the European sturgeon (Acipenser sturio). Federal Ministry for the Environment Nature Conservation and Nuclear Safety (BMU), Bonn

Guti G (2008) Past and present status of sturgeons in Hungary and problems involving their conservation. Fundam Appl Limnol Arch Hydrobiol Large Rivers 18:61–79

Heckel J, Kner R (1858) Die Süßwasserfische der östreichischen Monarchie. W. Engelmann, Leipzig

Hensel K, Holčík J (1997) Past and current status of sturgeons in the upper and middle Danube river. Env Biol Fish 48:185–200

Hochleithner M, Gessner J (2012) The sturgeons and paddlefishes of the world – biology and aquaculture. Aquatech Publications, Kitzbühel

Holčík J (1989) The freshwater fishes of Europe, Part II general introduction to fishes/acipenseriformes, vol 1. AULA-Verlag, Wiesbaden

Holčík J, Bastl I, Ertl M, Vranowsky M (1981) Hydrobiology and ichthyology of the Czechoslovak Danube in relation to predicted changes after the construction of the Gabcikovo-Nagymaros River barrage system. Práce Lab Rybar Hydrobiol 3:19–158

Hontz S, Iani M, Paraschiv M, Cristea A, Tănase B, Bâdiliță AM, Deák G, Suciu R (2012) Acoustic telemetry study of movements of adult sturgeon in the Lower Danube River (Km 175–375) during 2011. Book of abstracts, 39th IAD Conference, Szentendre

ICPDR (2015) Danube river basin district management plan—Update 2015

IUCN (2010) http://www.iucnredlist.org/search. Accessed 14 June 2016

Jahrl J (2013) Illegal caviar trade in Bulgaria and Romania, results of a market survey on trade in caviar from sturgeons (Acipenseridae). WWF-Austria\TRAFFIC, Vienna

Jarić I, Višnjić-Jeftić Z, Cvijanović G, Gačić Z, Jovanović L, Skorić S, Lenhardt M (2011) Determination of differential heavy metal and trace element accumulation in liver, gills, intestine and muscle of sterlet (Acipenser ruthenus) from the Danube River in Serbia by ICP-OES. Microchem J 98:77–81

Khodorevskaya G, Ruban GI, Pavlov DS (2009) Behaviour, migrations, distribution and stocks of sturgeons in the Volga-Caspian basin. World sturgeon conservation society, special publication n° 3, books on demand, Norderstedt

Kirschbaum F (2010) Störe—Eine Einführung in Biologie, Systematik, Krankheiten, Wiedereinbürgerung, Wirtschaftliche Bedeutung. Aqualog animalbook GmbH, Rodgau

Krisch A (1900) Der Wiener Fischmarkt. Carl Gerold's Sohn, Wien

Ludwig A, Lippold S, Debus L, Reinartz R (2009) First evidence of hybridization between endangered sterlets (*Acipenser ruthenus*) and exotic Siberian sturgeons (*Acipenser baerii*) in the Danube River. Biol Invasions 11:753–760

Poleksic V, Lenhardt M, Jarić I, Djordjevic D, Gačić Z, Cvijanovic G, Raskovic B (2010) Liver, gills, and skin histopathology and heavy metal content of the Danube sterlet (*Acipenser ruthenus* Linnaeus, 1758). Environ Toxicol Chem 29:515–521

Ratschan C, Zauner G, Jung M (2014) Grundlagen für den Erhalt des Sterlets. Interreg Projekt Bayern—Österreich (J00346). Bericht Projektsphase 2014. I. A. Amt der OÖ. Landesregierung

Reinartz R (2008) Artenhilfsprogramm Sterlet. Projekt 904, Abschlussbericht 2004–2007, I.A. des Landesfischereiverbandes Bayern e.V

Reinartz R (2015) Feasibility study—ex-situ measures for Danube River sturgeons (Acipenseridae). Conducted on behalf of the ICPDR and BOKU within the project "elaboration of pre-requisites for sturgeon conservation in the Danbue River basin"

Reinartz R, Slavcheva P (2016) Saving sturgeons—A global report on their status and suggested conservation strategy. WWF, Vienna

Reinartz R, Peterí A, Friedrich T, Sandu C (2016) Ex-situ conservation for Danube River sturgeons—concept, facts and outlook. Danube News 33(18):6–7

Rosenthal H, Bronzi P, Gessner J, Moreau D, Rochard E (eds) (2009) Action plan for the conservation and restoration of the European sturgeon, vol 152. Council of Europe, Nature and Environment, Strasbourg

Sandu C, Reinartz R, Bloesch J (eds) (2013) Sturgeon 2020: a program for the protection and rehabilitation of Danube sturgeons. Danube Sturgeon Task Force (DSTF) & EU Strategy for the Danube River (EUSDR) Priority Area (PA) 6—Biodiversity

Schmall B, Friedrich T (2014) Das Schicksal der großen Störarten in der Oberen Donau. Denisia 33:423–442

Suciu R, Lenhardt M, Oakland F, Nichersu I, Onără D, Hontz S, Paraschiv M, Holostenco D, Trifanov C, Iani M (2015) Monitoring strategy of sturgeon behaviour to ensure functionality of future fish pass: the iron gate 2 case. Book of abstracts, fish passage 2015, international conference on river connectivity best practices and innovations, Groningen

Wolfram G, Mikschi E (2007) Rote Liste der Fische (Pisces) Österreichs. In: Zulka H (ed) Rote Listen gefährdeter Tiere Österreichs, Grüne Reihe des Lebensministeriums, vol 14/2. Böhlau Verlag, Wien, pp 61–198

Chapter 27
Healthy Fisheries Sustain Society and Ecology in Burkina Faso

Andreas Melcher, Raymond Ouédraogo, Otto Moog, Gabriele Slezak, Moumini Savadogo, and Jan Sendzimir

27.1 Introduction

27.1.1 The Challenges of Sustaining a Fishery in Burkina Faso

Burkina Faso (BF) is a Sahelian country located in West Africa on the arid southern rim of the Sahara. In this region all work and movement revolve around water and its availability, whether in nature or society. The aquatic ecosystems responsible for storing and replenishing the quantity and quality of water are vital for the productivity

A. Melcher (✉)
Institute of Hydrobiology and Aquatic Ecosystem Management, University of Natural Resources and Life Sciences, Vienna, Austria

Centre for Development Research, University of Natural Resources and Life Sciences, Vienna, Austria
e-mail: andreas.melcher@boku.ac.at

R. Ouédraogo
Institute for Environment and Agricultural Research (INERA), Ministry of Higher Education, Scientific Research and Innovation, Ouagadougou, Burkina Faso

O. Moog · J. Sendzimir
Institute of Hydrobiology and Aquatic Ecosystem Management, University of Natural Resources and Life Sciences, Vienna, Austria
e-mail: otto.moog@boku.ac.at; jan.sendzimir@boku.ac.at

G. Slezak
Department of African Studies, University of Vienna, Vienna, Austria
e-mail: gabriele.slezak@univie.ac.at

M. Savadogo
IUCN International Union for Conservation of Nature, IUCN Burkina Faso Office, Ouagadougou, Burkina Faso
e-mail: moumini.savadogo@iucn.org

© The Author(s) 2018
S. Schmutz, J. Sendzimir (eds.), *Riverine Ecosystem Management*, Aquatic Ecology Series 8, https://doi.org/10.1007/978-3-319-73250-3_27

and food security of all flora and fauna as well as all society, whether nomadic herdsmen, sedentary farmers, or urban workers. The rising contribution of fish to provide protein in the diet has only increased the importance of managing fish and the aquatic landscapes they depend on. However, achieving sustainable fisheries is complicated by threats to productivity (more frequent droughts in a drying climate) and consumption (record population growth rates), as well as governance constraints.

Ranked in the bottom 5% of all developing countries (UNDP 2015), the urgent need for sustainable development in BF is challenged by severe socioeconomic and natural constraints. In the latter case, chronic water scarcity is interrupted by episodes of severe drought that continue for multiple years. However non-drought years can scarcely replenish the water deficit because too little of the meager rain that falls is retained long enough to help drive the processes supporting ecosystems and society. During the brief rainy season (4 months on average), precipitation occurs with very high variability in space and time and with such intensity that runs off often exceeds percolation, and groundwater levels are not replenished (Filippi et al. 1990; Wang et al. 2010; Pavelic et al. 2012).

Socioeconomic challenges constrain development with multiple factors: one of the highest population growth rates in a nation that already has seven times as many people as a century ago (INSD 2006) and poor access to either financial or human capital, e.g., almost half the people at poverty level and only 31% of children complete primary school (MEF 2004). As a result, famine is recurrent, and chronic malnutrition affects 44.5% of 5-year-old children and 13% of the women of child-bearing age. This makes food security central to national development policies and strategies (DGPSA 2007). Fisheries are also linked to wider security issues in that political instability and associated terror attacks from northern neighbors, e.g., Mali and Niger, curtail efforts to monitor and manage aquatic ecosystems, especially along the northern tier of BF.

27.1.2 Addressing Those Challenges Through the SUSFISH Project

In 2011 the SUSFISH project funded by the Austrian Development Agency was launched to build the basis in science and policy for sustainably managing natural and man-made aquatic systems in BF. This involved building scientific capacity to monitor and assess the dynamics of ecosystem services (fish, water, self-purification capacity) provided by aquatic ecosystems, the educational capacity to train scientists and technicians in these concepts and methods, and institutional capacities in management and policy formulation, all of which are linked with research and education in the sphere of water and fisheries in BF (Melcher 2015; Ouedraogo et al. 2015; Slezak et al. 2015).

The SUSFISH project established the capacity to manage fisheries by applying the latest scientific methods in a joint partnership between Austrians and Burkinabe.

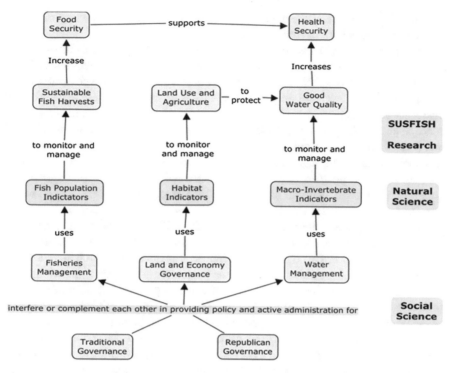

Fig. 27.1 SUSFISH project integration of research and policy goals in support of sustainable management of fisheries and aquatic ecosystems in BF (after Ouedraogo et al. 2015)

Global standards in statistical modeling were applied to rigorously establish bioindicators as management tools that link sets of species (fish and benthic invertebrates) with water quality parameters (Fig. 27.1), implemented by stakeholders involved in the SUSFISH project. However, these sustainability challenges are wider than ecology. For example, artificial aquatic ecosystems like reservoirs are novel landscape elements that alter the social, cultural, and economic features of local communities and their approaches to natural resources management. In Burkina resource users pursue fish conservation with little governmental support (Butterworth et al. 2010), but little is known as to how far local people can assume responsibility to manage waters and fish (Ouedraogo 2010). Fishing communities remain poor despite the emergence of fishing income to complement their many other economic activities. Furthermore, despite research over the past decade, e.g., the Sustainable Fisheries Livelihoods Programme in West Africa (FAO IDAF; e.g., Fabio et al. 2003), much research remains to understand their livelihoods and food security strategies as professional groups, households, and individuals. In addition, development research in Africa has a tragic history of narrowly focusing on technical methods whose use terminates with the end of each project and never become integrated into the policy and practice of society (Raynaut 1997; Batterbury and Warren 2001). Even if the capacity to monitor fisheries does become established, the reasons why fisheries became unproductive and might remain so may originate from a range of sources both natural and social.

To identify and explore the barriers and bridges to that integration, a range of social and systems sciences were applied to examine the effectiveness of "republican," e.g., national, based on the French system, and traditional forms of governance and the potential to harmonize them. In addition, exercises in scenario development allowed managers and planners to explore potential paths of policy development. Finally, conceptual modeling afforded a systems analysis of the ecological, economic, and social factors that can individually or by interaction create opportunities or barriers to sustainable fisheries management. The results were used to make recommendations for fish and water policies, in education (universities and governmental agricultural professional schools), and will have practical relevance for food security and health care (Melcher et al. 2013; Ouedraogo et al. 2014; Sendzimir et al. 2015).

This chapter pursues the central question of the SUSFISH project, e.g., what is the long-term potential to establish sustainable fisheries and aquatic ecosystems on which they depend in BF? SUSFISH embraced this question as a challenge wider than simply transferring the technical means to monitor fisheries but in instilling and awakening a culture of healthy fisheries that is broadly supported across Burkinabe society as well as in government policy. We begin by describing the scientific advances that the project used in application to establish bioindicators and monitoring methods as the technical base of fisheries management. We then consider the many factors, biophysical as well as economic, political, and cultural, which potentially can influence the trajectory of Burkinabe fisheries toward sustainability. We conclude with recommendations based on lessons learned.

27.2 Description of the Aquatic Resources in BF and Its History

BF is located in the heart of West Africa in the sub-Saharan region (12°16′ N, 2°4′ W) (Fig. 27.2). The climate is tropical semiarid, with ambient air temperature averaging around 28.8 °C and temperature extremes vary between 24 and 40 °C (Ly et al. 2013). The region is marked by a south to north gradient of increasing aridity as evapotranspiration (1700–2400 mm/year) exceeds annual precipitation (400–1200 mm) in each of three eco-regions (MECV 2007). Precipitation is extremely variable in space and time over the rainy season (May/June to September), and often such intense bursts run off rather than percolate into the groundwater. As a result, some water courses are intermittent, lying dry and dormant for weeks to months at a time (Ouedraogo 2010).

To meet the growing demand for water amid chronic water scarcity, more than 1400 reservoirs were built since 1950, making Burkina a leading country in water resource development (Ouedraogo 2010) and governance (Niasse et al. 2004). Reservoirs account for 82% of the surface water (SP/CONAGESE 2001) and are

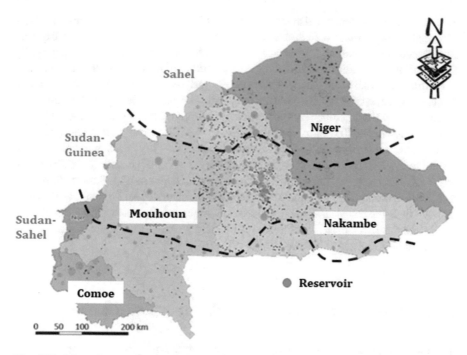

Fig. 27.2 Map of reservoirs and their size in all four river basins of BF, Comoe, Mouhoun, Nakambe and Niger and the three eco-regions, Sahel, Sudan Guinea and Sudan Sahel, reflecting increasing latitude of the country (Adapted from Cecchi et al. 2009)

used mainly for agriculture, livestock breeding, and fishing. Reservoir construction spurred 15-fold growth in fisheries harvests since 1950, employing more than 30,000 fishermen and the several thousand women involved in processing and selling fresh fish (Zerbo et al. 2007). Hence, fishing is an emerging socioeconomic activity that contributes significantly to household livelihoods. Typically a fisherman earns 40% of his livelihood from fishing and the remaining 60% from agriculture (rain fed and irrigated) and livestock breeding (Ouedraogo 2010). A woman fish processor earns about US$17/month (Zerbo et al. 2007), which is also above the absolute poverty threshold.

The fisheries and the aquaculture sectors in BF (MRAH 2013; Stratégie Nationale de Développement Durable de la Pêche et de l'Aquaculture à l'Horizon 2025 "National Strategy for Sustainable Development of Fisheries and Aquaculture") are increasingly productive. The total fish capture in BF is estimated at 12,000 tons and aquaculture at 400 T. The sector employs 32,700 fishermen, 5700 sellers, and 3000 fish processors and contributes to less than 1% to the GDP. BF imports more than 44,400 T of fish annually.

However, high water demands and low management capacity had led to overuse of surface waters. Rising water demand combined with severe sedimentation rates are depleting reservoir water volumes to the point where some reservoirs may disappear within 25 years (Ouedraogo 2010). In addition, rising fishing pressures have resulted in overfishing that depletes fish stocks in terms of total population number, biodiversity, and average fish size. Fish size is a critical indicator of fish reproductive capacity and hence the sustainability of fish populations, which is further threatened by declining water quality as expanding urban centers dispose of waste in rivers and their tributary creeks and canals. This is particularly the case for the nation's capital, Ouagadougou, whose rapid expansion outstrips treatment capacity. Hence, all waste is thrown into the catchment of the second most important river in BF, the Nakambe River, which hosts 40% of the dams and 35–40% of the population. These pressures increase food security risks in river systems (Cook et al. 2009) leading to multiple fish extinctions (Ouedraogo 2010). According to the IUCN Red list settings, 26.3% of the West African freshwater fish species face extinction because of pollution, deforestation, sedimentation, mining, and agriculture (Smith et al. 2009). To establish the basis to track how the status of aquatic resources changes in response to these threats, the SUSFISH project carried out literature and field surveys to assemble historic and current data into a national database (Mano 2016).

27.3 Diversity and Conservation Status of Aquatic Species

The inland waters of Western Africa support a high diversity of aquatic species with high levels of endemism. Estimated numbers of inland water-dependent species by major taxonomic group in West Africa: 563 fishes, 90 mollusks, 287 odonates, and 35 crabs representing 2–5% of global described species (Smith et al. 2009; IUCN 2016). More than 14% of species across the region are currently threatened, and future levels of threats are expected to rise significantly due to a growing population and the corresponding demand of natural resources. Threatened species, comprising critically endangered (endangered and vulnerable), are estimated at 26%, 9%, 18%, and 40% for fishes, mollusks, odonates, and crabs, respectively.

Fish

At the national level, under the SUSFISH project, we collated a total of 152 species from 28 families described so far, from which 145 have been assessed according to the criteria of IUCN Red List (Ouedraogo et al. 2016). The results indicate that 20.7% of the 145 evaluated species are threatened, of which 1.4% are critically endangered, 4.8% are endangered, and 14.5% are vulnerable. These proportions are comparable to the West African evaluation results (26%) from the IUCN Red List (Smith et al. 2009). One key finding is the vital role of protected sites as refuges for most of the threatened species. This highlights the importance of sustaining the health of protected areas, by protecting the surface water streams that irrigate them. Another finding is overall conservation assessment undermined by the large fraction of species with data deficiency (28.3%). Most such species have rarely been seen since their description in the 1960s by Roman (1966), so their conservation status remains uncertain. Finally, given the intensification of the different human-related and climatic pressures, the appeal for more rigorous conservation-related actions is in no way diminished by the relatively high (40.7%) proportion of species of least concern.

Fig. 27.3 Number of exclusive and shared fish species between four main sampling sites (after Meulenbroek 2013)

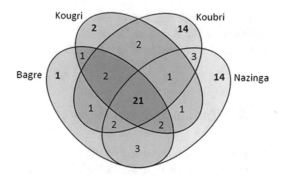

Within the SUSFISH project, 68 species, 41 genera, and 19 families were collected in seven sampling areas in the Upper Volta catchment. Cyprinidae, Alestidae, Mormyridae, and Mochokidae were the most represented families in terms of species and individual number. *Oreochromis niloticus, Tilapia zillii, Brycinus nurse, Clarias anguillaris, Sarotherodon galilaeus*, and *Barbus macrops* were the constant species. One part, species richness increased from downstream to upstream with, respectively, 14 and 2 exclusives species. Fig. 27.3 illustrates how local fish species diversity reflects low (Nazinga, Koubri) and high (Bagre, Kougri) human-induced impacts.

Benthic Invertebrates

The current knowledge of benthic macroinvertebrates in African rivers is comparatively fragmentary and restricted to a few countries, primarily in Eastern and Southern Africa. In West Africa, only a few studies treat the macroinvertebrate fauna, e.g., Aggrey-Fynn et al. (2011) in Ghana, and Camara et al. (2012) and Edia et al. (2013) in Ivory Coast. Some initial efforts have tested the relationship of benthic invertebrate diversity to pollution in Ghana (Thorne and Williams 1997) and in Gabon (Vinson et al. 2008). Recent SUSFISH project publications increased the knowledge base of benthic invertebrates in BF (Trauner et al. 2013; Koblinger and Trauner 2014; Kaboré 2016).

Benthic invertebrate communities in the running waters of BF exhibit a taxonomic gradient that is characterized as follows: insects > mollusks > annelids > crustacea > arachnids. According to Kaboré et al. (2016a), the rivers in the Volta and Comoé catchment are dominated by insects (relative abundance of 95%), represented mostly (80%) by midges and flies (Diptera). The mayflies (Ephemeroptera) and caddies flies (Trichoptera) made up 7% of the abundance. The Plecoptera as well as the Bivalvia, Ostracoda, and Arachnida were found in frequencies lower than 0.5%. With respect to the taxonomical composition, a total of 132 taxa was recorded, and the large majority of these (103 taxa) belonged to 57 families from eight orders of insects.

In a study on the water beetles of BF, Kaboré et al. (2016b), a total of 38 species of diving beetles (11 Noteridae and 27 species of Dytiscidae) and 22 species of water scavenger beetles Hydrophilidae could be detected from 18 lentic and lotic water bodies. The fact that out of these 60 water beetle species 24 species have been reported for the first time in BF indicates clearly that the taxonomic knowledge of benthic invertebrates is at its infancy in BF.

27.4 Human-Induced Impacts

New knowledge generated by SUSFISH field research serves as a foundational database that can inform the formulation and implementation of policy for managing aquatic resources. It provides benchmark data from which to measure progress and set target performance levels for policy and practice. For example, not only has SUSFISH research generated the most current and comprehensive species lists for BF, it has identified the significance of the relative scarcity of some of its species: a significant fraction (56%) of fish species in Burkina is threatened. SUSFISH has established the data basis to identify the multiple sources of those threats and quantify their impacts on aquatic species. Broadly, in BF the presence, diversity, trophic level, density, and biomass of certain fish and benthic invertebrate genera and species respond negatively to a range of anthropogenic pressures (Ouedraogo 2010; Melcher et al. 2012; Stranzl 2014; Kaboré et al. 2016a; Kaboré 2016; Mano 2016).

Using Biological Indicators to Distinguish Impacted and Nonimpacted Areas
Approaches that follow Moog and Stubauer (2003), Nijboer et al. (2004), or Pont et al. (2006) to identify "a priori criteria" from distinct areas were applied in BF to understand human impacts on aquatic resources and to describe reference conditions based on physicochemical features, hydro-morphology and in-stream structures, and land use. We found that protected areas can reasonably be considered as credible reference sites as far as they show low impact levels. Benthic invertebrates as well as fish taxa respond not only to threats and pressures but also to landscape and habitat parameters. As such, certain genera and species can be useful as bioindicators of water body typology and river morphology and structure in BF catchments as well as land use-land cover and habitat-type parameters. Kaboré (2016) investigated the benthic macroinvertebrate communities of 66 areas at Sahel Rivers and found that a multimeric index approach could be developed to assess the ecological quality of running water bodies.

The practical implications of SUSFISH research are that it provides specific knowledge about the sensitivity of certain fish and benthic invertebrate taxa to specific pressures and/or clusters of pressures that offer the data basis for monitoring the presence and impacts of pressures. Overall, SUSFISH surveys demonstrate that such parameters as fish size, abundance, and diversity are related to the quality of fisheries and habitat management. Therefore, both fisheries and water resources can be better managed based on science that rigorously monitors and manages multiple levels: the aquatic taxa, the water column, the habitat quality and surrounding land uses, and the human activities that generate pressures impacting these aquatic and terrestrial habitats. SUSFISH data indicate that fish management must be informed

Fig. 27.4 Increasing number of human-induced impacts (Global pressure index) show the dramatic decline of the number of fish species for each of the main sampling sites in terms of number of fish species (after Stranzl 2014)

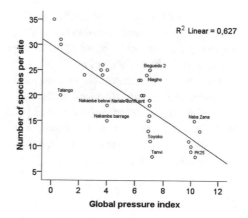

by data at scales larger than landscape, since fish biodiversity is related to their mobility and, ultimately, to water network connectivity (Ouedraogo 2010; Melcher et al. 2012; Stranzl 2014; Kaboré et al. 2016a; Kaboré 2016; Mano 2016).

In general, fish taxa exhibit lower numbers and body mass as human-induced impacts increase (Fig. 27.4). More specifically, fish taxa can be used as bioindicators of the impacts of anthropogenic pressures either in the positive or negative sense. For example, Stranzl (2014) and Mano (2016) concluded that *Auchenoglanis* gen. and *Hydrocynus* gen. could be used as sentinel genera because they are not found in areas of high anthropogenic impact. *Clarias* sp. and *Sarotherodon* sp. increase with pressures, unlike other species, e.g., *Alestes* sp. and *Schilbe* sp., which are sensitive and decrease in number as pressures rise. *Hemichromis* and, especially, *Tilapia* correlate positively with hydro-morphological pressure, but they respond negatively to chemical impacts.

27.5   Adaptive Governance of BF Fisheries

The SUSFISH project took the first concrete steps toward establishing a governance framework that can adaptively reestablish and manage sustainable fisheries in BF. These steps provided first the tools to identify the status of aquatic ecosystems in general and their component fish and benthic invertebrate fauna in particular. Application of these tools demonstrated that these fauna could be used as reliable bioindicators of aquatic ecosystem conditions and anthropogenic pressures impacting those ecosystems. Therefore, the first phase of SUSFISH proved that such tools can help to measure progress toward sustainability in managing fisheries.

The second phase of SUSFISH worked to identify which factors in nature and society help or hinder the said progress toward food security in BF. That means that beyond the technical tools, what needs to be done to establish a "shared understanding of sustainable fisheries in society" in BF? Surveys were conducted at two levels. In the field, practitioners whose livelihoods depended on the fisheries food chain in BF were interviewed. In work centers of science and policy in Ouagadougou and Bobo-Dioulasso, expert scientists and policy makers met in workshops to identify what scenarios of future development of aquatic ecosystems are of concern to management and to explore the variables and relationships that might determine the trajectory of these scenarios.

27.5.1 Scenarios of Declining Fishery Sustainability

Most of the potential pathways of fishery decline envisioned by experts resulted from interactions between biophysical factors. The salient trajectory of concern is that overfishing has driven fisheries into a trap that only extreme policies, e.g., prolonged shutdown of fishing, could liberate it from. In this scenario, overfishing drives down the average size of adult fish and, hence, the productivity of the whole fishery, since egg production is proportional to body mass. Sustained overfishing of an underproductive fishery could eliminate any chance of rebounding to its original productive potential. A number of scenarios explore potential consequences of excessive or abusive use of resources and habitats in and around aquatic ecosystems. Of particular concern is water extraction due to (gold) mining, which often is not strictly regulated because of its elite status as the main source (48%) of export earnings (Ouedraogo 2010; IMF 2014). At the end of the dry season, water volumes can decline below thresholds critical to fish capacity to survive and reproduce in the wet season.

The most ubiquitous human pressure is agriculture. It is the national economic mainstay found in every catchment whether in the arid Sahelian steppe in the north (animal husbandry) to the Northern Sudanian savanna that exists as a band across the nation's center [cotton, groundnuts, karite (shea nuts), and sesame] to the moister Southern Sudanian savanna in the southwest (sugar). Both agriculture and mining practices can degrade habitats in and around aquatic ecosystems, reducing the capacities to filter runoff or provide services that sustain biodiversity. Not only does agriculture drive water demand, it also utilizes pesticides and fertilizers that can bioaccumulate in aquatic ecosystems and degrade productivity (especially the reproductive capacity of commercial fish) or shift the balance of phytoplankton to species, e.g., cyanobacteria, unpalatable to fish or increase the proliferation of macrophytes and thereby reduce the capacity to catch fish. However, in the latter case, macrophyte proliferation might also help reduce overfishing and provide more nursery habitat in aquatic ecosystems.

Evidence from field research makes several of the scenarios associated with habitat degradation appear likely. Increasing intensity of anthropogenic pressures leads to declines in (1) diversity, (2) some (potential indicator) species and also families, (3) trophic level (the trophic level dropped from 3 to 2.5 at highly impacted sites.) based on a trophic level scale by (Pauly et al. 2000), and (4) density and biomass of intolerant species. The number of fish genera was found to decline with the number of pressures. Anthropogenic pressures on fisheries mostly occur in multiple forms (clusters) and correlate among each other to create "cumulative effects." Agricultural pressures were present in 87% of our sites. 70% of all sites exhibited bad water quality (expert opinion). 50% had connectivity pressures (GIS); stream morphology is mostly still ok (SS). In the Nakambe catchment, only 13% of the area remained as natural vegetation. 76% is cultivated land, and 11% is bare soil, and water holding capacity has decreased 33% in 30 years.

Two scenarios explore how the productive potential for reservoir fisheries can be limited by water volume lost to sedimentation or degradation of infrastructure, e.g., dikes and spillways. Lower water volumes have less capacity to buffer temperature increases or inputs of toxic chemicals and are more likely to cause fisheries productivity to decline. Productivity could be limited also if fish population declines due to harvesting could not be replenished by migration of fish from regional meta-populations. Fish migrations could be blocked when regional hydrographic networks of rivers are disrupted by overbuilding of dams without infrastructure to allow fish to bypass. This latter scenario is supported by evidence showing that study sites that are under anthropogenic pressure but have intact connectivity, e.g., fish can migrate to other parts of the water network, have a higher diversity than fragmented ones. This may be related to our observation that about 50% of all caught species are potamodromous and therefore normally migrate for spawning. Connectivity is severely impacted by hydrograph modification and reservoir dams. More than 90% of the annual discharge in the Nakambe basin is held back by dams, causing massive hydrographical modification. Approximately 89% of all sites in this study were regarded as under influence of hydrographical changes.

Several scenarios illustrate the potential for social, cultural, and political factors to limit fisheries productivity. One posits that fish harvests and habitat maintenance are substandard because most fishermen have inadequate skill sets. There may have been insufficient time to develop institutions for training, since fishing has only become common in recent decades. In our field data, learning fishing methods over generations contextualized in the environmental challenges was found only in a few cases. In addition, fisheries may not be productive enough to support enough fishermen full time, so fishing is mostly a part-time job for farmers and herdsmen. Another scenario posits that indigenous fisheries have been neglected because international markets have very quickly filled the majority (80%) of national demand. Similarly, failure to develop functioning institutions at all levels may arise when neglect from the national level and decline at the local level erode any trust that fisheries governance can successfully be established.

27.5.2 Socioeconomic Factors Influencing Fishery Sustainability

As previously noted, SUSFISH was founded in recognition of the history of failure of development projects grounded only on technical and/or scientific advances. In BF alone we found examples of failure to adopt modernizing technology in the form of abandoned equipment and infrastructure (fish ponds, refrigerators, fish weighing scales, fish shops). For that reason SUSFISH sponsored research into the social, economic, and political barriers and bridges the gap to sustainable fisheries. Our research indicates that, while some encouraging examples exist, there are abundant barriers to sustainable fisheries provided by challenges of governance at multiple levels in BF.

Our results confirm that fish size and fish community diversity are associated with the degree and quality of management, both of fisheries directly and of the habitat surrounding the fishery. For example, the Nazinga site has relatively unimpacted habitat (land management) and has a closed fishing season that is well-regulated (fisheries management), and it has significantly larger fish and a higher share in adult fish than any other sampling site. Furthermore, both fish and BMI diversity are higher in protected areas, e.g., Nazinga, Mare aux Hippo than in others. Just as important to biodiversity conservation as the legal content of policy is its execution at the appropriate level (subsidiarity) by well-organized and led local actors. For example, in Moussodougou the fisheries are directly controlled by a local association that effectively enforces rules. Situations with effective management are associated with an increased biomasses and abundance with large fish specimens.

A prominent overarching challenge is that it appears that fisheries management is not equally applied all over BF but is concentrated in a few large reservoirs of "national economic interest," e.g., Ziga, Bagre, Kompienga, and Sourou. Fishing in these large water bodies is dominated by commercial fishermen, who are regulated and in good communication with government officers. However, elsewhere in BF communication is not so good for management organs devoted to smaller reservoirs, except for four fishing concessions given to the associations of the local fishermen (Bapla, Moussodougou, Tandjari, and Lera). Aside from these few examples of successful organization of local management capacity, for the most part, there are gaps between national and lower levels of governance. Briefly, a governance system that effectively functions from the central, national level out to the regional and local levels has yet to be established. Often the link between law and practices to monitor fisheries is missing (law is not adopted in practice). Therefore there is little effective police monitoring or enforcement of fishing practices at the lower levels, e.g., smaller-scale fisheries.

While efforts to decentralize management authority have been underway for years, the failure to comprehensively bridge institutions from national to local levels is hampered by the frequency of shifts of governance responsibilities (institutional nomadism) for fisheries management at the national level. One salient example of the poor communication that results from such "administrative flux" is the general lack of expertise that is regionally or locally available for fishermen who need expert consultation. As a result, progress in improving fishing methods is blocked by lack of capacity to learn or to organize.

In the face of poor inter-level communication and sporadic or absent monitoring, the use of illegal equipment and fishing practices only mounts. It is hard to imagine how trust in governance can be built to strengthen compliance with laws and policies under such conditions, and evidence of this eroding trust is that in some areas local fishermen have swung their allegiance from republican to traditional authorities. Traditional authorities still constitute legitimate local sources of governance. Traditional institutions play a vital role in reaffirming the identity of communities reliant on aquatic ecosystems and thereby broadly influence water and fish management. However, the current governance structure has failed to link and harmonize republican and traditional sources. And efforts to decentralize have been poorly implemented, e.g., local management committees lack the funding to even meet regularly, or have been taken over by special economic or political interests, i.e., elite capture.

The governmental bodies responsible for the fisheries sector are unaware of women's specific role in the fisheries management. Consequently they did not consider females adequately as crucial actors in their strategic and political programs. SUSFISHs sociological research on fish as an important income-generating resource shed light on women as important preserving stakeholders (e.g., *system-preserving functions*) in the economic, nutrition, and health domains. Interdisciplinary work revealed important cross-sectoral activities, interrelated power relations, and hindering factors that play key roles in the value chain issue of the resource fish in BF. However, notwithstanding their important economic role, since women are excluded from decision-making processes on local levels, the focus of future analysis should be oriented toward the impeding factors emerging from incomplete or misguided education, structures of associations, and power asymmetries. These findings resulted in the draft of a strategy for the integration of these aspects in the fisheries management policies, which was developed in SUSFISH.

By law there are two kinds of status of fisheries based on management type: concession and PHIE (Perimetre Halieutique d'Intérêt Economique), i.e., a fishery that impacts the national economy. But actually, there are three categories of fisheries management: very large reservoirs that never dry out and fishing continues for much of the year [PHIE; Bagré, Kompienga, Sourou, Ziga, Toécé, Douna, Yakouta, Sirba (eight reservoirs) and Lake Bam] and concessions and "others" that have no legal status. Note: subsistence fisheries exist in all three categories mentioned above. Besides this, by law there are four categories of fisheries defined according to the use of the catch: commercial, subsistence, sport, and scientific. According to SUSFISH findings, this categorization does not reflect the status quo. Besides subsistence fisheries all other categories are not represented significantly.

Into what category a fishery falls depends on how national policies are prioritized—the PHIE are "nationally important" reservoirs, whereas the latter are more "subsistence level" fishing for local markets (Fig. 27.5). For the former, management is organized at a professional level: most of the fishermen involved are professionals

532   A. Melcher et al.

Fig. 27.5 Sustainable fishing in the protected area of Nazinga, close to the border to Ghana (Photo: A. Melcher)

and are aware of the legislation or rules. However, there is a link between education and fisheries management. As illiteracy is common among fishermen and fish processors, at a lower, more local, level awareness of regulations as well as access to information on improving fishing methods can be blocked. At the local levels (subsistence fisheries) when officials ignore rather than engage (no monitoring or enforcement) local problems, the lack of engagement gives local fishermen no opportunity to learn about fisheries policies through responding to them. Thus they remain largely uninformed about legislation and administrative policy. They do know some of the rules, but they rarely if ever observe them. Lower level agents perceive no functional links (e.g., communication) that tie GDFA staff with any of the decentralized lower level government layers. Therefore, most local fishermen are unaware of the GDFA or what its functions are. National legislation and policies are not known by any of the decentralized lower level government layers. In closing we must caution that this data was gathered during a transitional period (shifting responsibilities between people in different organizations) such that anyone interviewed would not claim responsibility.

27.6   Recommendations Based on Lessons Learned

Management of fisheries requires policies in place at the appropriate governance levels as well as reference data against which to compare policy performance over the long term. Steps must be taken to fill two gaps that prevent such management. First, there are no plans for almost all (more than 1000) minor fisheries, and only a few of the major ones have a management plan. Second, the means (equipment, training, and protocols) must be put in place and applied in as many fisheries as possible. The means to establish monitoring programs for reservoir and river fisheries all over BF have come from SUSFISH project's provision of equipment and training for surveying and monitoring as well as field data that confirm that fish and benthic macroinvertebrates can be used as bioindicators of conditions in and around aquatic ecosystems. On this basis, managers can monitor the status of fisheries and surrounding habitats as well as trends in anthropogenic impacts on aquatic fauna. Education programs for students and government agents need to be established and integrated with university research programs.

In BF, there are a number of ways to improve the governance of aquatic resources, especially fisheries, and the sectors of society and surrounding landscapes that impact fisheries (Fig. 27.6). At the national level, efforts should be made to forge a vision of sustainable fisheries that is communicated to and made workable at all levels of society and governance. National level policy makers have failed to make a national future perspective of fisheries operative either at the federal, regional, or local levels. A national "vision" has been recorded, but it remains theory on paper, not a practical, working vision that informs policy and implementation. Failure to lead and provide a unifying vision means that local as well as regional and national actors have no paradigm of fisheries development to rally around and use as a baseline against which to measure policy performance. While indicators are lacking, evidence of failure accumulates. Infrastructure (fish ponds, refrigerators, fish weighing scales, fish shops) has been installed but either not maintained and allowed to decay or converted for private use. Such patent failures undermine trust in the government institutions that establish and execute programs, projects, and policies to develop infrastructure.

National governance of fisheries could be improved by harmonizing policies of different ministries such that their implementation reinforces each other. Fisheries sustainability depends on integration of policies governing a range of diverse activities, e.g., water, agriculture, forests, mining, and tourism. Institutional nomadism,

Fig. 27.6 Sustainable fishing in the very dry north of Burkina at the swamps of Dori, close to the border to Mali (Photo: A. Melcher)

the unpredictable shifting of responsibilities between ministries, contributes to this frequent failure to create compatible policies or integrate them. It also promotes interagency conflicts (e.g., agricultural staff vs. foresters) by keeping the boundaries of agency responsibility vague and undefined. Without a clear mandate, field officers are not inclined to monitor fisheries. Some fishermen agree and claim such responsibility shifts prevented surveillance that would have hindered them from using smaller net mesh sizes, a trend driving down fish size. The implications of such unpredictable shifts in responsibility are general feelings of frustration, helplessness, and a lack of trust in governance, which are sustained when such nomadism perpetuates a state of ignorance by those officially responsible for management.

Better harmony can be achieved between national and local levels if local knowledge is better integrated in forming and implementing the visions and resulting policies for sustainable fisheries. So far such input has been blocked by lack of local capacity to constructively participate. The reasons for this lack of capacity are multiple: lack of funding to support participation, lack of experience or training in participation (people), and lack of effective processes of participative democracy (governance). This lack of local organizing capacity hinders efforts for bottom-up leadership in fisheries management to fill the gap left by failure from the top, the national level. Very few, if any local sustainability initiatives are started, is often blocked by the perception that they lack resources, usually money. This lack of inclusiveness is reinforced by chronic failure to evaluate policies periodically and improve them. Such periodic policy review requires rigorous measurement of policy impacts and could identify where lack of local involvement harms policy creation or administration. Also such periodic review could be done with mutual contributions from national and local levels, such that policy reform reflects experience from different governance levels.

In the vacuum between national and local governance levels, such governance breakdowns as elite capture are especially harmful and result in resistance of local communities to new regulations. Furthermore, this vacuum isolates and hinders efforts to harmonize the "republican," e.g., European-based democratic, with the traditional institutions. Both make relatively important contributions to the governance of water and fish resources. Despite the ongoing development and refinement of Republican institutions, there has been some resurgence of traditional practices (sacrifice by fishermen, traditional fishing). However, this trend is not universal and in some cases the opposite is true. Traditional stress [conflict] management seems to change or slip away leaving some room for modern management. Traditional institutions reinforce the sense of aquatic ecosystems as "sacred space" at the heart of the life of surrounding communities. "Water bodies occupy an important place in the history of the study area. They are loaded with symbolism and contain very often places of worship."

This failure to harmonize policies across different governance levels is linked to or ineffective efforts to decentralize power to regions while maintaining a functioning governance structure that works across all levels. When officials choose to ignore rather than engage local problems, the lack of engagement gives local fishermen no opportunity to learn about fisheries policies through responding to them. Thus they remain ignorant of legislation administrative policy and other matters in which knowledge is required for proper management. However, this is also a question of adequate training modules for the target groups of managers and practitioners. For instance, some training courses have been developed, but they are not specialized for fishing. Rather they are framed for agriculture in general with a very few sessions for fishing and aquaculture. Furthermore, this training is not generally available across all BF and not comprehensive or detailed enough to really make a difference.

Long-term sustainability of ecosystems and society in BF is challenged by population growth rates that exceed economic development and environmental carrying capacity. Lowering population growth rates requires the second phase of the demographic transition, which in turn depends on elevating the capacity of women in education and business. The elaboration of a strategy for the integration of a gender-sensitive approach in projects and policies for fisheries will support the participation of female actors in the sector. But as findings showed, the consideration of women's needs and strategic interests is often linked to other policy sectors such as health, nutrition, water management, and education (Fig. 27.7). The coordination and integration of research findings and policy recommendations between the involved ministries are crucial for the development of a comprehensive development plan.

Fig. 27.7 Sustainable fisheries management needs a lot of women power, especially in processing, marketing, and trading fish, like at the market of Koubri close to the capital city Ouagadougou (Photo: A. Melcher)

References

Aggrey-Fynn J, Galyuon I, Aheto DW, Okyere I (2011) Assessment of the environmental conditions and benthic macroinvertebrate communities in two coastal lagoons in Ghana. Ann Biol Res 2(5):413–424

Batterbury S, Warren A (2001) The African Sahel 25 years after the great drought: assessing progress towards new agendas and approaches. Glob Environ Chang 11(1):1–8

Butterworth DS, Bentley N, De Oliveira JA, Donovan GP, Kell LT, Parma AM, Punt AE, Sainsbury KJ, Smith ADM, Stokes TK (2010) Purported flaws in management strategy evaluation: basic problems or misinterpretations? ICES J Mar Sci 67(3):567–574

Camara IA, Diomande D, Bony YK, Ouattara A, Franquet E, Gourène G (2012) Diversity assessment of benthic macroinvertebrate communities in Banco National Park (Banco Stream, Côte d'Ivoire). Afr J Ecol 50:205–217

Cecchi P, Gourdin F, Kone S, Corbin S, Etienne J, Casenave A (2009) Les petits barrages du Nord de la Côte d'Ivoire: Inventaire et potentialités hydrologiques. Sécheresse 20(11):112–122

Cook SE, Fisher MJ, Giordano M, Andersson MS, Rubiano J (2009) Water, food and livelihoods in river basins. Water Int 34:13–29

DGPSA (2007) Production d'un atlas dynamique sur la sécurité alimentaire du Burkina Faso. Document de projet. Direction Générale des Prévisions et des Statistiques Agricoles. Ministère de l'Agriculture de l'Hydraulique et des Ressources Halieutiques, Février 2007, 8 p

Edia OE, Bony KY, Konan KF, Ouattara A, Gourène G (2013) Distribution of aquatic insects among four costal river habitats (Côte d'Ivoire, West-Africa). Life Sci 2(8):68–77

Fabio P, Braimah LI, Bortey A, Wadzah N, Cromwell A, Dacosta M, Seghieri C, Salvati N (2003) Poverty profile of riverine communities of southern lake Volta. Sustainable Fisheries Livelihoods Programme in West Africa, Cotonou, 70 p, SFLP/FR/18

Filippi C, Milville F, Thiéry D (1990) Evaluation de la recharge naturelle des aquifères en climat Soudano-Sahelien par modélisation hydrologique globale: application a dix sites au Burkina Faso. Hydrol Sci J 35(1):29–48

IMF (2014) Burkina Faso. Selected issues. IMF (International Monetary Fund) country report no. 14/230. July 2014. 35 p

INSD (2006) Institut National de la Statistique et de la Démographie. Statistiques de l'environnement. Available online http://www.insd.bf/fr/

IUCN (2016) IUCN red list status. Available online http://www.iucnredlist.org/initiatives/freshwa ter/westafrica/rlwa

Kaboré I (2016) Benthic invertebrate assemblages and assessment of ecological status of water bodies in the Sahelo-Sudanian area (Burkina Faso, West Africa). Doctoral thesis, Doctoral thesise, Institute of Hydrobiology and Aquatic Ecosystem Management (IHG), University of Natural Resources and Life Sciences, Vienna, Austria (BOKU), p 242

Kaboré I, Moog O, Alp M, Guenda W, Koblinger T, Mano K, Ouéda A, Ouedraogo R, Trauner D, Melcher A (2016a) Using macro-invertebrates for ecosystem health assessment in semi-arid streams of Burkina Faso. Hydrobiologia 766(1):57–74

Kaboré I, Jäch MA, Ouéda A, Moog O, Guenda W, Melcher A (2016b) Dytiscidae, Noteridae and Hydrophilidae of semi-arid waterbodies in Burkina Faso: species inventory, diversity and ecological notes. J Biodivers Environ Sci 8(4):1–14

Koblinger S, Trauner D (2014) Benthic invertebrate assemblages in water bodies of Burkina Faso. Master thesis, Institute of Hydrobiology and Aquatic Ecosystem Management (IHG), University of Natural Resources and Life Sciences, Vienna, Austria (BOKU), p 156

Ly M, Traore SB, Agali A, Sarr B (2013) Evolution of some observed climate extremes in the West African Sahel. Weather Clim Extrem 1:19–25

Mano K (2016) Fish assemblages and fish based assessment of the ecological integrity of river networks in Burkina Faso. Doctoral thesis, Institute of Hydrobiology and Aquatic Ecosystem Management (IHG), University of Natural Resources and Life Sciences, Vienna Austria (BOKU), p 244

MECV (2007) Rapport de l'inventaire National des sources de production, d'utilisations et de rejets du mercure dans l'environnement au Burkina Faso Eds: Dir. Gen. Env. Cadre de Vie, Burkina Faso, 84 p

MEF (2004) Politique nationale de population au Burkina Faso. Conseil National de Population, Ministère de l'Economie et des Finances, Burkina Faso

Melcher A (2015) Sustainable management of water and fish resources in Burkina Faso. The project—SUSFISH. In: Obrecht AJ (ed) APPEAR. Participative knowledge production through transnational and transcultural academic cooperation. Böhlau, Wien, p 319. ISBN 978-3-205-79690

Melcher A, Ouedraogo R, Schmutz S (2012) Spatial and seasonal fish community patterns in impacted and protected semi-arid rivers of Burkina Faso. Ecol Eng 48:117–129

Melcher A, Ouedraogo R, Savadogo M, Sendzimir J (2013) Current questions in sustainable water management and higher education in Burkina Faso. The SUSFISH consortium book of abstracts, West Africa Symposium, Vienna, Austria, 23–28 June 2013. https://doi.org/10.13140/2.1.1970.8967

Meulenbroek P (2013) Fish assemblages and habitat use in the Upper Nakambe Catchment, Burkina Faso. Master thesis, Institute of Hydrobiology and Aquatic Ecosystem Management (IHG), University of Natural Resources and Life Sciences, Vienna, Austria (BOKU), p 60

Moog O, Stubauer I (2003) Adapting and implementing common approaches and methodologies for stress and impact analysis with particular attention to hydromorphological conditions. Final report, UNDP/GEF DANUBE REGIONAL PROJECT. Strengthening the implementation capacities for nutrient reduction and transboundary cooperation

MRAH (2013) Stratégie nationale de développement durable de la pèche et de l'aquaculture à l'horizon 2025. Ministère des Ressources Animales et Halieutiques. Adoptée le 18 décembre 2013, Burkina Faso, 35 p

Niasse M, Alejandro I, Amidou G, Olli V (2004) La gouvernance de l'eau en Afrique de l'Ouest: aspects juridiques et institutionnels—Water Governance in West Africa: legal and institutional aspects. UICN, Gland, Suisse et Cambridge, Royaume Uni. xxiv + 247 p

Nijboer RC, Jhonson RK, Verdonschot PFM, Sommerhäuser M, Buffagni A (2004) Establishing reference conditions for European streams. Hydrobiologia 516:91–105

Ouedraogo R (2010) Fish and fisheries prospective in arid inland waters of Burkina Faso, West Africa. Doctoral thesis, Institute of Hydrobiology and Aquatic Ecosystem Management (IHG), University of Natural Resources and Life Sciences, Vienna, Austria (BOKU), p 232

Ouedraogo R, Oueda A, Savadog M, Melcher A (2014) Sustainable management of water and fish resources in Burkina Faso—Susfish Péche Eau. The SUSFISH consortium book of abstracts—Recueil des Résumés, Symposium, Ouagadougou 15–16 July 2014, Ouagadougou, Burkina Faso. https://doi.org/10.13140/2.1.3805.9048

Ouedraogo R, Savadogo M, Kabore C, Kabre G, Oueda A, Nianogo A, Peloschek F, Sendzimir J, Slezak G, Toe P, Zerbo H, Melcher A (2015) The SUSFISH project—a trans-disciplinary approach to integrating people, fishery, socio-economy and higher education. In: Obrecht AJ (ed) APPEAR. Participative knowledge production through transnational and transcultural academic cooperation. Böhlau, Wien, p 319. ISBN 978-3-205-79690

Ouedraogo R, Oueda A, Kabore AW, Savadogo M, Zerbo H, Seynou O, Kabore Zanbsore C, Dibloni TO, Laleye P (2016) Liste rouge des poissons du Burkina Faso. UICN Burkina Faso (in print)

Pauly D, Christensen V, Froese R, Palomares MLD (2000) Fishing down aquatic food webs. Am Sci 88:46–51

Pavelic P, Giordano M, Keraita B, Ramesh V, Rao T (2012) Groundwater availability and use in Sub-Saharan Africa: a review of 15 countries. International Water Management Institute (IWMI), Colombo, p 274. https://doi.org/10.5337/2012.213

Pont D, Hugueny B, Beier U, Goffaux D, Melcher A, Noble R, Rogers C, Roset N, Schmutz S (2006) Assessing river biotic condition at a continental scale: a European approach using functional metrics and fish assemblages. J Appl Ecol 43(1):70–80

Raynaut C (1997) Societies and nature in the Sahel, SEI global environment and development series. Routledge, London

Roman B (1966) Les poissons des Haut-Bassins de la Volta. Musée Royal de l'Afrique Centrale, Tervuren

Sendzimir J, Slezak G, Ouedraogo R, Savadogo M, Cecchi P, Kabore C, Kabre G, Magnuszewski P, Oueda A, Nianogo A, Moog O, Peloschek F, Savadogo L, Toe P, Zerbo H, Melcher A (2015) Sustainable management of water and fish resources in burkina faso (SUSFISH)—a synthetic overview of the susfish project—society meets ecology. Editor: The SUSFISH Consortium, p 80. https://doi.org/10.13140/RG.2.1.5015.2482

Slezak G, Oueda A, Cecchi P, Kabre GB, Moog O, Ouedraogo R, Peloschek F, Savadogo LGB, Schmutz S, Sendzimir J, Toe P, Waidbacher H, Melcher A (2015) The SUSFISH project—joint research and scientific exchange for higher education. In: Obrecht AJ (ed) APPEAR. Participative knowledge production through transnational and transcultural academic cooperation. Böhlau, Wien, p 319. ISBN 978-3-205-79690

Smith KG, Diop MD, Niane M, Darwall WRT (2009) The status and distribution of freshwater biodiversity in Western Africa Gland, Switzerland and Cambridge. IUCN, UK, x + 94 p + 4 p cover

SP/CONAGESE (2001) Communication nationale du Burkina Faso. Convention-Cadre des Nations Unies sur les Changements Climatiques. Adoptée par le gouvernement en novembre 2001, 126 p

Stranzl S (2014) Quantification of human impacts on fish assemblages in the Upper Volta catchment, Burkina Faso. Master thesis, Institute of Hydrobiology and Aquatic Ecosystem Management (IHG), University of Natural Resources and Life Sciences, Vienna, Austria (BOKU), p 90

Thorne R, Williams P (1997) The response of benthic macroinvertebrates to pollution in developing countries: a multimetric system of bioassessment. Freshw Biol 37:671–686

Trauner D, Koblinger T, Huber T, Moog O, Melcher A (2013) Benthic invertebrate assemblages in the Upper Nakambe Basin. In: SUSFISH Consortium Appear Project [56], Current questions in sustainable water management and higher education in Burkina Faso. Book of abstracts. http://susfish.boku.ac.at/. https://doi.org/10.13140/2.1.1970.8967

UNDP (2015) Human development report 2015. Work for human development. United Nations Development Programme 1 UN Plaza, New York, NY 10017, USA

Vinson MR, Dinger EC, Kotynek J, Dethier M (2008) Effects of oil pollution on aquatic macroinvertebrate assemblages in Gabon wetlands. Afr J Aquat Sci 33:261–268

Wang L, Dochartaigh OB, Macdonald D (2010) A literature review of recharge estimation and groundwater resource assessment in Africa. British Geological Survey Internal Report, IR/10/051, 31 p

Zerbo H, Ouattara DC, Soubeiga Z, Kabore K, Kabore C, Bado E, Goumbri BA, Yerbanga RA, Ouedraogo N, Baro S (2007) Analyse de la filière pêche au Burkina Faso. Projet d'Appui au Renforcement des Capacités d'Analyse des Impacts des Politiques Agricoles. Direction Générale des Ressources Halieutiques. Ministère de l'Agriculture, de l'Hydraulique et des Ressources Halieutiques. Août 2007, 63 p

Chapter 28
The Tisza River: Managing a Lowland River in the Carpathian Basin

Béla Borsos and Jan Sendzimir

At 156,000 km^2 the Tisza river is one of the largest tributaries of the Danube river. Historically, almost the entire Tisza river basin (TRB) was under one administration (the Austro-Hungarian Empire), but management has become far more complex after World War I, when the basin was split among five newly formed countries (Hungary, (Czecho)Slovakia, Ukraine, Romania and Serbia). The river exhibits extreme dynamics due to its particular geomorphology: a very short, steep fall from the Carpathian mountains suddenly turns into the very flat lowland expanse of the Hungarian Great Plain. The arc-like shape of mountains around the basin amplifies the flood peak by causing stormwater received from the tributaries to converge on the main river channel in near unison. The resulting impoundment of high water in the main bed backs water up into the tributaries, threatening the neighbouring floodplain communities. The mountains receive 3–4 times the amount of precipitation that falls on the plains (2000 vs. 600 mm/year). These combined factors make the Tisza naturally "flashy," with flow rates varying by a factor of 50 or more, accompanied by sudden (in 24–36 h) and extreme (up to 12 m) rises in river stage (Lóczy 2010).

Increasing variation in nature (climate) and accelerating socio-economic processes in society (urbanisation, agriculture) challenge all aspects of water management. Rising trends in precipitation extremes have increased the dramatic variations in flows: 100-fold differences between the highest and the lowest stage often occur, and the stage can rise as much as 4 m within 24 h (Bodnár 2009). Additionally, the temporal pattern of the flow regime increasingly varies across the seasons. Spring tides issue from snow melt in the high mountains, while the summer flood is usually

B. Borsos (✉)
Institute of Geography, University of Pécs, Pécs, Hungary
e-mail: dioliget@bckft.hu

J. Sendzimir
Institute of Hydrobiology and Aquatic Ecosystem Management, University of Natural Resources and Life Sciences, Vienna, Austria
e-mail: jan.sendzimir@boku.ac.at

a result of sudden and torrential rainfall early in June. Then, 2 months with little or no rainfall follows, leaving the river with an annual minimum in autumn and a serious drought in the valley by the end of the summer. Another feature of the physical geography in the plains is that since the whole lowland river basin sits on an alluvial cone, and no rock bed exists up to a certain depth, the soil easily conducts groundwater, which emerges on the surface during high water stages. This, accompanied by high rainfall and snowmelt, saturates the soil and may cause extended water logging on the plains, with limited runoff due to the low natural gradient. In fact, on the 270 km upper reach of the Tisza up to Tiszabecs, the river falls 1577 m, while on the remaining Great Plain stretch of close to 700 km, it falls only 32 m.

This chapter describes the main river management problems in the TRB, including a historical background, and then discusses contrasting management strategies that currently contend for control of the vision guiding further development in the TRB.

28.1 Historical River Management

In its natural state, the Tisza was very meandering river, changing its bed quite often and leaving many side arms and oxbows. Centuries of river engineering along the Tisza have made this natural state a distant memory. The continuous work of the Hungarians who settled here after 900 AD shaped the landscape, transforming the Great Plain into a cultivated region where the natural, periodic inundations of the floodplain would temporarily cover an area up to 30,000 km^2 (Somogyi 1994, p. 22).

Sometime during the Medieval period (ca. 1100–1200), early water works called the "fok" management or the *fok* system of dikes (with sluices) were developed to control inundation of floodwaters onto specific areas of the floodplain. An extensive floodplain economy was practiced both along the Danube (Andrásfalvy 1973) and the Tisza river (Molnár 2009; Fodor 2002), including their respective tributaries, such as the Bodrog (Borsos 2000). This economy took advantage of the several-metre-high, and sometimes many-hundred-metre-long, flat natural levees built by the rivers on the floodplain during recurrent floods. Water was conducted onto the deeper-lying floodplain areas in small channels with the help of incisions ("fok" in Hungarian), cut into these natural formations. There were also natural gaps where side arm streams feeding permanent water surfaces in the river valley started. However, most of the smaller *foks* were human-made or altered and acted as outlets to deep bed canals branching off from the middle-stage water bed of the main channel, where the direction of water flows was dependent on the water level in the main river bed. During high stage flooding, the incisions discharged water from the river onto the floodplain. By discharging water slowly against the general gradient of the landscape, the *foks* gently inundated the floodplain. As the main river channel ebbed, the same structures drained floodwater back into the river.

The shallow floodplain "backswamp" ponds and major oxbow lakes played an important role in the local economy during late Medieval times—in addition to serving as natural water reservoirs (Bellon 2003). The ecological potential of the floodplains with the help of the *foks* was exploited through a wide variety of means ranging from fishing, fruit orchards and livestock management to reed harvesting and logging, and, occasionally on the higher elevations, tillage. The channels even provided convenient transport routes for timber, reed and hay, while water flows in them were used by mills (Rácz 2008). Despite the fact that the floodplain was inundated more frequently during this period, the inundations were shallower, and the settlements would not, in fact, have been inundated, since they were built on high natural terraces (relicts of depositional features of an older floodplain).

During the Ottoman rule in the seventeenth and eighteenth century, some areas were deliberately converted into marshland for military purposes, to provide better strategic defences for border castles seated in the river corners (Hamar 2000). The fok system was neglected because the prolonged conflict dispersed the population, and, after the expulsion of the Turks, mislaid water mills, which let water out onto the fields, aggravated waterlogging of the area further. Additionally, deforestation in the upper, mountainous, portion of the catchment triggered much bigger runoff events (Andrásfalvy 2009), causing really dangerous floods in the eighteenth and nineteenth century. This—and the quest of landlords for plough land to produce cash crops like wheat—triggered much large-scale river engineering efforts in the late nineteenth century. All these factors combined to redefine water as a threat, whereas prior cultures had used it to drive their regional economy.

At the close of the nineteenth century, the full force of the industrial revolution was brought to bear in reshaping rivers all over Europe. The large-scale river training works—called the Vásárhelyi Plan—were implemented with the aim to reduce the length of the Tisza by shortcutting meandering bends, cutting off and draining the floodplain with earthen embankments—dikes—that prevented river channel water from entering the large areas formerly inundated periodically. As a result, river velocity increased, incising the channel and, thereby, increasing the gradient of the river, thus shortening the water's travel time. The average gradient of the river bed rose from 3.7 to 6 cm/km, and it became more balanced, i.e. uniform between the upper and lower reaches of the river (Lászlóffy 1982). The pre-industrial, full length of the river on the plains was 1419 km, which regulation reduced by 32% to 966 km by the time the works were completed. All in all 114 crosscuts were made to eliminate 589 km of meanders, the total length of the cuts ranging up to 136 km. Later on it turned out that water caught on the floodplain has to be drained artificially, forcing the construction of a draining canal system as an auxiliary measure. Currently, a 2700-km-long line of dikes "protects" 17,300 km^2 of land along the Tisza within Hungary. In total, dikes within the Tisza river valley extend for 4500 km and have reduced the area of the active floodplain by 90% (Bellon 2004).

28.2   Current Management Issues

28.2.1   Faster Flows in a Land Without Buffers

The well-meant engineering interventions of the nineteenth and early twentieth centuries triggered grave consequences for the ecological functioning and the local economies of river basins. Throughout Europe prior to the Industrial Revolution, man and environment coexisted in river valleys through economies and technologies with much smaller impacts. The application of these pre-industrial lifeways of society was less extreme in scale, extensive in space or consistent in time. The emergence of a market economy teleconnected the Tisza river to unprecedented economic and political forces over a much wider region than the TRB: all of Europe. Exposure to these forces precipitated huge social and psychological changes as well as shifts in the ownership structure. As a consequence, the frequency, degree and extent of human technical interventions have changed dramatically, leaving permanent marks on the physical geography and the dynamic equilibrium of river systems, including the Tisza.

Vegetation cover and structure in the entire river basin was altered by mass conversion from a semi-forested polyculture of orchards, meadows and ponds to grain-dominated monocultures. The rising demand for wheat as a cash crop producing income for landlords and used to feed cavalry horses (wars) and urban populations (industrial concentrations) drove this conversion from polyculture to monoculture. Dikes were built to prevent flooding of wheat fields and settlements. During the eighteenth and nineteenth centuries, these landscape conversions profoundly changed the boundary conditions (water retention capacity of the plain, discharge and river dynamics), depleted the buffer capacities and damaged certain subsystems such as the gallery forests and wetland habitats. Consequently, the functional integrity of the river valley systems was gradually eliminated. The sponge effect, i.e. the catchment's and the floodplain's capacity to retain excess water, was lost, and the landscape became barren. In the wake of this change, the runoff of surface waters was accelerated, triggering a reinforcing feedback effect by increasing erosion and, hence, the bed loads in rivers, shifting the ratio to floating sediment derived from the washed off forest soil.

As the flood control works were implemented from the second half of the nineteenth century on, the hydrodynamic processes triggered by the alterations on the river dynamics resulted in siltation of the floodway between the dikes, incision of the low stage river bed in the main channel, draining the floodplain of groundwater in times of low water and water stagnation in open fields on the floodplain in times of high water or intensive rainfall or snow melt. These factors—reinforced by other interdependent changes in the basin upstream, such as the increasing amount of paved surfaces, reduced vegetation cover and strong water erosion—gave rise to ever-growing flood crests (Lászlóffy 1982). The habitual reaction was to raise the height of the dikes (Fig. 28.1).

From 1860 to 2000, in seven, separate, consecutive stages, the dikes along the Tisza were expanded and raised to strengthen flood defences. Today, the dikes tower 4–6 m above the mean river bed—and the surrounding terrain. The seven stages were prompted by at least two reasons: (1) the headwater regions in the mountains were further deforested, leading to less storage of water in the uplands and more and faster runoff, and (2) the floodway within the dikes gradually silted up over time due to sedimentation and could not contain the larger volumes of flood water. The latter process has continued as a positive feedback until the dikes (earth embankments) reached their physical limits, and now they cannot be raised any further (as evidenced

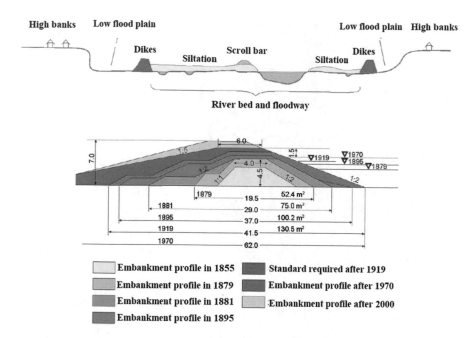

Fig. 28.1 Increasing the height of flood control levees, adapted from Schweitzer (2009)

by dike breaks becoming more frequent). Over time an onion-like structure was formed which reached the limits of its structural strength by the end of the twentieth century. Further heightening of the dikes would entail the risk of bursts due to the hydrostatic pressure of the water and the limited resistance of the earthen material. Also, a dike is only as strong as its underlying substrate. At one point a flood can "blow out" a dike from underneath. This also sets the limits of dike height. Additionally, it was also recognized that the mathematical models used to predict design flood levels were flawed, as they could only make forecasts based on past experience but are unable to take into account expected—or unexpected—future processes (Koncsos et al. 2000). One of these newly recognized unexpected and unpredictable factors is the local impact of increasingly variable climatic events which will definitely make—or indeed, has made—historical data obsolete (Nováky 2000). Another unpredictable factor is the management of the upstream basin, which belongs to the national territory of other countries—one of them, Ukraine not even a Member State of the EU—and hence, beyond the influence of the Hungarian water administration.

In spite of heavy engineering, especially the confinement of the natural floodplain to 5–10% of its former area, the geomorphology of the Tisza valley did not change much: higher and lower elevations on the now inactive floodplain remained intact. Figure 28.2 above shows a section of the Hungarian reach of the river on a schematic diagram indicating the lower elevations of the former floodplain and the high banks that can still be clearly distinguished by the naked eye. The difference in elevation

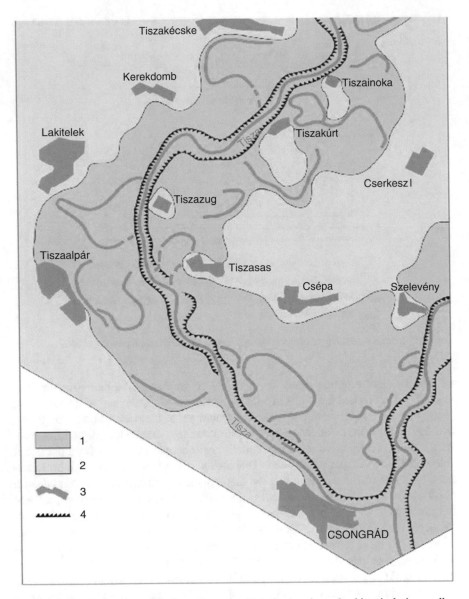

Fig. 28.2 Settlements are still situated on the high banks along the historical river valley. (1. lowland 2. high banks 3. settlements 4. dikes). Adapted from Schweitzer (2009)

between the parts formerly inundated regularly by the river and the parts considered to be safe and at low risk of floods is more than a metre. It is also clear that infrastructure still follows more or less the aforesaid distinction, and most settlements have been and are still being built on high banks, relatively safe from floods.

Figure 28.2 also reveals that former river branches—now cut off from the main river bed and the floodway by dikes—can be clearly distinguished on the plain as deeper depressions on the flatland. The pooling of water due to poor drainage (water stagnation) is most severe on these parts (Schweitzer 2009). Such stagnation can be extensive and costly to farmers in terms of productivity lost when prolonged inundation kills biological activity in the soil. Often it can take years to re-establish such bioactivity. On 15 January 2011, a total of 380,000 ha of arable land was covered by water upwelling (stagnation) for several weeks to months (Vízügy 2011, website of the national water administration). Compensation payments for agricultural losses due to stagnant water in 2013 ranged up to HUF 9 billion (ca. 28.8 million euros) nationwide (Szeremlei 2013). Unfortunately, however, recent urbanisation and the dominance of industrial agricultural practices resulted in a situation when today ~34.23 billion euros (agricultural production and municipal/industrial infrastructure) are at risk of damage by floods. Over the past 20 years, the rising trend of flood stages has meant that high waters have increasingly overtopped the dikes. The largest and most damaging flood was in 2010. In a single county, Borsod-Abaúj-Zemplén, the costs of disaster management exceeded HUF 2 billion (6.45 million euros) (KSH 2011).

The long-term sustainability of communities in the Tisza river valley is severely challenged by a range of outcomes from river engineering. In addition to the increasing potential for devastating floods, the faster flows in the river channel have degraded (lowered) its bed, thus lowering the water table during dry periods. On the other hand, the dikes contain many large flows in the active floodway, and thus raise the water table during wet periods. Because of these processes, and because of the spatially varying capacity of the floodway to transmit water, there might be areas found within the Tisza valley flooded and other areas in the state of drought at the same time, or the same areas suffer both flood, water stagnation and drought, respectively, in different periods of the year.

28.3   Competing Concepts of River Management

28.3.1   Business as Usual

The "hard" path (*sensu* Gleick 2003) is driven by a technocratic focus on controlling water flows through geo-engineering approaches and still dominates the agenda of the Hungarian water management administration. Failure to re-examine this attitude despite mounting evidence of its drawbacks is an excellent example of the concept of *Path Dependence* (see Chap. 16). This path rigidly adheres to the industrial vision of a river valley as a transport (river channel) and production (floodplain) resource delivery system. The principal elements of this approach always revolve around the same responses to flooding: further strengthening of the dike system, clearing of the floodway, stabilisation of embankments and creating concrete canals to increase hydraulic throughput. A parallel arm of the "hard" path addresses water scarcity through construction of barrages to retain water in big reservoirs within the course of the river and mitigate drought by artificial irrigation schemes. The rigidity of such hard infrastructures precludes any innovations that might flexibly connect and integrate these two arms (flood and drought protection). This hobbles the capacity of managers or communities to adapt and greatly increases vulnerability to climatic variation.

The same conservative view is seen in the field of urban planning. Szolnok, for instance, the largest city in the middle section of the Hungarian reach, considered the river as a fixed part of the infrastructure and not as a dynamic part of the landscape, which requires room to flood and move, i.e. shift the channel bed. The confrontation of the dynamic (a trend of increasing flood crest elevations) with the static (fixed dike elevation and location) resulted in numerous near failures of the dikes during the serious floods of the last 20 years. The "soft" option for the cities to pay countryside communities to open their dikes and store floodwater on meadowland cannot be implemented currently due to a combination of incoherent legal and psychological barriers (Sendzimir et al. 2008, 2010).

Instead, expensive river engineering schemes are in the planning pipeline. In Szeged, downstream of Szolnok and close to the Hungarian–Serbian border, the river passes through the downtown of the city. The river channel is in the grip of concrete walls that must be raised further every now and then to address rising flooding trends. One recent strategic concept addresses those trends with a mobile, aluminium quay embankment on top of the current abutment. This retention method would boost flood crest levels by up to 5½ m above the average ground level of the city (Kozák 2011), increasing river velocity and greatly increasing the damages should the embankment fail. One alternative does not seem to be much more cost efficient: a dry river bed to be constructed afresh on fertile land as a greenfield investment just to bypass the city in times of high floods (Rigó 2013).

Dependence on "hard path" solutions is reinforced by paradigms that view river dams as beneficial in terms of both flood control (as storage reservoirs) and drought (as sources of irrigation water) (Gleick 2003). Since such paradigms influence how you interpret and filter data, a number of conclusions can be drawn from the same set of facts. So far, there is only one such scheme in operation on the Hungarian stretch of the river: the Kisköre dam and the so-called Tisza Lake, the impoundment behind the barrage. This is considered to be a great success, both in terms of water governance of the river and as a social benefit. Recreational opportunities, fishing, bird watching and the like are mentioned most frequently. However, such rigid nature conservation measures and approaches do not facilitate the dynamic systems thinking needed to adapt in increasing variability of climate and water flows. Tisza Lake is praised for its role in boosting biodiversity, but it actually stifles the biodiversity that previously emerged from water level dynamics. The "lake", actually a reservoir, is a stagnant water body that disrupts the dynamic pattern of floods and low water stages in the middle of a living water course (Teszárné Nagy et al. 2009, see Chap. 6). The complete eutrophication of the lake can only be avoided by permanent anthropogenic manipulation.

Despite these problems there are still planning schemes to build more dams on the lower Tisza stretches at Csongrád to provide irrigation water to a part of the plains named Homokhátság, which is morphologically higher than the adjacent river floodplains. This expensive project increases the danger of waterlogging from water stagnation while doing little against flooding. River dams—whether or not producing electricity—are a logical consequence of the previous phase of classical river training works: the dams slow the river down just 100 years after it was accelerated by channel straightening (Balogh 2014).

To protect the ill-planned build-up of vulnerable assets (community, industrial and agricultural) on the floodplain, management has been trapped in a series of expensive stages to shore up the "hard" path infrastructure. While economics dictates this, it is ironic that the costs of the current system—including the disaster relief operations in times of floods—far exceed the value of the assets that might be protected by them (Koncsos 2006). There are less expensive alternatives that might break us out of such path dependence. Compared to conventional flood control wisdom, there are two distinct and, to some extent, related design schemes (VTT and ILD) designed to overcome the flood problem by discharging surplus flood water onto lower-lying deep floodplain areas on arable land on the former natural floodplain.

28.3.2   Advancement of the Vásárhelyi Plan

The water management establishment considers this concept as its "softer" alternative, because for the first-time agricultural land on the open floodplain is used conceptually for emergency water storage in state-of-the-art artificial reservoirs outside the dike system. It is a *flood reduction and mitigation system consisting of engineering structures and reservoirs dedicated to the controlled discharge and eventual return of floods into the river as necessary (or transferring surplus onto areas in shortage of water[1])*.

The new program was named in remembrance of the original river training concept envisaged by the short-lived but influential water engineer Pál Vásárhelyi in the nineteenth century. The selection of the revered historical name gives the program a political "spin" to increase its acceptance. However, it also reveals how questionable the development following the Vásárhelyi vision has been. Problems emerging from the original Vásárhelyi plan have ongoing effects on the life of the Tisza valley up to date. The first and main result of the Vásárhelyi plan—which was implemented poorly and incompletely anyway, even within the theoretical framework of the technocratic approach of the time—was that engineers and developers are trapped now in the need for ever newer interventions into the system, as explained in the previous section. Therefore one can reasonably ask whether this initiative will "clean up the mess" or simply extend problems inherent in the whole concept.

---

[1] Act No LXVII of 2004 on the Advancement of the Vásárhelyi Plan.

The VTT proudly boasts of a change in attitudes, even a paradigm shift. And indeed, the focus is moved from defence (and a military-like organisation) to regulation, control and prevention, and a long-term sustainable solution with ecological considerations in mind. The most important change in the approach was the idea of retaining water instead of draining it from the plains, which could be one step toward integrating ideas of flood and drought management. However, as conceived, such a technical solution does not really reflect the kind of paradigm shift the name suggests. The published program still states that the key objective was to enhance flood security in the Tisza valley, and not the implementation of integration of land management and development practices. Such integrative, alternative practices disarm floods by lowering crest elevation and velocity, and then use their storage to lower drought risk. This renders the very concept of risk, danger and exposure to floods irrelevant.

Instead, there are three major segments in the program, of which only the second one is a relatively new idea; the other two are business as usual methods:

1. Improvement of the water carrying capacity in the high water stage river bed on the Tisza (in other words: clear the floodway)
2. Construction of a flood detention emergency reservoir system with a total storage capacity of 1.5 billion m^3 (10–12 reservoirs)
3. Development of the existing flood control works and structures.

Later, the VTT concept was broadened to involve infrastructure development in the settlements concerned (excess water drainage in the built-up areas, sewage systems, waste water treatment plants, replacement and construction of byroads, bicycle paths) and implementation of husbandry methods driven by natural conditions (landscape management). Yet the actual solutions treat only the symptoms. For instance, as part of the flood control measures, the bank protection works at the bottleneck in Kisar were reinforced, but nothing was done to overcome the bottleneck itself.

Cost cuts and funding difficulties resulted in mistranslation and piecemeal implementation of the original concept. As an incomplete and imperfect edition of the complex system of water storage bodies originally intended by the VTT, these current reservoirs are now prone to functional inaptitude. The first structure to be inaugurated was the Cigánd reservoir in the Bodrogköz in 2008. The second structure, the Tiszaroff reservoir, was completed in 2009 with the expectation that it will be used only once every 30 or 40 years. Conceived as an infrequently used "emergency reservoir", it obviously would not make society and ecosystems adaptive to the mounting pressures of increasing climatic variability. Additionally, the poor design of both structures does not follow the natural depressions of the floodplain. Today, 6 of the 11 reservoirs are operational, and in the period between 2014 and 2020, an additional 50 billion euros worth of European Union funding is earmarked for the completion of the series of projects (MTI 2015). This expensive system partially addresses only one problem: floods. It does not help with waterlogging or drought. Also, as it turned out, it is of not much use in the case of icy floods, striking last time in February 2017 (VG/MTI 2017).

The VTT also has structural flaws that mainly result from a combination of institutional and legal barriers and a conservative engineering approach. Poor design features are reflected in the following aspects:

- Functional landscape features are not exploited in storing or moving water.
- The river floodway already lies higher than the floodplain itself because of the accumulation from decades of siltation.
- Design is subject to rigid artificial and legal constraints. For instance, a 60 m protective zone surrounding public roads means that some new dike sections had to be built on the deepest lying land.
- Inlet structures are oversized and with high threshold level, so they can only be opened at very high water stages.
- Reservoirs are considered to be rigid structures dedicated for flood control only, and hence, barriers to agricultural production.
- The system is paradoxical and self-contradictory: during the flood of 2010, water was discharged into the Tiszaroff reservoir to skim off the peak flows and protect Szolnok, but regional water authorities upstream pumped excess surface water into the river at the same time to drain open fields from stagnating water.

Overall, in the view of the authors and based on lessons learnt from former technocratic approaches, the VTT does not offer sufficient capacity to cope with or adapt to the impacts of increasing climatic variability.

28.3.3 The Integrated Land Development Concept

The integrated land development concept (ILD) adapts human practices and infrastructure such that they balance with the provisions of the natural environment (climate, the hydrological cycle). Rather than developing and maintaining massive and expensive engineering to tame environmental dynamics, it aims to use ecosystem services to enhance adaptability to diverse sources of uncertainty, e.g. variance in climate, water, economy, etc. It is a concept developed from multiple perspectives, including engineers, social and natural scientists, NGOs and environmentalists. It starts from a comprehensive goal to simultaneously build resilience to floods, drought and waterlogging by changing the space/time dimensions of the water regimes. Put simply, that means slowing water movement to the point where its excess does less damage and can be accumulated to sustain ecology and economy when water is scarce. Restoring the original dynamic equilibrium of water in the landscape offers safe flood control and the replenishment of missing precipitation. This can be done by setting up *land use patterns* that accommodate nature (biodiversity and ecosystem services) as well as society (husbandry that exploits those services to sustain local economies). For example, converting *cropland* to *grassland* can reliably transform the more extreme water dynamics outside the dikes into animal products for food and consumption. Such land uses make both human and natural communities more adaptable and resilient to variability of climate.

An ILD landscape is a mosaic of different land uses that allows multiple uses in parallel. Such a multi-use system consists of various agricultural practices like horticulture, orchards, livestock management and cropland production supplemented with a variety of other activities related to land use, many of them conventionally not qualified as part of modern agriculture. Such activities include fisheries, forest management, industrial crops like hemp or reed, hunting, apiculture, alternative transportation means (rafting), energy generation facilities (water mills) and direct water use for drinking, washing, watering, cooking, other domestic water needs, and so on. Such a complex land use system supports local self-sufficiency by providing a diversity of functions that work in a wide variety of circumstances.

To establish a robust land use and water management system and make it work requires experimentation in land use innovations in areas denied for these purposes since the late nineteenth century: the floodplain. The current Tisza valley must be assessed first from a geomorphologic point of view in order to determine those areas that can be flooded by "natural" water movement (Fig. 28.3). As a key design principle, efficiency is achieved by conserving and enhancing natural processes that deliver ecosystem services, not working against them. To apply such principles, one recent modelling project (Koncsos 2006) systematically surveyed the left and the right bank of the Tisza for sites that were morphologically feasible for water storage. A total of 19 such deep floodplains—polders—were identified, the inundation of which could result in significant reduction of the river water level during flooding. Only deep

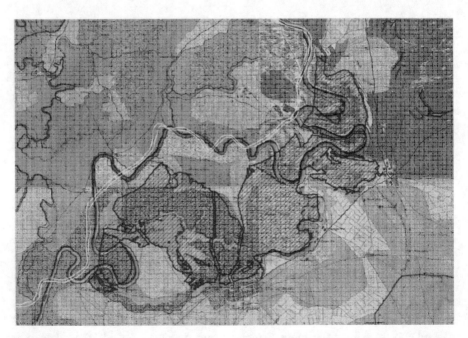

Fig. 28.3 Red lines indicate the borders of potential deep floodplain polders fit for water retention under the ILD concept in the Middle Tisza region. The yellow line shows the current path of the river drawn on a map of the region before river regulation (Koncsos 2011)

floodplains with a retention capacity of at least 50 million m³ were considered, while the storage capacity of the largest area measured exceeded 200 million m³. Total storage capacity of the deep polders assessed exceeds 2 billion m³. That is a buffer volume that would have rendered most of the floods of the past century harmless by slowing the speed of the flood wave and lowering its elevation. The VTT (in full completion) is expected to lower the flood crest by 1 m. Deep floodplain inundation has twice that potential. Designing flooding of deep-lying floodplain areas is not a simple job. Quantitative and temporal conditions of water replenishment, the impact of local water steering canal system and the alternatives of water steering must all be investigated (Koncsos 2006). The size of the area shown by the model as potential candidate for flood control is several times larger than the area of the reservoirs finally approved for construction, yet the need for actual construction works—once the delicate design process has been completed—would be a lot less than in the case of the VTT.

A strategic methodology to implement a sustainable landscape management strategy should build on the lessons learnt from traditional floodplain husbandry just as much as on modern scientific achievement of water and land management, data collection and processing, remote sensing, GIS, topographic surveys and precisions earthworks. It consists of the following elements:

1. Connectivity between floodplain and river channel created by primary notches ("fok") and a set of secondary incisions allowing communication with the floodplain behind the levees bordering the river banks.
2. Carefully controlled water discharge onto the riverine floodplain by allowing floodwater to enter through the *fok* incisions and "back up" the secondary channels against the general gradient of the basin.
3. A lock at the mouth of the notch to regulate water levels on the plain as a function of time, water volumes and discharge as well as drainage operations.
4. Careful design with due observance of natural contour lines in order to allow for both discharge and return gravitationally, thus avoiding the need for external energy use.
5. Areal inundation by actuating the lock at the main outlet site and by raising low embankments along the channels to govern water.
6. Different geographic locations for different water uses. Moving water for productive use, stagnant water bodies for fish ponds, reservoir for irrigation or recreation. Aquatic communities are preserved until the next inundation/replenishment.
7. Assist infiltration where water is needed or drying out where ploughing is intended to be done. Excess water is drained back to the main river bed when the water level in the mean stage river bed dropped to a lower relative elevation than that on the plains.
8. Water governance can be achieved by locks as well as bottom sills at strategic points of the water transportation network. Locks are more expensive but can be used to proactively retain the water on either side, wherever it happens to be higher, while bottom sills guide water gravitationally when it reaches their design height.
9. Water thus can be managed wisely without forced hydromorphological alterations in the riverine system. It is not simply a reconnection of the floodplains but a method preserving or restoring to a great extent the original functions of the landscape.

The ILD strategy requires a serious "paradigm shift" in current water and landscape management principles and practices. It acknowledges that flooding is not a risk to get rid of; it is rather an opportunity to take advantage of. The Tisza valley as a whole has no "excess water". On the contrary, it is a naturally arid landscape where missing water was supplemented under pristine conditions by periodic floods of its river. If you want to design a long-term sustainable landscape management strategy, you have to understand the landscape properly. The design should take the contours and land relief into account and land use and, hence, the water supply of the land should be adjusted to the relief and not the other way round.

Depending on the local conditions and morphology, inundation of the flooded areas in the floodplain can either be natural or managed by human interventions (Fig. 28.4).

- Natural flooding: means a system where water only follows the native depressions and brooklets of the landscape formed by the dynamics of the river and its floodplain.
- Assisted flooding: water movements can be governed by bottom sills at strategically important locations and some man-made infrastructure needs to be protected by dikes.
- Artificial water steering: in situations where flooding is restricted, water is led between low levees along wide channels. To drain excess "stagnant" water that wells up from below and rests on the surface, these channels are currently deeply dredged. Sustainable land development would reverse this process by broadening these channels. The flooding of the surrounded areas would be controlled by side locks.

In any framework of managing and developing the functions of a landscape, a sustainable water management system ensures replenishment of water bodies in the land and—in times of need—careful drainage of excess inland water and waterlogged fields. It should be set up as a complex whole of natural beds, bottoms and depressions, combined with man-made system components—existing channels and road networks—as well as freshly built structures constructed for the purpose of water governance.

Flooding of the plains can be started by opening the main lock at the flood control line when water levels in the main river bed reach a desirable height, e.g. the elevation of the lock bottom. The natural hydrostatic pressure of the rising tide would drive water from the river through the freshly established notches to the former excess water drainage canals. While the primary locks along the system's main branches are open to assist flooding, secondary or side locks can be manipulated in accordance with the water needs of the surrounding areas. As soon as water has penetrated up to the highest point of the system and the landscape, the main lock and the primary locks in the canals are closed. This way no overspill will occur, and once water levels in the main river bed subside, the water discharged onto the plains can be retained as applicable and necessary.

The possibility of gravitational reverse flooding—that is, inundation of an area started from relatively lower elevations along the river course and filling the

Fig. 28.4 Conceptual
illustration of the VTT
versus the ILD concept
(original drawings by Péter
Balogh)

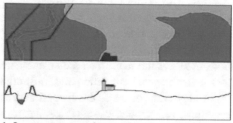

1. Low water stage (current state)

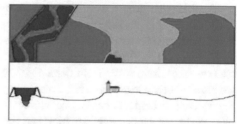

2. High water stage (current state)

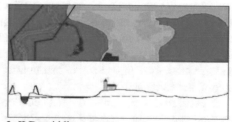

VTT

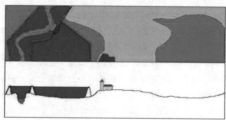

3. ILD, middle water stage

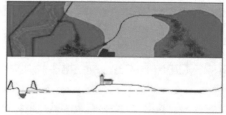

4. ILD, high water

floodplain upwards—can be realized along the mid-Tisza reach once mean stage highs occur, which is the case quite frequently (that is, several times a year). This strategy would prevent more extreme high stages from ever occurring. For the purposes of design, the historical water flow patterns need to be consulted and the bottom sill of the main lock gate established at a level that allows use of relatively low water stages. Penetration and infiltration rates need to be taken into account, so that the amount of water discharged addresses needs such as replenishment of soil moisture and groundwater tables. Historical figures supplemented with climate change forecasts will also provide an insight into the temporal patterns of flooding possibilities that in turn would help agricultural production planning.

When water levels in the mean stage river bed retreat, then the main gate lock has to be opened as soon as possible to drain water from the main canals where it stands above the level of the surrounding terrain. Any other locks need to be opened afterwards to drain water from fields into the canals. For most purposes, a couple of weeks of inundation at a time is the maximum length of time which can be tolerated by the vegetation, land and field crops without damage or deformation. This is especially so when water temperatures are high and the oxygen concentration is low.

Draining is theoretically possible down to the level of the bottom sill at the main gate lock, but it is advisable to retain some more water in the land for the purposes of infiltration and to make up for losses through evaporation. At the same time, this level ought to be low enough to allow for drainage of the fields. If the system is properly designed, drainage is possible gravitationally, without the need for any pumping. Again, consulting historical data of water level dynamics during pulse floods may help. Since high water can stand no longer than the land's submergence tolerance period, one must carefully judge the time between opening the locks and subsidence of the flood in the main bed below the bottom sill. Meeting the specific conditions for gravitational drainage minimizes flood and drought risk and avoids waterlogging. Such methods are cheaper than conventional geo-engineering. However, one must overcome significant barriers in the minds of people and the legal and administrative systems as well as certain parts of the above ground (power lines) and underground (gas pipelines) infrastructure. However, most of the latter can be accomplished by skilful design.

A detailed description of the ILD concept, theoretical and practical, geographic, legal, social, institutional and psychological opportunities and barriers, constraints and difficulties in the way of its implementation are set forth in a book compiling the outcomes of a UNDP financed international project (Borsos 2014).

28.4   Climate Change and Possible Future Paths

Current scientific evidence strongly suggests that climate change is a fact, not a possibility. Therefore, the need for adaptation to a changing climate and the consequential alterations in many of the large biogeochemical cycles of the Earth shall become a compelling driver to reconsider current management practices, including surface and underground water regimes. Forecast scenarios as to the probable impacts of the change may vary to a large extent globally, but converge pretty much in the case of the Carpathian basin (Bartholy et al. 2011): drought, less precipitation in summer, more rain and less snow in winter, with the two transient seasons (autumn and spring) shortened. Specifically, it seems that the south of the Great Hungarian Plain will occasionally receive as little as 100 mm precipitation in the summer season, which corresponds to a quite arid, almost desert climate (Kis et al. 2014). Even more worrisome is the prediction that precipitation in the higher mountain ranges of the Eastern Carpathians, where rain and snow fall in the winter period, will increase by 10% or more over the current—already high and torrent—levels (Jurek et al. 2014). However, higher temperatures mean that less water will be stored in ice and snow buffers to be more slowly released as spring arrives. This means that the temporal pattern of water availability in the lowland rivers of the Hungarian plain will be even more extreme: while summers are expected to be dryer than ever, spring snowmelt accompanied by occasional torrent rainfall will greatly increase the risk of flash floods.

From the perspective of flood control, the most visible and worrying signs are the appearance of sudden, high-intensity rainfall events, mainly in the Carpathian section of the Tisza, in Ukraine. Torrential outflows from these unprecedented events cannot be attributed to deforestation alone but also to changing weather patterns and altered temporal and spatial distribution of precipitation. The local hydrological cycle, which had previously provided relatively even rainfall distributions, now appears dangerously concentrated. In certain parts of the Carpathian basin, for instance, in the Kárpátalja, over several days rainfall equalling half a year of precipitation fell on forests too denuded to prevent massive runoff. The rain arrived at the beginning of November, where the river bed was already full and the catchment area saturated, with no sponge effect left to retain runoff water (Bodnár 2009). The hydrological balance between individual river basins has been shifted as well. For instance, while the Danube river basin used to be more humid in the past, the Tisza catchment receives more rain these days (Borhidi 2009). Clearly, a strategy balancing this inhomogeneous supply is of paramount importance.

Modern societies are not a bit less exposed to extreme weather events than their forebears but are a lot less adaptable. Our human and industrial capital was designed and calibrated under more predictable conditions, and therefore the rigid technical systems designed to protect fields, crops and assets do not perform very well in emergency situations. A shift toward more integrated land management concepts becomes increasingly attractive as one recognizes how it increases our adaptability to stress and shock.

The full potential of any adaptation strategy is realized when it is understood and applied both from the top (technocrats, government) and the bottom (NGOs, local practitioners). Tools to visualize how climate change occurs as well as its expected outcomes can help broaden that understanding. For instance, geographic projections of precipitation and temperature distribution patterns are less comprehensible to illustrate the expected changes in vegetation distribution than life zone maps, as a recent investigation in Hungary showed (Szelepcsényi et al. 2013). This more directly conveyed the likely impacts of climatic change to inhabitants of the Tisza valley. Public understanding of how a problem arises can be a key to their support of the implementation of potential strategies in the future, especially if these strategies are experiments. Once people understand the impossibility of current practices, they will be more easily convinced to switch to other cash crops or even deeper changes such as converting cropland to pasture or forest.

Currently in the Tisza valley, the conventional infrastructure and practices of water management as applied to agricultural, communal and industrial water use are not adaptive to future uncertainty associated with climate variability. Of the three strategies presented above, ILD is arguably the most comprehensive and flexible candidate for successful adaptation. Unfortunately, a variety of factors combine to trap current management in path dependence, such that VTT continues to be implemented. This can provide a temporary water storage capacity of 1.5 billion m^3, a fair amount to reduce the crest of flood waves but a far cry from the system theoretical needs of the region. Its very expensive and resource intensively operated structures cannot do anything else but skim the flood crests at the price of ruining agriculturally productive land. Once the flood is there, they reduce the crest level to an extent ranging from 10–12 cm up to 30–40 cm along the river, depending on the exact geographic location (OVF 2014).

For a truly adaptive strategy, the temporal aspects of the water regime ought to be handled in a holistic manner, taking into account drought and water stagnation, floods and underground resource management as a single whole. In fact, human presence, infrastructure and activities need to be adapted to a changing landscape and not the other way round. To date it has not been encouraging to see how decision-makers in Hungary are slow to ask the right questions and experimentally test them or to react to scientific evidence with adaptive policies. For example, after decades of ignoring water stagnation, only recently has the first attempt been made in Hungary to mitigate the consequences of the expected higher water stagnation levels and rising groundwater table by modelling extreme precipitation cases—ala, not in the Tisza, but in the Danube basin on a pilot project in Tát, Hungary (Bauer 2015). It remains in question whether such modelling results can be applied to experimentally test policies to mitigate water stagnation and then apply them in river basins throughout the nation and beyond.

References

Andrásfalvy B (1973) A Sárköz és a környező Duna-menti területek ősi ártéri gazdálkodása és vízhasználatai a szabályozás előtt [Ancient floodplain economy and water utilisation schemes in the Sárköz and the neighbouring areas along the Danube river]. In: Zsigmond K (ed) Vízügyi Történeti Füzetek [Historical papers on water management], vol 6. Vízdok, Budapest
Andrásfalvy B (2009) A gazdálkodás következtében végbement földfelszín változások vizsgálata a Kárpát medencében [Assessment of the changes on the surface of the earth in the Carpathian-basin due to husbandry]. In: Bertalan A, Gábor V (eds) Antropogén ökológiai változások a Kárpát-medencében [Anthropogenic ecological changes in the Carpathian-basin]. L'Harmattan, Budapest, pp 9–19

Balogh P (2014) Vízlépcsőlátás: megújuló energiákkal a folyó ellen [Narrow minded water dams: using renewable energy resources against the river]. http://greenr.blog.hu/2014/09/14/balogh_peter_vizlepcsolatas_megujulo_energiakkal_a_folyok_ellen

Bartholy J, Bozó L, Haszpra L (2011) Klímaváltozás—2011. Klímaszcenáriók a Kárpát-medence térségére [Climate change—2011. Climate scenarios for the Carpathian-basin region]. MTA, Budapest, p 287

Bauer Z (2015) Adaptation to climate change. Newsletter of the Regional Environmental Centre No 5, April 2015

Bellon T (2003) A Tisza néprajza. Ártéri gazdálkodás a tiszai alföldön. [Ethnography of the Tisza. Flood plain management on the Tisza lowlands]. Timp Kiadó, Budapest. 230 p

Bellon T (2004) Living together with nature: farming on the river flats in the valley of the Tisza. Acta Ethnogr Hungar 49(3–4):243–256

Bodnár G (2009) A Tiszába érkező vizekről. A Felső Tisza vízmegtartása [On waters discharged into the Tisza River—water retention in the Upper Tisza]. Conference presentation at the conference: the future is in our hands. All drop of water is a value, Budapest 7th May, 2009

Borhidi A (2009) Régebben még volt hová menni [Earlier on, you had a place to go]. Népszabadság, 5th September, 2009, Weekend supplement page 1

Borsos B (2000) Három folyó között. A bodrogközi gazdálkodás alkalmazkodása a természeti viszonyokhoz a folyószabályozási munkák idején (1840–1910) [Among the three rivers: adaptation of farming practices to the natural conditions at the time of river training works (1840–1910)]. Akadémiai Kiadó, Budapest

Borsos B (2014) A practical guide to integrated land management methods. Methods intended to improve land use and water management efficiency in the floodplains of the Tisza Basin. Lap Lambert, Saarbrücken. 212 p

Fodor Z (2002) A Tisza-menti fokok tájhasznosítási szerepe az újkori folyószabályozások előtt [The role of 'fok' along the Tisza river before the river regulations in the new age]. Falu, Város, Régió, A vidékfejlesztés hírei 2002/4:14–17

Gleick P (2003) Global freshwater resources: soft-path solutions for the 21st century. Science 302:1524–1528

Hamar J (2000) Lesznek-e még folyóink? [Will we have any more rivers left?]. In: Gadó PG (ed) A természet romlása, a romlás természete: Magyarország [The deterioration of nature and the nature of the deterioration]. Föld Napja Alapítvány, Budapest, pp 67–93

Jurek M, Crump J, Maréchal J (eds) (2014) Future imperfect: climate change and adaptation in the Carpathians. GRID-Arendal, a centre cooperating with UNEP, Arendal. 40 p

Kis A, Rita P, Judit B (2014) Magyarországra becsült csapadéktrendek: hibakorrekció alkalmazásának hatása [Projected trends of precipitation for hungary: the effects of bias correction]. Légkör 59(3):117–120

Koncsos L (2006) A Tisza árvízi szabályozása a Kárpát-medencében [Flood control of the Tisza in the Carpathian basin]. Magyar Természetvédők Szövetsége, 32 p

Koncsos L (2011) Természetközeli árvízvédelem [Nature-like flood control]. Presentation at the workshop entitled A nemzeti vidékfejlesztési stratégia, a Vásárhelyi Terv Továbbfejlesztése és az integrált tájgazdálkodás eredményeinek integrációs lehetőségei [National rural development strategy, advancement of the Vásárhelyi Terv and integrated land development: potential for integration], 18–19 November, Nagykörü, Hungary

Koncsos L, Reimann J, Vágás I (2000) Matematikai-statisztikai módszerek árvízvédelmi feladatok elemzéséhez [Mathematical-statistical methods for analysing flood control tasks]. In: Somlyódy L (ed) A hazai vízgazdálkodás stratégiai kérdései [Strategic issues in Hungarian water management]. MTA, Budapest

Kozák M (2011) Szeged belvárosi árvízvédelmi rendszer fejlesztése [Development of Szeged downtown flood control system]. Presentation, TICAD—Development of the Tisza river basin, international conference 24–25 March 2011, Szeged

KSH (2011) A 2010. évi árvíz Borsod-Abaúj-Zemplén megyében. Központi Statisztikai Hivatal (Central Statistical Office), June 2011

Lászlóffy W (1982) A Tisza [The Tisza]. Akadémiai Kiadó, Budapest, 609 p

Lóczy D (2010) Flood-hazard in Hungary: a re-assessment. Cent Eur J Geosci 2:537. https://doi.org/10.2478/v10085-010-0029-0

Molnár G (2009) Ember és természet. Természet és ember [Man and nature. Nature and man]. Kairosz Kiadó, Budapest, 172 p

MTI (2015) 150 milliárdot tolnak árvízvédelmi fejlesztésekbe [HUF 150 billion earmarked for flood control development projects] MA.hu, 11 March 2015, http://www.ma.hu/belfold/243323/150_milliardot_tolnak_arvizvedelmi_fejlesztesekbe

Nováky B (2000) Az éghajlatváltozás vízgazdálkodási hatásai. A hazai vízgazdálkodás stratégiai kérdései [Impacts of climate change on water management. Strategic issues in national water management]. MTA Stratégiai Kutatások Programja, Budapest

OVF (2014) Press release of the National Chief Directorate of Water Management on 12 November 2014

Rácz L (2008) Magyarország környezettörténete az újkorig [Environmental history of Hungary up to the new age]. MTA Institute for Historical Sciences, Budapest, 261 p

Rigó M (2013) Mindenek felett és előtt: Szeged megóvása az árvíztől [Before and beyond everything: protect Szeged from the flood] Manuscript, Szeged, 9 January 2013

Schweitzer F (2009) Strategy or disaster. Flood prevention related issues and actions in the Tisza river basin. Hungar Geogr Bull 58(1–4):3–17

Sendzimir J, Magnuszewski P, Flachner Z, Balogh P, Molnar G, Sarvari A, Nagy Z (2008) Assessing the resilience of a river management regime: informal learning in a shadow network in the Tisza River Basin. Ecol Soc 13(1):11. http://www.ecologyandsociety.org/vol13/iss1/art11/

Sendzimir J, Pahl-Wostl C, Kneiper C, Flachner Z (2010) Stalled transition in the upper Tisza River Basin: the dynamics of linked action situations. Environ Sci Pol 13(7):604–619

Somogyi S (1994) Az Alföld földrajzi képének változásai (XVI–XIX. század) [Changes in the geographical appearance of the Great Plain (16th–19th centuries)]. Nyiregyháza

Szelepcsényi Z, Anna K, Nóra S, Hajnalka B (2013) Életzóna-térképek alkalmazása az éghajlatváltozás vizualizációjára [Applying life zone maps to visualize climate change]. LÉGKÖR 59:111–116

Szeremlei Közösségi Önkormányzat (2013) Lakossági tájékoztató 15(3):2, April 2013, Spring

Teszárné Nagy M, Péter V, István B, Szilágyi Enikő K, Anikó AR, Pál K (2009) 30 éves a Kiskörei tározó [Kisköre reservoir is 30 years old]. Hidrológiai Közlöny 89(6):68–71

VG/MTI (2017) Tokajnál pusztított a jeges árvíz. Világgazdaság, 13 February 2017 http://www.vg.hu/kozelet/kornyezetvedelem/tokajnal-pusztitott-a-jeges-arviz-482877

Vízügy (2011) https://www.vizugy.hu/print.php?webdokumentumid=280

Part IV
Summary

Chapter 29
Landmarks, Advances, and Future Challenges in Riverine Ecosystem Management

Stefan Schmutz, Thomas Hein, and Jan Sendzimir

Science and society are interlinked systems as research topics are defined by societal needs and research outputs trigger societal development. This was particularly the case in the environmental sciences within recent decades: the "Environmental Movement" emerged as a powerful social phenomenon in twentieth-century society via different pathways. Pioneers of the movement were protesters against large infrastructure projects such as hydropower dams or massive pollution of rivers. Green parties took up the momentum and provided political platforms for green thinking. Environmental legislation was implemented, and science contributed to a more sustainable management of aquatic ecosystems via the so-called triangle of sustainability linking environment, society, and economy.

Two environmental events in the 1980s exemplify this paradigm shift in Europe. In 1984, about 3000 activists occupied the Danube floodplains downstream of Vienna, Austria, to protest the construction of the Hainburg hydropower dam. The protest finally ended with withdrawal of the construction plans, and in 1996, the same reach of the Danube and its floodplains that hosted the protest became a national park. In 1986, during the Sandoz accident in Basel (Switzerland), 20 tons of a toxic pesticide mix flowed unhindered with the fire water into the Rhine. In the following 2 weeks, the spill spread more than 400 km downriver, destroying practically the entire eel population and other fish species in its path. As a consequence,

S. Schmutz (✉) · J. Sendzimir
Institute of Hydrobiology and Aquatic Ecosystem Management, University of Natural Resources and Life Sciences, Vienna, Austria
e-mail: stefan.schmutz@boku.ac.at; jan.sendzimir@boku.ac.at

T. Hein
Institute of Hydrobiology and Aquatic Ecosystem Management, University of Natural Resources and Life Sciences, Vienna, Austria

WasserCluster Lunz Biological Station GmbH, Lunz am See, Austria
e-mail: Thomas.hein@boku.ac.at

© The Author(s) 2018
S. Schmutz, J. Sendzimir (eds.), *Riverine Ecosystem Management*, Aquatic Ecology Series 8, https://doi.org/10.1007/978-3-319-73250-3_29

environmental safety regulations were improved, risk management was established, and since then water pollution in the Rhine has decreased significantly.

In the USA, already a decade earlier, environmental targets had been implemented in *legal frameworks*, such as the *Clean Water Act* (1972) and the *Endangered Species Act* (1973), supporting environmental planning and conservation. In Europe, comprehensive environmental legislation did not become effective until the formation of the *European Union* (*EU*) in 1992 enabling the release of a series of directives enforceable over the entire union: the Nitrate *Directive* (1991), *Urban Wastewater Directive* (1991), *Birds & Habitat Directive* (1992), *SEA Directive* (2001), Water Framework Directive (*WFD*, 2000), and the *Floods Directive*, although some environmental directives (e.g., EIA Directive 1987) had been already implemented under the umbrella of the *European Economic Community* (*EEC*), established in 1957. These regional environmental developments were also reflected in international agreements, such as the *UN Conventions on Water, Sustainability and Biodiversity* (1992). Nowadays, legal frameworks consist of a complex network of international agreements, EU-wide directives for EU countries, and national legislations (see Chaps. 17 and 18).

Although not explicitly dedicated to aquatic ecosystems, a number of other *international initiatives* contributed to increase awareness of conservation needs. For example, the Millennium Assessment Report clearly pinpointed a 50% decline in diversity as measured by the Living Planet Index for freshwater vertebrate species (based on 323 species) between 1970 and 2000. This trend provides the factual justification for classifying freshwater ecosystems as the most threatened ecosystems on earth (www.millenniumassessment.org). Within freshwater ecosystems, running water species belong to the most endangered group of species as about half of running water species (vertebrates and crayfish) are threatened or their status is unknown (Collen et al. 2014).

Global threats are increasingly identified and addressed by *internationally agreed development targets*, such as the UN Sustainable Development Goals (2015) focusing on the protection and restoration of water-related ecosystem (goal 6.6). Assessing ecosystem status is nowadays seen as the fundamental basis for ecosystem management. For example, the continuous assessment of climate change and its potential impacts on ecosystems via the Intergovernmental Panel on Climate Change (reports from 1990, 1995, 2001, 2007, 2014) has raised public awareness and brought the climate change debate to the top political level. A similar approach is envisaged by the Intergovernmental Science-Policy Platform on Biodiversity and Ecosystem Services (IPBES) founded in 2012 and currently supported by 125 countries. The first findings of research performed under the aegis of IPBES are expected in 2018.

The most comprehensive *monitoring of surface waters* worldwide has been undertaken within the implementation of the WDF in EU countries. Overall, 108,000 stations have been monitored in surface waters and groundwaters within the first monitoring cycle of the WFD by 2009 (see Chap. 19). Algae (phytoplankton and benthos), macrophytes, invertebrates, and fish have been surveyed in lakes and rivers (57,000 monitoring stations, EC 2009). The results indicate that the majority

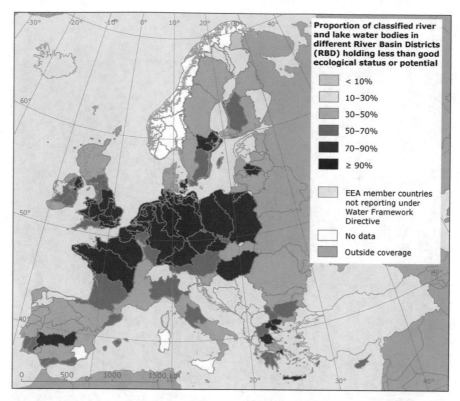

Fig. 29.1 Ecological status of surface waters in EU countries (www.eea.europa.eu, accessed 21 October 2016)

of rivers and lakes fail the WFD objectives of good status or potential, and significant restoration efforts are required (Fig. 29.1).

Hand in hand with the environmental movement, advances in environmental legislation and international initiatives have supported the expansion of the scope of scientific inquiry to include environmental issues, both in terms of theoretical ecosystem understanding and research application (Fig. 29.2). Sound *conceptual understanding of the functioning of natural ecosystems* is a prerequisite for developing effective restoration and management strategies. For a long time in riverine science, river ecosystems were perceived as a sequence of more or less isolated "river zones" (Thienemann 1925; Huet 1949; Illies and Botosaneanu 1963). The *River Continuum Concept* (RCC; Vannote et al. 1979) was the first concept linking important processes (such as P/R ratios, organic matter input, and functional diversity of organisms) along the longitudinal gradient of river catchments. This concept was complemented by subsequent concepts that integrated longitudinal irregularities (*Serial Discontinuity Concept*, Ward and Stanford 1995) and spatial/temporal dynamics as intrinsic features of running water ecosystems, e.g., *Flood Pulse Concept* (FPC) (Junk et al. 1989). The FPC introduces the role of flood

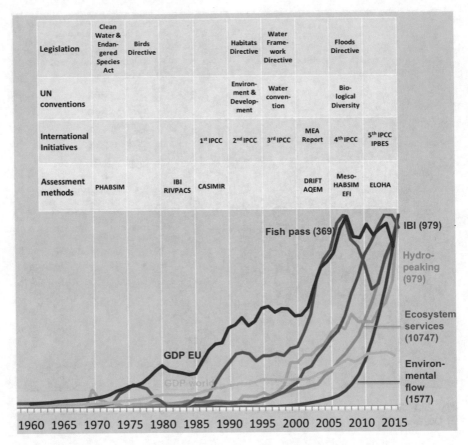

Fig. 29.2 Development of selected landmarks in riverine ecosystem management and economic development between 1960 and 2015. GDP: relative gross domestic product development. Other parameters: relative number of citations for keywords as indicated in figure (absolute number of citations in brackets). IBI: Index of Biotic Integrity (based on www.scopus.com, accessed 15 November 2016)

pulses in connecting aquatic and terrestrial environments at the landscape scale as measured by their changes over time. Together with its extension for temperate systems (Tockner et al. 2000), the FPC describes the role of discharge variability (flow and flood pulses) for ecosystem processes. The *Nutrient Spiraling Concept* uses the continuum perspective addressed in the RCC to develop a model of how elements (nutrients) interact with organisms in running water systems. It provides a framework for studying how transport and transformation processes interact and, thus, provides a basis to parameterize nutrient dynamics in river systems (Newbold et al. 1981). The *Riverine Productivity Model* (Thorp and Delong 1994; Thorp and Delong 2002) emphasizes the importance of autochthonous (aquatic) in-stream production for riverine food webs. This adds an important contribution to the energy balance of riverine food webs as the RCC and the FPC emphasize the importance of terrestrial subsidies for the riverine food webs.

Besides processes, structures play a major role in understanding ecosystem complexity. The *Concept of the Four-Dimensionality* of running waters (Ward 1989) links the three spatial scales (longitudinal, lateral, vertical) with the time scale. Static views of habitats, such as the Multidimensional Niche Concept (Hutchinson 1957), were replaced by dynamic concepts, such as the *Patch Dynamics Concept* (PDC, Townsend 1989) and the *Shifting Habitat Mosaic Concept* (SHMC, Stanford et al. 2005). The *Riverine Landscapes Concept* (Wiens 2002) extended the spatial scope to river-influencing, land ecosystems. While the PDC described the general role of distinct landscape units and their temporal variability and interactions, the SHMC addresses the high heterogeneity in riverine systems, a mosaic of diverse habitats at different successional stages driven by geomorphic dynamics. This allows the coexistence of a high diversity, e.g., species number, based on the connectivity between habitat patches with a regime of episodic disturbances that lead to periodical resets at different locations within the riverine landscape.

The *River Ecosystem Synthesis Concept* (RES, Thorp et al. 2006) represents an integrated model derived from aspects of other aquatic and terrestrial models, combining the view on distinct geomorphic river sections and various ecosystem properties to functional process zones (FPZs) and other aspects of riverine biocomplexity.

Theoretical concepts of ecosystem functioning triggered also *methodological developments in river science and application*. While river habitat was measured by labor-intensive field assessments and analyzed by simple 1D models in the past, nowadays, new instrumentation (e.g., acoustic Doppler current profiler) produces high-resolution data on bathymetry and flow velocity that can then be processed in 3D models (see Chap. 3). The new technologies reduce data acquisition costs and enable the analysis of high-resolution, spatial/temporal, habitat dynamics. When linked to GIS data on land use, those approaches can be integrated into catchment-scale analyses (see Chap. 13). Habitat analyses at various scales are linked to biotic assessments that range from microhabitat preference models for distinct species and life stages to biotic community assessment at reach scale up to species distribution models at catchment and continental scale. Micro- and mesohabitat models are nowadays standard models for assessing environmental flow and habitat improvement (see Chap. 7). Today, biotic community assessment methods such as the Index of Biotic Integrity are standard methods used for river management that are increasingly supported by modeling approaches and developed from site-specific to river-type, national and supranational assessment tools (e.g., European Fish Index; see Chap. 19).

Developments of scientific methodologies are closely linked to *data availability* (see Chap. 20). The integration of dispersed biodiversity data at species and site level into supranational databases, the collection of standardized WFD monitoring data across the EU, and satellite-based earth observation, just to mention a few, create data sources of unprecedented breadth and resolution. New data management systems and data mining techniques are required to handle those data. A network on International Long-Term Ecological Research (ILTER) sites, founded in 1993 in the USA, is gradually developing. ILTER is one of the few hosts of long-term data that are indispensable for analyzing and predicting long-term trends, which are key to understanding, for example, climate change impacts. However, at the moment in terms of rivers, it covers only 58 sites mainly located in Europe.

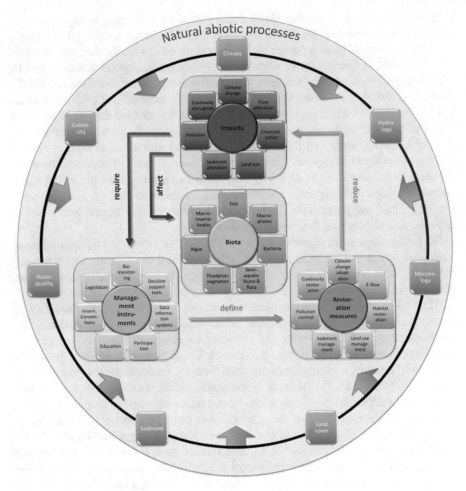

Fig. 29.3 Riverine ecosystem management: effects of natural abiotic processes, human-induced impacts, and restoration measures on biota

As described by a number of river concepts (see above), *essential elements of riverine ecosystems* are *river morphology hydrology, sediment, continuity, water quality, and biota*. Understanding the interrelation between abiotic system elements and biota and food webs is vital for assessing human impacts and developing effective restoration and mitigation strategies. Although treated in separate chapters in this book for practical purposes (see Chaps. 2–13), it is evident that the linkages among system elements are as important as processes within each system element. It is likely that while human stressors may alter each system element individually, interactions among stressors strongly affect overall system behavior and response (Fig. 29.3).

Finally, managers are facing a very complex system of natural and human-induced processes making it impossible to find simple solutions for restoring degraded rivers. Disentangling effects of multiple stressors, developing stressor-specific restoration strategies, and, lastly, integrating those again into *effective management programs* are a challenging task. Legacy effects have to be considered as human interventions may go back for several 100 years (see Chap. 2). Humans have polluted rivers since the first larger population agglomerations were established. Sewage treatment plants are known to combat pollution efficiently. However, diffuse inputs from agriculture require alternative land management strategies and protection of riparian vegetation and floodplains (see Chaps. 10 and 13). Toxicants still represent a relevant stressor in river ecosystems, despite major improvements of the situation over the last decades, at least in regions with a strong governmental regulation. Intelligent strategies are required to deal with how exposure to toxicants is complicated by mixtures, multiple stressors, and other features of the environmental context that all influence the magnitude of potential negative effects (see Chap. 12).

Scale dependencies and upstream/downstream effects are evident in all riverine ecosystem elements. Long-term flow alterations may be as critical as short-term alterations (see Chaps. 4 and 5). Sediment seems to be as important as flow, but understanding of sediment processes lags behind, making it difficult to identify effective and sustainable mitigation measures (see Chap. 8). While building fish passes is supposed to be an adequate mitigation measure for continuity disruptions, their efficiency, in particular, for downstream migration, is still questionable. Furthermore, the overall response of fish communities to multiple stressors in the context of multi-fragmented, river systems has to be explored in more detail (see Chap. 9). By restoring habitats, providing environmental flow, and building fish passes, we can improve habitat quality and connectivity. However, some impacts are not reversible or are hard to mitigate. Large dams cause such altered ecosystems since dams fundamentally change former riverine ecosystems into lake-type or hybrid systems, and restoration measures are, in general, limited here (see Chaps. 6 and 24). In addition to the challenges described above, our current understanding of causal relationships underpinning ecological processes becomes increasingly obsolete over time and needs revision, as the environmental conditions drift away from "natural" conditions due to climate change (see Chap. 11).

Many methods and successful applications of *river restoration* at local scales are presented in this book. However, comprehensive restoration of running water is lacking in most of the rivers worldwide. We are increasingly running the risk of losing the necessary free space to experimentally explore alternative restoration interventions. For example, river widening, as one of the favorite morphological restoration measures, requires space that is increasingly becoming the limiting factor, particularly in areas with natural constraints and high levels of development, such as the Alps. But even if land is available, land owners have to be convinced, subsidies provided, and implementation procedures developed to be successful. Strategic planning at catchment and even larger scales is necessary to cope with conflicting interests and new infrastructure development. However, even with best intensions and maximal support, we might not be able to significantly restore heavily impacted rivers in all cases due to how society (acceptance of existing land uses) and ecosystems have shifted over time and become irreversibly locked into a new regime. Accepting this fact may redirect our efforts into restoring and protecting less-impacted rivers in order to successfully provide other ecosystem services (see Chap. 15).

Apparently, where, when, and how to restore rivers are not trivial questions that should be answered by gut decisions but require scientifically based knowledge of ecosystem functioning, underlying mechanisms of restoration processes, and involvement of all relevant stakeholders. While the need for inter- and transdisciplinarity has been invoked so many times that they have become worn-out buzzwords, they remain as concepts that are rarely applied in routine river management. As solutions are not straightforward and may only evolve over time, we cannot manage only based on certainty. We must manage by learning along the way what needs to be done. In that context, the principle of adaptive management seems an appropriate way to manage under very uncertain conditions, e.g., climate change, by using comprehensive monitoring of the human/ecosystem to periodically challenge our science and policies within structured learning cycles (see Chap. 16).

One might say that restoring rivers for ecological purposes is a luxury only rich countries can afford. However, when it comes to providing essential or well-appreciated services for human beings (e.g., water supply, fisheries, or recreation), the functioning of riverine ecosystems becomes the foundation of sustainable development (see Chaps. 14, 27 and 28). Therefore, the concept of ecosystem services is very helpful to make the linkage between ecosystem functioning and ecosystem services better understandable for stakeholders and the public and apply it more widely in river management (Chap. 21). The importance of ecosystem services will only increase as society adapts to climate change by lowering the use of fossil fuels, which were used initially to replace ecosystem services. If fossil fuels can no longer provide alternative services, then ecosystem-driven services must be enhanced to replace them.

Environmental movements triggered societal transformations and redirected the research agenda. However, environmental movements come and go, and economic crises can easily remove environmental tasks from the political agenda, as seen recently. Therefore, the *institutionalized involvement of stakeholders and interested public in decision processes* is very important to keep the dialog on environmental issues alive. NGOs play a critical role in this process, but openness of all participating actors (decision makers, administration, stakeholders, NGO, science) is required to elaborate sound and widely accepted solutions (see Chap. 23). Finally, adequate capacity building and educational programs taking up the challenges in riverine ecosystem management are required to guarantee sustainable development of riverine ecosystems in the long run (see Chap. 22).

Aquatic ecosystems such as rivers will become increasingly important in sustaining and improving our quality of life by providing such services as food, water, transport, and the aesthetics that define a region. The science of understanding and managing rivers cannot provide the luxury of high certainty anymore, which we now understand was an illusion in the first place. However, in an increasingly unpredictable world, it can provide the best questions, methods, and potential solutions to test as society adapts to uncertainty. As improved restoration efforts return the natural beauty of river valleys to many regions of the world, river science will become a vital bridge between the integrity of nature and of society.

References

Collen B, Whitton F, Dyer EE, Baillie JEM, Cumberlidge N, Darwall WRT, Pollock C, Richman NI, Soulsby A-M, Böhm M (2014) Global patterns of freshwater species diversity, threat and endemism. Glob Ecol Biogeogr 23:40–51

EC (2009) Report from the commission to the European Parliament and the council in accordance with article 18.3 of the Water Framework Directive 2000/60/EC on programmes for monitoring of water status. Brussels

Huet PM (1949) Apercu des relations entre la pente et les populations piscicoles des eaux courantes Station de Recherches des Eaux et Forets, Groenendaal (Belgique), pp 332–351

Hutchinson GE (1957) Concluding remarks. In: Paper presented at the Cold Spring Harbor Symp Quant Biol 22: 415–427

Illies J, Botosaneanu L (1963) Problèmes et méthodes de la classification et de la zonation écologique des eaux courantes considerées surtout du point de vue faunistique. Internationale Vereinigung für theoretische und angewandte Limnologie 12:1–57

Junk WJ, Bayley PB, Sparks RE (1989) The flood pulse concept in river-floodplain systems. Can Spec Publ Fish Aquat Sci 106:110–127

Newbold JD, Elwood JW, O'Neill RV, Van Winkle W (1981) Measuring nutrient spiralling in streams. Can J Fish Aquat Sci 38(1755):860–863

Stanford JA, Lorang MS, Hauer FR (2005) The shifting habitat mosaic of river ecosystems. SIL Proc 29(1):1922–2010

Thienemann A (1925) Die Binnengewässer Mitteleuropas. Eine limnologische Einführung. Die Binnengewässer. Schweizerbart'sche Verlagsbuchhandlung, Stuttgart

Thorp JH, Delong MD (1994) The riverine productivity model: an heuristic view of carbon sources and organic processing in large river ecosystems. Oikos 70(2):305

Thorp JH, Delong MD (2002) Dominance of autochthonous autotrophic carbon in food webs of heterotrophic rivers. Oikos 96(3):543–550

Thorp JH, Thoms MC, Delong MD (2006) The riverine ecosystem synthesis: biocomplexity in river networks across space and time. River Res Appl 22(2):123–147

Tockner K, Malard F, Ward JV (2000) An extension of the food pulse concept. Hydrol Process 14 (16–17):2861–2883

Townsend CR (1989) The patch dynamics concept of stream community ecology. J N Am Benthol Soc 8(1):36–50

Vannote RL, Minshall GW, Cummins KW, Sedell JR, Cushing CE (1979) The river continuum concept. Can J Fish Aquat Sci 130:137

Ward JV (1989) The four-dimensional nature of lotic ecosystems. J N Am Benthol Soc 8:2–8

Ward JV, Stanford JA (1995) The serial discontinuity concept of lotic ecosystems. Regul Rivers Res Manag 10:159–168

Wiens JA (2002) Riverine landscapes: taking landscape ecology into the water. Freshw Biol 47:501–515

Printed in the United States
By Bookmasters